85.00

555

5122773231

32-3

SYNTHETIC OILS AND ADDITIVES

FOR LUBRICANTS

SYNTHETIC OILS
AND ADDITIVES
FOR LUBRICANTS

Advances Since 1977

Edited by M. William Ranney

NOYES DATA CORPORATION
Park Ridge, New Jersey, U.S.A.
1980

Copyright © 1980 by Noyes Data Corporation
    No part of this book may be reproduced in any form
    without permission in writing from the Publisher.
Library of Congress Catalog Card Number: 79-24355
ISBN: 0-8155-0781-X
Printed in the United States

Published in the United States of America by
Noyes Data Corporation
Noyes Building, Park Ridge, New Jersey 07656

Library of Congress Cataloging in Publication Data

Ranney, Maurice William, 1934–
    Synthetic oils and additives for lubricants.

    "Contains all new advances since ... Lubricant
additives--recent developments, published in 1978."
    Includes indexes.
    1. Lubrication and Lubricants--Additives--Patents.
2. Lubricating oils--Patents. I. Title.
TJ1077.R265   621.8'9   79-24355
ISBN 0-8155-0781-X

To
Laurel Susan
my lovely daughter
and
world traveler

FOREWORD

The detailed, descriptive information in this book is based on U.S. patents, issued since January 1978, that deal with synthetic oils and additives for lubricants. This title contains all new advances since our previous title, *Lubricant Additives—Recent developments,* published in 1978, as well as new material relating to synthetic oils.

This book serves a double purpose in that it supplies detailed technical information and can be used as a guide to the U.S. patent literature in this field. By indicating all the information that is significant, and eliminating legal jargon and juristic phraseology, this book presents an advanced, technically oriented review of the latest advances in synthetic oils and additives for lubricants.

The U.S. patent literature is the largest and most comprehensive collection of technical information in the world. There is more practical, commercial, timely process information assembled here than is available from any other source. The technical information obtained from a patent is extremely reliable and comprehensive; sufficient information must be included to avoid rejection for "insufficient disclosure." These patents include practically all of those issued on the subject in the United States during the period under review; there has been no bias in the selection of patents for inclusion.

The patent literature covers a substantial amount of information not available in the journal literature. The patent literature is a prime source of basic commercially useful information. This information is overlooked by those who rely primarily on the periodical journal literature. It is realized that there is a lag between a patent application on a new process development and the granting of a patent, but it is felt that this may roughly parallel or even anticipate the lag in putting that development into commercial practice.

Many of these patents are being utilized commercially. Whether used or not, they offer opportunities for technological transfer. Also, a major purpose of this book is to describe the number of technical possibilities available, which may open up profitable areas of research and development. The information contained in this book will allow you to establish a sound background before launching into research in this field.

Advanced composition and production methods developed by Noyes Data are employed to bring these durably bound books to you in a minimum of time. Special techniques are used to close the gap between "manuscript" and "completed book." Industrial technology is progressing so rapidly that time-honored, conventional typesetting, binding and shipping methods are no longer suitable. We have by-passed the delays in the conventional book publishing cycle and provide the user with an effective and convenient means of reviewing up-to-date information in depth.

The table of contents is organized in such a way as to serve as a subject index. Other indexes by company, inventor and patent number help in providing easy access to the information contained in this book.

15 Reasons Why the U.S. Patent Office Literature Is Important to You —

1. The U.S. patent literature is the largest and most comprehensive collection of technical information in the world. There is more practical commercial process information assembled here than is available from any other source.

2. The technical information obtained from the patent literature is extremely comprehensive; sufficient information must be included to avoid rejection for "insufficient disclosure."

3. The patent literature is a prime source of basic commercially utilizable information. This information is overlooked by those who rely primarily on the periodical journal literature.

4. An important feature of the patent literature is that it can serve to avoid duplication of research and development.

5. Patents, unlike periodical literature, are bound by definition to contain new information, data and ideas.

6. It can serve as a source of new ideas in a different but related field, and may be outside the patent protection offered the original invention.

7. Since claims are narrowly defined, much valuable information is included that may be outside the legal protection afforded by the claims.

8. Patents discuss the difficulties associated with previous research, development or production techniques, and offer a specific method of overcoming problems. This gives clues to current process information that has not been published in periodicals or books.

9. Can aid in process design by providing a selection of alternate techniques. A powerful research and engineering tool.

10. Obtain licenses — many U.S. chemical patents have not been developed commercially.

11. Patents provide an excellent starting point for the next investigator.

12. Frequently, innovations derived from research are first disclosed in the patent literature, prior to coverage in the periodical literature.

13. Patents offer a most valuable method of keeping abreast of latest technologies, serving an individual's own "current awareness" program.

14. Copies of U.S. patents are easily obtained from the U.S. Patent Office at 50¢ a copy.

15. It is a creative source of ideas for those with imagination.

CONTENTS AND SUBJECT INDEX

INTRODUCTION. 1

DISPERSANTS AND DETERGENTS. 3
  Succinic Anhydride-Based Ashless Compositions 3
    Polyamino Cycloaliphatic Compounds 3
    Amine Condensation Products 6
    Alkylene Polyamine Reaction Products. 7
    Arylamine-Aldehyde Products. 9
    Substituted Pyrimidine and Triazine Ring Compounds. 11
    Polyalkenyl Oxazoline and Polyamine. 16
    Oxazoline Adducts 18
    Polyesters Containing Carboxypyrrolidone 22
    Alkenylsuccinimides Based on N,N,N',N'-Tetrakis(3-Aminopropyl)-
      Ethylenediamine. 24
    Borated Oxazolines. 25
    Boric Acid Treated Compounds 25
    Hydrocarbon-Soluble Alkyl Lactone Polyol Esters 26
    Dimercaptothiadiazole-Dispersant Reaction Products. 30
    Demulsifier Additive System. 30
  Modified Polyolefin Ashless Compositions 31
    Alkyl-Guanidino-Heterocyclic Compounds. 31
    Chlorinated Polyisobutylene-Pyridine Reaction Products 33
    Oxazoline-Containing Polymers. 34
    Copolymer, Polyamine and Polycarboxylic Acid Condensation Products . .36
    Olefin-Thionophosphine Sulfide Reaction Products 38
    Polybutyl Phenoxyacetamides. 38
    Catalyst for Mannich Condensation Reaction of Oxidized Olefinic
      Polymers. 40
  Phenol-Based Ashless Compositions 42
    Aminophenol-Detergent Combinations. 42
    Haloalkyl Hydroxy-Aromatic Condensation Products. 45
    Esters of Alkyl-Substituted Salicylic Acids 46
    Amides of Alkyl-Substituted Salicylic Acids. .47

Phenyl Glycidyl Ether-Diamine Reaction Products 48
Mannich Reaction Products. 53
Sulfurized Mannich Condensation Products 56
Boronated Amide Product. 57
Other Ashless Compositions 60
Terpolymer Polymethacrylates Based on 4-Vinylpyridine. 60
Aminoguanidine Sulfonate 63
Tetrahydropyrimidyl-Substituted Compounds 64
Metal Sulfonates, Phenates and Naphthenates. 66
Metal Carbonate Overbased Metal Sulfonates 66
Calcium Oxide with Low Reactivity Towards Water 70
Single Stage Low Temperature Addition of Carbon Dioxide. 71
Metal Salts of Substituted Ethylsulfonic Acids 72
Alkaline Earth Metal Sulfonates and Dispersants. 75
Magnesium-Containing Complexes 77
Overbased Magnesium Phenates. 78
Overbased Metal Phenates. 84
Improved Filtration Properties for Overbased Metal Naphthenates 85
Promoter System for Alkyl Phenol Sulfide Reactions 86
Phenoxide-Halocarboxylic Acid Condensates 87
Metallic Compounds of Polyarylamine Sulfides. 88
Other Compositions 90
Solid Graphite Particles. 90
Overbased Calcium Sulfonates in Diester Lubricating Oils. 93

VISCOSITY INDEX IMPROVERS AND OTHER ADDITIVES 96
Viscosity Index Improvers. 97
Styrene and tert-Butylstyrene Block Copolymers 97
Styrene-Isoprene Block Copolymers and Polybutene 100
Block Copolymers of Styrene and Methacrylate Components. 101
Styrene, Butylstyrene and Vinyl Benzyl Ether Terpolymers 105
Hydrogenated Block Copolymers of Butadiene and Isoprene 108
Hydrogenated and Sulfonated Block Copolymers of Styrene and
    Isoprene 110
Hydrogenated Copolymers of Butadiene and Styrene 110
Hydrogenated Star-Shaped Polymer 110
High Viscosity Index Gear Oil Formulations. 111
Addition of Oil-Soluble Acid to Prevent Hazing 112
Pour Point Depressants 113
α-Olefin Copolymers. 113
Polyalkylacrylate and Ethylene-Propylene Copolymers. 115
Vinyl Acetate, Ethylene and Vinyl Chloride Terpolymer 118
Alkylated Polybenzyl Polymers. 119
Viscosity Index Improvers and Dispersants 119
Epoxidized Terpolymers. 119
Oxidized Copolymer and Acrylonitrile 122
Aminated Oxy-Degraded Copolymers 124
Aminated Interpolymers. 128
Polyamine-Hydroperoxidized Ethylene-Propylene Copolymer Reaction
    Products 131
Ethylene, Propene and Cyclic Imide Terpolymers. 134
Ethylene, Propene and N-Vinylimidazole Terpolymers 135
Copolymer, Amine Compound and Oxygen Reaction Products 138

Nitrogen-Containing Copolymers from Lithiated Copolymers. 140
Grafted Copolymers 142
Stabilized Imide Graft of Ethylene Copolymers 144
Polyolefin Copolymers with Maleic Anhydride Grafting Reactions. 148
Star-Shaped Isoprene-Divinylbenzene Polymers 150
Multistep Polymerization for Polymer-in-Oil Solution. 152
Multifunctional Viscosity Index Improvers. 153
Sulfone Copolymers 153
Carboxylic Acid Grafting 157
Electrophilically Terminated Ethylene-Alpha Olefin Copolymers. 160
Polymers Containing Postreacted Phenol Antioxidant Functionality 163
Dithiophosphorylated Copolymers of Aziridineethyl Acrylates 164
Polyoxyalkylene Phenothiazines for Ester Lubricants. 166

LOAD-CARRYING ADDITIVES. 167
Mineral Oils. 167
Ammonium Salts of Oxa- and Thiaphosphetanes 167
Amine Salt of Dialkyldithiophosphate 170
Imidazoline-Phosphonate Reaction Products. .171
(N,N-Diorganothiocarbamyl) Phosphorothioites. 173
Phosphorus Acid Compounds-Acrolein-Ketone Reaction Products. 175
Dimethyl Octadecylphosphonate 177
Mixture of Zinc Salts of Dialkyldithiophosphoric Acid. 178
Metal Dithiophosphates. 181
Zinc Alkyl Dithiophosphate and Substituted Succinic Anhydride. 183
Phosphorothionate Derivatives of Alkylphenol Sulfides 185
Chlorinated Wax-Trihydrocarbyl Phosphite Reaction Products. 188
Phosphorus Trihalide-Multifunctional Alcohol Reaction Products 189
Monoaryl Phosphonates 190
Organophosphorus Derivatives of Hydroxycarboxylic Acids. 192
Lubricant Composition. 194
Substituted 1,2,3-Triazoles 196
Benzotriazole and Substituted Succinic Anhydride. 197
1,2,4-Thiadiazole Polymers. 200
Triazole Derivatives. 200
N-Butoxydodecenylsuccinimides. 202
Azole Aminopolysulfides 205
Sulfurization of Alcohol-Epichlorohydrin Reaction Products 206
Sulfurized Tallow Oil 207
Sulfurization of Dicyclopentadiene and Alloocimene 208
Sulfurized Fatty Acid Esters. 211
Sulfur-Containing Molybdenum Dihydrocarbyldithiocarbamate
   Compounds. 212
Mercapto-Substituted Boron-Containing Compounds 215
Alkyl Beta-Thiopropionic Acid Esters. 217
Alkene-Sulfur Monochloride Reaction Products 218
Wax Esters of Vegetable Oil Fatty Acids. 220
Silicone Composition for Lubricating Copper and Bronze. 221
Hydbrid Lubricant Containing Colloidal Dispersion of Polytetrafluoro-
   ethylene 222
Metal Flakes and Molybdenum Disulfide Particles. 223
Ester Lubricants 224
Organic Sulfonic Acid Ammonium Salts. 224
Organoamine Salts of Phosphate Esters. 227

OXIDATION AND CORROSION INHIBITORS

OXIDATION AND CORROSION INHIBITORS. 229
  Antioxidants and Corrosion Inhibitors 229
    Substituted Triazine-Mercapto Compound Reaction Products 229
    Benzotriazole-Allyl Sulfide Reaction Products 231
    Polydisulfide and Arylamines 232
    Methylene Bis(Dibutyldithiocarbamate) and 4-Methyl-2,6-Di-tert-
      Butylphenol 232
    Sulfur-Based Antioxidant Compositions 233
    Adducts of Benzotriazole Compounds and Alkyl Vinyl Ethers.........236
    Alkenyl Succinimide Dispersant and Aminobenzoic Acid Compounds ...237
    Phenylated Naphthylamine, Sulfoxide Compound and Copper.........238
    Antioxidants and Halogen-Containing Compounds..................241
    Mercaptan-Carbonyl Compound Reaction Products 244
    Sulfur-Bridged Hydrocarbon Ring Compounds.....................246
    Sulfurized Dihydroxybenzene Compounds........................249
    Alkaline Earth Metal Sulfonate and Phosphonate-Phenate Sulfide 252
    Condensation Products of Dithiophosphoric Acid Esters and Styrene....253
    Alkyl Phenol, Aromatic Amine and Triesters of Dithiophosphoric Acid..254
    Bispiperazidophosphorus Compounds.............................256
    Substituted Trialkanolamines and Zinc Phosphates.................259
    Titanate Dithiophosphate Compositions 261
    Antioxidant and Phosphorus-Containing Compound.................262
    Triketone 264
  Synthetic Ester Compositions.......................................265
    Dibenzothiophene and Other Sulfur-Containing Additives 265
    Phosphorus Pentasulfide Adduct of Polycyanoethylated Keto Fatty
      Esters 266
    Esters of Arylaminophenoxyalkyl Carboxylic Acids..................266
  Other Fluids and Processes 268
    Fluorinated Polyether Fluids.....................................268
    Phosphate Ester Turbine Lubricant Formulation 269
    Organosulfur-Containing Nickel Complexes as UV Stabilizers.........271

METALWORKING LUBRICANTS. 272
  Water-Based Lubricants...272
    Mixture of Salts of Alkylaryl Sulfonic Acids.......................272
    Complex Phosphate Surfactants 274
    Methacrylic Acid Grafted Polyoxyalkylene.........................276
    Amino-Amides Derived from Polymeric Fatty Acids.................278
    Phosphate Esters and Sulfur Compounds 279
    1,3,5-Tris(Furfuryl)Hexahydro-s-Triazine.........................282
    Disodium Monocopper(II) Citrate................................282
    Zinc Phosphate, Dimer Acid and Alkenyl Succinic Anhydride 284
    Sulfurized Molybdenum Compound and Water-Soluble Fatty Ester.....285
    Modified Triglycerides...288
    Pasty Lubricant for Nonchip Metal Forming.......................291
    Forging Lubricant..292
    Lubricating Composition Applied over Primer Coat 293
  Oil-Based Lubricants...294
    Organic Hydroperoxides and Sulfur 294
    Sulfurized Oil and Polyalkenylsuccinic Anhydride-Polyglycol Reaction
      Product..295
    Partially Neutralized Aluminum Acid Alkylorthophosphates 297

Hot Melt Drawing Lubricant.....................................298
Drawing Lubricant 299
Oilless Fluid for Scoring Glass.................................301

SOLID LUBRICANT COMPOSITIONS 302
 Solid Lubricants 302
  Zinc-Based Polysulfides..302
  Finely Divided Metal Carbonate and Halogenated Compounds........303
  Polyphenylquinoxaline Antifriction Fillers and Modifiers............305
  Molybdenum Disulfide, Hydroxyethyl Cellulose and Silicate 307
  Self-Lubricating Bearings.......................................308
  Self-Lubricating Antifriction Material 310
  Molten Glass Applications......................................310

GREASE COMPOSITIONS.....................................312
 Thickeners..312
  Benzimidazobenzisoquinolinone Ureas 312
  Triazine-Urea Compounds.......................................314
  Polyurea Composition..316
  Alkali Metal Salts of Gamma-Keto Acids..........................317
 Extreme Pressure Additives......................................319
  Sodium and Potassium Borates 319
  Alkali Metal Triborate..321
  Phosphorus and Sulfur Compounds...............................322
 Other Additives and Grease Compositions..........................323
  Amine Salt of Half Ester of Substituted Succinic Acid as Antifatigue
   Additive 323
  Acylated Polyamides as Rust Inhibitors 324
  Lithium Base Grease Containing Polyisobutylene for Water Resistance...326
  Preparation of Clay-Based Compositions..........................326
  Biodegradable Grease 328
  Aluminum Fatty Acid Soap Thickened Polyisobutylene Composition ...330
  Brake Grease Composition with Polyglycol Base....................331
  Perfluorinated Polyether Greases................................334

SPECIALTY LUBRICANTS AND RELATED PROCESSES..............337
 Fiber and Textile Lubricants.....................................337
  Oxidation Stable Polyoxyalkylene Compounds.....................337
  Silicone and Perfluoropolymer Composition.......................343
  2,2,4-Trimethyl-1,2-Dihydroquinoline Polymers...................343
  Flame-Retardant Composition Containing Bromooctadecane.........346
  Hot Drawing Lubricant...347
 Tire Lubricants 349
  Green Tire Lubricants..349
  Internal Tire Lubricant 350
  Band-Ply Lubricant Concentrate.................................352
 Recording Substrates and Plastics................................354
  Dialkanolamine Derivatives and Halogenated Hydrocarbon Carrier.....354
  Tetrafluoroethylene Telomeric Compositions 356
  Methyl Alkyl Siloxanes for Video Disc Lubricant...................357
  Soaps and Esters from Alpha-Olefin Acids for PVC Lubricant 359
  Lubricant for Thermoplastic Materials 363
  Magnetic Field Reactor for Preparation of Lubricant Dispersions.......363

Specialty Lubricants and Fluids 366
   Railway Lubricating Oil 366
   Bearing Lubricant 369
   Temporary Rust-Proofing Lubricant for Steel Plates.................369
   Centrifugal Compressor Refrigeration Lubricant Based on Castor Oil....371
   High Density Metal-Containing Lubricants for Oil Well Drilling 372
   Nuclear Service Lubricant....................................373
   Penetrating Oil Composition..................................374
   Gas Bearing Gyroscopes 375
   Lubricant for Processing of Molten Glass 376
   Automatic Transmission Fluid 378
   Boron Dispersant for Automatic Transmission Fluid................379
Hydrotreating and Processing of Specialty Oils......................380
   Upgrading Lubricating Oil Stock...............................380
   Simultaneous Dewaxing and Hydrogenation of Feedstock 382
   Hydrogenation of Low Molecular Weight Polyisoprene..............384
   Hydrogenation of Olefin Polymers 386
   Two-Step Alkylation Process..................................387
   Thermal Inhibition of α-Oligomer Oils 388
   Processing of Sulfonic Acids 389
Synthetic Ester Compositions....................................393
   Mixtures of Complex and Monomeric Esters.....................393
   Trimethylolpropane-Dimerized Fatty Acids 396
   Trimethylolpropane-Isopalmitic Acid-Alkanoic Acid 398

COMPANY INDEX..400
INVENTOR INDEX...402
U.S. PATENT NUMBER INDEX 406

INTRODUCTION

Mineral oil stocks are the prime source of lubricants for an almost endless list of applications. Nearly all of the lubricants are formulated with a variety of additives. Lubricant additives generally are defined simply as materials which enhance or impart desirable properties to a mineral-base oil. The high quality of modern lubricants results for the most part from the use of additives.

Lubricating oils and related hydraulic and transmission fluids for present-day machinery, and particularly for present-day internal-combustion engines and other uses contain a wide variety of additives. The additives usually are classified according to their intended function such as dispersant; oxidation, corrosion and rust inhibitor; viscosity index improver; pour point depressant; antiwear agent; antifoam agent; and many more.

The advent of high-speed automotive engines in particular, coupled with increased engine operating temperatures and increased complexity of antipollution devices associated with such engines, has resulted in substantial increases in additive quantities in automotive lubricating oils to meet a continuing demand for improved properties and results. The quantities of additives employed in some uses have been approaching quantities so large as to affect negatively the primary mission of the lubricating oil: to lubricate.

What is needed is a single additive which will provide a multiple function to satisfy at least some of the basic requirements of individual additives for lubricating and other oils now presently satisfied by package of several additives. With such an additive, the quantity of overall additives employed in the lubricating oil potentially could be substantially reduced, permitting a single effective unit quantity to fulfill multiple requirements.

Modern technology is currently supplying the general public and the process industries with machinery which is designed to operate under a wider range of temperatures and under greater loads than previously available. In addition, many of the newer machines are designed to operate at extremely high speeds. Many of these machines require certain specific lubricating properties which

1

are not available in the conventional lubricants. Thus modernization of high-speed and high-temperature equipment has strained the petroleum industry for the development of a second generation of lubricants capable of satisfying the requirements of the new machines. Recently, for example, there has been an increased demand for lubricants capable of performing well at temperatures above 300°F in high-speed bearings and gears for periods in excess of 500 hours. In addition, with the further development of the high-speed sealed bearings, the lubricant must be able to endure for the life of the bearing.

Metalworking operations, for example, rolling, forging, hot-pressing, blanking, bending, stamping, drawing, cutting, punching, spinning and the like generally employ a lubricant to facilitate the same. Lubricants greatly improve these operations in that they can reduce the power required for the operation, prevent sticking and decrease wear of dies, cutting bits and the like. In addition, they frequently provide rust-inhibiting properties to the metal being treated.

The worldwide research and development activity in lubrication and lubricant additive technology continues to accelerate as increasing performance demands must now be related to rapidly escalating raw material costs and limited long-term supply. This book, based on the patent literature of the United States since January 1978, provides over 260 processes, comprising synthetic techniques, formulation technology and test results. These processes represent much of the latest technology advances and trends not only in the United States, but in Europe, Japan and other countries.

DISPERSANTS AND DETERGENTS

The major problem associated with crankcase lubricants employed in internal combustion engines is the presence in the lubricant of foreign particles, such as dirt, soot, water and decomposition products resulting from a breakdown of lubricating oil. Similarly, problems are encountered with the presence of foreign materials in the fuel burned in the internal combustion engine. In order to prevent a buildup of sludge or other undesirable constituents within the lubricating oil, or fuel, it is necessary that such foreign particles be maintained in suspension within the lubricating oil or fuel so that they can readily be removed or passed through the internal combustion engine without a buildup of such foreign matter within the lubricating oil or the fuel tank.

Numerous additives have been proposed to prevent the buildup of sludge and particulate matter within the crankcase oil of the internal combustion engine or within the fuel tank supplying the fuel to the internal combustion engine. For example, to obviate the problem or reduce the problem, the approach has generally been to employ either known detergents, such as metal phenates and sulfonates, or oil-soluble nitrogen-containing compositions. Many of the prior art additives are the reaction products of relatively high molecular weight carboxylic acid acylating agents and certain specified amines, alcohols, and combinations of alcohols.

However, problems have been encountered in providing additives which are stable and functional to suspend the foreign material in both the lubricating oil and the fuel, not only at high temperatures, but also at low temperatures. Thus, new and improved dispersants for lubricants and fuels are constantly being sought.

SUCCINIC ANHYDRIDE-BASED ASHLESS COMPOSITIONS

Polyamino Cycloaliphatic Compounds

E.W. Kluger, J.W. Miley and T.K. Su; U.S. Patent 4,153,567; May 8, 1979; assigned to Milliken Research Corporation describe improved fuel and lubricant

3

additive compositions produced by the process comprising simultaneously reacting a polyamine substituted cycloaliphatic compound having the general structure

$$NH-CH_2CH_2CH_2NH_2$$
$$NHR$$

where R is hydrogen or $CH_2CH_2CH_2NH_2$, with at least one substituted aliphatic polycarboxylic acid acylating agent selected from the group consisting of carboxylic acids and their corresponding halides, anhydrides, and esters, the carboxylic acid acylating agents having at least 12 carbons per molecule. More specifically, the oil-dispersible and/or oil-soluble additives have the general structure

where R is H, $-CH_2CH_2CH_2NH_2$ or

and R_1 is an alkyl moiety containing at least about 8 carbons or a substantially saturated olefin polymer having an average molecular weight of from about 200 to about 5,000.

The compositions are useful as dispersing agents in lubricants and fuels, especially fuels employed in internal combustion engines and lubricants intended for use in the crankcase of internal combustion engines, gears, and power transmitting units. The following examples illustrate the process.

Example 1: In a 250 cc three-necked flask equipped with a magnetic stir bar, Dean Stark trap, and heating mantle was placed 21.9 grams (0.08 mol) of dodecenylsuccinic anhydride which had been dissolved in 50 cc of toluene. To this solution was added 18.2 grams (0.08 mol) of tetramine, N,N'-di-(3-aminopropyl)-1,2-diaminocyclohexane,

which also had been dissolved in 50 cc of toluene. An exothermic reaction occurred on mixing. The reaction mixture was then refluxed for 6 hours and water was collected in the Dean Stark trap. The toluene was removed by evaporation to give 99.3% yield of the crude viscous adduct. An infrared spectrum indicated that no anhydride was present. The compound produced from the above reaction can be represented structurally as

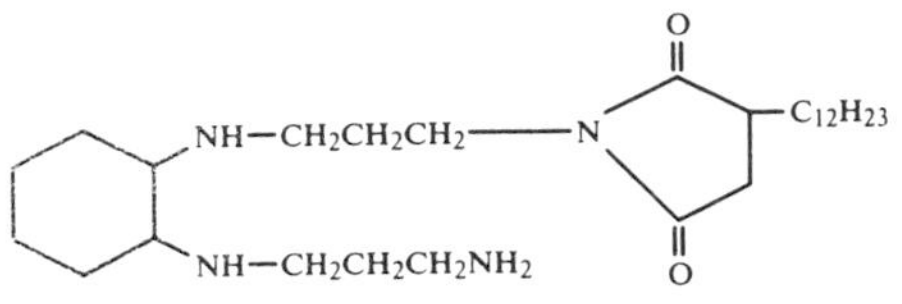

Example 2: In a 250 cc three-necked flask equipped with a magnetic stir bar, Dean Stark trap, and heating mantle was placed 31 grams (0.087 mol) of alkenylsuccinic anhydride having an average molecular weight of 356 in 50 cc of toluene. To this solution was added 19.9 grams (0.087 mol) of the tetramine, N,N'-di-(3-aminopropyl)-1,2-diaminocyclohexane:

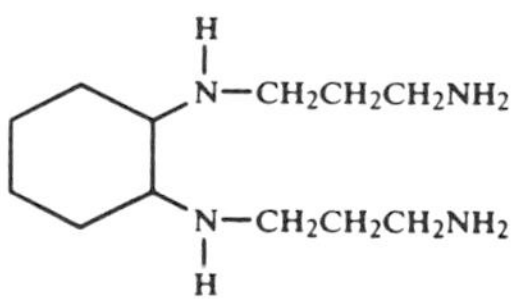

An exothermic reaction occurred on mixing the solutions. The reaction mixture was then refluxed for 14 hours and the resulting water was collected in the Dean Stark trap. The toluene was removed by evaporation to give 97% yield of the crude viscous adduct. An infrared spectrum indicated that no anhydride was present. The compound produced from the above reaction can be represented structurally as

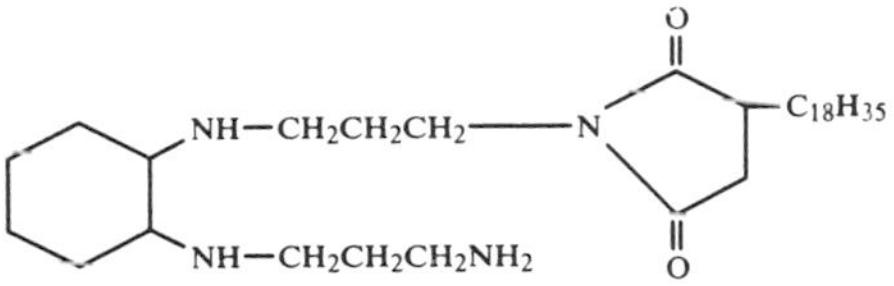

Example 3: To mineral oil with viscosity of 22 cp at 23°C, the alkenylsuccinic anhydride tetramine adduct (product produced in Example 2) was introduced to make a mixture at room temperature. This adduct was completely dissolved in the mineral oil at 1% and partially dissolved in it above 5%. The infrared spectrum of the mineral oil rich phase shows the new absorption bands at 177 and 1,700 cm^{-1}. This indicates that this particular adduct (imide structure) is miscible with mineral oil in amounts of up to about 5%.

Example 4: One-tenth gram of carbon black having a particle size of 24 μ was introduced into 10 grams of mineral oil solution containing 0 to 1% of the product produced in Example 2. After mixing thoroughly, these oil mixtures were each placed in a transparent centrifuge tube and centrifuged for one minute at the speed of 3,000 rpm. The oil-carbon black mixture became almost substantially clear with the precipitation of carbon black in the bottom; but the mineral oil-carbon black admixture of the product of Example 2 appeared to have a uniform dispersion of carbon black.

Amine Condensation Products

F. Fossati, A. Peditto and V. Petrillo; U.S. Patent 4,127,492; November 28, 1978; assigned to Liquichimica Robassomero SpA, Italy describe a dispersing additive for lubricating oils consisting of a mixture of polyfunctional compounds of the type resulting from this process according to which an olefin polymer is condensed with an unsaturated carboxylic acid, and the condensation product is reacted wtih an amine. The process is improved in that the condensing step is carried out in the presence of an olefin derivative having a relatively short chain and the condensation product, before being reacted with the amine, is added with a fatty acid and then partially esterified with a polyhydroxylated compound. The following examples illustrate the process.

Example 1: 0.5 mol of a polybutene having molecular weight of 110 was heated to 105°C and subjected to a partial chlorination with a chlorine amount of 0.25 mol.

The temperature was allowed to rise up to 110°C, and then 0.05 mol of the product resulting from the polymerization of propylene and having a distillation range of between 180° and 220°C at 760 mm Hg, which is usually known as the propylene tetramer, was added, the added olefin being allowed to react with the chlorine dissolved into the polybutene.

The chlorine excess was then removed by purging with nitrogen and the product was thereafter supplemented with an amount of 0.65 mol of maleic anhydride.

The condensation was carried out at a temperature of 230°C until completed. The unreacted excess maleic anhydride was then removed together with the last traces of chlorine residue by purging with nitrogen. The analysis of the thus obtained product showed a saponification number of 105. To the thus obtained mixture of alkenylsuccinic anhydrides, an amount of 0.05 mol of oleic acid, and then an amount of 0.1 alcoholic equivalents of a polyoxyethylene glycol, having molecular weight of 400, were added.

The mixture was then heated for 2 hours to 180°C to promote the esterification. Then residue amine ("amine bottom") containing 35% of nitrogen in an amount corresponding to 0.85 basic equivalents was added and the condensation reaction was initially carried out at 140°C, and then up to 200°C, the little residue water being removed by purging with nitrogen.

The product was then diluted with a solvent refined paraffinic oil, having a viscosity of 150 SUS at 100°F, so as to bring the total nitrogen content to 1.3%. The thus obtained product was used in a 2% by weight lubricating composition, together with 0.75% of a calcium superbasic sulfonate and 0.75% of a zinc alkyldithiophosphate, dissolved in a SAE 30 solvent refined paraffinic oil. The thus prepared composition was tested in a Petter AV 1 engine test run, and an average merit rating of piston cleanness of 9 was obtained. In In comparative tests carried out with known prior art succinimide dispersing agents the average merit rating was 8.2.

Example 2: Polybutene (0.5 mol) having molecular weight of 1,100 was heated to 105°C and partially chlorinated with a chlorine amount of 0.25 mol. The temperature was allowed to rise to 110°C and then an amount of 0.05 mol of a fraction of n-olefins, having an internal double bond and with between 11 and 14 carbon atoms was added, this fraction being the product of the dehydrogenation of n-paraffins obtained according to the Isosiv process, the mixture being allowed to react with the chlorine dissolved in the polybutene for about 30 minutes.

The chlorine excess was then removed by purging with nitrogen and the product was then supplemented with an amount of 0.63 mol of maleic anhydride. The condensation was carried out at a temperature of 230°C until completed. The unreacted excess maleic anhydride was then removed together with the last traces of residue chlorine by purging with nitrogen. The analysis of the obtained product revealed a saponification number of 120.

To the resulting mixture of alkenylsuccinic anhydrides an amount of 0.05 mol of a synthetic fatty acid was added, the fatty acid being prepared, through the well-known Oxo process, starting from a cut of internal n-olefins, having a number of carbon atoms of between 15 and 18 and being the product of the dehydrogenation of the corresponding n-paraffins obtained by the process known as Isosiv.

The thus obtained mixture was reacted with an amount of 0.15 alcoholic equivalents of the sorbitan oleic monoester. The mixture was then heated to 180°C for 4 hours to promote the esterification. An amine residue containing 35% of nitrogen in an amount corresponding to 0.8 basic equivalents was then added and the condensation reaction was carried out at an initial temperature of 140°C, and thereafter at 200°C, the residue water being removed by purging with nitrogen.

The product was then diluted with a solvent refined paraffinic oil, having a viscosity of 150 SUS at 37.8°C, until the total nitrogen content was 1.3%. The resulting product was used in a 2% by weight lubricating composition, together with 0.75% of zinc alkyldithiophosphate and 0.75% of calcium superbasic sulfonate, dissolved into a SAE 30 solvent refined paraffinic oil. The thus obtained composition was tested in a Petter AV 1 engine test run, and the resulting average merit rating of piston cleanness was 9.5.

Alkylene Polyamine Reaction Products

J.C. Nnadi and M.H. McIntyre; U.S. Patent 4,086,173; April 25, 1978; assigned to Mobil Oil Corporation have found that certain multifunctional additives may be incorporated in lubricant compositions for imparting improved dispersancy and antirust properties employing polybutenylsuccinic anhydrides, for preparing alkenylsuccinimide dispersants and antirust improvers, in an amount two or more times greater than those conventionally employed for making alkenylsuccinimide dispersants.

These products are obtained by reacting an alkenylsuccinic anhydride having from about 20 to 500 carbon atoms in the alkenyl group with a member of the group consisting of: (a) an alkylene polyamine having from about 2 to 10 carbon atoms in the alkylene group and having the following structure, $H_2N-(C_nH_{2n}NH)_x-C_nH_{2n}-NH_2$ in which n is a whole number from 2 to 4, x is a whole number from 1 to 10, the anhydride being present in an amount sufficient to react with all amino groups in the alkylene polyamine; (b) a mono- or a bis-alkenylsuccinimide of an alkylene polyamine having up to 500 carbon atoms in the alkylene group; and (c) a glycol ester, aminoglycol ester, imidoester, amidoester or oxazoline ester of an alkenylsuccinic anhydride having from about 20 to 500 carbon atoms in the alkylene group.

Example 1: 170 grams 61% active bis-succinimide (made from polybutenyl-succinic anhydride MW 1,000 and tetraethylene pentamine) 130 grams 56% active polybutenylsuccinic anhydride (MW 1,450) and 60 grams process oil were heated to and kept at 170°C for 4 hours and at the same temperature under house vacuum for one-half hour. The yield of filtered product was quantitative. Found: 0.91% N; 0.30% basic N; and acid number (ASTM D664-5) 4.16. The above represents the case of a 1:1 molar ratio of succinimide to succinic anhydride.

Example 2: The procedure of Example 1 was repeated except that the molar ratio of the succinimide to the anhydride was 1:2. The yield of filtered product was quantitative. Found: 0.65% N; 0.14% basic N; and acid number 8.3.

Example 3: The procedure of Example 1 was repeated except that the molar ratio of the succinimide to the anhydride was 1:3. Found: 0.50% N; 0.11% basic N; and acid number 12.4.

Example 4: The procedure of Example 2 was repeated except that the alkenyl group of the succinic anhydride had a MW of 2,000. Found: 0.45% N; 0.08% basic N; and acid number 6.3.

Example 5: The procedure of Example 1 was repeated except that

was used in place of the succinimide. The molar ratio of the triazine crosslinked succinimide to the polybutenylsuccinic anhydride (polybutene, MW 1,350) was 1:1. Found: 0.49% N; 0.20% basic N; and acid number 4.9.

Example 6: 7 grams bis-alkenylsuccinimide (made from tetrapropenylsuccinic anhydride and tetraethylene pentamine), 250 grams polybutenylsuccinic anhydride (the alkenyl group having a MW of 2,000) and 50 grams process oil were reacted as in Example 1 to obtain a quantitative yield of product. This example employs a molar ratio of 3:1 of anhydride to bis-succinimide and illustrates the case where the alkenyl group of the bis-succinimide is both propylene derived and low molecular weight.

Examples 7 through 14: From 7 grams bis-succinimide of Example 6, 130 grams alkenylsuccinic anhydride (the alkenyl group having a MW of 1,350) and 20 grams process oil, reacted as in Example 6, quantitative yield of product was obtained.

The dispersancy improving and antirust products of this process were subjected to a series of tests to evaluate their utility in lubricating oils. In the following humidity chamber rust test the base oil employed was a 150 SUS at 210°F base stock lubricating oil. To this oil are added 2% by weight of the products of this process, and the days required to show rust were recorded.

The rust test is a general purpose and rather severe test. It utilizes a humidity chamber operated at 120°F and 97 to 98% relative humidity with an air circulation rate of 150 cubic feet per minute.

The test panels are 2" x 4" x ⅛" polished steel plates of SAE 1010 steel of a 10 micron finish. The test is performed by first cleaning a new panel in naphtha, absolute methanol and xylene in that order. The air dried panel is then dipped in a test formulation for one minute and then "drip-dried" for 2 hours prior to insertion into the chamber. The panels are suspended in a vertical position within the chamber and can be continuously monitored through the glass dome of the chamber.

The severity of the test can be judged by the rapid rusting rate (1 hour) of a panel coated only with a base stock compared to complete rust inhibition for periods up to 7 days when utilizing an effective rust inhibitor in concentrations of 0.5 to 4.0%.

As will be apparent from the table below, the presence of the products of this process in the base oil results in a longer period of time for rust to occur (Examples 10 through 14) than those for the base oil containing only the succinimide of Example 1 (Example 9) or the base oil alone (Example 8).

Rust Test in Humidity Chamber

Example		Days to Show Rust
8	Base oil	<1
9	Base oil plus 2% succinimide of Example 1	1
10	Base oil plus 2% product of Example 2	4
11	Base oil plus 2% product of Example 3	>7
12	Base oil plus 2% product of Example 4	>7
13	Base oil plus 2% product of Example 6	>7
14	Base oil plus 2% product of Example 7	>7

Arylamine-Aldehyde Products

A process described by *S. Chibnik; U.S. Patent 4,153,564; May 8, 1979; assigned to Mobil Oil Corporation* is concerned with a class of products made by reaction of an alkenylsuccinic anhydride with arylamine-aldehyde moieties.

As an example the product may be made by: (1) reacting an alkenylsuccinic anhydride with a resin prepared from an aromatic amine and an aldehyde; and (2) reacting the material from (1) with an aromatic triazole and an aldehyde. The following examples illustrate the process.

Example 1: Forty-one parts of a mixture of products having the following formula

where x is either 0, 1 or 2, was reacted with 412 parts of polyisobutenylsuccinic anhydride (prepared using 1,300 MW polyisobutene) at 150°C for 3 hours under a 5 mm vacuum. The product was diluted with a light mineral oil and was filtered.

Example 2: 12 parts of the reactant shown in Example 1 and 130 parts of the same polyisobutenylsuccinic anhydride were reacted for 3 hours at 150°C under a 2 mm vacuum. The molar ratio of anhydride to resin was 1:1. The product was diluted with 150 parts of light oil and was cooled to 95°C. 8 parts of tolyltriazole was added and a total of 3 parts of paraformaldehyde was added in equal portions over a 1-hour period. The reaction was completed by removal of volatiles at 150°C for 1 hour at 2 mm of pressure.

Example 3: Polyisobutenyl (MW 500) succinic anhydride (100 parts) and 31 parts of the resin shown in Example 1 were reacted, and this product was reacted with 19 parts of benzotriazole and 6.3 parts of paraformaldehyde. Reaction conditions were similar to those of Examples 1 and 2.

Example 4: Under conditions similar to those outlined in Examples 1 and 2, a product was made using 130 parts of polyisobutenyl (MW 1,300) succinic anhydride, 15 parts of the Example 1 amine-aldehyde resin, 8 parts of tolyltriazole and 3 parts of paraformaldehyde.

Test for Emulsification — The products described above were examined for emulsibility by dissolving 1.3% of each in a 100" solvent refined, paraffinic, neutral oil and testing in accordance with ASTM D-1401. There was no separation of oil or water with either of the Example 1 and 2 products.

Diesel Oil Test — 3.2 parts of the respective products (on an oil-free basis) was compounded with 1.6 parts of calcium sulfonate, 0.4 part of calcium phenate and 1 part of zinc alkyldithiophosphate in 98.3 parts of solvent refined SAE 30 grade lubricating oil. The base fluid and the same base fluid containing the individual additives were next subjected for evaluation in a diesel oil test. This test was developed to produce deposits from the oxidation of lubricating oil under conditions which closely approximate those found in the piston zone of a diesel engine.

The test consists of an aluminum cylinder heated by radiant energy from an internal heater. The surface temperature of the heater is maintained at 575°F during the test period (140 minutes). The shaft turns slowly (2 rpm) and into an oil sump where it picks up a thin film of oil. This thin film is carried into the oxidation zone where heated gases (moist air at 350°F is typically employed, however, nitrogen oxides, sulfur oxides and other mixtures can be used) to form oxidation deposits.

These deposits can be affected by the detergent as the test cylinder rotates into the sump. The efficiency of the detergent is rated by the color and intensity of the shaft at the end of the test. The comparative results obtained employing this test, are shown in the table below.

Antioxidant Test — In this test, the additive is placed in an oil, and then the composite is heated to 375°C. Dry air is passed through it at the rate of 10 pounds per hour. There are also present in the composite iron, copper, aluminum and lead. After 24 hours the neutralization number for each oil composition is obtained according to ASTM Method D741-1. The effectiveness of the additive is revealed by comparison of the control of viscosity increase and control of acids with the additive-free oil or standard. Results of viscosity determinations at 100°F are shown below.

	Example 1	Example 2	Standard*
Diesel Oil Test			
(100 = clean)	74.0	68.0	50
Antioxidant Test			
Viscosity (cs)	175.2	32.3	323.8
Sludge	Trace	Trace	Heavy

*Solvent refined SAE 30 mineral oil

Substituted Pyrimidine and Triazine Ring Compounds

J.C. Nnadi and I.J. Heilweil; U.S. Patents 4,113,725; September 12, 1978 and 4,116,875; September 26, 1978; both assigned to Mobil Oil Corporation describe a class of multifunctional additives for industrial fluids, the compounds having the following general formulas:

(a) 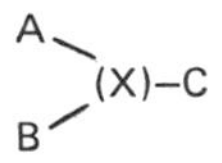 $\begin{array}{c} A \\ \diagdown \\ \diagup \\ B \end{array}(X)-C$ and   (b) 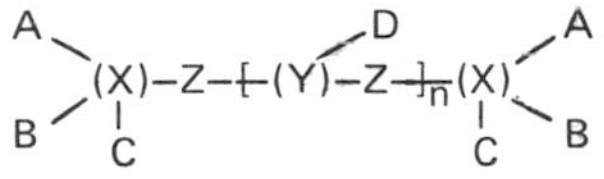 $\begin{array}{c} A \\ \diagdown \\ \diagup \\ B \end{array}(X)-Z-\!\!\left[(Y)-Z-\right]_n(X)\begin{array}{c} D \diagup \diagdown A \\ | \\ C \diagdown B \end{array}$

in which each X and Y represents a heterocyclic nitrogen radical and may be the same or different for each occurrence of X and Y; Z is a basic nitrogen-containing radical; n is 0 or an integer of at least 1, preferably 1 to 10; A, B, C and D are linked groups derived from compounds which may provide desired functions, such as detergent, antioxidant, and antiwear properties, or indirectly useful functions, such as adsorbency.

At least one of A, B, C or D is amino or anilino or is derived from an alkenyl-succinimide or an alkyl lactam or tetrahydropyrrolidine, or alkyl-substituted Mannich base, having at least 8 carbons in the alkenyl or alkyl radical, or combinations of any of these. The reaction between these products and metal compounds, particularly alkali and alkaline earth metal compounds, provides more improved properties.

The compounds may be prepared generally by reacting a halogenated heterocyclic compound, such as di- or trichloropyrimidine or cyanuric chloride, with the reactants necessary to supply the desired substituents, at least one of which is a basic nitrogen compound.

To produce (b)-type compounds, ammonia and primary amines, diamines and higher polyamines are used to obtain additional heterocyclic linkages. This substitution reaction is preferably carried out at a temperature between 70° and 250°C over a period of from 0.5 to 15 hours. To illustrate preparation of typical compounds of this process using a chlorinated pyrimidine as the heterocyclic compound, one of the basic nitrogen atoms of an alkenylsuccinimide amine (the terminal nitrogen atom) having the structure

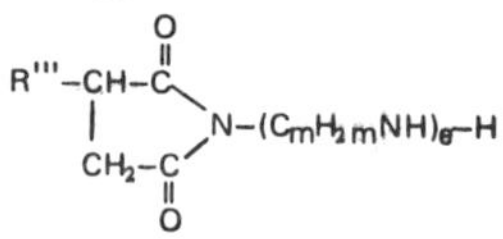

$$R'''-CH-\overset{\overset{\displaystyle O}{\|}}{C} \diagdown \underset{\underset{\displaystyle O}{\|}}{\underset{CH_2-C}{\diagup}} N-(C_mH_{2m}NH)_{g}-H$$

or a bis(alkenylsuccinimide)amine (one of the inner nitrogen atoms) having the structure

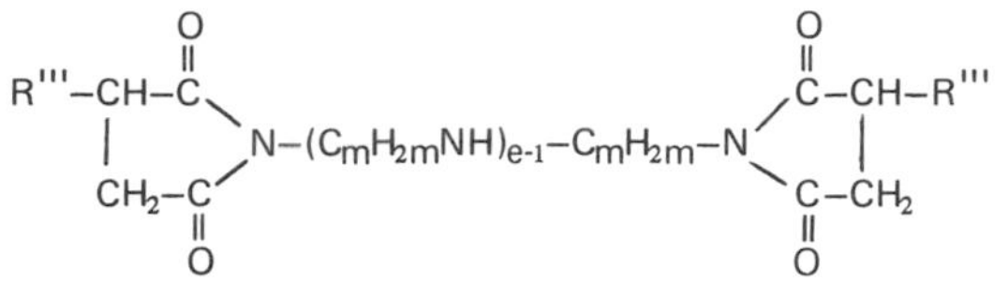

$$R'''-CH-\overset{\overset{O}{\|}}{C}\diagdown\underset{CH_2-\underset{\overset{\|}{O}}{C}}{}N-(C_mH_{2m}NH)_{e-1}-C_mH_{2m}-N\diagup\overset{\overset{O}{\|}}{C}-CH-R'''$$

where R''' is an alkenyl group, m is 1 to 3, and e is 1 to 10, becomes linked to the pyrimidine upon evolution of hydrogen chloride. The molar ratios of the two reactants may be varied to replace all of the chlorine atoms to produce the (a)-type compound. The corresponding alkyl lactam or bis(alkyl lactam)amines and other amino or anilino substituents may be added in a similar manner. Compounds having mixed substituents, for example, a mono-amino-di-succinimideamino pyrimidine, may be obtained by varying the amine-type reactant. One of the most preferred products of this process consists of a pyrimidine or triazine ring substiuted by three bis(polyalkenylsuccinimide)-tetraethylene pentamine groups. The following examples illustrate the process.

The products produced in these examples are analyzed by gel permeation chromatography using Water Associates Permeation Chromatograph, Model 200. The procedure has been described in literature as, e.g., *Journal of Chemical Ed.,* volume 43, page A567 (1966).

In these examples the essential ingredient may be referred to by structure or by name. The naming or depicting of these products is for convenience only in describing the type of molecule believed to be produced.

Example 1: In a suitable reactor equipped with an agitator and condenser were added 18.2 grams (0.1 mol) of 2,4,6-trichloropyrimidine and 52 grams (0.2 mol) of dodecylaniline. The mixture was heated to 150° to 180°C for 90 minutes during which period hydrogen chloride evolution ceased.

To the di(dodecylanilino)chloropyrimidine was added 20 grams (0.1 mol) of tetraethylenepentamine. The reaction mixture was heated to a temperature of 190° to 210°C for 3 hours, then 135 grams (0.1 mol) polybutenylsuccinic anhydride, obtained by reacting a polybutene having a molecular weight of 1,350 with maleic anhydride, was added. The reaction mixture was stirred at 150° to 180°C for 5 hours, during which time water condensed out. The reaction was stopped after 1½ hours at 150°C under house vacuum and nitrogen atmosphere after water condensation ceased. The yield of final reaction product was 170 grams containing a compound having the presumed structure below.

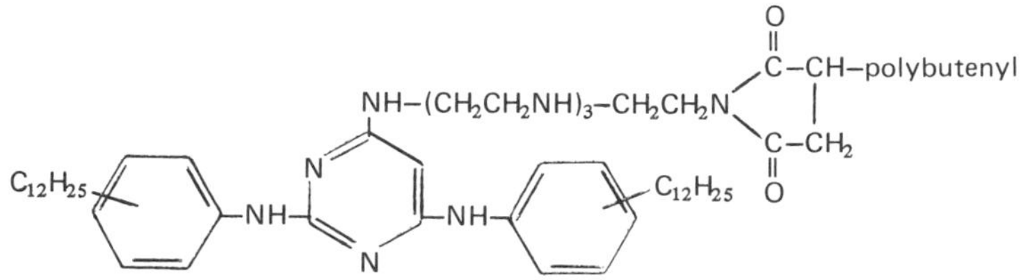

The product was dissolved in about 500 cc toluene and washed once with 50 cc 5% NaOH solution and twice with 500 cc H_2O. n-Butanol was used to break up the resulting emulsion. Upon distilling off the toluene-alcohol from the organic layer and holding the residue at 170°C under house vacuum for 2 hours, the yield was 160 grams. Found: 4.25% N. The reaction mixture may also contain a minor amount of the corresponding half amide.

Example 2: To the same apparatus used in Example 1 were added 30 grams (0.165 mol) of 2,4,6-trichloropyrimidine and 1,050 grams (0.50 mol) of the bis-polybutenyl succinimide of tetraethylene pentamine, the polybutenyl groups having about 900 MW. The reaction mixture was heated at 150° to 180°C for 7 hours, cooled and dissolved in 1,500 cc toluene and washed twice with 200 cc of 12.5% NaOH solution and twice with 250 cc distilled water; n-butanol was used to break the emulsion during the washes.

The washed material was distilled to 150°C under vacuum and nitrogen for 2 hours. The yield of final reaction product was 940 grams (96% theoretical), a major component of which is believed to be 2,4,6-trisubstituted pyrimidine, the substituent obtained from the bis-succinimide reactant. Found: 2.26% N; 0.97% basic N; and 0.17% Cl.

Example 3: To a reactor were added 8.5 grams (0.05 mol) of 2-amino-4,6-dichloropyrimidine and 300 grams (0.1 mol) of the bis-polybutenylsuccinimide of Example 2. The reaction mixture was heated at 200° to 220°C for 6 hours during which hydrogen chloride evolved. The resulting reaction mixture was treated as in Example 2. The yield of the remaining reaction product, believed to contain primarily the corresponding 4,6-disubstituted-2-aminopyrimidine, was 270 grams (96% theoretical). This reaction product has the following analysis: Found—2.36% N; 0.98% basic N; and 0.15% Cl.

Example 4: In a reactor similar to that of Example 1, 32 grams (0.2 mol) of a 2-amino-4,6-dichloropyrimidine was reacted with 550 grams (0.2 mol) of the bis-polybutenylsuccinimide of Example 2 at 170°C for 2 hours. To the resulting product was added 20 grams (0.1 mol) of tetraethylene pentamine, and the mixture heated at 175° to 180°C for a period of 5 hours. The reaction mixture was treated in the same manner as in Example 2.

The yield of reaction product, containing primarily bis-substituted-pyrimidinyl-amine of the (b)-type (in which n is 0, A is $-NH_2$ and B is derived from the bis-succinimide), was 565 grams (over 96% theory). The product has the following analysis: Found—3.23% N; 2.34% basic N; and 0.32% Cl.

Example 5: Using the same procedure as in Example 3, 1 mol of aminodichloro-triazine, made by reacting cyanuric chloride with ammonia gas, is reacted with 2 mols of the bis-polybutenylsuccinimide reactant of Example 2. The resulting product, consisting primarily of an aminotriazine bearing two succinimide amino substituents, is obtained in the yield of 96% theoretical and has the following analysis: Found—2.20% N; and 0.81% basic N.

Example 6: Comparison — Following the procedure of U.S. Patent 3,623,985, two products were prepared as follows: (a) One mol of tetraethylenepentamine was reacted with one mol of polybutenylsuccinic anhydride obtained from a polybutene having a molecular weight of about 900 and maleic anhydride.

3 mols of the resultant product was reacted with 1 mol of cyanuric chloride. The product was washed with sodium bicarbonate solution and then with water. (b) The same conditions and procedure as in (a) were followed, except the polybutene precursor had a molecular weight of 1,350.

Two further examples show the reaction of substituted triazines with NaOH.

Example 7: Comparison — In a suitable reactor, 100 grams of the product of Example 6 is reacted with 2 cc of a solution of 50% by weight of sodium hydroxide in water. The reaction is carried out by heating the above in solvent mixture consisting of 50 cc toluene and 50 cc isopropyl alcohol to remove all the toluene-H_2O-alcohol overhead and holding the residue at 170°C under house vacuum for 1½ hours and filtering. Analysis: Found, 3.5% N; and 0.52% Na.

Example 8: Comparison — One mol of cyanuric chloride is reacted with 3 mols of the polybutenylsuccinimide of Example 2 under the conditions of Example 2, to produce the corresponding trisubstituted traizine. The resulting product (100 g) was reacted with 4 cc of NaOH solution as in Example 7. The yield of product was quantitative. Analysis: Found, 2.36% N, and 0.98% Na.

Evaluation of Products — The additives of this process are tested in a series of tests to indicate their utility in lubricating oils. The sulfuric acid and pyruvic acid tests indicative of detergent properties are described in U.S. Patent 3,368,972. The test oil consists of a blend of solvent refined mineral oils (SUV at 210°F of 64.1) with 1% by weight of a zinc dialkyl phosphorodithioate. To the oil is added 3% by weight of a compound of this process. In the sulfuric acid test, the lower the result the better the additive. In the pyruvic acid test, the higher the result the better the additive. The following results are obtained.

Test Oil	Sulfuric Acid Test	Pyruvic Acid Test (%)
Alone	0.102	58.6
Example 1 product	0.002	99.6
Example 2 product	0.006	99.9
Example 3 product	0.004	99.9
Example 4 product	0.004	99.9

The compounds of this process have also been tested as lubricant additives in oxidation stability or antioxidant tests. The test procedure consists of mixing air, flowing at a controlled rate, with a second controlled flow stream of nitrogen oxides and sulfur dioxide in a mixing tank. The gas mixture is saturated with water by passing it through a fritted glass bubbler and then through a preheater. The heated stream is introduced into a reactor at a controlled rate. Samples of test oil blends (similar to that used in the previous test except the SUV at 120°F is 86.1) containing compounds of this process are also preheated and pumped into an oil reservoir of the reactor.

An aluminum shaft, equipped to rotate at constant speed, is immersed in the oil reservoir while a portion thereof is exposed to the water-air-gas mixture. The shaft is maintained at a temperature of about 575°F. Thus, the oil from the reservoir coats the shaft as it rotates and becomes exposed to the upper portion of the reaction chamber which is filled with air and oil vapor as a thin film. The duration of the test is seventy minutes. The rating of this test is based on the amount of oil oxidation-degradation products, such as lacquer, which become deposited on the aluminum surface of the shaft. The rating is made visually by classifying the deposits as follows:

1 = clean aluminum surface or extremely light deposit
2 = moderately light or iridescent surface
3 = light or golden deposit and transparent
4 = medium or brown and translucent
5 = heavy or brown and opaque
6 = very heavy black or brown and rough

Oil Composition	Concentration (% by wt)	Rating
Oil blend	0	4.0, 4.3, 3.7
Bis-succinimide reactant*	10	4.5
Example 2 product	5	1.0, 1.1
Example 3 product	5	1.0
Example 4 product	5	1.0
Example 5 product	6	1.3
Example 6 product (a)	5	5.2
Example 6 product (b)	5	5.0
Example 7 product	5	1.3
Example 8 product	5	1.0

*Used in Example 2

An oil containing the bis-pyrimidine compound of Example 4 is tested in a standard 4-ball wear test. The oil medium is the same mineral oil blend used in the sulfuric and pyruvic acid tests. The additive concentration is 5% by weight.

| | . . . . Wear Scar Diameter (mm). . . . . | | |
| | . . . . 40 kg . . . . | | 80 kg |
Lubricant Composition	300°F	400°F	500°F	300°F
Alone	0.745	0.785	0.895	0.908
Example 4 product	0.531	0.690	0.844	0.807

The composition of Example 3 is tested in the Caterpillar 1-G Engine Test. The oil composition used in the test consists of the same mineral oil blend used in the sulfuric acid and pyruvic acid tests, containing 1.3% magnesium alkyl benzene sulfonate, 1.2% of a zinc dialkyl phosphorodithioate, 1.0% of barium phosphosulfonate of polypropylene and 2.5% of the product produced in Example 3. The test engine is a single cylinder, 4-cycle Caterpillar engine operated under the following conditions:

Speed	1,000 rpm
Brake load	19.8 hp
Oil temperature	150°F
Jacket temperature	150°F
Fuel	Diesel*

*Containing 1% sulfur

The engine is operated for 480 hours, ratings are made periodically. These ratings consists of: piston deposits (100% is clean), lacquer demerits (0 is clean) and percent top groove packing deposits (0 is clean). The results obtained are shown on the following page.

	120 hours	240 hours	480 hours
Piston rating	90.9	90.2	82
Lacquer demerits	3.7	4.3	9.6
Top groove	69.0	68.0	92.0

Using the same lubricant formulation but with the bis-polybutenyl succinimide reactant of Example 3 instead of the final product, the piston rating after 120 hours is 84; after 240 hours, 68. The lacquer demerits are 9.6 and 21.4; top groove packing 35 and 51, both for 120 and 240 hours, respectively.

Polyalkenyl Oxazoline and Polyamine

T.F. Lonstrup, D.W. Brownawell, E. Goletz, Jr. and S.J. Numair; U.S. Patent 4,113,639; September 12, 1978; assigned to Exxon Research and Engineering Company have found that the combination of an oil-soluble polyalkenyl oxazoline compound ($\overline{M}n$ = 1,000 to 3,300), preferably polybutenyl succinic-bis-oxazoline and an oil-soluble acyl nitrogen compound ($\overline{M}n$ = 1,300 to 8,000), preferably polybutenyl succinimide exhibits synergistic behavior in dispersancy and/or varnish inhibition when employed in a ratio of one part per weight of the oxazoline compound to preferably 1 to 3 parts by weight of the acyl nitrogen compound and the combination is present in at least a dispersing amount in a lubricating oil.

Example 1: A mixture of 500 grams (0.4 mol) of polyisobutenyl succinic anhydride having a saponification number of 89 and a ($\overline{M}n$) of 980, 500 ml of mineral lubricating oil (Solvent Neutral 150) as solvent, 4 grams of zinc acetate dihydrate as a promoter and 96.8 grams (0.8 mol) of THAM was charged into a glass reactor fitted with thermometer, stirrer and a Dean-Stark moisture trap, and heated. Heating at about 180°C for 4 hours gave the expected quantity of water, i.e., about 1.1 mols of water in the trap. After filtration and rotoevaporation, the concentrate (50 wt % of the reaction product) analyzed for 1.00 wt % nitrogen, and 0.06 wt % zinc. The product had a ($\overline{M}n$) of about 1,400.

The polyisobutenyl succinic anhydride used herein (also used in Example 2) was prepared by conventional technique, namely the reaction of chlorinated polyisobutylene having a chlorine content of about 3.5 wt %, based on the weight of chlorinated polyisobutylene, and an average of 70 carbons in the polyisobutylene group, with maleic anhydride at about 200°C.

Example 2: A borated derivative of the reaction product of polyisobutenyl succinic anhydride and an alkylene polyamine was prepared by first condensing 2.1 mols of polyisobutenyl succinic anhydride, having a saponification number of 89 and a ($\overline{M}n$) of 980, dissolved in Solvent Neutral 150 mineral oil to provide a 50 wt % solution with 1 mol of tetraethylenepentamine (hereafter noted as TEPA). The polyisobutenyl succinic anhydride solution was heated to about 150°C with stirring and the polyamine was charged into the reaction vessel over a 4-hour period which was thereafter followed by a 3-hour nitrogen strip. The temperature was maintained from about 140° to 165°C during both the reaction with the TEPA and the subsequent stripping.

While the resulting imidated product was maintained at a temperature of from about 135° to about 165°C a slurry of 1.4 mols of boric acid in mineral oil was added over a 3-hour period which was followed by a final 4-hour nitrogen strip.

After filtration and rotoevaporation, the concentrate (50 wt % of the reaction product) contained about 1.5 wt % nitrogen and 0.3 wt % boron. The product had a ($\overline{M}n$) of about 2,420.

Example 3: In the same manner as Example 1, 2.1 mols of polyisobutenyl succinic anhydride (saponification number of 103 and an $\overline{M}n$ of about 1,300) was utilized in place of the polyisobutenyl succinic anhydride of Example 2. The resulting concentrate (50 wt % active ingredient) analyzed for 1.46% nitrogen and 0.32% boron.

Example 4: The general process of Example 1 was used, however, in this instance, one mol of polyisobutenyl succinic anhydride (having a saponification number of 103 and a ($\overline{M}n$) of 1,300) dissolved to 50 wt % in Solvent Neutral 150 mineral oil was heated with 0.036 mol of zinc acetate dihydrate and 1.9 mols of THAM at a temperature of from 168° to 174°C. At the end of the THAM addition, the reaction mixture is sparged with nitrogen at 177°C for 10 hours. After rotoevaporation, the concentrate (50% wt % active ingredient) analyzed for 1.0 wt % nitrogen and 0.1 wt % zinc. The product had a ($\overline{M}n$) of about 1,700.

Example 5: The general process of Example 2 was used, however, 1.3 mols of polyisobutenyl succininc anhydride was used and boration was not undertaken. The ($\overline{M}n$) of the product was about 1,520.

Evaluation of Combinations in Varnish Inhibition Test — Each test sample consisted of 10 grams of lubricating oil containing 0.07 gram of the addition concentrate (50% active) which results in a total of 0.35 wt % additive present in the test sample. The test oil to which the additive is admixed was 9.93 grams of a commercial lubricating oil obtained from a taxi after 2,000 miles of driving with the lubricating oil. Each 10 gram sample was heat soaked overnight at about 140°C and thereafter centrifuged to remove the sludge.

The supernatant fluid of each sample was subjected to heat cycling from about 150°C to room temperature over a period of 3.5 hours at a frequency of about 2 cycles per minute. During the heating phase, the gas containing a mixture of about 0.7 volume percent SO_2, 1.4 volume percent NO and balance air was bubbled through the test samples and during the cooling phase water vapor was bubbled through the test samples.

At the end of the test period, which testing cycle can be repeated as necessary to determine the inhibiting effect of any additive, the wall surfaces of the test flasks in which the samples were contained are visually evaluated as to the varnish inhibition. The amount of varnish imposed on the walls is rated at values of from 1 to 7 with the higher number being the greater amount of varnish.

It has been found that this test correlates with the varnish results obtained as a consequence of carrying out an MSVC engine test. The results which are recorded in the table below indicate that combinations of the oxazoline reaction product and the acyl nitrogen compound exhibit enhanced behavior when their weight ratios range from about one part by weight of the acyl nitrogen compound to from 0.2 to 3 parts by weight of the oxazoline compound; with a synergistic result when the polyalkenyl substituent of each has a ($\overline{M}n$) of about 1,300 and about 3 parts by weight of the acyl nitrogen compound is combined with 1 part by weight of the oxazoline reactant.

Weight Percent of Additive Added to Test Oil

Test Sample	Additive of Example 2	Additive of Example 3	Additive of Example 4	Additive of Example 5	VIB Rating
1	0.35	–	–	–	6
2	0.26	–	0.09	–	5
3	0.18	–	0.18	–	4
4	0.09	–	0.26	–	5
5	–	–	0.35	–	7
6	–	0.35	–	–	5.5
7	–	0.26	0.09	–	3
8	–	0.18	0.18	–	4
9	–	0.09	0.26	–	5
10	–	–	0.35	–	6
11	–	–	–	0.35	7
12	–	–	0.09	0.26	7
13	–	–	0.18	0.18	5
14	–	–	0.26	0.09	5
15	–	–	0.35	–	6

Oxazoline Adducts

J. Ryer, J. Zielinski, H.N. Miller and S.J. Brois; U.S. Patents 4,102,798; July 25, 1978 and 4,153,566; May 8, 1979; both assigned to Exxon Research and Engineering Company have found that oil-soluble oxazoline reaction products of hydrocarbon substituted dicarboxylic acid, ester, or anhydride, for example, polyisobutenyl succinic anhydride, with 2,2-disubstituted-2-amino-1-alkanols, such as tris(hydroxymethyl)aminomethane (THAM), are useful additives in oleaginous compositions, such as sludge dispersants for lubricating oil or gasoline.

Example 1: A bis-oxazoline of polyisobutenyl succinic anhydride and tris(hydroxymethyl)aminomethane (THAM) was prepared as follows. 280 grams (0.5 equivalent) of polyisobutenyl succinic anhydride was charged into a laboratory glass one-liter reaction flask, equipped with a bottom draw-off, a thermometer, a charging funnel, a nitrogen bleed, and an overhead condenser equipped with a Dean-Stark water trap. The flask was heated in an oil bath. The anhydride was then heated to about 200°C under a blanket of nitrogen. While stirring at this temperature, 0.5 mol (60.5 grams) of THAM was added in a series of portions of about 5 grams each, over an hour period with stirring.

Thereafter, the reaction was continued with stirring at 200°C for 2 hours while collecting water from the condenser. The flask was then allowed to cool and a liter of hexane was added to the flask to dissolve the reaction product, which was then drained from the flask and filtered through filter paper to remove any solids. The hexane solution was then washed three times with 250 ml portions of methanol. Thereafter the hexane layer was placed in a rotoevaporator at 90°C for about 2 hours to evaporate off the hexane.

Then an equal weight of a neutral mineral lubricating oil having a viscosity of about 150 SUS at 100°F (Solvent Neutral 150) was added with stirring to give an oil concentrate consisting of about 50 wt % of the oxazoline reaction product in about 50 wt % mineral lubricating oil. The infrared spectrum of this concentrate product featured a strong absorption band at about 6.0 microns as expected for the bis-oxazoline. The product (50% active ingredient in Solvent

Neutral 150 oil) analyzed for 0.78 wt % nitrogen and 2.30 wt % oxygen. The observed oxygen to nitrogen (O/N) ratio of 2.95 is in excellent agreement with the theoretical O/N ratio of 3. The product showed a total acid number (ASTM-D664) of 0.03.

The polyisobutenyl succinic anhydride used above had been prepared by conventional techniques, namely the reaction of chlorinated polyisobutylene having a chlorine content of about 3.8 wt %, based on the weight of chlorinated polyisobutylene, and an average of 70 carbons in the polyisobutylene group, with maleic anhydride at about 200°C. The resulting polyisobutenyl succinic anhydride showed a saponification number of 80 mg KOH/gram.

Example 2: A mixture of 500 grams (0.78 equivalent) of polyisobutenyl succinic anhydride having a saponification number of 87, 500 ml of tetrahydrofuran (THF) as solvent, 4 grams of zinc acetate dihydrate as a catalyst and 96.8 grams (0.8 mol) of THAM was charged into the previously described glass reactor and heated. When the reaction temperature had risen to 72°C, the THF solvent distilled off. Further heating at about 200°C for four hours gave the expected quantity of water, i.e., about 1.1 mols of water in the trap. After filtration, the reaction product analyzed for 1.99 wt % nitrogen, and 0.12 wt % zinc. The product was drawn from the flask and diluted with an equal weight of the Solvent Neutral 150 mineral lubricating oil for testing in the Sludge Inhibition Bench (SIB) test. The polyisobutenyl group of the succinic anhydride averaged about 70 carbons.

Example 3: A mixture of 500 grams (0.78 equivalent) of the polyisobutenyl succinic anhydride of Example 2, 96.8 grams (0.8 mol) of THAM and 4.0 grams of zinc acetate dihydrate were charged into the glass reactor previously described. The mixture was heated in the oil bath at about 200° to 220°C for about three hours, until water ceased to evolve from the reactor. Approximately 18.0 grams (1 mol) of water collected in the trap.

The infrared spectrum of the reaction product drawn from the flask showed a strong absorption band at 6.0 microns showing the oxazoline structure had formed. Elemental analysis showed that the final product of 50 wt % of the reaction product dissolved in 50 wt % Solvent Neutral 150 oil, contained 1.08% nitrogen and 0.058% zinc.

Example 4: A mixture of 60.57 pounds (44.0 mols) of polyisobutenyl succinic anhydride of Example 2, 11.73 pounds (27.5 mols) of THAM, and 0.48 pound (1 mol) of zinc acetate dihydrate as catalyst, were charged into a small pilot plant stirred reactor equipped with a nitrogen purge, stirrer, and overhead condenser with a water trap. The reaction mixture was heated to 427°F at a rate of 122°F per hour and held on temperature until 3.35 pounds (about 84.49 mols) of water of reaction was produced.

Thereafter, the reaction contents were cooled and diluted with the aforesaid Solvent Neutral 150 oil to give a 50 wt % solution of the reaction product in 50 wt % oil. This oil concentrate showed a hydroxyl number of 37.0 and contained 0.92 wt % nitrogen and 0.05 wt % zinc, based on the weight of the concentrate.

Examples 5 through 23: Using the same general procedure as described in Example 3, various polyisobutenyl succinic anhydrides (PIBSA) of polyisobutylene having number average molecular weights of 980, 2,300 and 18,000 were reacted with 2 mol equivalents of various amino alcohols to form bis-oxazolines, including 2-amino-2-methyl-1-propanol (AMP), 2-amino-2-methyl-1,3-propane-diol (AMPD) and 2-amino-2-(hydroxymethyl)-1,3-propanediol. The reactants, proportions, and analyses of the products formed in Examples 1 through 23 are summarized in the table shown on the following page.

A number of the additives of this process were subjected to a Sludge Inhibition Bench (SIB) test which has been found, after a large number of evaluations, to be an excellent test for assessing the dispersing power of lubricating oil dispersant additives.

Using this test, the dispersant action of oxazoline additives of the process was compared with the dispersing power of a commercial dispersant referred to as PIBSA/TEPA. The PIBSA/TEPA was prepared by reaction of 1 mol of tetra-ethylenepentamine with 2.8 mols of polyisobutenyl succinic anhydride obtained from polyisobutylene of about 1,000 number average molecular weight. The PIBSA/TEPA dispersant was used in the form of an additive concentrate containing about 50 wt % PIBSA/TEPA in 50 wt % mineral lubricating oil.

This PIBSA/TEPA additive concentrate analyzed about 1.14% nitrogen, indicating that the active ingredient, i.e., PIBSA/TEPA per se, contained about 2.28% nitrogen. Sufficient quantities of all the additive concentrates tested below were used in making the test blends to furnish the 1.0, 0.5 and 0.3 wt % of actual additive. The test results are given in the table below.

Sludge Dispersancy Test Results

Additive of Example	. . . . .Milligrams Sludge per 10 Grams Oil at.		
	1.0 wt %	0.5 wt %	0.3 wt %
10	1.6	6.6	—
15	3.5	—	—
14	3.8	—	—
12	2.9	5.9	7.3
11	2.7	6.1	7.3
21	2.6	5.9	—
22	2.8	6.3	—
23	2.0	4.1	—
PIBSA/ TEPA	5.2	7.5	7.7

It will be noted from the table above that the dispersants of the process were more effective than the commercial PIBSA/TEPA dispersant which is in widespread use in crankcase lubricating formulations. Specifically 1.0 wt % of the PIBSA/TEPA per se (i.e., 2 wt % of its 50 wt % concentrate) gave 5.2 mg of new sludge precipitated, per 10 grams of crankcase oil.

On the other hand, the bis-oxazoline products of this process shown in the table above were more effective as sludge dispersants since they stably suspended a larger proportion of the new sludge as shown by the fact that less sludge precipitated down during the centrifugation. Similar results are shown at the 0.5 wt %, and 0.3 wt % active ingredient levels.

Oxazoline Adducts of Polyisobutenylsuccinic Anhydride and 2,2-Disubstituted-2-Amino Ethanols

Example	PIB (MW)***	PIBSA Sap. No.	(g)	Amino Alcohol Type	Amino Alcohol (g)	Zinc Salt** (g)	Analyses of Concentrate* N (wt %)	Zn (wt %)	O_2 (wt %)
1	980	80	280	THAM	60.5	0	0.78	0	2.30
2	980	87	500	THAM	96.8	4.0	1.99	0.12	–
3	980	87	500	THAM	96.8	4.0	1.08	0.058	–
4	980	84	27,500	THAM	5,300	220	0.92	0.05	–
5	980	80	500	AMP	71.2	4	0.89	0.07	–
6	980	112	500	AMP	100.0	0	1.16	0.01	–
7	980	112	150	AMP	26.8	4	0.90	–	1.10
8	980	112	150	AMP	28	0	0.95	–	1.60
9	980	80	350	AMPD	79	1.4	–	–	–
10	980	80	500	THAM	96.8	4.0	0.99	0.12	–
11	980	80	350	THAM	60.5	2.7	0.93	0.017	–
12	980	80	350	THAM	60.5	0	1.15	–	–
13	980	112	150	THAM†	36.3	0	1.18	–	3.58
14	980	112	500	THAM†	121	0	1.2	–	3.31
15	980	112	2,000	THAM†	484	0	1.38	–	3.57
16	980	112	600	THAM†	145.2	4	1.17	–	–
17	980	112	500	THAM†	121	4	1.19	0.01	–
18	980	112	500	THAM	121	4	1.28	0.06	–
19	980	112	520	THAM	121	0	1.43	–	–
20††	980	80	350	THAM	30.3	1.31	1.18	–	–
				AMP	24.51	–			
21	18,000	1.9	200†††	THAM	0.86	0.07	–	–	–
22	18,000	1.9	200†††	THAM	0.73	0.08	–	–	–
23	18,000	1.9	2,000†††	THAM	7.3	0.8	–	–	–

*Analyses on 50 wt % solution of the oxazoline reaction product in 50 wt % Solvent 150 Neutral mineral lubricating oil.

**Zinc diacetate dihydrate.

***Molecular weight of the polyisobutenyl group (PIB) by Vapor Pressure Osmometry (VPO).

†THF used to dissolve THAM.

††30.3 g of THAM added over 1 hour, then 24.51 of AMP added over 1 hour.

†††20 wt % solution in S 150 N.

Polyesters Containing Carboxypyrrolidone

According to a process described by *J.S. Elliott, B.T. Davis and S. Norman; U.S. Patents 4,070,370; January 24, 1978; assigned to Edwin Cooper and Company Limited, England and 4,127,493; November 28, 1978; assigned to Ethyl Corporation* polyesters suitable for use as lubricant additives are prepared by reacting a dicarboxylic acid or anhydride having a branched chain alkyl or alkenyl substituent containing at least 30 carbons with a compound having the formula:

(A)

$$\begin{array}{c} O \\ \parallel \\ C-R^1 \\ | \\ CH\!-\!\!-\!CH_2 \\ | \qquad | \\ CH_2 \quad C\!=\!O \\ \diagdown \; \diagup \\ N \\ | \\ R \end{array}$$

where R and R^1 are specified substituents which between them possess from 2 to 6 free hydroxyl groups.

In a particularly preferred form of the process a polybutenyl succinic anhydride of molecular weight from 900 to 1,500 is reacted with a carboxypyrrolidone of formula (A) in which R is a hydroxyalkyl or hydroxyalkylaminoalkyl group, R^1 is $-NHR^2$ and R^2 is hydroxyalkyl or hydroxyalkylaminoalkyl.

The esterification may be carried out using well-known techniques. An esterification catalyst, such as p-toluene sulfonic acid is preferably used. The esterification is preferably carried out in the absence of a solvent but an inert organic solvent, particularly, a water-entraining solvent such as toluene or xylene optionally with a polar solvent such as dimethylformamide, or mineral oil, may be used if desired.

Example 1: (a) 122.2 grams (0.2 mol) ethanolamine was added slowly with stirring to 130.1 grams (1.0 mol) itaconic acid in 200 ml water at 80°C. The solution was refluxed for 1 hour and then stripped under vacuum (~20 mm Hg) at 190°C. The product at this stage had a total acid number (TAN) of 43.8 (0 calculated). 10 grams ethanolamine was added and the mixture heated up to 200°C. Water formed in the reaction was distilled out. The acid number had dropped to 28.1. A further 30 grams of ethanolamine was added and the mixture heated at 190°C for 1 hour. The mixture was finally vacuum stripped (5 mm Hg) at 240° to 250°C. The product was found to contain 13.3 wt % nitrogen (13.0 wt % calculated) and to have a total acid number of 4.9 and a total base number (TBN) of 36.1 (0 calculated).

(b) 942.9 grams (0.7 mol) polyisobutene (MW 1,000) succinic anhydride was dissolved in 700 ml xylene. To the solution was added 151.3 grams product (a) and 1.1 grams p-toluene sulfonic acid. The resultant mixture was refluxed for 11 hours to azeotrope off the water formed (10 ml of water collected). 600 ml of solvent was then distilled off and refluxing was continued for a further 4 hours (1.2 ml of water collected). A further 2 hour reflux produced no more water. The product was cooled, dissolved in toluene, and washed with 200 ml water. The toluene solution was dried over anhydrous magnesium sulfate, filtered and stripped under reduced pressure to remove solvent. Paraffinic SAE 5 lubricating oil (100 grams) was stirred into the product and the resultant concentrate was found to contain 1.6 wt % nitrogen, 6.0 wt % oxygen and to have a TAN of 2.9, a TBN of 10.1 and a saponification number of 78.3.

Example 2: (a) 48.7 grams (0.3 mol) N-(3-aminopropyl)-diethanolamine was added slowly to a solution of 39.0 grams (0.3 mol) itaconic acid in 80 ml of water at 80°C. The resultant solution was refluxed for 1½ hours and then stripped under vacuum (~20 mm Hg) at 60°C for 2 hours and 100°C/5 mm Hg for 1 hour. The product was a very viscous yellow liquid and was found to contain 9.3 wt % nitrogen (10.2 wt % calculated) and to have a TAN of 187 (204.5 calculated) and a TBN of 226.2 (204.5 calculated).

(b) 68.6 grams of the product of (a) was heated to 120°C and 40.6 grams (0.25 mol) N-(3-aminopropyl)-diethanolamine was added in four portions distilling out any water formed after each addition. After the final addition the mixture was heated at 150° to 180°C for 2 hours and finally stripped at 210°C under reduced pressure (4 mm Hg). The product was found to contain 12.5 wt % nitrogen and to have a TAN of 3.4 and a TBN of 28.1.

(c) A mixture of 242.5 grams (0.18 mol) polyisobutene (MW 1,000) succinic anhydride, 75.3 grams of the product of (b) and 0.3 gram of p-toluene sulfonic acid was stirred and heated at 200° to 220°C in a stream of nitrogen (to remove water as it was formed) for 10 hours.

The product was cooled, dissolved in toluene and washed with 1:1 water/methanol (100 ml). The toluene solution was dried over anhydrous magnesium sulfate, filtered and stripped under reduced pressure to remove solvent. The product was found to contain 3.4 wt % nitrogen, 6.90 wt % oxygen; and to have a TAN of 5.9, a TBN of 82.1 and a saponification number of 61.7.

Example 3: (a) 107.2 grams (1.0 mol) benzylamine was added slowly with stirring to 130.1 grams (1.0 mol) itaconic acid at room temperature. The mixture was heated slowly to 170°C and water formed during the reaction was distilled out. Heating was continued at 170°C until no more water was evolved. The mixture was allowed to cool and then ground to a fine buff colored powder. The product was found to contain 6.4 wt % nitrogen (6.4 wt % calculated) and to have a TAN of 246.9 (255.8 calculated), a saponification number of 262 (225.8 calculated) and a TBN of 0 (0 calculated).

(b) 87.7 grams of the product of (a) was heated with 51.0 grams (0.44 mol) trimethylol propane oxetane at 200°C for 8 hours and then stripped under vacuum (~20 mm Hg) at 220°C. The product was a viscous brown liquid and was found to contain 4.1 wt % nitrogen and to have a TAN of 0.3, a saponification number of 159.6 and a TBN of 0.2.

(c) 212.0 grams (0.3 mol) polyisobutene (MW 450) succinic anhydride was dissolved in 150 ml xylene. To the solution was added 100.6 grams of the product of (b) and 0.3 gram p-toluene sulfonic acid. The resultant mixture was refluxed for 12 hours to azeotrope off the water formed (3.2 ml of water collected). The solvent was then distilled off and the mixture heated at 200° to 210°C for 7 hours. The product was cooled, dissolved in 80/100 petroleum ether and washed with 1:1 water/methanol (100 ml). The petroleum ether solution was dried over anhydrous magnesium sulfate, filtered and stripped under reduced pressure to remove solvent. The product was a dark brown extremely viscous liquid and was found to contain 1.3 wt % nitrogen and to have a TAN of 8.1, a TBN of 0 and a saponification number of 130.1.

Throughout the examples, total acid numbers (TAN), total base numbers (TBN) and saponification numbers are in units of mg KOH/g. Suitability of the products of this process for use as ashless dispersants in lubricants was shown by MSVC and Petter AV-B engine tests, by Panel Coker Tests and by Spot Tests.

Alkenylsuccinimides Based on N,N,N',N'-Tetrakis(3-Aminopropyl)Ethylenediamine

G. Soula and P. Duteurtre; U.S. Patent 4,094,802; June 13, 1978; assigned to Société Orogil, France describe lubricant additives comprising compositions based on alkenylsuccinimides containing at least one alkenylsuccinimide of the formula:

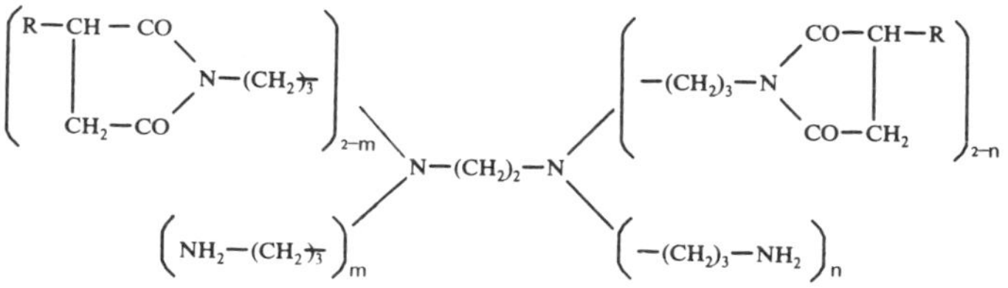

$$\left[\begin{array}{c} R-CH-CO \\ | \\ CH_2-CO \end{array} \right\rangle N-(CH_2)_3 \left. \right]_{2-m} N-(CH_2)_2-N \left[-(CH_2)_3-N \left\langle \begin{array}{c} CO-CH-R \\ | \\ CO-CH_2 \end{array} \right]_{2-n} \right.$$

$$\left(NH_2-(CH_2)_3 \right)_m \qquad \left(-(CH_2)_3-NH_2 \right)_n$$

where R is an alkenyl group of about C_{20-200}, m is an integer equal to 0, 1 or 2, and n is an integer equal to 0 or 1. These compositions can be prepared by condensation of an alkenylsuccinic anhydride with N,N,N',N'-tetrakis(3-aminopropyl)-ethylenediamine.

The N,N,N',N'-tetrakis(3-aminopropyl)ethylenediamine starting material can be prepared by cyanoethylation of ethylenediamine with acrylonitrile in a molar ratio of acrylonitrile to amine of about 4, followed by hydrogenation of the nitrile obtained.

The additive compositions impart detergent/dispersing, antirust, and antifoam properties to lubricating oils.

G. Soula and P. Duteurtre; U.S. Patent 4,081,388; March 28, 1978; assigned to Orogil, France also describe similar compositions, based on alkenylsuccinimides which impart useful dispersing, detergent, antirust and antifoam properties when employed as an additive in a lubricating oil. The compositions are alkenylsuccinimides obtained by condensing an alkenylsuccinic anhydride with at least one polyamine of the formula shown below in a molar ratio of amine:anhydride of <1.

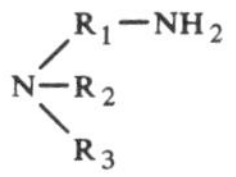

$$N \begin{array}{l} \nearrow R_1-NH_2 \\ - R_2 \\ \searrow R_3 \end{array}$$

R_1 is a $-R-O-R'-$ radical, R_2 is one of the radicals $-R-O-R'-NH_2$, $-R'-NH_2$, or $-R-OH$, R_3 is one of the radicals $-R-O-R'-NH_2$, $-R'-NH_2$, $-R-OH$,

C_{1-4}-alkyl or phenyl. R represents an optionally branched C_{2-3} alkyl, preferably ethyl or isopropyl and R' represents propyl or isobutyl.

The polyamines preferentially employed for preparing the compositions are, for example, tris-(3-oxa-6-amino-hexyl)-amine, N-ethyl-N,N-bis(3-oxa-6-amino-hexyl)-amine, N-(3-aminopropyl)-N,N-bis-(2-methyl-3-oxa-6-amino-hexyl)-amine, tris-(2-methyl-3-oxa-6-amino-hexyl)-amine, tris-(2,5-dimethyl-3-oxa-6-amino-hexyl)-amine, tris-(3-oxa-5-methyl-6-amino-hexyl)-amine, and N-(2-hydroxyethyl)-N,N-bis-(3-oxa-6-amino-hexyl)-amine.

Borated Oxazolines

S.J. Brois, A. Gutierrez and E.D. Winans; U.S. Patent 4,116,876; September 26, 1978; assigned to Exxon Research and Engineering Company describe borated derivatives of: (a) hydrocarbyl substituted mono- and bis-oxazolines obtained as a reaction product of hydrocarbyl substituted dicarboxylic acid, ester, or anhydride, e.g., polyisobutenyl succinic anhydride with from 1 to 2 molar equivalents of a 2,2-disubstituted-2-amino-1-alkanols, such as tris-(hydroxymethylamino)methane (THAM); and (b) lactone oxazolines obtained as a reaction product of hydrocarbyl substituted lactone carboxylic acids, e.g., polybutyl lactone carboxylic acid, with 2,2-disubstituted-2-amino-1-alkanols, such as tris-(hydroxymethyl)aminomethane (THAM), and their derivatives.

These products are useful additives in lubricating oils since both the sludge dispersant and/or varnish inhibiting properties of the oil are enhanced.

Thus, it has been found that a hydrolytically stable borated oxazoline lubricating oil additive which has enhanced varnish-inhibition activity can be realized by condensing a boron compound, e.g., boric acid, with the hydroxy alkyl groups of the oxazoline ring of an oil-soluble hydrocarbyl substituted oxazoline material when from about 0.1 to 2.0, preferably 0.3 to 1.0 wt % of boron as a borate ester, is present in the material.

Although the boron can sometimes be introduced readily at temperatures of up to about 200°C by transesterification of the oxazoline material, undesirable viscosity increases and gel formation can occur. Moreover, introduction of the boron via an acidic compound, such as boric acid, at such high temperatures converts the oxazoline material by destruction of the oxazoline ring into undefined mixtures of boron containing imide/amide products of reduced sludge dispersant and/or varnish-inhibition activity.

It has been discovered that boric acid can be used with little if any destruction of the oxazoline ring when esterification is effected at a temperature of not greater than 120°C, preferably from about 60° to 100°C, thus preserving the enhanced dispersant activity resulting from the presence of the intact borated oxazoline structure.

Boric Acid Treated Compounds

N. Okamoto: U.S. Patents 4,071,548; January 31, 1978 and 4,120,887; October 17, 1978; both assigned to Toa Nenryo Kogyo KK, Japan describes polyether additives and polyether-boron additives which have excellent antioxidative and anticorrosive effects in addition to dispersing effect and have an excellent thermal stability superior to that of conventional, commercial, ashless detergent-dispersants comprising succinic imide or hydroxybenzylamine.

The ashless detergent-dispersants (lubricating oil additives) can be represented by the following general formulas:

(1)
$$R-CH-\overset{\overset{\displaystyle O}{\|}}{C}-ZR'$$
$$CH_2-\underset{\underset{\displaystyle O}{\|}}{C}-OR''$$

(2)
$$\left[\begin{array}{c} R-CH-\overset{\overset{\displaystyle O}{\|}}{C}-ZR' \\ CH_2-\underset{\underset{\displaystyle O}{\|}}{C}-OR'' \end{array} \right]_{1-3} \cdot B$$

and

(3)
$$\left[\begin{array}{c} R-CH-\overset{\overset{\displaystyle O}{\|}}{C}-(OR''')_n \\ CH_2-\underset{\underset{\displaystyle O}{\|}}{C}-OR'''' \end{array} \quad \overset{(R'''O)_n}{\underset{N}{\diagdown}} \quad R'''-N\overset{(R'''O)_n}{\diagup} \right]_{3/2} \cdot B$$

where R represents an alkyl or alkenyl group of more than 40 carbons inclusive, R' represents a group of general formula $-(R'''OH)_2$ when Z is nitrogen or a group of general formula $-(R'''O)_m H$ when Z is oxygen, R'' represents a group of general formula $-R'''NH(R'''OH)$ or $-(R'''O)_m H$, R''' represents an alkylene group of 2 or 3 carbons, R'''' represents $-H$ or a group of general formula $-R'''NH(R'''OH)$, Y represents an alkyl group, having from 1 to 20 carbons, n represents a number from 3 to 8, and m represents a number from 5 to 20.

The compounds of formula (1) are obtained by reacting a product obtained from a polyalkenyl succinic anhydride and a polyalkylene glycol with a secondary alkanolamine.

The compounds of formula (2) are mixtures of boric acid esters obtained by treating the compounds of formula (1) with boric acid or anhydride. These esters are assumed to have a structure in which 1 to 3 mols of compound (1) are combined with 1 mol of boron.

The compounds of formula (3) are obtained by treating with boric acid or anhydride products obtained by treating polyalkenyl succinic anhydrides with N,N,N'-tris(polyoxyalkylene) alkylalkylenediamines or further with a secondary alkanolamine.

It is to be noted that by using the compounds of general formulas (1) through (3) of this process, the same effect as that obtained by dissolving a polyalkylene glycol in a mineral oil can be obtained. It has been known that boron is added to various petroleum products because of its antioxidative action and detergent-dispersing action. In this connection, according to this process, effective amounts of both polyether and boron can be incorporated in the form of just a single compound in petroleum products by employing compound (2) or (3).

Hydrocarbon-Soluble Alkyl Lactone Polyol Esters

A process described by *S.J. Brois and A. Gutierrez; U.S. Patent 4,123,373; October 31, 1978; assigned to Exxon Research and Engineering Company* concerns hydrocarbon-soluble alkyl lactone polyol esters, their method of preparation and the utility of the lactone polyol esters preferably in hydrocarbon fuel and lubricating systems as stable sludge dispersants and/or varnish inhibiting additives.

Thus, it has been found that saturated hydrocarbon, preferably long carbon chain, substituted structures which feature vicinal lactone and polyol ester functions can be so constructed using synthetic methods whereby a highly stable additive of enhanced dispersancy, enhanced viscosity properties and/or antirust properties is obtained. Moreover, further functionalization of the lactone polyol ester system via the process with vicinal hydroxyl, thiyl and sulfo groups can engender other desirable properties, such as antioxidation and anticorrosion activity. This class of additives can be represented in part by the formula:

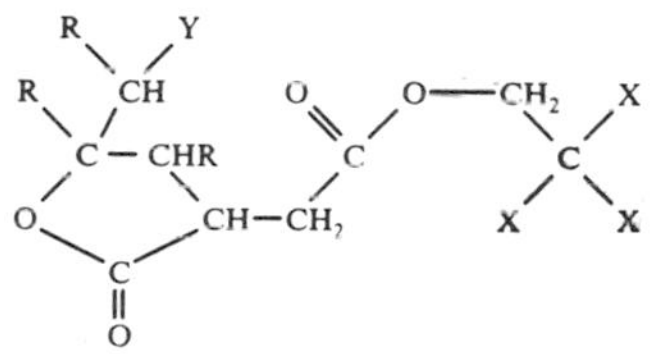

where R is selected from the group consisting of hydrogen, hydrocarbyl and substituted hydrocarbyl containing from 1 to 10,000 or more carbons with the restriction that at least one R has at least about 8 carbons and preferably from about 16 to about 400 carbons, X is selected from the group consisting of hydrogen, alkyl, hydroxyalkyl, $-OCH_2C(CH_2OH)_3$, $-(CH_2)_nOH$ or the radical $-(CH_2OCH_2CH_2O)_nH$ where n is 1 to 3 and with the restriction that at least one X contains a hydroxy moiety and Y is selected from the group consisting of hydrogen, hydroxyl, sulfo, alkylthio (TS−), alkyldithio (TSS−), and a sulfur bridge, e.g., −S− and −S−S−, joining two lactone polyol ester units together as depicted below where z is a number ranging from 1 to 4 and T is defined as containing 1 to 50, preferably 2 to 20 carbons.

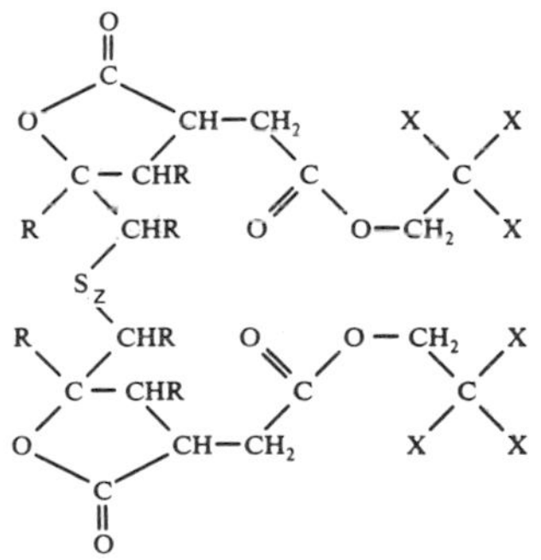

Preferred herein is polyisobutyl lactone polyol ester of number average molecular weight ($\overline{Mn}$) ranging from about 400 to about 140,000 prepared by the reaction of equimolar proportions of polyisobutyl lactone carboxylic acid with pentaerythritol at a temperature from about 100° to 240°C, preferably 150° to 180°C until one mol of H_2O per mol of reactant is removed from the reaction.

Example 1: Polybutyl [($\overline{Mn}$) of 960] lactone acid is prepared as follows. 120 grams of polybutenyl succinic anhydride (PIBSA) of ($\overline{Mn}$) ≈ 960 and having a saponification number of 92 were diluted in 100 ml of tetrahydrofuran (THF). 2 grams of water were added and the resulting mixture was heated to reflux temperature for about 2 hours. Infrared analyses of the mixture showed that the anhydride was fully converted to the succinic acid analog. The THF solvent was boiled off and 1 ml of concentrated sulfuric acid was added to the mixture at about 110°C. Heating for 2 hours at 120°C effected the conversion of the polybutenyl succinic acid to the desired lactone acid product. Infrared analyses showed the presence of strong absorption bands at about 6.5 to 8.5 microns which is characteristic of lactone acids.

The mixture was diluted in 200 ml of hexane, washed twice with 200 ml of water and subsequently concentrated by rotoevaporation for 2 hours at 80°C. Infrared analysis of the lactone acid product treated with diethylamine featured an infrared spectrum with a strong lactone carbonyl absorption band at 5.64 microns.

Example 2: Polybutyl [($\overline{Mn}$) of 960] lactone acid is prepared as follows. 600 grams of about a 51 wt % solution of the PIBSA (as described in Example 1) dissolved in Solvent Neutral 150 oil, 4 grams of water and 2 grams of Amberlyst 15 catalyst were heated at about 100°C for about 8 hours, and then at 130°C for 2 hours. Infrared analysis showed the presence of lactone acid by the presence of strong absorption bands at about 6.5 to 8.5 microns. The product was diluted with hexane, filtered, and rotoevaporated at 80°C for 4 hours. The residue upon treatment with an excess of diethylamine featured an infrared spectrum with an intense lactone carbonyl absorption band at 5.64 microns.

Example 3: Polybutyl [($\overline{Mn}$) of 1,400) lactone acid is prepared as follows. A mixture of 52 grams (about 0.04 mol) of polyisobutenyl succinic anhydride of ($\overline{Mn}$) ≈ 1,400 and having a saponification number of 80, 2 grams of water and 5 grams of Amberlyst 15 catalyst were heated for about 8 hours at 105°C. Infrared analysis indicated that the anhydride was directly and completely converted to the desired lactone acid as evidenced by the strong carbonyl absorption bands at from 5.63 to 5.84 microns. THE IR spectrum of the amber concentrate treated with an excess of diethylamine showed a strong lactone carbonyl absorption at 5.64 microns.

Example 4: 100 grams (~0.05 mol) of the polyisobutyl lactone acid prepared as in Example 2 and 6.8 grams (0.05 mol) of pentaerythritol were mixed and heated gradually to 200°C. The reaction mixture was stirred at 200°C for 3 hours, then filtered and analyzed. The filtrate featured an infrared spectrum with hydroxyl absorption at 2.9 microns and intense lactone and ester carbonyl absorptions in the 5.65 to 5.85 micron region. The product analyzed for 4.76% oxygen and was found to have a hydroxyl number of 77.9.

Example 5: 25 grams (~0.018 mol) of polyisobutyl lactone acid prepared according to the procedure outlined in Example 3, and 2.5 grams (0.018 mol) of pentaerythritol were mixed and heated to 150°C for an hour. The reaction temperature was raised to 200°C and the stirred mixture was maintained at this temperature for 2 hours. Solvent Neutral 150 oil was added to the residue to give a 51 wt % oil solution of the lactone polyol ester. The filtered product solution featured an infrared spectrum with prominent absorption bands at 2.9,

5.65 and 5.8 microns consistent with the presence of hydroxyl, lactone and ester functionality. The product solution featured a hydroxyl number of 51.9.

Example 6: Pentaerythritol Ester of PIBSA — 200 grams (~0.1 mol) of a 51 wt % solution in Solvent Neutral 150 oil of PIBSA as described in Example 2 and 13.6 grams (0.1 mol) of pentaerythritol were mixed and heated to 200°C. The reaction mixture was stirred at 200°C for about 3 hours and then filtered. The filtrate (50% ai) featured an infrared spectrum with a strong ester carbonyl absorption band at 5.8 microns and analyzed for 5.04% oxygen. The hydroxyl number for the ester product in solution (50 wt % ai) was determined to be 57.4.

Example 7: The reaction product of pentaerythritol and alkenyl chlorolactone acid according to U.S. Patent 3,755,173 was prepared as follows. 130 grams of PIBSA having a ($\overline{M}n$) of about 960 and a saponification number of 84 was mixed with 10 cc of methanol and 80 ml of benzene. Chlorine gas was bubbled through the mixture for one hour while the temperature ranged between 26° and 49°C. The reaction mixture was sparged with nitrogen for 2 hours and then rotoevaporated at 80°C for about 3 hours. The residue (135 grams) was analyzed and found to have a chlorine content of 3.9 wt %.

130 grams of the chlorine-containing product were combined with 15.9 grams of pentaerythritol and stirred under nitrogen for 12 hours at approximately 180°C. The product was then diluted with Solvent Neutral 150 oil to 50% ai and filtered through Celite 503. The filtrate (50% ai) analyzed for 2.04 wt % chlorine.

Evaluation of Additives in Varnish Inhibition Test — Each test sample consisted of 10 grams of lubricating oil containing 0.07 gram of the additive concentrate (50% active) which results in a total of 0.35 wt % additive present in the test sample. The test oil to which the additive is admixed was 9.93 grams of a commercial lubricating oil obtained from a taxi after 2,000 miles of driving with the lubricating oil. Each ten gram sample was heat soaked overnight at about 140°C and centrifuged to remove the sludge.

The supernatant fluid of each sample was subjected to heat cycling from about 150°C to room temperature over a period of 3.5 hours at a frequency of about 2 cycles per minute. During the heating phase, the gas containing a mixture of about 0.7 volume percent SO_2, 1.4 volume percent NO and balance air was bubbled through the test samples and during the cooling phase water vapor was bubbled through the test samples. At the end of the test period, which testing cycle can be repeated as necessary to determine the inhibiting effect of any additive, the wall surfaces of the test flasks in which the samples were contained are visually evaluated as to the varnish inhibition.

The amount of varnish imposed on the walls is rated at values of from 1 to 7 with the higher number being the greater amount of varnish. It has been found that this test correlates with the varnish results obtained as a consequence of carrying out an MSVC engine test. The results which are recorded in the table below indicate that the lactone polyol ester reaction products of this process have superior varnish control over that activity shown by a halo-lactone polyol ester produced according to U.S. Patent 3,755,173 and the conventional pentaerythritol ester of polyisobutenyl succinic anhydride as in U.S. Patent 3,381,022, Example 20.

In the reaction of the lactone acid or ester reagent, it is preferred to have about 1 mol of polyol per carboxy group of the hydrocarbyl substituted lactone acid or ester reagent.

Test Sample	Additive of Example	VIB Rating	Chemical Name of Additive
1	4	4	Polybutyl [($\overline{M}$n) of 960] lactone pentaerythritol ester
2	5	5	Polybutyl [($\overline{M}$n) of 1,400] lactone pentaerythritol ester
3	6	7	Reaction product of pentaerythritol and alkenyl chlorolactone acid
4	7	6	Pentaerythritol ester of polyiso-butenyl succinic anhydride

Dimercaptothiadiazole-Dispersant Reaction Products

According to a process described by *K.E. Davis; U.S. Patents 4,136,043; January 23, 1979 and 4,140,643; February 20, 1979; both assigned to The Lubrizol Corporation* nitrogen- and sulfur-containing compositions are prepared by reacting an oil-soluble dispersant with a dimercaptothiadiazole such as 2,5-dimercapto-1,3,4-thiadiazole and subsequently reacting the intermediate thus formed with a carboxylic acid or anhydride containing up to about 10 carbons and having at least one olefinic bond. The preferred carboxylic acid or anhydride is maleic anhydride. The compositions thus obtained are useful in lubricants as dispersants, extreme pressure agents, corrosion inhibitors and inhibitors of copper activity and "lead paint" deposition.

Demulsifier Additive System

E.J. Friihauf; U.S. Patent 4,129,508; December 12, 1978; assigned to The Lubrizol Corporation describes lubricant and fuel compositions which are characterized by improved demulsifying properties. These properties are contributed substantially by an additive comprising a mixture of: (a) one or more reaction products of a hydrocarbon-substituted succinic acid or anhydride with one or more polyalkylene glycols or monoethers thereof; (b) one or more organic basic metal salts; and (c) one or more alkoxylated amines.

Example 1: A mixture of 133 parts (0.5 mol) of a tetrapropenyl substituted succinic anhydride, 375 parts (0.5 mol) of a commercially available methoxy polyoxyethylene glycol (Carbowax 750) having an average formula molecular weight of about 750, and 200 parts of toluene is heated to 100°C. The reaction mixture is held at 100° to 120°C for 8 hours, then stripped at 120°C under vacuum for 1 hour. The reaction mixture is filtered to yield the filtrate as the desired acid ester product.

Example 2: A mixture of 466 parts (1 equivalent) of the acid ester prepared in Example 1 and 114 parts of a basic magnesium sulfonate (Hybase M-400) is heated at 120° to 130°C for 3 hours. The residue is the desired product.

These compositions are useful as lubricant additives in which they function primarily as demulsifier additives. They can be used in a variety of lubricants based on diverse oils of lubricating viscosity.

MODIFIED POLYOLEFIN ASHLESS COMPOSITIONS

Alkyl-Guanidino-Heterocyclic Compounds

C. Cohen and B. Sillion; U.S. Patents 4,159,898; July 3, 1979; 4,113,637; Sept. 12, 1978 and 4,071,459; January 31, 1978; all assigned to Institut Francais du Petrole, France describe alkyl-guanidino-heterocyclic compounds which have sufficient thermal stability and surfactant properties to result in a detergent action both in crankcase oils and in gasolines, even in contact with hot surfaces.

The alkyl-guanidino-heterocyclic compounds are defined as resulting, as a general rule, from the alkylation, with a halogenated hydrocarbon RX_n, of guanidino-heterocyclic compounds of the general formula:

(A)

In the formula RX_n of the halogenated hydrocarbon, R is a saturated or unsaturated substantially hydrocarbon radical containing about 10 to 200 carbon atoms, and which may further contain polar atoms or groups in proportions which do not alter its hydrocarbon nature, for example, oxygen atoms $-O-$, carbonyl groups or halogen atoms; X is a halogen atom, for example, chlorine, bromine or iodine and n is an integer, generally 1 or 2.

In the formula of the guanidino-heterocyclic compound (A), Y may be:

an oxygen atom $-O-$; in this case the heterocycle is a benzoxazole;

a sulfur atom $-S-$; in this case the heterocycle is a benzothiazole;

a $-NH-$ group; in this case the heterocycle is a benzimidazole;

a $-CONH-$ group in which the carbon atom is directly bound to the benzene ring; in this case the heterocycle is a 4-quinazolone;

R' and R'' may be a hydrogen atom, or an alkyl radical having from 1 to 10 carbon atoms; when they are in ortho positions, R' and R'' may also form together a ring fused to the benzene ring, this ring being either aliphatic or aromatic; R' and R'' may also represent a $-Z-Ar$ group in which Z is an oxygen atom, a sulfur atom or a divalent aliphatic group, such as, $-CH_2-$; $-(CH_2)_4-$ or

$$-\overset{\overset{\displaystyle CH_3}{|}}{\underset{\underset{\displaystyle CH_3}{|}}{C}}-$$

and Ar is a simple aromatic radical such as a phenyl radical or a substituted aromatic radical such as an alkyl-phenyl radical in which the alkyl group contains from 1 to 10 carbon atoms.

The halogenated hydrocarbons RX_n are preferably derived from substantially saturated petroleum cuts, or from polymers of α-olefins or of internal olefins.

Example 1: In a glass reactor, provided with a stirrer, shaped as an anchor, and heated in an oil bath, various compounds of the process (designated under references Ia to Ig), are prepared by reacting, under nitrogen scavenging, chlorinated polyisobutenes with 2-guanidino benzimidazole (A) in the presence of sodium carbonate, at the conditions reported in Table 1 below.

Table 1

| | | Chlorinated . Polyisobutene . . | | (A) | | . . . . Reaction. . . . . | | Final . . . Product . . . | |
| | | Cl | | | | | | | |
Ref.	M*	% (wt)	Weight (g)	Weight (g)	Carbonate (g)	Duration (hr)	Temperature (°C)	H % by wt	Cl % by wt
Ia	950	3.66	10	4.37	1	4	190	2.7	0.4
Ib	1,280	2.7	12.8	4.37	1	4	200	2.3	0.45
Ic	480	8.04	18	8.74	2	4	200	5.1	0.6
Id	980	4.3	10	4.37	1	4	200	3.2	0.4
Ie	950	3.66	10	2.2	1	8	200	3	0.35
If	950	3.66	2	0.84	0.2	2	250	3	0.4
Ig	960	0.4	460	200	48	4	200	1.4	0.2

*M is the molecular weight of the chlorinated polyisobutene.

Products Ia to If have been diluted with hexane after reaction, filtered with a filtration aid and then evaporated.

Product Ig has been diluted with 470 g of a 100 N oil at the end of the reaction and then filtered. The nitrogen and Cl contents mentioned for Ig have been determined on the obtained solution.

Example 2: By operating as in Example 1, there is reacted 255 g of chlorinated polyisobutene having a molecular weight of 950 and containing 3.66% by weight of chlorine, with 112 g of 2-guanidino benzoxazole and 25 g of sodium carbonate for 4 hours at 200°C. The reaction medium is diluted with hexane, filtered and hexane is evaporated therefrom. The resulting product (product II) contains 3.4% by weight of nitrogen and 0.5% by weight of chlorine.

Example 3: While operating as in Example 1, there is reacted 70 g of chlorinated polyisobutene having a molecular weight of 950 and a chlorine content of 3.66% by weight, with 30 g of 2-guanidino benzothiazole and 7.4 g of sodium carbonate, for 4 hours at 200°C.

The reaction medium is diluted with hexane, filtered and hexane evaporated therefrom. The resulting product (product III) contains 5.25% by weight of nitrogen and 0.2% by weight of chlorine.

Example 4: While operating as in Example 1, there is reacted 300 g of chlorinated polyisobutene having a chlorine content of 3.66% by weight and a molecular weight of 950, with 152 g of 2-guanidino-(3H)-4-quinazolone and 32 g of sodium carbonate for 4 hours at 200°C. The reaction medium is diluted with hexane, filtered, and hexane evaporated therefrom. The resulting product (product IV) contains 4.5% by weight of nitrogen and 0.6% by weight of chlorine.

Example 5: The dispersing power of lubricating compositions containing certain additives prepared as described in the preceding examples have been tested by examination of spots obtained on a filter paper after deposition thereon of a drop of used mineral oil containing 2% by weight of additive.

In strictly identical operating conditions there was determined, in each case, the ratio of the average diameter of the black spot to the average diameter of the oil aureole. The higher this ratio, the better the dispersing power of the additive. The results obtained are reported in Table 2 below, in which are also mentioned the results of identical tests conducted without additive and with 2% of an additive available in the trade (alkenylsuccinimide).

Table 2

Temperature (°C)	None	. . . . Additive						
		Commercial	Ib	Ic	Ig	II	III	IV
25	0.4	0.60	0.67	0.54	0.65	0.56	0.62	0.58
150	0.30	0.48	0.54	0.46	0.58	0.43	0.51	0.50

The detergent property of product Ig, added in a proportion of 2% by weight to a lubricating oil, has been tested on a Petter AV1 engine according to method AT4 and on a Petter AV1 engine according to method DEF. A merit over 100 of 94 and a merit over 10 of 9.4 have been obtained.

Chlorinated Polyisobutylene-Pyridine Reaction Products

A process described by *J.M. Larkin and H. Chafetz; U.S. Patent 4,100,086; July 11, 1978; assigned to Texaco Inc.* relates to lube oil dispersants prepared by reacting chlorinated polyolefins with a nitrogen heterocycle in the presence of alkali metal or alkaline earth metal salts.

The compositions are synthesized as illustrated by the following general equation:

$$RX \; + \; \underset{N}{\bigcirc}^{R^1} \; + \; MY \; \xrightarrow[80°-160°C]{2-20\ hr} \; dispersant \; + \; MX$$

R in this equation can be derived from a polymer of 500 to 5,000 molecular weight such as polypropylene, polybutylene, polyethylene, etc., or copolymers of these materials. X is a halogen, preferably chlorine. MY is a metal salt such as potassium, sodium, magnesium, or calcium oxides, hydroxides, carbonates, bicarbonates, phosphates, nitrates, alkanolates, phenolates, etc. The nitrogen heterocycle can be pyridine, substituted pyridines such as the picolines, nitromethylpyridines, ethyl pyridines, etc., quinolines, isoquinolines, phenazines, purines, pyrimidines, etc. Pyridine and 4-picoline are preferred amines. Sodium carbonate, sodium bicarbonate, sodium hydroxide, and sodium gluconate (or mixtures of these salts) are preferred as MY.

A solvent is not necessary, but the use of 0.1 to 10% of a polar component such as water or a low molecular weight alcohol is preferred.

The aromatic heterocycle is generally used in excess (1.5 to 30 mols per mol of halogenated polymer) and the metal salt (MY) is generally present in quantities ranging from 0.2 to 3.0 mols per mol of RX.

The dispersant is isolated by removal of the aromatic heterocycle under reduced pressure and filtration to remove MX and unreacted MY. An extraction step may be employed by partitioning the dispersant between a hydrocarbon such as heptane, gasoline, or diluent oil and methanol. (The methanol portion is discarded or the methanol and the materials it contains can be recycled.) After distillation of the hydrocarbon, the dispersant remains.

In the oils contemplated herein, the additive compositions normally constitute between about 0.1 to 10 wt % of the composition, preferably between about 0.5 and 5 wt %. Also contemplated are the concentrates thereof wherein the dispersant product content is between about 10 and 50 wt %. Concentrates are formulated for ease of handling, storage and transportation. The finished lubricating oil compositions are prepared from the concentrates via dilution with additional base oil.

The hydrocarbon base oils employed in major amounts in the finished lubricating oil compositions are derived from a wide variety of hydrocarbon base oil materials such as naphthenic base, paraffinic base and mixed base mineral oils or other hydrocarbon products such as synthetic hydrocarbon oils, e.g., polyalkylenes such as polypropylene, polyisobutylene of a molecular weight of about 250 to 2,500. Advantageously, the base oil employed in the finished lubricating compositions have an SUS viscosity at 100°F of between about 50 and 2,000, preferably between 75 and 375.

In the finished lubricating oil compositions of the process, additional additives such as supplementary detergent-dispersants, oxidation inhibitors, corrosion inhibitors, antifoamants, etc., may be employed in addition to the dispersant of the process.

Oxazoline-Containing Polymers

W.L. Smith, J.S. Kelyman and L. Jones; U.S. Patent 4,120,804; October 17, 1978; assigned to The Dow Chemical Company have discovered that oil-soluble polymers corresponding to the formulas:

$$R-X\{N-CH_2CH_2\}_n Y \quad \text{or} \quad R\{N-CH_2CH_2\}_n Y$$
$$\qquad\quad\; R_1 \qquad\qquad\qquad\quad R_1$$

$$(1) \qquad\qquad\qquad (2)$$

wherein

  (a) R is a saturated or substantially saturated polyolefin having a molecular weight of at least about 250 (preferably from about 500 to 5,000, and more preferably from 700 to 3,000);

  (b) R_1 is hydrogen or R_2CO- where R_2 is hydrogen or alkyl of from 1 to 18 carbon atoms with the proviso that at least 60% of the R_1 groups are R_2CO-, (preferably all or substantially all of the R_1

groups are R_2CO- and more preferably all of the R_1 groups are
R_2CO- wherein R_2 is hydrogen or a lower alkyl of 1 to 4 carbon
atoms);

(c)   X is $>C=O$ or $-SO_2-$;

(d)   Y is a terminal organic or inorganic group (preferably chloro,
bromo, iodo or hydroxyl); and

(e)   n is an average value of from 2 to 15, preferably from 3 to 10.

The polymers represented by formulas (1) and (2) are oil-soluble materials which
are normally viscous liquids or low-melting solids. The compounds are conveniently prepared by reacting a polyolefin bearing a displaceable halo group (e.g.,
chloro, bromo or iodo) or a polyolefin end capped with a carboxylic or sulfonic
acid group, a carboxylic or sulfonic acid halide group, or an anhydride of a dicarboxylic ethylenically unsaturated carboxylic acid (e.g., maleic or succinic
acid or anhydride) with an oxazoline or a polyoxazoline corresponding to formulas (3) and (4), respectively.

(3)   (4)

If a polyoxazoline reactant is used, it must be first partially hydrolyzed so that
there are secondary amino groups available for reaction with the acylating reagent, displaceable halide, etc., in the hydrophobic polymer reactant.

Example: Polyisobutylene sulfonyl chloride was prepared by reacting polyisobutylene (average molecular weight 920) with chlorosulfonic acid and then treating the reaction mixture with thionyl chloride. Residual hydrochloric acid was
removed by vacuum stripping. The polyisobutylene sulfonyl chloride thus produced was blended with 3 equivalents of 2-ethyl-2-oxazoline in a sealed citrate
bottle and heated at 90°C for a period of approximately 8 hours. The functionalized polymer thus produced was isolated as a viscous liquid by dissolving the
reaction product in hexane, isolating the hexane layer, and stripping away the
hexane under reduced pressure.

The isolated product was analyzed by nuclear magnetic resonance spectroscopy
to determine the proportion of oxazoline to polyisobutylene and by liquid
chromatography to determine the amount of unreacted polyisobutylene. These
analytical data showed the reaction product to be the desired grafted "three mer"
with three ring-opened oxazoline units attached to the polyisobutylene chain
through the $-SO_2-$ linkage.

The above material was evaluated as an ashless dispersant in a "spot dispersancy
test." This test is conducted by adding a quantity of the above polymer to a
5-dram bottle along with 5 g of a heavily sludged automobile crankcase petroleum-based motor oil. The bottle was capped, warmed to 300°F, and shaken for
one hour to insure proper dissolution and mixing of the components. The mixture was then heated overnight in a closed container at 320°F. After the heat
treatment, the bottle was removed from the oven and allowed to cool to room
temperature. Six drops of the oil were dropped onto the center of a 4 inch by

5 inch piece of standard white blotter paper. After 24 hours, the diameter of the sludge spot and oil spot were measured. Dispersancy is reflected by the ability of the oil to keep the sludge in suspension. In this test, dispersancy is reflected by the difference in diameters of the sludge and oil spots.

$$SDT = \frac{\text{Diameter sludge spot}}{\text{Diameter oil spot}} \times 100$$

A heavily sludged oil (with no test additive) will yield a SDT rating of 40 or less. An excellent dispersant will yield SDT ratings of from 70 to 80 when added at a 1 to 3% by weight of the base oil.

In this test method, the above oxazolinated polymer gave SDT ratings of 50.2, 80.1 and 81.7 at additive levels of 1, 2 and 3%, respectively.

Copolymer, Polyamine and Polycarboxylic Acid Condensation Products

According to a process described by *R.E. Malec; U.S. Patent 4,120,803; Oct. 17, 1978; assigned to Ethyl Corporation* there is provided an ashless dispersant for lubricating oil which is made by copolymerizing a mono-alkene, an aliphatic conjugated diene and an unsaturated fatty acid or ester thereof to form an intermediate copolymer which is amidated by reaction with a hydrocarbyl polyamine. The product is then reacted with an aliphatic or aromatic polycarboxylic acid to form inter- and intra-molecular bridges between residual amine groups of the polyamine.

Example 1: Preparation of Copolymer — A solution of 68.6 g (0.25 mol) mono-alkene and 18.5 g (0.063 mol) of methyl oleate was cooled to 0° to 5°C and 10.1 g (0.188 mol) of 1,3-butadiene was added to it. The monoalkene used was a 19.5 wt % paraffin and 78.6 wt % monoolefin which consisted mainly of C_{10}-C_{26} (mainly C_{12}-C_{18}) monoolefins which were 12.6 mol % linear alpha-olefins, 35.7 mol % branched chain vinylidene olefin and 51.7 mol % internal olefin. A second solution of 15 g of $AlCl_3$ in 300 g of ethyl chloride was prepared at 0°C.

The two solutions were added concurrently to a stirred reaction vessel. Addition took 13 minutes during which time external heat was applied to assist in vaporizing ethyl chloride. The boiling ethyl chloride controlled the reaction temperature (BP 12.3°C). The temperature was 24°C at the end of the addition. Then 400 ml of a 50/50 isopropanol/hexane mixture was added and the mixture stirred for 30 minutes. It was then washed with water and the aqueous phase removed. The organic layer was washed with a solution of 35 ml concentrated HCl in 265 ml water.

Finally, the organic layer was washed with water and the product dried over anhydrous magnesium sulfate. Volatiles were stripped from the product at 60°C at 2 mm Hg leaving 91.3 g of oily copolymer.

Amidation of the Copolymer — In a reaction vessel was placed 80.8 g of the above copolymer and 9.8 g (0.052 mol) of tetraethylene pentamine. The mixture was blanketed with nitrogen and stirred for 3.25 hours at 243° to 250°C during which time a small amount (ca 2 ml) of volatiles distilled out and was collected. The resultant amidated copolymer was blended with mineral oil to give 140 g of a solution containing 50 wt % amidated copolymer. The remainder was the oil diluent and paraffin carried through from the original monoolefin mixture.

Coupling of the Amidated Copolymer — The 140 g of amidated copolymer solution was placed in a reaction flask. To it was added 10 g (0.068 mol) of adipic acid and the resultant mixture stirred 2 hours at 200° to 210°C under vacuum (water aspirator). The resultant coupled product weighed 148.5 g and was diluted with an additional 7.5 g of SAE-7 mineral oil to maintain a 50% active product.

Other coupled amidated copolymers can readily be prepared following the above general procedure. The following examples illustrate this.

Example 2: 140 g of amidated copolymer, made as in Example 1, was placed in a reaction flask. To it was added 18.1 g of tetrapropenyl succinic acid and the mixture was then heated at 200°C for two hours to obtain a coupled copolymer.

In like manner, an aromatic polycarboxylic acid coupled amidated copolymer can be made by using 14.5 g of dimethyl phthalate in place of the tetrapropenyl succinic acid.

Similarly, any of the other aliphatic dicarboxylic compounds or aromatic polycarboxylic compounds can be substituted in Example 1 for adipic acid in the manner previously discussed to obtain a coupled amidated copolymer.

The coupled copolymers of this process are very effective ashless dispersants in lubricating oil. They can be used in both mineral oil and synthetic oils. Examples of such synthetic oils are alkylated benzenes such as octadecyl benzene, olefin oligomers such as decene-1-trimer, synthetic ester lubricants such as di(2-ethylhexyl) adipate, esters of trimethylolpropane with octanoic acid, complex condensation products of pentaerythritol, sebacic acid and methanol.

Example 3: 10,000 gal of SAE-5 solvent refined mineral oil was placed in a blending vessel. To this was added 408 kg of a commercial zinc dialkyldithiophosphate (8.3% zinc), 850 kg of a commercial over-based calcium alkaryl sulfonate (300 base number), and 4,767 kg of a commercial ethylene-propylene copolymer V.I. improver and 1,702 kg of the coupled copolymer of Example 1. The mixture was stirred until homogeneous, filtered and packaged for use in internal combustion engines.

Tests have been carried out which show that the coupled copolymer is an effective dispersant in lubricating oil. In this test, a mineral lubricating oil was formulated to contain a commercial zinc dialkyldithiophosphate, a calcium alkaryl sulfonate, a commercial viscosity index improver and other conventional additives. The commercial ashless dispersant usually included in the formulation was replaced with 2.4 wt % of the additive of Example 1.

The oil was placed in the sump of a gasoline engine and subjected to a low temperature dispersancy test. In this test, the engine was operated on a controlled cycle and at intervals it was opened up so that various internal parts could be rated for cleanliness on a scale from 0 to 10 (10 = clean). Parts were rated for both sludge and varnish contamination. The test's criteria are the hours to a No. 9 rating on both sludge and varnish. In this test, the sludge rating of the Example 1 additive, was 92 hours and the varnish rating was 83 hours which are good ratings demonstrating that the additive is an effective dispersant.

Olefin-Thionophosphine Sulfide Reaction Products

According to a process described by *S.J. Brois; U.S. Patent 4,100,187; July 11, 1978; assigned to Exxon Research & Engineering Co.* adducts are formed by a method which involves the reaction of an olefin, including olefin polymers, with a dimeric thionophosphine sulfide. Further derivatives may be formed by reacting the resulting olefin-thionophosphine sulfide adduct products with amines, e.g., alkylene polyamines, aziridines, phosphines or metal salts, e.g., zinc salts; or, alkylene polyamine- or aziridine-treated olefin-thionophosphine sulfide reaction products may be further reacted with metal salts, e.g., zinc acetate.

A representative material can be prepared by reacting polyisobutylene with an arylthionophosphine sulfide, and then further reacting the resulting cyclic bis(phosphinodithioic acid) anhydride product with an alkylene polyamine or aziridine, which, in turn, may be further modified by reaction with zinc acetate.

The oil-soluble olefin-thionophosphine sulfide products, as well as its oil-soluble derivatives outlined above, may be used as additives for lubricating oil, and petroleum fuels, e.g., gasoline or fuel oil.

Polybutyl Phenoxyacetamides

According to a process described by *J.S. Elliott, B.T. Davis and R.M. Howlett; U.S. Patent 4,083,791; April 11, 1978; assigned to Edwin Cooper and Co. Ltd., England* amides of C_{50-200} alkylphenoxy-substituted aliphatic carboxylic acids made by reacting a C_{50-200} alkylphenol with a halocarboxylic ester followed by displacement of the ester group with an amide group have a much higher content of amide than a product made using a halocarboxylic acid reactant in place of the ester.

Particularly preferred additives are the polybutyl phenoxyacetamides derived from polybutenes of molecular weight 900 to 2,800 and polyethylene polyamines having an average of from 4 to 6 amino groups, prepared by reacting polybutyl phenoxyacetic acid or an ester thereof with the polyethylene polyamine in a molar ratio of 2:1 to 1:5.

The lubricating oil used in the lubricating compositions of the process, may be any of the well-known synthetic ester oils, such as dioctyl sebacate. The preferred oils, however, are mineral oils of lubricating viscosity of well-known type.

Example 1: In this example, the amide dispersant was prepared going through an alkylphenoxy acetic ester intermediate. In a reaction vessel was placed 580 g (0.5 mol) of polyisobutylphenol and 300 ml petroleum ether. To this was added a solution of 11.5 g (0.5 mol) sodium in 120 ml methanol. The mixture was heated at reflux for 30 minutes and then solvents distilled out at 200°C/40 mm for one hour.

To the above phenate was added 300 ml petroleum ether and 120 ml methanol. While stirring at 70°C, 82.8 g (0.55 mol) of butyl chloroacetate was added. The mixture was stirred at reflux for 3 hours. The product was washed 3 times with 90% aqueous methanol. Following this, the solvent was distilled out yielding 588 g of an intermediate containing butyl polyisobutylphenoxyacetate. Saponification value of the mixture was 36.5 mg KOH/g. Theoretical saponification value if all of the polyisobutylphenol had been converted to butyl polyisobutylphenoxyacetate is 44 mg KOH/g. Thus, conversion of polyisobutylphenol to intermediate ester was 83%.

To 480 g of the ester intermediate was added 35.4 g of tetraethylenepentamine (1.2 mols tetraethylenepentamine per equivalent of ester) and the mixture stirred at 200°C for 5 hours. The resultant product was diluted with 86.9 g of process oil to give 542 g of an 85% concentrate analyzing 2.08 wt % nitrogen and having a total base number of 51.8 mg KOH/g.

Example 2: Preparation of n-Butyl Polyisobutylphenoxyacetate — To a solution of polyisobutylphenol (107.0 g, 0.1 mol), prepared by alkylation of phenol with 1,000 molecular weight polyisobutylene using a BF_3/phenol complex as catalyst, and n-butyl chloroacetate (22.6 g, 0.15 mol) in xylene (100 ml) was slowly added, over about one hour, a solution of sodium methoxide (8.1 g, 0.15 mol) in anhydrous methanol (40 ml). The addition was carried out at 100°C and on completion the solution was heated at this temperature for a further hour.

The solution was washed with 10% hydrochloric acid (50 ml) followed by 3 x 80 ml portions of aqueous methanol (1:4). After being dried over magnesium sulfate the solution was stripped of solvent. Yield, 102 g.

A sample of this product was saponified with excess aqueous potassium hydroxide, acidified with hydrochloric acid and then washed with portions of aqueous methanol (1:4) until acid free. The acid value of the thus-formed polyisobutylphenoxy acetic acid (36.5 mg KOH/g) indicated that a conversion of 73% had been obtained.

Reaction of n-Butyl Polyisobutylphenoxyacetate Tetraethylenepentamine — A mixture of polyisobutylphenoxy butyl acetate (50 g) and tetraethylenepentamine (3 g) was heated at 200°C for 4 hours under an atmosphere of nitrogen. After dilution with mineral oil the mixture was filtered to give a clear product. % N 1.8 (calculated 1.8); acidity 5 mg KOH/g; TBN 37 mg KOH/g.

Example 3: Preparation of n-Butyl Polyisobutylphenoxyacetate — A solution of sodium methoxide, prepared by dissolving sodium (43.0 g, 1.87 mols) in dried methanol (600 cc) was added to a stirred solution of polyisobutylphenol (1,943 g, 1.7 mols, prepared by boron trifluoride catalyzed alkylation of phenol with 1,000 molecular weight polyisobutylene) and n-butyl chloroacetate (281.4 g, 187 mols) in xylene (850 cc) at 100°C. The addition took 1¾ hours and the solution was then stirred at 69°C for 2 hours.

The solution was stirred with hydrochloric acid/water (1:4, 225 cc) and allowed to separate. The organic layer was then washed with water/methanol (1:4, 3 x 500 cc), petroleum spirit (BP 62° to 68°C) being added to assist separation. The solution was dried over magnesium sulfate, filtered and stripped to 170°C under vacuum. 1,734 g (79%) of the intermediate was thus obtained. Saponification value: 40.4 mg KOH/g.

Example 4: Preparation of Tetraethylenepentamine Amide of Polyisobutyl-phenoxyacetic Acid — The product of Example 3 (1,500 g 1.03 mols) was stirred with tetraethylenepentamine (119 g, 0.63 mol) at 200°C, in an atmosphere of nitrogen for 4 hours. Evolved n-butanol was permitted to escape during the reaction. The product was cooled, diluted with mineral oil (276 g) and filtered in petroleum ether solution. Removal of the solvent gave the amide (1,650 g, 91%). % N 2.2 (calculated 2.2%); Anhydrous TBN 5.5 mg KOH/g.

Suitability of the products of the process for use as ashless dispersants in lubricants was shown by MS VC and Petter AV-B Engine tests, by Panel Coker Tests and by Spot Tests.

Catalyst for Mannich Condensation Reaction of Oxidized Olefinic Polymers

C.T. West; U.S. Patent 4,131,553; December 26, 1978; assigned to Standard Oil Company (Indiana) describes a process for and products of the Mannich condensation reaction of oxidized olefinic polymers. The Mannich reaction is carried out in the presence of about 0.01 to 40.0% by weight of an oil-soluble sulfonic acid based on the neat polymer.

The sulfonic acid catalyzes the reaction producing the Mannich modified oxidized polymer, increases the resistance of the Mannich modified oxidized polymer to oxidation and varnish formation, increases the detergent-dispersing properties of the viscosity-index improver, does not appreciably harm the color of the product and increases the performance of the additive in engine tests. The Mannich reaction may be carried out with polymer oxidized in the absence of oil-soluble sulfonic acids.

Example 1: (A) Oxidation of the Copolymer — 7.0 g of one ethylene-propylene (about 40 mol % propylene) copolymer, molecular weight about 30,000, in 100 g of 5W 100 N oil was placed in a flask fitted with a stirrer and means to sparge a 50/50 mixture of air and nitrogen through the contents. A 0.14 g portion of an over-based magnesium polypropyl benzene sulfonate, molecular weight about 900, oxidation catalyst was added and the flask was heated to 395°F. Nitrogen and air were bubbled through the mixture until oxidation and polymer degradation reduced the viscosity of the mixture to about 2,100 Saybolt Universal seconds at 210°F. The reaction was complete in 2½ hours and the product was then cooled to room temperature.

(B) The Mannich Reaction — 0.42 g of hexamethylene diamine (0.0036 mol), 0.82 g of a 39.5% in oil solution of polypropyl benzene sulfonic acid (0.0036 mol) Mannich reaction catalyst, molecular weight about 900, and 0.50 g formalin (44%, 0.0073 mol) were added to 100 g of the 7% by weight oil solution of the oxidized copolymer at 390°F. The mixture was maintained at 390°F and the reaction was continued for 1 hour. The mixture was sparged with nitrogen at 390°F for 1 hour.

Example 2: Example 1A was repeated with the addition of 0.66% by weight based on the polymer oil solution of a 45.0% by weight, based on the oil, solution of a polypropyl benzene sulfonic acid, equivalent weight 807, in place of the 0.15 g of the magnesium sulfonate. In Example 1B no additional sulfonic acid was added.

Example 3: Example 1 was repeated with the mol ratio of formalin to hexamethylene diamine 1:1.

Example 4: Example 1 was repeated except with a mol ratio of formalin to hexamethylene diamine of 1.1:1.

Example 5: Example 1 was repeated with 0.68 g of alkyl benzene sulfonic acid (0.0003 mol) and the ratio of formalin to hexamethylene diamine is 1.5:1.

Example 6: Example 1 was repeated except with 0.68 g of alkyl benzene sulfonic acid (0.0003 mol) and the mol ratio of formalin to hexamethylene diamine was 2:1.

Example 7: Example 1 was repeated with 0.68 g of alkyl benzene sulfonic acid (0.0003 mol) and the mol ratio of formalin to hexamethylene diamine was 1:1.

Example 8: Example 1 was repeated except 0.68 g alkyl benzene sulfonic acid (0.0003 mol) was used and the mol ratio of formalin to hexamethylene diamine was 3:1.

Example 9: Example 1 was repeated except with the formalin to hexamethylene diamine ratio of 1:1 without the alkyl benzene sulfonic acid.

Example 10: Example 1 was repeated except with the formalin to hexamethylene diamine ratio of 2:1 without the alkyl benzene sulfonic acid.

These experiments were carried out with polymer oxidized in the presence of the magnesium polybutyl benzene sulfonate. The results of the following tests point out the absence of the desired increase in performance when the sulfonate oxidized polymer is reacted in a Mannich process without a sulfonic acid Mannich catalyst. The increased performance is attained only when the sulfonic acid is added during the Mannich reaction. The sulfonate present during the oxidation step does not inherently produce the desired Mannich reaction catalysis.

The Spot Dispersancy Test gives a measure of the oil's ability to disperse sludge and varnish. In the Spot Dispersancy Test, a dispersant is mixed with an amount of Ford VC sludge oil and is incubated at 300°F for 16 hours and 3 to 10 drops of this mixture are dropped onto a standard white blotter paper producing a sludge-oil spot which separates into a sludge spot and an oil spot. After 24 hours, the diameter of the sludge and the oil rings are measured. Dispersancy is reflected by the ability of an oil to keep sludge in suspension. Thus, dispersancy will be reflected by the difference in diameters of the slude and oil rings.

A rating (SDT) is given by the diameter of the sludge ring divided by the diameter of the oil ring, and multiplied by 100. A high numerical rating indicates good dispersancy.

The Hot Tube Test is a determination of the oxidation and varnish resistant properties of an oil package. A measured quantity of oil is metered into a 2 mm heated glass tube through which hot air or nitrogen dioxide is blown through the tube. The oil is consumed in the test and the deposits in the tube are measured. The tubes are rated from 0 through 10, 0 being a heavy black, opaque deposit, and 10, perfectly clean.

The table below demonstrates that the use of the sulfonic acid catalyst and high mol ratios of HCHO/HMDA produces a superior additive.

Example	Mol Equivalents of Sulfonic Acid	HCHO/HMDA Ratio	IR*	Spot Dispersancy Test	Hot Tube . . Test. . . Air	NO$_x$
1	0.10	2.0	0.22	83	2.5	1.5
2	0.10	2.0	0.18	83	–	–
3	0.10	1.0	0.05	59	3.5	1.8
4	0.10	1.5	0.16	86	3.5	2.0
5	0.083	1.5	0.17	88	–	–
6	0.083	2.0	0.22	83	–	–
7	0.083	1.0	0.06	59	–	–
8	0.083	3.0	0.24	86	–	–
9	0.0	1.0	0.10	55	2.7	1.6
10	0.0	2.0	0.13	60	3.5	1.5

*IR 1660–1680 cm^{-1} a per 0.5 mm cell path length.

The ASTM color of the product of Example 1 was 3.5 and that of Example 2 was greater than 8. Both determinations were made on products diluted to 15% by weight in xylene.

PHENOL-BASED ASHLESS COMPOSITIONS

Aminophenol-Detergent Combinations

D.L. Clason, J.F. Pindar and J.M. Cohen; U.S. Patent 4,100,082; July 11, 1978; assigned to The Lubrizol Corporation describe combinations of aminophenols, wherein the phenols contain a substantially saturated hydrocarbon substituent of at least 10 aliphatic carbon atoms, and one or more detergent/dispersants selected from the groups consisting of:

(a) neutral or basic metal salts of an organic sulfur acid, phenol or carboxylic acid;

(b) hydrocarbyl-substituted amines wherein the hydrocarbyl substituent is substantially aliphatic and contains at least 12 carbon atoms;

(c) acylated nitrogen-containing compounds having a substituent of at least 10 aliphatic carbon atoms; and

(d) nitrogen-containing condensates of a phenol, aldehyde and amino compound.

Fuels and lubricants containing such combinations as additives are particularly useful in two-cycle (two-stroke) engines.

The following specific illustrative examples describe how to make the aminophenols and detergent/dispersants which comprise the compositions of this process.

Example 1: (A) A mixture of 4,578 parts by weight of a polyisobutene-substituted phenol prepared by boron trifluoride-phenol catalyzed alkylation of phenol with a polyisobutene having a number average molecular weight of approximately 1,000 (vapor phase osmometry), 3,053 parts by weight of diluent mineral oil and 725 parts by weight of textile spirits is heated to 60°C to achieve homogeneity.

After cooling to 30°C, 319.5 parts by weight of 16 M nitric acid in 600 parts by weight of water is added to the mixture. Cooling is necessary to keep the mixture's temperature below 40°C. After the reaction mixture is stirred for an additional two hours, an aliquot of 3,710 parts by weight is transferred to a second reaction vessel. This second portion is treated with an additional 127.8 parts by weight of 16 M nitric acid in 130 parts by weight of water at 25° to 30°C. The reaction mixture is stirred for 1.5 hours and then stripped to 220°C at 30 torrs. Filtration provides an oil solution of the desired intermediate (1A).

(B) A mixture of 810 parts by weight of the oil solution of the 1A intermediate described in (A), 405 parts by weight of isopropyl alcohol and 405 parts by weight of toluene is charged to an appropriately sized autoclave. Platinum oxide catalyst (0.81 part by weight) is added and the autoclave is evacuated and purged with nitrogen four times to remove any residual air. Hydrogen is fed to the autoclave at a pressure of 29 to 55 psig while the content is stirred and heated to 27° to 92°C for a total of 13 hours. Residual excess hydrogen is removed from the reaction mixture by evacuation and purging with nitrogen four times. The reaction mixture is then filtered through diatomaceous earth and the filtrate stripped to provide an oil solution of the desired aminophenol. This solution contains 0.578% nitrogen.

Example 2: A mixture of 906 parts by weight of an oil solution of an alkyl phenyl sulfonic acid (having an average molecular weight of 450, vapor phase osmometry), 564 parts by weight mineral oil, 600 parts by weight toluene, 98.7 parts by weight MgO and 120 parts by weight water is blown with carbon dioxide at a temperature of 78° to 85°C for 7 hours at a rate of about 3 ft^3 of carbon dioxide per hour. The reaction mixture is constantly agitated throughout the carbonation. After carbonation, the reaction mixture is stripped to 165°C at 20 torrs and the residue filtered. The filtrate is an oil solution of the desired over-based magnesium sulfonate having a metal ratio of about 3.

Example 3: A polyisobutenyl succinic anhydride is prepared by reacting a chlorinated polyisobutene (having an average chlorine content of 4.3% and an average of 82 carbon atoms) with maleic anhydride at about 200°C. The resulting polyisobutenyl succinic anhydride has a saponification number of 90. To a mixture of 1,246 parts by weight of this succinic anhydride and 1,000 parts by weight of toluene, there is added at 15°C, 76.6 parts by weight of barium oxide. The mixture is heated to 115°C and 125 parts by weight of water is added dropwise over a period of one hour. The mixture is then allowed to reflux at 150°C until all the barium oxide is reacted. Stripping and filtration provides a filtrate having a barium content of 4.71%.

Example 4: A mixture of 1,500 parts by weight of chlorinated polyisobutene (molecular weight of about 950 and having a chlorine content of 5.6%), 285 parts by weight of an alkylene polyamine having an average composition corresponding stoichiometrically to tetraethylenepentamine and 1,200 parts by weight of benzene is heated to reflux. The mixture's temperature is then slowly increased over a 4-hour period to 170°C while benzene is removed. The cooled mixture is diluted with an equal volume of mixed hexanes and absolute ethanol (1:1). This mixture is heated to reflux and a one-third volume of 10% aqueous sodium carbonate is added to it. After stirring, the mixture is allowed to cool and the phases separated. The organic phase is washed with water and stripped to provide the desired polyisobutenyl polyamine having a nitrogen content of 4.5%.

Example 5: A mixture of 140 parts by weight of toluene and 400 parts by weight of a polyisobutenyl succinic anhydride (prepared from the polyisobutene having a molecular weight of about 850, vapor phase osmometry) having a saponification number of 109 and 63.6 parts by weight of an ethylene amine mixture having an average composition corresponding in stoichiometry to tetraethylenepentamine, is heated to 150°C, while the water/toluene azeotrope is removed. The reaction mixture is then heated to 150°C under reduced pressure until toluene ceases to distill. The residual acylated polyamine has a nitrogen content of 4.7%.

Example 6: To 1,133 parts by weight of commercial diethylene triamine heated at 100° to 150°C is slowly added 6,820 parts by weight of isostearic acid over a period of two hours. The mixture is held at 150°C for one hour and then heated to 180°C over an additional hour. Finally, the mixture is heated to 205°C over 0.5 hour; throughout this heating, the mixture is blown with nitrogen to remove volatiles. The mixture is held at 205° to 230°C for a total of 11.5 hours and then stripped at 230°C at 20 torrs to provide the desired acylated polyamine as a residue containing 6.2% nitrogen.

The nitrogen-containing compositions of this process are particularly useful in formulating lubricating oils for use in two-cycle engines. Typical compositions contain about 90 to 60% oil. The preferred oils are mineral oils and mineral oil-synthetic polymer and/or synthetic ester oil mixtures. Polybutenes of molecular weights of about 250 to 1,000 (as measured by vapor phase osmometry) and fatty acid ester oils of polyols such as pentaerythritol and trimethylol propane are typical synthetic oils used in preparing these two-cycle oils.

These oil compositions contain about 2 to 30%, typically about 5 to 20%, of at least one aminophenol as described above and about 1 to 30%, typically 2 to 20% of at least one detergent/dispersant. Other additives such as viscosity index (V.I.) improvers, lubricity agents, antioxidants, coupling agents, pour point depressing agents, extreme pressure agents, color stabilizers and antifoam agents can also be present.

The table below describes several illustrative two-cycle engine oil lubricant compositions of this process.

Composition	Amino-phenol* of Example 1	..Detergent/Dispersant*.. Example	Amount	Oil** Amount (pbw)
A	6	2	2	92
B	3	2	1	96
C	10.6	6	2.1	87.3
D	7.5	4	3.5	89
E	6	3	2	92
F	15	5	3	82

*Parts by weight of oil solution described in indicated Examples.
**The same base oil is used in each blend; a 650 neutral solvent extracted paraffinic oil cut with 20% by vol Stoddard solvent and containing 9 pbw/hundred parts of final blend of a bright stock having a viscosity of 150 SUS at 100°F.

Haloalkyl Hydroxy-Aromatic Condensation Products

D.E. Ripple; U.S. Patent 4,108,783; August 22, 1978; assigned to The Lubrizol Corporation has found that condensation products made by reacting an α-haloalkyl hydroxy-aromatic compound also having at least one nonfused hydrocarbyl substituent with at least one olefinic nitrile, carboxylic acid or carboxylic acid derivative are useful as additives for fuels and lubricants. The number of carbon atoms in the aromatic hydrocarbyl compound's substituents are each about 25, while the haloalkyl group contains from 1 to 36 carbons. The acid or nitrile reactant usually contains 3 to 40 carbons.

Products made from halomethylalkyl-substituted phenols and α,β-olefinic diacid derivatives such as maleic anhydride are particularly useful. Similarly useful products can be made from these condensation products by further reacting their acid, acid derivative or nitrile groups with alcohols, polyols, monoamines, polyamines, metal salts or metals.

Example 1: (A) A mixture of 1,412 parts by weight of phenol and 1,090 parts by weight of benzene is heated to 50° to 55°C; then 283 parts by weight of a boron trifluoride phenol complex is added over 20 minutes. Following this addition, 5,000 parts by weight of a polyisobutene having a $\overline{M}n$ of about 1,000 is added. The mixture is stirred for two hours at 55° to 60°C and then 645 parts by weight of ammonium hydroxide is added. Stirring is continued for an additional hour. The mixture is heated to 160°C for four hours while an azeotrope of phenol, water and benzene distills from it. Stripping the mixture to 220°C at 10 mm Hg and filtering it through diatomaceous earth provides the desired alkylated phenol which has a $\overline{M}n$ of 1,047 and an infrared spectrum consistent with its structure as an alkylated phenol.

(B) A mixture of 4,549 parts by weight of the alkylated phenol described in (A), 540 parts by weight of paraformaldehyde and 2,500 parts by weight of petroleum naphtha boiling between 96° and 102°C is heated to 55° to 60°C for two hours to effect homogenization. Gaseous hydrogen chloride is then bubbled into the reaction mixture at a rate of 2 cfh through a glass tube whose orifice is located below the mixture's surface for a total of twelve hours. The mixture is stirred for an additional 3.5 hours and blown with nitrogen at a rate of 1.5 to 2 cfh for an additional eight hours. The mixture is filtered through filter aid and the filtrate stripped to 90°C at 25 mm Hg to provide the final product which has a chlorine content of 2.1%, and an infrared spectrum consistent with its structure as a chloromethylated alkylated phenol.

(C) A mixture of 4,302 parts of the chloromethyl alkylated phenol described in (B) and 274 parts of maleic anhydride is heated to 210° to 215°C for 7.5 hours while being nitrogen blown. Excess maleic anhydride is removed by stripping the mixture to 210°C at 10 mm Hg and the reaction mixture diluted with 2,973 parts by weight of a diluent oil. Filtration with filter aid provides a 40% solution of the desired product.

Example 2: To 2,028 parts by weight of the 40% oil solution described in Example 1(C) at 140° to 145°C is slowly added 65 parts by weight of a polyethylene polyamine having an average of three to seven amino groups per molecule. The reaction mixture is heated at 175° to 180°C for two hours and the mixture then filtered through diatomaceous earth to provide a 40% oil solution of the desired product.

Examples 3 through 7: These examples are all carried out in substantially the same fashion using the following procedure: A mixture of haloalkyl-substituted phenol and unsaturated compound is heated under nitrogen to a temperature of 200° to 240°C for 4 to 6 hours. The mixture is stripped to approximately 200°C at 10 to 25 mm Hg to obtain a residue which is then diluted with an approximately equal volume of diluent oil and filtered through diatomaceous earth to yield a filtrate which is a solution of the desired product. Details as to the reactants and proportions used in these examples are summarized in the table below.

Example No.	. . . . α-Haloalkylphenol	mols	. . . Unsaturated Reactant . . .	mols
3	Example 1(B)	0.245	2-Ethylhexyl acrylate	0.25
4	Example 1(B)	0.245	Di(n-butyl)fumarate	0.25
5	Example 1(B)	0.32	Itaconic acid	0.32
6	Example 1(B)	0.2	4-Cyclohexene-1,2-dicarboxylic acid anhydride	0.2
7	Alkyl-substituted chloromethyl phenol*	2.55	Maleic anhydride	2.8

*Made by alkylating phenol with a mixture of C_{15-18} 1-olefins according to general procedure described in Example 1(A).

As previously indicated, the condensation products of this process are useful as additives in preparing lubricant compositions where they function primarily as detergents and dispersants, particularly where the oil is subjected to high temperature environments or to cyclic stresses such as those encountered in stop and go automobile driving. Many such compositions are particularly useful in dispersing engine sludge and reducing engine varnish.

In related work *C.P. Bryant; U.S. Patent 4,108,784; August 22, 1978; assigned to The Lubrizol Corporation* has found that condensation products made by reacting an α-hydroxyalkyl hydroxy-aromatic compound also having at least one non-fused hydrocarbyl-substituent with at least one olefinic nitrile, carboxylic acid or carboxylic acid derivative are useful as additives for fuels and lubricants. The total number of carbon atoms in the nonfused hydrocarbyl substituents is at least about 7 while the α-hydroxyalkyl group contains from 1 to 36 carbon atoms and the olefinic acid or nitrile reactant usually contains 2 to 40 carbon atoms.

Products made from hydroxymethyl alkyl-substituted phenols and α,β-olefinic acid derivatives such as maleic anhydride are particularly useful. Similarly useful posttreated products can be made from these condensation products by further reaction with alcohols, amines, metal salts or metals.

Esters of Alkyl-Substituted Salicylic Acids

C.W. Stuebe; U.S. Patent 4,098,708; July 4, 1978; assigned to The Lubrizol Corporation has found that esters of alkyl-substituted hydroxy-aromatic carboxylic acids (especially alkyl-substituted salicylic acids) and the like, in which the alkyl substituent contains at least 10 carbon atoms, are useful as dispersant additives for lubricants and fuels. Particularly useful are esters in which the substituent is derived from propylene, 1-butene or isobutene and has a molecular weight of about 150 to 1,750, and in which the alkyl moiety is derived from pentaerythritol. Such esters are preferably prepared by the reaction of the corresponding organic hydroxy compound with a vic-hydroxyalkyl ester of the salicylic acid.

The preparation of the esters of this process is illustrated by the following examples. All percentages and parts are by weight. Molecular weights are number average molecular weights ($\overline{M}$n) as determined by vapor phase osmometry, acid number or hydroxyl analysis.

Example 1: A mixture of 5,720 parts (5.5 equivalents) of an alkyl-substituted phenol ($\overline{M}$n 885) formed by alkylating phenol with tetrapropene and 399 parts (6.05 equivalents) of potassium hydroxide is heated at 250° to 260°C for 12 hours and dried under nitrogen. At 210°C 2,000 parts of mineral oil is added. Carbon dioxide blowing is started at 180°C and the reaction mixture is allowed to cool to 145°C. Carbon dioxide blowing is continued at 145° to 155°C until the reaction is complete. The reaction mixture is cooled to 85°C and 589 parts of concentrated hydrochloric acid is added. Then 800 parts of toluene and 500 parts of water are added and the mixture is heated under reflux for 2 hours, stripped and filtered to yield a substituted salicylic acid.

A mixture of 437 parts (0.23 equivalent) of the substituted salicylic acid prepared as described above, 10 parts of p-toluenesulfonic acid and 300 parts of xylene is heated at 140° to 145°C for 3 hours. Pentaerythritol, 17 parts, is added and the mixture is heated to 205°C to remove volatiles and maintained at 200° to 205°C for 3 hours under nitrogen. The mixture is filtered to yield the desired pentaerythritol ester (75% solution in mineral oil).

Example 2: A mixture of 6,417 parts (3.38 equivalents) of the substituted salicylic acid of Example 1 and 15 parts of lithium carbonate is heated to 145° to 155°C. Propylene oxide, 232 parts (4 equivalents), is added over 3.5 hours and the mixture is stripped at 135° to 145°C under vacuum. The desired 2-hydroxypropyl ester (76% solution in mineral oil) is obtained by filtration.

A mixture of 980 parts (0.5 equivalent) of the 2-hydroxypropyl ester and 37.5 parts (1.1 equivalents) of pentaerythritol is heated to 225° to 230°C while propylene glycol is continuously removed by nitrogen blowing. After stripping, the mixture is filtered to yield 871 parts of the desired pentaerythritol ester (76% solution in mineral oil).

The esters of this process are useful as additives for lubricants and normally liquid fuels, where they function primarily as dispersants; that is, they maintain accumulated sludge, dirt and other insolubles in suspension.

Amides of Alkyl-Substituted Salicylic Acids

D.I. Hoke; U.S. Patent 4,090,971; May 23, 1978; assigned to The Lubrizol Corporation has found that amides of alkyl-substituted hydroxy-aromatic carboxylic acids (especially alkyl-substituted salicylic acids) and the like, in which the alkyl substituent contains at least about 10 carbon atoms, are useful as dispersant additives for lubricants and fuels. Particularly useful are salicylamides in which the substituent is derived from propylene, 1-butene or isobutene and has a molecular weight of about 150 to 1,750, and in which the amide moiety is derived from an alkylene polyamine. Such amides are prepared by the reaction of the corresponding nitrogen-containing compound with a vic-hydroxyalkyl ester of the salicylic acid.

The preparation of the amides of this process is illustrated by the following examples. All percentages and parts are by weight. Molecular weights and equivalent weights are number average figures as determined by vapor phase osmometry, acid number or hydroxyl analysis.

Example 1: A mixture of 1,830 parts (5 equivalents) of an alkyl-substituted salicylic acid formed by alkylating phenol with tetrapropene and subjecting the resulting alkyl phenol to the Kolbe reaction, 207 parts (5 equivalents) of a commercial ethylene polyamine mixture containing about 3 to 7 nitrogen atoms per molecule, and 500 parts of mineral oil is heated to 170°C for 3 hours as water is removed by distillation. It is then heated to 210°C for 2 hours with the removal of additional water, and is filtered with the addition of a filter aid material. The filtrate is the desired amide (80% solution in mineral oil) and contains 2.94% nitrogen.

Example 2: A mixture of 3,400 parts of an alkyl phenol ($\overline{M}n$ 262) similar to that of Example 1, 500 parts of xylene, 1,000 parts of mineral oil and 944 parts of potassium hydroxide is heated under reflux for 6 hours and then stripped to 255°C under nitrogen. Carbon dioxide blowing is started at 207°C and the reaction mixture is allowed to cool to 145°C. Carbon dioxide blowing is continued at 145° to 150°C until the reaction is complete.

The mixture is then cooled to 90°C. Toluene, 1,000 parts, is added, followed by 1,440 parts of concentrated hydrochloric acid over 4.5 hours. The mixture is then heated under reflux at 100° to 103°C for 4.5 hours as water is removed by azeotropic distillation. The product is filtered, stripped to 141°C under vacuum and filtered again to yield 4,809 parts of a substituted salicylic acid.

A mixture of 1,414 parts of the substituted salicylic acid and one part of lithium hydroxide hydrate is heated to 103° to 105°C. Ethylene oxide, 200 parts, is added over 4.5 hours. The mixture is stripped at 106°C under vacuum and filtered, yielding a 2-hydroxyethyl ester (73% solution in mineral oil).

A mixture of 550 parts (1.0 equivalent) of the 2-hydroxyethyl ester and 105 parts (1.0 equivalent) of diethanolamine is heated at 195° to 200°C for 12 hours as ethylene glycol is removed by blowing with nitrogen. The mixture is stripped to 116°C under vacuum and filtered to yield the desired amide (75% solution in mineral oil, containing 2.48% nitrogen).

Example 3: A mixture of 319 parts of an ester prepared as described in Example 2 and 15 parts of ethylene diamine is heated at 110° to 118°C for 4.5 hours, stripped at 170°C under vacuum and filtered. The filtrate is the desired amide (73% solution in mineral oil, containing 2.24% nitrogen).

The amides of this process are useful as additives for lubricants and normally liquid fuels, where they function primarily as dispersants; that is, they maintain accumulated sludge, dirt and other insolubles in suspension.

Phenyl Glycidyl Ether-Diamine Reaction Products

W.H. Machleder and J.M. Bollinger; U.S. Patents 4,147,641; April 3, 1979 and 4,134,846; January 16, 1979; both assigned to Rohm and Haas Company have demonstrated that mixtures of (1) the reaction products of certain substituted phenols, epichlorohydrin and amines, and (2) a polycarboxylic acid ester show excellent carburetor, induction system and combustion chamber detergency and, in addition, provide effective rust inhibition when used in hydrocarbon fuels at low concentrations, i.e., about 60 to 100 ppm of (1) and about 200 to 300 ppm of (2).

Component (1) of the multipurpose additive of the process is the reaction product of (a) a glycidyl ether compound (I) of the formula:

(I)

$$(R^6)_{\overline{m}} \overline{} \text{—OCH}_2\text{CH} \underset{O}{\overset{\triangle}{\longrightarrow}} \text{CH}_2$$

where R^6 is an aliphatic hydrocarbon group containing at least 8 carbon atoms, m is 1 to 3, and (b) a primary or secondary monoamine or polyamine, that is, an amine having at least one amino group having at least one active hydrogen atom.

The glycidyl ether compound (I) is conveniently prepared by condensing a metal alkoxide of a phenol having 1 to 3 aliphatic hydrocarbon substituents (R^6) with an excess of epichlorohydrin. The carbon content and number of aliphatic hydrocarbon substituents are chosen to provide the required degree of solubility of the final glycidyl ether compound/amine adduct in hydrocarbon fuels or lubricating oils.

A preferred class of amines is given by the formula (II):

(II)
$$R^1 N \overline{(R^4 N)_n} - R^4 - NH$$
$$\underset{R^2}{} \quad \underset{R^5}{} \quad \underset{R^3}{}$$

where R^1, R^2 and R^3 independently are hydrogen, C_1-C_6 alkyl substituted by $-NH_2$ or $-OH$, R^4 is a C_1-C_6 divalent hydrocarbon radical (alkylene or phenylene), R^5 is hydrogen or C_1-C_6 alkyl, and n is 0 to about 5. These amines include amines wherein the amino groups are bonded to the same or different carbon atoms. Some examples of diamine reactants where the amine groups are attached to the same carbon atoms of the alkylene radical R^4 are N,N-dialkylmethylenediamine, N,N-dialkanol-1,1-ethanediamine, and N,N-di(aminoalkyl)-2,2-propanediamine.

A preferred component (1) of the additive mixtures of the process is N,N'-bis[3-(p-H35-polyisobutylphenoxy)-2-hydroxypropyl]ethylenediamine, shown by the structural formula (III) where R^6 is hydrogen or PIB$_{H35}$:

(III)

$$\underset{\text{PIB}_{H35}}{} \quad \text{O—CH}_2\overset{\text{OH}}{\underset{|}{\text{CHCH}_2}}\text{NHCH}_2\text{CH}_2\text{NHCH}_2\overset{\text{OH}}{\underset{|}{\text{CHCH}_2}}\text{—O}$$

PIB is an abbreviation for a polyisobutene generically of any molecular weight. H35 is the commercial designation for Amoco Chemical Company's polyisobutene having a number average molecular weight ($\overline{M}n$) of about 670.

As indicated in the general description above, the preferred component (1) can be a mixture of structure (III) and structure (IV) set forth below or it can be (III) or (IV) taken singly. In other words, on a parts per 100 parts basis, (III) can vary from 1 to 99 parts and (IV) can vary from 99 to 1 part; or there can be 100 parts of (III) or 100 parts of (IV), all parts being on a weight basis.

(IV)

$$PIB_{H35} - \underset{R^6}{\bigcirc} - O-CH_2CHCH_2 \quad (OH) \quad NCH_2CH_2NH_2 \quad PIB_{H35} - \underset{R^6}{\bigcirc} - O-CH_2CHCH_2 \quad (OH)$$

where R^6 is as defined in structure (III).

The polycarboxylic acid ester component (2) is used primarily for cost considerations and lowers treating costs. However, component (2) also promotes deposit reduction as shown in results of the Induction System Test in the table below.

Component (2) preferably is a mixed ester of a di- or tricarboxylic acid, although polycarboxylic acid esters, wherein the ester groups are the same, are also useful. An especially preferred polycarboxylic acid ester is the mixed adipate diester comprising the monoisodecyl, monooctyl phenoxy polyethoxy ethanol (containing an average of 5 mols of condensed ethylene oxide) mixed ester of adipic acid made by a conventional acid esterification process.

In the table, the comparison primarily is with Chevron F-310. The essential component in Chevron F-310 is believed to be a polybutene amine as described in U.S. Patent 3,438,757. As can be seen from the table, the mixture of the amine adduct component (1) and the polycarboxylic acid ester component (2) compares very favorably with component (1) and with the Chevron F-310, when the latter two additives are used alone. It should also be noted from the table, that the Chevron F-310 is used at a much higher level (1,000 pounds per thousand barrels of gasoline) than is either the amine adduct component or the mixture of component (1) and component (2). Descriptions of test procedures follow the table below.

Additive	Treating Level (lb/1,000 bbl of Gasoline)	Carburetor Detergency Blowby Test % Deposit Reduction Leaded Gasoline (MS-08)	Induction System Test Single Cylinder % Deposit Reduction Unleaded Gasoline*
Control base gasoline	—	0	0
Chevron F-310	1,000	96	~99

(continued)

Additive	Treating Level (lb/1,000 bbl of Gasoline)	Carburetor Detergency Blowby Test % Deposit Reduction Leaded Gasoline (MS-08)	Induction System Test Single Cylinder % Deposit Reduction Unleaded Gasoline*
Component (1)	75	95	94
Additive mixture**	25/50	84, 90, 83	~81
Additive mixture**	25/75	94	~98

*Phillips Reference Fuel.
**Mixture of Component (1)/Component (2) as follows: Component (1), adduct of Formula III or IV where R_1 is H or PIB_{H35}; and Component (2), monoiodododecyl, monooctyl phenoxy polyethoxy ethanol (containing an average of 5 mols of condensed ethylene oxide) mixed ester of adipic acid.

The Blowby Carburetor Detergency Keep Clean Engine Test (BBCDT-KC) measures the ability of a gasoline additive to keep clean the carburetor throttle body area, and is run in a 1970 Ford 351 CID V-8 engine equipped by means of a special Y intake manifold with two one-barrel carburetors, which can be independently adjusted and activated.

The Induction System Deposit Test (ISDT) which is used to evaluate the ability of gasoline additives or mixtures of additives to control induction system deposits, is run using a new air-cooled, single-cylinder, 4-cycle, 2.5 hp Briggs and Stratton engine for each test. The engine is run for 150 hours at 3,000 rpm and 4.2 ft-lb load, with a 1-hour shutdown every 10 hours to check the oil level. Carbon monoxide exhaust emission measurements are made each hour to insure that a constant air to fuel (A/F) ratio is being maintained.

Upon completion of a test run, the engine is partially disassembled, and the intake valve and port are rated and valve and port desposits are collected and weighed.

MS-08 gasoline is used in the Blowby Carburetor Detergency Keep Clean Engine Test; Phillips J Reference Fuel (an unleaded fuel), is used in the Induction System Test.

In the following examples all parts and percentages are by weight unless otherwise noted, and R^6 is hydrogen or the same as the other hydrocarbon substituent on phenol.

Example 1: (A) Polyisobutene H35 Phenol —

To a 5-liter 3-necked flask equipped with a thermometer, mechanical stirrer, and reflux condenser with Dean-Stark trap was charged 1,920 g (2.9 mols) of polyisobutene H35 (Amoco) of about 660 molecular weight, 564 g (6 mols) of phenol, 200 g of Amberlyst 15 acid catalyst, and 550 ml of hexane. The stirred mixture was heated at reflux (pot temperature 100° to 107°C) under a nitrogen atmosphere for 24 hours, during which time 5.4 ml of water had separated.

After cooling 60° to 80°C, the mixture was filtered to remove the resin beads, the latter being washed with hexane, and the filtrate subjected to vacuum concentration with a pot temperature of 160°C. There was obtained 1,971.4 g of product residue having an oxygen content of 2.92% (theoretical, 2.12%) and molecular weight of about 754. The product is actually a mixture of alkylated phenols with an average molecular weight of 548 based upon the oxygen and 556 calculated from UV spectral parameters, i.e., in some of the products R^6 is hydrogen and in others R^6 is PIB$_{H35}$.

(B) 1,2-Epoxy-3(p-H35-Polyisobutylphenoxy)Propane —

To a 5-liter 3-necked flask fitted with a thermometer, mechanical stirrer, addition funnel and reflux condenser was charged 973 g (1.75 mols based upon 2.92% oxygen) of the polyisobutene H35 phenol obtained in (A), 72 g (1.75 mols based upon 97.4% assay) of sodium hydroxide pellets, and 450 ml of toluene. The stirred mixture was heated under a nitrogen atmosphere at 84° to 90°C for one hour to effect the dissolution of the base. Epichlorohydrin (161.9 g, 175 mols) was then added dropwise at 60°C during 2.5 hours, followed by a hold period at 70°C. The reaction mixture was then cooled, filtered, and the salt (107 g dry) washed with toluene. The filtrate was stripped (100°C at 15 mm) to give 1,075.3 grams of product residue having a molecular weight of about 810.

(C) N,N'-Bis[3-(p-H35-Polyisobutylphenoxy)-2-Hydroxypropyl] Ethylenediamine —

A mixture of 1,018.4 g of the epoxide product of (B), 122.6 g (2.04 mols) of ethylenediamine, and xylene (700 ml) was heated at reflux (131° to 136°C) with stirring under a nitrogen atmosphere for 18 hours. After vacuum stripping (18 mm, pot temperature of 120°C), there was obtained 1,053.4 g of turbid residue which was filtered through a bed of Celite 545 in a steam-heated Buchner funnel to give a clear, yellow viscous product of molecular weight about 1,680. The mixture obtained in this synthesis may contain the N,N' diadduct, the N,N diadduct and some N or N' monoadduct. The product prepared in this way had 1.26% basic nitrogen (1.67% theory) and 5.26% oxygen (3.81% theory).

Example 2: (A) Polyisopropylphenol — To a 5-liter, 4-necked round-bottomed flask fitted with a stopcock on the bottom, a condenser, a stirrer, a thermometer, and an addition funnel, were charged, under nitrogen, 1,150 g (2.0 mols) Ampol C_{20} polypropylene. The reaction was heated to 70°C and 236 g (2.5 mols) of phenol were added followed by the dropwise addition (10 minutes) of 102.4 g (0.4 mol) BF_3 phenol complex. The reaction mixture was heated to 95°C and held there 5 hours. The reaction mixture was then cooled to 70°C, diluted with 600 cc toluene, and a solution of 131.4 g (1.24 mols) Na_2CO_3 in 1,050 cc water was slowly added. The mixture was heated to 80°C and the layers were allowed to separate. After discarding the aqueous layer, the organic layer was washed with 100 cc water. The organic layer was then vacuum stripped (180°C, 0.25 mm) to afford 1,260 g (94%) polyisopropylphenol ($\overline{M}n$ ~737).

(B) Polyisopropylphenyl Glycidyl Ether — To a 5-liter, 4-necked round-bottomed flask fitted with a condenser, addition funnel, stirrer, and thermometer was charged 1,260 g (1.71 mols) polyisopropylphenol from (A). A 50% NaOH solution (137.3 g, 1.71 mols) was then added and the mixture heated with stirring to reflux (118°C) and held there for 0.5 hour. The mixture was vacuum stripped at 100°C (0.5 mm) to remove water, recharged with 50 g toluene and restripped (105°C, 0.2 mm) to azeotropically remove the last traces of water. The reaction was then cooled to 65°C and 792 g (8.55 mols) epichlorohydrin was added and the reaction heated to reflux (~120°C) for three hours. The excess epichlorohydrin was then vacuum stripped at 120°C (0.05 mm) to yield ~1,450 g of the crude glycidyl ether.

(C) N,N'-Bis[3-(p-Polyisopropylphenoxy)-2-Hydroxypropyl] Ethylenediamine — (C) of Example 1 was repeated in all essential respects except for substitution of the polyisopropyl (PIP) glycidyl ether adduct of (B) above for the polyisobutene phenol/epichlorohydrin adduct of (B) of Example 1. The product may also contain N,N diadduct and N or N' monoadduct.

Mannich Reaction Products

R.E. Karll and R.J. Lee; U.S. Patent 4,142,980; March 6, 1979; assigned to Standard Oil Company (Indiana) describe a process for making certain compositions and lubricating oils. The process utilizes a substituted phenol comprising the reaction product of an alkyl phenol, the alkyl substituent containing about 50 to 20,000 carbon atoms, and an aliphatic unsaturated carboxylic acid containing about 3 to 100 carbon atoms. Also described are oil-soluble Mannich reaction products of the substituted phenol, an amine having at least one reactive nitrogen, and a formaldehyde-affording reactant.

The method of alkylating an alkyl-substituted phenol having the formula:

OH

R

where R is an alkyl substituent containing from about 50 to 20,000 carbon atoms, with an aliphatic unsaturated carboxylic acid having from about 3 to 100 carbon atoms with no substantial degradation of the R substituent generally comprises reacting one equivalent weight of the alkyl-substituted phenol with one equivalent weight of the aliphatic unsaturated acid in the presence of a catalytically effective amount of BF_3 and HCOOH.

This process can be used to manufacture disubstituted phenol having the formula:

OH
$C-(CH_2)_rCOOH$
$(CH_2)n$
CH_3
R

where R is an alkyl substituent containing from about 50 to 20,000 carbon atoms, and n and r are integers such that $n + r \geqslant 5 \leqslant 15$.

These substituted phenols are extremely useful for reaction with an amine having at least one reactive nitrogen and a formaldehyde-affording reactant in a Mannich condensation. This oil-soluble condensation product is a highly effective detergent or dispersant in lubricating oils.

Commonly, the oil-soluble Mannich reaction product comprises the reaction product of (a) the reaction product of an alkyl phenol, the alkyl-substituent containing about 50 to 20,000 carbon atoms, and an aliphatic unsaturated carboxylic acid containing about 3 to 100 carbon atoms; (b) an amine having at least one reactive nitrogen and containing less than about 100 carbon atoms; and (c) a formaldehyde-affording reactant. The molar ratio of a:b:c is about 1:0.7-1.0:1.5-2, preferably about 1:1:1.5

Example 1: For the preparation of high molecular weight disubstituted phenol, 870 g of polybutylene phenol, equivalent to 0.25 mol was diluted with 500 ml of hexane. The molecular weight of the polybutyl substituent was about 3,500 and contained an average number of carbon atoms of about 240. To the solution was added 80 g of $BF_3 \cdot HCOOH$ catalyst, equivalent to 0.5 mol. Under vigorous agitation, 72 g of oleic acid, equivalent to 0.25 mol, was introduced dropwise over a period of about one-half hour. A slight exothermic reaction occurred, raising the reaction temperature by 6°F. The reaction was held at room temperature for 24 hours after which the catalyst layer was removed. The oil layer was washed with methanol and then water.

The hexane solution was then distilled to a pot temperature of 250°F to remove all solvent and water. The residual disubstituted phenol reaction product was obtained as a clear distillation residue. The product consists essentially of 8-(1-hydroxy-4-polybutyl)phenol stearic acid in mineral oil.

Example 2: To the entire amount of the disubstituted phenol prepared in Example 1, was added 48 g (0.25 mol) of tetraethylenepentamine. The reaction mixture was heated to 350°F and held at this temperature for 2 hours. The reaction mixture was cooled to 150°F and 40.5 g of 36% aqueous formaldehyde was added dropwise thereto. The mixture was then slowly heated to 300°F with nitrogen sparging to remove water. The reaction mixture was then diluted with SAE 5W oil and filtered over Celite clay. The filtrate comprised a solution containing 30% of Mannich condensation product in SAE 5W oil. The nitrogen content of the solution was 1.41%.

Example 3: A solvent-extracted mineral oil solution of 0.66 mol of 1,600 molecular weight monosubstituted polybutyl phenol was reacted with 0.61 mol of tetraethylenepentamine, 1.2 mols formaldehyde and 0.33 mol of oleic acid and the reaction mixture heated to 300°F for 3 hours with nitrogen gas purging to remove water. The Mannich reaction product was then filtered through diatomaceous earth to yield a crystal clear filtrate of 1.4% nitrogen content and a SSU viscosity at 210°F of 1,070. The function of the oleic acid was to prevent haze formation during processing and in storage. No substantial amount of this acid, if any, becomes attached to the phenol.

The use of the Mannich product of Example 2 as a dispersant addition agent in lubricating oils is demonstrated in the Spot Dispersancy Test and in the Ford Sequence V-C Test. In the Spot Dispersancy Test, a measured amount of the Mannich product is thoroughly mixed with a measured amount of used test oil (crankcase oil resulting from a Lincoln Engine Sequence V Test run for 384 hours), and the stirred mixture is heated to 360°F for 16 hours. The used crankcase oil from the Lincoln Test contains sludge to the extent that the original dispersant was no longer capable of keeping the sludge suspended in the oil.

As a control for purposes of comparison, the same volume of used crankcase oil is likewise heated and stirred at 360°F for 16 hours without the addition of any further dispersant. Equal aliquot portions of each of the test oils are deposited on different marked areas of a large sheet of blotter paper. The blotter paper is held at room temperature for 3-hour development (low severity) and for 27-hour development (high severity). The spots develop into two separate concentric rings: the inner ring is the sludge ring and the outer ring is the sludge-free oil ring. The diameters of these rings are measured and the ratio of the diameter of the sludge ring (D_s) to the diameter of the oil ring (D_o) times 100 provides an indication of the dispersancy function of the test compound. Ideally the ratio should be 100.

Spot Dispersancy Test

Dispersant	Concentration	$D_s/D_o \times 100$ Low Severity (3-hr development)	High Severity (27-hr development)
		%	
Example 2	3	100	89.5
Example 2	1.5	90	79.5
Example 3	4.0	90	72
Control	—	55	45

With 5% of the product of Example 2 in a conventional crankcase oil blend, the following results were obtained in the Ford Sequence V-C Test. A score of 10 is perfect.

Sludge	
Average of 10 engine parts	9.7
Required for passing	8.5
Varnish	
Average of 10 engine parts	8.6
Required for passing	8.3
Piston skirt	8.6
Required for passing	8.2

From the data shown above, it is clear that the compositions of this process provide superior dispersancy properties in lubricating oils.

Sulfurized Mannich Condensation Products

A process described by *K.E. Davis; U.S. Patent 4,161,475; July 15, 1979; assigned to The Lubrizol Corporation* provides a nitrogen-containing Mannich condensation product containing about 0.1 to 20% sulfur by weight, based on the total weight of the improved product. The sulfur is introduced into the product by sulfurizing with elemental sulfur a conventional nitrogen-containing Mannich condensation product useful as an additive for lubricants and normally liquid fuels.

In the following examples, all number average molecular weights ($\overline{M}n$) are determined by vapor pressure osmometry and all weight average molecular weights ($\overline{M}w$) by gel permeation chromatography unless expressly stated to the contrary. All "mm's" refer to millimeters of mercury vacuum.

Example 1: (A) Benzene (217 parts) is added to phenol (324 parts, 3.45 mols) at 38°C and the mixture is heated to 47°C. Boron trifluoride (8.8 parts, 0.13 mol) is added to the mixture over a one-half hour period at 38° to 52°C. Polyisobutene (1,000 parts, 1.0 mol) derived from the polymerization of C_4 monomers predominating in isobutylene is added to the mixture at 52° to 58°C over a 3.5-hour period. The mixture is held at 52°C for one additional hour. A 26% solution of aqueous ammonia (15 parts) is added and the mixture heated to 70°C over a 2-hour period. The mixture is then filtered and the filtrate is the desired crude polyisobutene-substituted phenol. This intermediate is stripped by heating 1,465 parts to 167°C and the pressure is reduced to 10 mm as the material is heated to 218°C in a 6-hour period. A 64% yield of stripped polyisobutene-substituted phenol ($\overline{M}n$ is 885) is obtained as the residue.

(B) A commercial mixture of ethylene polyamines (41 parts, 1.0 equivalent) corresponding in empirical formula to penta(ethylene)hexamine is added to a mixture of the substituted phenol (400 parts, 0.38 equivalent) described in (A) and diluent oil (181 parts) at 65°C. The mixture is heated to 93°C and paraformaldehyde (12 parts, 0.4 equivalent) added. The mixture is heated from 93° to 140°C over a 5-hour period and then held at 140°C for 4 hours under nitrogen. The mixture is cooled to 93°C and additional paraformaldehyde (12 parts, 0.4 equivalent) is added. The mixture is heated from 93° to 160°C for a total of 12 hours. The total amount of distillate collected is 13.2 parts. An additional amount of diluent oil (119 parts) is added to the mixture which is then filtered. The filtrate is a 40% oil solution of the desired Mannich condensation product containing 1.87% nitrogen.

(C) To 1,850 parts (1.0 equivalent) of the Mannich condensate described in (B) is added sulfur flowers (64 parts, 2.0 equivalents) at 80°C. The mixture is heated to 160°C over a 10-hour period removing the hydrogen sulfide evolved (35 g). The mixture is then filtered. The filtrate is a 40% oil solution of the desired sulfurized product containing 1.79% nitrogen and 1.43% sulfur.

Example 2: The procedure of Example 1 is repeated except the polyisobutene-substituted phenol used has a $\overline{M}w = 4084$, a $\overline{M}n = 1292$. To the substituted phenol (1,400 parts, 0.75 equivalent) and diluent oil (374 parts) is added the ethylene polyamine of Example 1(B) (77 parts, 1.85 equivalents) at 80°C. The mixture is heated to 96°C and sulfur flowers (42.7 parts, 1.33 equivalents) and paraformaldehyde (22.5 parts, 0.75 equivalent) is added. The mixture is heated to

150°C over 3 hours under nitrogen. A total of 5 parts distillate is removed. The mixture is cooled to 120°C and additional paraformaldehyde (22.5 parts, 0.75 equivalent) is added. The mixture is held at 120° to 125°C for one hour and then heated to 165°C for 5 hours. An additional 12 parts distillate is removed. The mixture is filtered to provide as a filtrate a 20% oil solution of the desired product containing 0.83% nitrogen and a 0.27% sulfur.

Example 3: A 37% aqueous solution of formaldehyde (55 parts, 0.68 equivalent) is added to Ethyl Antioxidant 733 (148 parts, 0.68 equivalent), a commercially available isomeric mixture of butyl-substituted phenols available from the Ethyl Corp., a 25% aqueous solution of dimethylamine (122 parts, 0.68 equivalent) and isopropanol (148 parts) as solvent. The mixture is heated to 75°C and held for 1.67 hours. The mixture is allowed to stand and separate into two layers. The aqueous layer is removed and the organic layer washed twice with water. Sulfur flowers (44 parts, 2.36 mols), along with dimethylformamide (148 parts) used as a sulfurization promoter, is added and the mixture heated to 150°C over 3 hours. The mixture is then filtered. Benzene is added and the filtrate water washed 3 times. The filtrate is then stripped to 90°C at 15 mm. The residue is the desired product containing 2.8% nitrogen and 13.98% sulfur.

As previously indicated, the sulfurized Mannich condensate compositions of this process are useful as additives for lubricants, in which they function primarily as sludge dispersants. They can be employed in a variety of lubricants based on diverse oils of lubricating viscosity, including natural and synthetic lubricating oils.

Boronated Amide Product

A process described by *B.T. Davis and M.F. Crook; U.S. Patent 4,119,552; Oct. 10, 1978; assigned to Edwin Cooper and Company Limited, England* provides a boronated amide of an alkylaryloxy-substituted alkanoic acid suitable for use as a lubricating oil additive. The process comprises reacting an amide of an alkylaryloxy-substituted alkanoic acid where the alkyl group contains at least about 30 carbon atoms with a boronating agent, e.g., a boron compound selected from the group consisting of boron acids, esters of boron acids, salts of boron acids with weak bases, boron oxides, boron halides, and boron salts of oxy acids, preferably at a temperature of about 50° to 300°C, the amount of the boronating agent being such that the boronated amide has a boron:nitrogen atom ratio of at least about 0.01:1.

The following examples illustrate the manner by which the initial alkylphenoxy-substituted aliphatic carboxylic acid and esters can be prepared.

Example 1: Preparation of PIB Phenoxy Acetic Acid — (A) Preparation of Sodium PIB Phenate: A p-PIB-substituted phenol (equivalent weight 1,130) was prepared by alkylation of phenol, in the presence of a boron trifluoride/phenol complex, with a PIB of molecular weight 1,000. To a solution of the resulting PIB phenol (79 g, 0.07 mol) in petroleum ether (30 ml, BP 100° to 120°C) was added a solution of sodium methoxide in methanol, prepared from sodium metal (1.61 g, 0.07 mol) and anhydrous methanol (25 ml). After stirring for 30 minutes, the product vacuum stripped to 150°C.

(B) Reaction of Sodium PIB Phenate with Chloroacetic Acid: To a solution of the sodium PIB phenate prepared in (A) (75 g, 0.65 mol) in petroleum ether (50 ml, BP 100° to 120°C) was added chloroacetic acid (7 g, 0.074 mol) and a solution of sodium methoxide in methanol prepared from sodium metal (1.7 g, 0.074 mol) and anhydrous methanol (20 ml). The mixture was heated, with stirring, under nitrogen, at 100°C for 3 hours. After allowing to cool, the product was further diluted with petroleum ether, washed with 200 ml of dilute hydrochloric acid, followed by three 200 ml portions of water, dried over anhydrous magnesium sulfate, vacuum stripped to 170°C and finally filtered. Acidity 19 mg KOH/g.

Example 2: Preparation of n-Butyl Polyisobutylphenoxyacetate — To a solution of PIB phenol (107.0 g, 0.1 mol), prepared by alkylation of phenol with 1,000 molecular weight polyisobutylene using a BF_3/phenol complex as catalyst, and n-butyl chloroacetate (22.6 g, 0.15 mol) in xylene (100 ml) was slowly added, over about one hour, a solution of sodium methoxide (8.1 g, 0.15 mol) in anhydrous methanol (40 ml). The addition was carried out at 100°C and on completion the solution was heated at this temperature for a further one hour. The solution was washed with 10% hydrochloric acid (50 ml) followed by 3 x 80 ml portions of aqueous methanol (1:4). After being dried over magnesium sulfate the solution was stripped of solvent. Yield 102 g.

A sample of this product was saponified with excess aqueous potassium hydroxide, acidified with hydrochloric acid and then washed with portions of aqueous methanol (1:4) until acid free. The acid value of the thus-formed PIB phenoxy acetic acid (36.5 mg KOH/g) indicated that a conversion of 73% had been obtained. Example 3 illustrates the conversion of the initial alkylphenoxy aliphatic carboxylic acid ester to the amide intermediate.

Example 3: A mixture of 50 g of polyisobutylphenoxy butyl acetate from Example 2 and 3 g of tetraethylenepentamine was heated to 200°C and stirred at that temperature for 4 hours. Butanol which formed during the amidation was allowed to distill out. After dilution with mineral oil, the mixture was filtered to obtain a clear product analyzing: N, 1.8%; acidity, 5 mg KOH/g; and total base number (TBN), 37 mg KOH/g.

Example 4: In a reaction vessel was placed 25.84 kg of n-butyl polyisobutyl (MW 1,000) phenoxy acetate. Over a one-hour period, 1.64 kg of tetraethylenepentamine was added while heating to 200°C. The mixture was stirred for 4 hours at 200°C at 50 mm while distilling out n-butanol. The product was diluted by adding neutral mineral oil to form a concentrate containing 15 wt % oil. Its analysis was: N, 1.75% and TBN, 39. Example 5 illustrates the boronation of the amide intermediate.

Example 5: In a reaction vessel was placed 800 g of the polyisobutylphenoxy acetamide of tetraethylenepentamine from Example 4. To the reaction vessel was added 61.8 g of boric acid (H_3BO_3). This mixture was stirred at 160°C for 4 hours while passing a slow stream of nitrogen through the vessel to aid in water removal. Following this, the product was filtered while at 100°C. The product analyzed: N, 1.7%; B, 0.31%; TBN, 28.9; giving a B:N atom ratio of 0.24:1.

Tests were carried out which demonstrate the properties of the boronated dispersants of this process compared to the unboronated material. These tests were the MS IIc and IIIc, and L-38 engine tests. In these tests, the oil used was a fully formulated SAE 10W/30 mineral oil containing conventional commercial additives including a metal sulfonate, zinc dialkyldithiophosphate, a V.I. improver, and the like, in which the dispersant normally used was replaced with either an alkylphenoxy acetamide of tetraethylenepentamine or its boronated counterpart.

The IIc and IIIc tests are standard engine tests described in ASTM Special Technical Publication 315-4. The IIc test is designed to evaluate the rusting and corrosion characteristics of motor oils. A standard test engine is operated continuously for 28 hours at moderate engine speed, partially warmed-up coolant and rich air/fuel ratio. The engine is then operated 2 hours at elevated coolant temperature and then shut down for 30 minutes. This is followed by 2 hours further operation at higher temperature and lean air/fuel ratio. Following this, the engine is disassembled and various parts visually inspected to determine the extent of rust and corrosion. The parts are rated on a scale of from 0 to 10, with 10 being clean. The overall rating is an average of the rating of the various parts.

A comparison was made of the IIc results obtained with a polyisobutylphenoxy acetamide of tetraethylenepentamine with the results obtained with the same type of product after boronation. Both dispersants were used at 3% concentration in a formulated motor oil in which they replaced the dispersant normally used. The results were: unboronated alkylphenoxyacetamide, a rust rating of 5.9, 6.8 and 7.1 (average 6.6); and boronated alkylphenoxyacetamide, a 6.9 rating. The results show that the boronation has somewhat improved the rust and corrosion protection provided by the oil.

The IIIc test evaluates the high-temperature oil thickening characteristics, sludge and varnish deposits as well as engine wear. A standard test engine is operated under noncyclic, moderately high speed, high load and temperature conditions for 64 hours. Oil samples are withdrawn at 8-hour intervals and viscosity measurements made. Following the test, the pistons are examined for ring-land face varnish and rated on a scale from 0 to 10, with 10 being clean. The test additives were used at 1.8% concentration in a formulated motor oil in which they replaced about one-half of a commercial succinimide dispersant normally used in this oil. The results of the tests are shown in the table below.

Additive	...Viscosity Increase...		Varnish
	40 Hours	64 Hours	
Unboronated alkylphenoxyacetamide	41%	1452%	4.7
Boronated alkylphenoxyacetamide	24%	189%	7.95

These tests show a sharp increase in the stability of the oil containing the boronated alkylphenoxyacetamide as evidenced by a much lower increase in oil viscosity during the test. The test also showed a very significant improvement in the piston ring-land face varnish due to the boronation.

The L-38 test is a standard engine test described in the Coordinating Research Council (CRC) Report No. 426 (1969). The test is designed to study the copper-lead corrosion characteristics of motor oils during engine operation. The test involves the continuous operation of a single cylinder CLR test engine under constant speed conditions for 40 hours. The bearing weight loss during the test is determined in mg and is a measure of the corrosivity of the oil. Results obtained in this test using 3% test additive in a formulated oil in which the test additive replaced the dispersant normally used were as follows: unboronated alkylphenoxyacetamide, a bearing weight loss of 51.71 mg; and boronated alkylphenoxyacetamide, 24.5 mg.

OTHER ASHLESS COMPOSITIONS

Terpolymer Polymethacrylates Based on 4-Vinylpyridine

R.I. Yamamoto, C.M. Cusano and F.J. Gaetani; U.S. Patent 4,081,385; March 28, 1978; assigned to Texaco Inc. describe a fully formulated crankcase lubricating oil composition containing between about 0.1 and 10 wt % of a tetrapolymer of 4-vinylpyridine (4-VP), a first alkyl methacrylate of the formula:

$$CH_2{=}\overset{\overset{\displaystyle CH_3}{|}}{\underset{}{C}}{-}\overset{\overset{\displaystyle O}{||}}{C}{-}OR$$

a second alkyl methacrylate of the formula:

$$CH_2{=}\overset{\overset{\displaystyle CH_3}{|}}{\underset{}{C}}{-}\overset{\overset{\displaystyle O}{||}}{C}{-}OR^1$$

and a third alkyl methacrylate of the formula:

$$CH_2{=}\overset{\overset{\displaystyle CH_3}{|}}{\underset{}{C}}{-}\overset{\overset{\displaystyle O}{||}}{C}{-}OR^2$$

where R is alkyl of from 1 to 6 carbons, R^1 is alkyl of from 10 to 15 carbons and R^2 is alkyl of from 16 to 20 carbons, of a molecular weight between about 25,000 and 2,500,000 having a component ratio of 4-VP:C_{1-6} alkyl methacrylate:C_{10-15} alkyl methacrylate:C_{16-20} alkyl methacrylate of between about 2:10:78:10 and 10:30:30:30 and between about 0.1 and 5 wt % of a calcium containing rust inhibitor. The composition has a sulfated metal ash content of between about 0.05 and 1 wt %.

The tetrapolymer is prepared by contacting a mixture of 4-vinylpyridine and first, second and third alkyl methacrylate monomers in the presence of a standard polymerization catalyst and transfer agent in an inert atmosphere, preferably in the presence of a diluent such as between about 20 and 60 wt % hydrocarbon oil. The reaction is advantageously conducted under conditions of agitation and at a temperature of between about 50° and 100°C. Under the preferred conditions, the monomers, transfer agent and a portion of the hydrocarbon oil diluent are first charged to the reactor and when the desired reaction temperature is obtained the polymerization catalyst is added.

Most advantageously, the catalyst is added in some two to five additions and additional diluent oil may be added during the second or later dose of polymerization catalyst, normally in the amount of between about 0 and 250 wt % of the reaction mixture. Polymerization is continued until all the monomers are essentially consumed, this latter occurrence is signified by the refractive index remaining constant.

Example 1: This example illustrates the preparation of the 4-vinylpyridine polymethacrylate terpolymer composition component.

To a 1-liter resin kettle equipped with a nitrogen inlet tube, stirrer, heater, cooling fan, thermister and thermocouple, the following materials were charged:

Materials	Grams	Mols
4-Vinylpyridine (4-VP)	40	0.38
Butylmethacrylate (BMA)	210	1.48
Neodol 25L Methacrylate (NMA)	500	1.79
Alfol 1620 SP Methacrylate (AMA)	250	0.77
1-Dodecanethiol	0.25	0.001
Mineral Oil A (~145 SUS at 100°F)	500	–

After purging the charged vessel with prepurified nitrogen for about a 15-minute
period with stirring, the reaction mixture was heated to 83°C. Stirring was conducted during the entire reaction period. When a temperature of 83°C was
reached, the initiator azobisisobutyronitrile (AIBN) was added in an amount of
2 g on a neat basis. Samples were taken at intervals and the refractive index determined at 54°C. When the refractive index became constant an additional
0.75 g of AIBN and 940 g of mineral oil (~100 SUS at 100°F) were added.
After 1 hour a third dosage of 0.75 g of AIBN was added. After an additional
hour the temperature was raised to 100°C for an hour and the formed terpolymer concentrate was allowed to cool and used for preparing blends. The product formed was characterized as a 41 wt % lube oil solution of a 4:21:50:25
weight ratio 4-VP:BMA:NMA:AMA polymethacrylate polymer in mineral oil.

The NMA and AMA monomers described above are respectively derived from
Neodol 25L and Alfol 1620 SP which are tradenames for technical grade alcohols respectively of Shell Chemical Company and Continental Oil Co. having
the following typical analysis:

	Typical Approx. Homolog Distribution, wt %
Neodol 25L (Synthetic lauryl alcohol)	
Lighter than $C_{12}OH$	4
$C_{12}OH$	24
$C_{13}OH$	24
$C_{14}OH$	24
$C_{15}OH$	15
$C_{16}OH$	2
Alfol 1620SP (Synthetic stearyl alcohol)	
$C_{14}OH$ and lighter	4
$C_{16}OH$	55
$C_{18}OH$	28
$C_{20}OH$	9

The resultant alkyl methacrylate monomers derived from the reaction of methacrylic acid with the above alcohols are in essence a mixture of C_{12-16} alkyl
methacrylates for those derived from Neodol 25L and C_{16-20} alkyl methacrylates
for those derived from Alfol 1620 SP with the same weight percent distribution
for a specific alkyl methacrylate as is found in the alcohol mixture. This same

weight distribution of the C_{12-16} and C_{16-20} methacrylate will also carry over into the interpolymer.

Example 2: This example illustrates the 4-vinylpyridine terpolymer containing lubricating oil compositions and the effectiveness of the 4-vinylpyridine in providing superior dispersancy to the formulation containing same without degrading the rust protection of the formulations below the critical standard specification requirements of the auto industry.

Seven fully formulated crankcase lubricating oil compositions were tested for rust protection in the Oldsmobile Sequence IIC Rust Test and for dispersancy in the Ford Sequence VC Test, both described in ASTM publication STP 315 F, "Multicylindered Test Sequence for Evaluating Automotive Engine Oils," Jan. 1973. For a given test the seven formulations employed were identical with the exception that the nitrogen containing component in the polymethacrylate ingredient was varied. Specifically, the representative polymethacrylate formulation employed is the 41 wt % lube oil formulation prepared and described in Example 1.

In the comparative formulations the terpolymer solutions were identical to the Example 1 formulation with the exception that the following monomer components were substituted for 4-vinylpyridine: 2-vinylpyridine, 2-methyl-5-vinylpyridine, dimethylaminoethyl methacrylate, diethylaminoethyl methacrylate, dimethylaminopropylmethacrylamide, and N-vinylpyrrolidone. The generic blends tested in the Oldsmobile Sequence IIC Rust Test and the Ford Sequence VC Test are as follows.

Table 1: Generic Formulation (Blend)

Ingredients	Blend A	Blend B
	. . . . . . (wt %)	
Mineral oil (~200 SUS at 100°F	89.10	89.65
Zinc (C_{3-8} dialkyldithiophosphate)	0.65	1.35
$CaCO_3$ overbased Ca sulfonate (~300 TBN)	1.00*	2.00*
Polyisobutylene (~1,200 MW) succinimide of tetraethylenepentamine	—	2.15
Ethyl substituted mono- and dinonylphenylamine	0.25	—
Poly(dimethyl silicone)	—	**
Polymethacrylate	9.00***	4.85***

  *45 wt % lube oil solution.
 **41 wt % lube oil solution.
***150 ppm.

Blend A gave a sulfated ash of 0.5 wt % and Blend B a sulfurized ash of 1 wt %. Evaluation of the generic lubricating oil compositions of Table 1 in which the ingredients are varied are reported in Table 2 below.

Table 2: Evaluation of Specific Crankcase Formulations

Nitrogen Component in Polymethacrylate of Blends A and B	Oldsmobile Sequence IIC Average Rust Rating* on Blend A	Ford Sequence VC Sludge Rating** on Blend B
4-Vinylpyridine	8.5	9.4
2-Vinylpyridine	8.6	5.4

(continued)

Table 2: (continued)

Nitrogen Component in Polymethacrylate of Blends A and B	Oldsmobile Sequence IIC Average Rust Rating* on Blend A	Ford Sequence VC Sludge Rating** on Blend B
2-Methyl-5-vinylpyridine	7.7	—
Dimethylaminoethyl methacrylate	5.9	9.0
Diethylaminoethyl methacrylate	5.1	—
Dimethylaminopropyl methacrylamide	5.3	—
N-vinylpyrrolidone	8.3	4.9

*8.4 minimum rating for a pass; 0.5% ash formulation.
**8.5 minimum rating for a pass; 1.0% ash formulation.

As can be seen from above, only the representative 4-vinylpyridine terpolymer polymethacrylate containing formulation meets the specification of both the Oldsmobile Sequence IIC Rust Test and the Ford Sequence VC Dispersant Test whereas the comparative polymethacrylates either degrade the formulation to a point of unacceptability in respect to antirust properties and/or do not have sufficient dispersant properties to meet the critical test requirements.

Aminoguanidine Sulfonate

According to a process described by *A. Abdul-Malek, C.G. Brannen and W.C. Edmisten; U.S. Patent 4,149,980; April 17, 1979; assigned to Standard Oil Company (Indiana)* ashless lubricant additive compositions which contain a small but effective amount of overbased aminoguanidine sulfonate, sufficient to impart acid neutralization, antirust, anticorrosive, dispersant, and detergent properties are formed by reacting greater than about one mol of aminoguanidine with one mol of a hydrocarbon sulfonic acid.

Aminoguanidine is a nitrogen-containing organic compound of the formula

$$NH_2-NH-\overset{\overset{\displaystyle NH}{\|}}{C}-NH_2$$

Aminoguanidine can be prepared by the zinc reduction of nitroguanidine, *Org. Synthesis*, Coll. vol. 3, 399 (1941), by the reaction of cyanamide at 20° to 50°C with hydrazine and carbon dioxide, German Patent 689,191, and other syntheses are described in an article by Lieber and Smith, *Chem. Rev.*, 25, 213 (1939). Aminoguanidine can be prepared in the free state, or in the form of a salt of an acid free aminoguanidine in a relatively unstable compound. A salt of aminoguanidine is commonly more stable, and more generally easily reacted and solubilized. In most applications the salt is the equivalent of the free aminoguanidine.

Example 1: 13.9 g aminoguanidine bicarbonate is dissolved in 20 ml water and placed in a round bottom flask under a blanket of nitrogen. 59 g of ammonium polybutyl sulfonic acid and 65 g of SX-5 oil is added to the aminoguanidine bicarbonate. The viscosity of the polymer at 100°F is 130 SUS. The equivalent molecular weight is about 450. This mixture is heated to 212°F for one hour to remove water. The water vapor is swept out of the flask with nitrogen. The mixture is then heated for two hours at 300° to 320°F. The mixture is cooled and ready for use.

Example 2: 36.8 g aminoguanidine bicarbonate is dissolved in 20 ml of water and placed in a round bottom flask under a blanket of nitrogen. 200 g of Bryton sulfonic acid and 113 g of SX-5 oil is added to the aminoguanidine bicarbonate. Bryton sulfonic acid is an alkyl aryl sulfonic acid purchased from Bryton Chemical Company which is a polypropyl benzene sulfonic acid, the equivalent molecular weight of which is about 460. The mixture is heated to 340°F while blanketed with nitrogen, and is reacted for 2 hours. When the mixture is cooled, it is ready for use.

Example 3: 16.3 g of aminoguanidine bicarbonate is dissolved in 20 ml of water and placed in a round bottom flask under a blanket of nitrogen. 156.5 g of benzene sulfonic acid and 59 g of SX-5 oil is added to the aminoguanidine bicarbonate. C_{20} benzene sulfonic acid is a polypropyl benzene sulfonic acid, the equivalent weight of which is about 680. The mixture is heated to a temperature of 340°F for two hours while blanketed with nitrogen. The mixture is then cooled and ready for use.

The aminoguanidine sulfonate displays antirust, detergent, anticorrosion, dispersion, and acid neutralization properties.

Tetrahydropyrimidyl-Substituted Compounds

B.W. Hotten; U.S. Patent 4,157,972; June 12, 1979; assigned to Chevron Research Company has discovered that tetrahydropyrimidyl-substituted compounds prepared from a C_3 to C_{50} amine containing a 1,3-diaminopropane group and ethylenediamine tetraacetic acid (EDTA) or nitrilotriacetic acid (NTA) are exceptionally superior ashless base additives for lubricating oil having good thermal stability as well as basicity.

The tetrahydropyrimidyl-substituted compounds of this process are prepared by reacting ethylenediamine tetraacetic acid or nitrilotriacetic acid with a compound of the following formula:

$$\begin{array}{cccc} R & R^1 & R^2 & R^3 \\ | & | & | & | \\ HN- & CH- & CH- & CH- & NH_2 \end{array}$$

wherein each of R, R^1, R^2 and R^3 is independently hydrogen or hydrocarbyl. The reaction is carried out at a temperature of 150° to 250°C for 10 to 100 hours. The reaction product may be used directly in lubricating compositions, or it may be purified by methods well known in the art to substantially isolate the primary polytetrahydropyrimidine product. In general, the use of the reaction product per se is preferred.

The compositions of this process are found to function as superior ashless additives for lubricating oil compositions in that they retain substantial alkalinity values under conditions of sustained high temperatures and they are highly rust-inhibitory.

Example 1: Into 600 ml of xylene were placed 57 g (about 0.3 mol) of nitrilotriacetic acid and 360 g (about 0.9 mol) of N-oleyl-1,3-diaminopropane. The mixture was held at about 150° to 200°C for about 27 hours and a total of 30 ml of water was evolved (calculated, 35 ml). The 431 g of product had 7.4 wt % of nitrogen and an alkalinity value of 160 mg KOH/g. The infrared spectrum showed the strong C=N band at 1640 cm⁻¹, and nuclear magnetic resonance (NMR) confirmed the presence of the methylene-ring hydrogens of the tetrahydropyrimidinyl group.

Example 2: N-oleyl-1,3-diaminopropane (2,400 g, about 6 mols) and nitrilotriacetic acid (382 g, about 2 mols) were mixed under nitrogen with stirring to 200°C over a 2-hour period. The mixture was maintained at this temperature for about 18 hours, stripped under vacuum and nitrogen to 150°C, and 2,661 g of product was recovered having an alkalinity value of 176 mg KOH/g. The product is tris-(3-oleyl-3,4,5,6-tetrahydro-2-pyrimidylmethyl)amine with some intermediate amides.

Example 3: Into 300 ml of xylene were mixed 146 g of ethylenediamine tetraacetic acid (about 0.5 mol) and 800 g of N-oleyl-1,3-diaminopropane. The mixture was heated at 150° to 200°C for about 48 hours, and 69 ml of water was evolved (72 ml calculated). The 904 g of product had an alkalinity value of 180 mg KOH/g and showed the infrared absorption at 1630 cm⁻¹ typical of C=N. The product is N,N,N',N'-tetrakis-(3-oleyl-3,4,5,6-tetrahydro-2-pyrimidylmethyl) ethylenediamine, mixed with some amido intermediates.

The polytetrahydropyrimidinyl products prepared by the process display satisfactory antivarnish detergency as additives in lubricating oils for the internal combustion engine as illustrated in the Ford V8 varnish test results of Table 1. In this test, a Ford V8 engine of 302 in³ displacement is operated in cycles of 500/2,500/2,500 rpm for periods of 45/120/75 minutes on a Chevron gasoline containing FCC heavy fraction (i.e., product of fluidized-bed catalyst cracking).

Table 1: Varnish Rating Using Ashless Base in Ford V8

	. . . . .Hours.			
	20	40	60	80
No base	8.9	8.0	7.7	—
Metallic base	9.7	9.4	9.1*	8.8*
Polytetrahydropyrimidine	9.6	8.9	8.7	8.3

*Mean value of two runs.

All oils contained 6 wt % polyisobutenyl succinimide of tetraethylene pentamine and 15 mmol/kg of zinc dialkyldithiophosphate in a neutral petroleum oil. The metallic base consisted of 30 mmol/kg of carbonated, sulfurized, calcium polypropylene phenate (9.25% calcium) and 30 mmol/kg of overbased calcium sulfonate (11.4% calcium). The polytetrahydropyrimidine was tris-(3-oleyl-3,4,5,6-tetrahydro-2-pyrimidylmethyl)amine at 2 wt % (63 meg/kg).

In the Ford V8 varnish test, the engine is disassembled at 20-hour intervals and the piston varnish is measured on a scale of 0 to 10, with 10 being completely clean. The polytetrahydropyrimidine ashless base is found to give antivarnish protection which is comparably satisfactory to the metallic base-containing, e.g., overbased, lubricating oil in present use.

The polytetrahydropyrimidines display excellent rust-inhibitory ability in the ASTM D1748 Humidity Cabinet Rust Test. In Table 2, various low-ash and ashless lubricating oil compositions have been tested in the Humidity Cabinet

Rust Test with and without the addition of 1 wt % of the product of Example 1.

Table 2: Rust Inhibition of Polytetrahydropyrimidine (1%)

| Composition | Humidity Cabinet Rust Life (hr) | |
	Without	With
Low ash*	<24	50
Low ash**	<24	90
Ashless***	24	800
Ashlest	40	700
Ashlesstt	130	>2,000

*6 wt % of polyisobutenyl succinimide of tetraethylpentamine
and 18 mmol/kg of zinc dialkyldithiophosphate.
**The composition of Footnote * + 0.2 wt % of tetrapropenylsuccinic acid.
***5 wt % of polyisobutenyl succinimide of triethylenetetramine, 1 wt % of
diisobornyldiphenylamine and 1% of bisalkylphenol sulfide.
tThe composition of Footnote *** + 0.2 wt % of tetrapropenylsuccinic
acid.
tt6 wt % polyisobutenyl succinimide of tetraethylenepentamine, 1 wt %
sulfurized wax, 3 wt % sulfurized alkylphenol and 1.5 wt % hindered
bisphenol (Ethyl 702).

METAL SULFONATES, PHENATES AND NAPHTHENATES

Metal Carbonate Overbased Metal Sulfonates

N. Bakker; U.S. Patent 4,137,184; January 30, 1979; assigned to Chevron Research Company describes sulfonates which are Group II metal carbonate overbased metal sulfonates of the formula $R-SO_3M$ where R is a substantially saturated aliphatic hydrocarbyl substituent containing from about 50 to 300 and preferably from about 50 to 250 carbon atoms. "Substantially saturated" means that at least about 95% of the carbon-to-carbon covalent linkages are saturated. Too many sites of unsaturation make the molecule more easily oxidized, degraded and polymerized. This makes the products unsuitable for many uses in hydrocarbon oils.

The especially useful polymers are the polymers of 1-monoolefins such as ethylene, propene, 1-butene, isobutene, 1-hexene, 1-octene, 2-methyl-1-heptene, 3-cyclohexyl-1-butene and 2-methyl-5-propyl-1-hexene. Polymers of olefins in which the olefinic linkage is not at the terminal position, such as 2-butene, 3-pentene and 4-octene, are also useful.

Example A is an illustrative preparation of a composition falling outside the scope of this process. Examples 1 through 6 illustrate the process.

Example A: Preparation of Calcium Polyisobutenyl Sulfonate — To a 10-gal glass-lined reactor are added 14,430 g of polyisobutylene of a number average molecular weight of 330 and an approximate average carbon number of 24, and 20,600 g of 1,2-dichloroethane. To this mixture is slowly added over a period of 1¼ hours 7,650 g chlorosulfonic acid. The reaction mixture is cooled continuously during the chlorosulfonic acid addition to maintain the temperature at 60°F. After the addition is completed, the reaction mixture is heated to 140°F. After maintaining the temperature of the reaction mixture at 140°F for 5½ hours, there is added slowly over a period of one hour a solution of 3,200 g NaOH in

6,400 ml methanol. The reaction mixture is then stripped to 196°F at atmospheric pressure, and one gallon of hydrocarbon thinner and 130 g NaOH in 260 ml methanol are added and the stripping operation continued to 248°F at atmospheric pressure. The contents of the reactor are cooled and transferred to a large reactor and sec-butyl alcohol and a solution of 6,300 g CaCl$_2$ in 32 liters of water is then added. This mixture is stirred at 100° to 120°F for 45 minutes.

After settling, the water layer is drained off and the metathesis repeated twice with 3,900 g CaCl$_2$ in 18 liters of water. The reaction mixture is then washed 3 times with approximately 4 gallons of water. 1 kg Ca(OH)$_2$ is added after the first water wash. After the water from the last wash is drained off, the supernatant product solution is filtered through diatomaceous earth. 3,000 g of diluent oil is added to the filtrate and the mixture stripped to 280°F and 60 mm Hg pressure to yield 17,070 g of calcium sulfonate concentrate containing 1.85% Ca, 4.57% S and 0.30% Cl. Neutral calcium as sulfonate, determined by Hyamine titration, a procedure published in *Analytical Chemistry*, Vol. 26, September 1954, pp 1492-1497, authors R. House and J.L. Darragh, is 1.81%.

Example 1: Preparation of Sodium Polyisobutenyl Sulfonate — To a 10-gallon glass-lined reactor are added 12,000 g of polyisobutylene having a number average molecular weight of 950 and an approximate average carbon number of 68, and 6,000 g of 1,2-dichloroethane. To this mixture is slowly added over a period of 1½ hours a solution of 2,100 g chlorosulfonic acid in 6,000 g butyl ether. The reaction mixture is cooled continuously to maintain the temperature at 40°F. After the addition is completed, the reaction mixture is warmed to 104°F. After maintaining the temperature of the reaction mixture at about 100°F for about 5 hours, there is added slowly over a period of 2 hours 3,810 ml of a 25% aqueous sodium hydroxide solution (approximately 1,150 g NaOH). 1,000 ml of hydrocarbon thinner is added and the reaction mixture is stripped to 195°F at atmospheric pressure. An additional 10,000 ml of hydrocarbon thinner is then added to yield 32,090 g of product.

Example 2: Preparation of Sodium Polyisobutenyl Sulfonate — The procedure of Example 1 is repeated with the exception that the reaction mixture is neutralized with a methanolic solution of sodium hydroxide prepared from 1,020 g NaOH and 4,300 ml of methanol. The product is 26,780 g of sodium polyisobutenyl sulfonate solution.

Example 3: Preparation of Calcium Polyisobutenyl Sulfonate — To the product solutions of Examples 1 and 2 are added half a volume of hydrocarbon thinner and half a volume of isobutyl alcohol and mixed thoroughly. This is the feed used in the continuous metathesis process.

The apparatus consists of a metathesis column 100 x 5 cm and a water-wash column 100 x 11.5 cm, both packed with ¼-inch Penn State packing and maintained at 40°C with heating tape.

The metathesis column is filled with 20% aqueous CaCl$_2$ solution. CaCl$_2$ solution and water are fed into the columns 20 cm from the top at 40 and 80 ml/min, respectively. The outlets are at the very bottom of the columns. The height of the CaCl$_2$ solution and the water level in the columns is controlled by raising or lowering the outlet of ⁵/₁₆ inch tubing connected to the bottom outlet of the columns and usually maintained 15 cm from the top.

The product feed solution is pumped into the metathesis column 20 cm from the bottom at 20 ml/min and taken off 2 cm from the top. Residence time of the product in the metathesis column is 20 minutes. The metathesized product is then pumped into the water-wash column 20 cm from the bottom at 20 ml/min and taken off 2 cm from the top.

To the water-washed product is then added enough $Ca(OH)_2$ to neutralize any acid product that may have formed and enough diluent oil to give a 70% concentrate after stripping off of the solvent. The stripped and filtered product contains by x-ray fluorescence analysis 1.31% calcium, 1.97% sulfur, 0.07% chlorine and 1.10% neutral calcium as sulfonate by Hyamine titration.

Example 4: Calcium Carbonate Overbased Calcium Polyisobutenyl Sulfonate — To a 1-liter, 3-neck flask with nitrogen sparge are added 410 g of calcium polyisobutenyl sulfonate solution containing 147 g calcium sulfonate wherein the polyisobutylene has a number average molecular weight of 950, and 100 g diluent oil. The mixture is stripped to 160°C bottoms under vacuum. To this mixture is added 500 ml of hydrocarbon thinner. The solution is transferred to a 2-liter, baffled, 3-neck flask and 50 ml methanol, 70 ml 2-ethyl-1-hexanol and 100 g calcium hydroxide are added. The mixture is carbonated for 3 hours at 34° to 40°C with 60 g of CO_2. The solution is then heated to 160°C at atmospheric pressure while solvent is removed overhead. After cooling, the mixture is filtered through diatomaceous earth. The filtrate is stripped to 170°C bottoms at 6 mm Hg to yield 337 g of calcium carbonate-overbased calcium polyisobutenyl sulfonate concentrate. The alkalinity value of this overbased sulfonate is 363, with 12.96% basic calcium. Hyamine titration indicated 0.68% neutral calcium as sulfonate. The base ratio for the product is 19:1.

Example 5: Calcium Carbonate Overbased Calcium Polyisobutenyl Sulfonate — To a 2-liter, baffled, 3-necked flask is added 799 g of calcium polyisobutenyl sulfonate solution straight from the water-wash column (see Example 3) containing 288 g calcium sulfonate wherein the polyisobutenyl group has a number average molecular weight of 950, 50 ml methanol, and 200 g calcium hydroxide. The reaction mixture is carbonated at ambient temperature for 160 minutes with 110 g of CO_2. During the carbonation the temperature increases to a maximum of 54°C. To the reaction mixture is added 500 ml of hydrocarbon thinner. Solvent is then distilled off to 150°C bottoms at atmospheric pressure.

The solution is cooled and filtered through diatomaceous earth and the filtrate is stripped to 170°C bottoms at 6 mm Hg. Shortly before all of the solvent has been removed, 160 g of diluent oil is added to yield 543 g of calcium carbonate overbased calcium polyisobutenyl sulfonate having an alkalinity value of 253 (9.04% basic calcium). The product contains 0.75% neutral calcium as sulfonate and has 9.26% calcium, 1.29% sulfur and 0.07% chlorine content as determined by x-ray fluorescence analysis. The base ratio is 12.0:1.

Example 6: Calcium Carbonate Overbased Calcium Polyisobutenyl Sulfonate — To a 1-liter, 3-neck flask is added 400 g of calcium polyisobutenyl sulfonate solution containing 144 g calcium sulfonate wherein the polyisobutylene has a number average molecular weight of 950, and 80 g of diluent oil. The mixture is stripped to 160°C bottoms under vacuum, yielding 249 g overhead. To the product in the flask are added 500 ml of hydrocarbon thinner, 50 ml methanol, 70 ml 2-ethyl-1-hexanol and 74 g of calcium hydroxide. This mixture is carbonated at ambient temperature with 41 g CO_2. The temperature increases to 39°C over a period of 130 minutes. The mixture is heated to 155°C bottoms at atmospheric pressure while taking off solvent overhead, and then is heated to 130°C under vacuum. The mixture is cooled, filtered through diatomaceous earth and the filtrate is stripped to 170°C at 6 mm Hg to yield 310 g of product in lubricating oil having an alkalinity value of 315 (11.26% basic calcium), a base ratio of 16.3:1 and a Hyamine titration value of 0.69% neutral calcium as sulfonate.

Example 7: General Motors Sequence IIC Rust Test — Lubricating oil formulations containing conventional commercial overbased sulfonates were evaluated in comparison to the sulfonate of Example 5 in the General Motors Sequence IIC Rust Test. The test oil is an R.I. Sun Puerto Rico SAE 30 lubricating oil. Formulation I contains 4.0% of a conventional succinimide dispersant, 24 mmol/kg of a calcium phenate, 18 mmol/kg of a zinc dihydrocarbyl dithiophosphate, and 24 mmol/kg of the overbased sulfonate being tested. Formulation II contains 8.0% of a conventional succinimide dispersant, 25 mmol/kg of a conventional, nonoverbased calcium sulfonate, 19 mmol/kg of a calcium phenate, 18 mmol/kg of a zinc dihydrocarbyl dithiophosphate and 18 mmol/kg of the overbased sulfonate being tested. The average engine rust (AER) is measured after 32 hours, with lower ratings indicating poorer performance. GM specifications require at least an AER rating of 8.4 to pass the test. The results are reported in Table 1.

Table 1: Sequence IIC Rust Test Performance

Composition Tested	AER Rating
Formulation I containing conventional overbased sulfonate	7.3
Formulation I containing Example 5 overbased sulfonate	8.4
Formulation II containing conventional overbased sulfonate	7.8
Formulation II containing Example 5 overbased sulfonate	8.4

Example 8: Caterpillar Engine Test — The lubricating oil compositions of this process were tested in a high-output diesel engine test to evaluate the ring-sticking performance of the lubricant. The test is carried out in a Caterpillar 1-G engine run until ring-sticking was recorded.

The base oil used in these tests is a Mid-Continent base stock SAE 30 oil containing 3% of a conventional succinimide dispersant, 7 mmol/kg of a zinc dialkylaryl dithiophosphate and 11 mmol/kg of a zinc dialkyl dithiophosphate. To this base oil is added 41 mmol/kg of the overbased sulfonate to be tested. For comparison, 41 mmol/kg commercially available overbased calcium sulfonate having a base ratio of about 9 was also tested. The sulfonate designated as "Z" in Table 2 is the commercially available sulfonate. The results of testing the lubricating oils of this process as well as the lubricating oils containing the commercially available sulfonate are set forth in Table 2 on the following page.

Table 2: Caterpillar Engine Test

Composition Tested	Hours to Ring Sticking
41 mmol/kg Z	6, 10
41 mmol/kg product of Ex. A	63*, 8, 6**
36 mmol/kg Z + 5 mmol/kg neutral product of Ex. 3	30, 27**
41 mmol/kg product of Ex. 5	72***
41 mmol/kg product of Ex. 6	45
41 mmol/kg product of Ex. 4	67***

  *Ring probably stuck at less than 20 hours, but the engine did not shut down.
 **Results of repeat tests.
***Ring was still free when test terminated.

From the data in Table 2, it can be seen that the overbased calcium sulfonates of this process are significantly better detergents than the commercially available sulfonate or the neutral sulfonate, even when combined with the commercially available sulfonate.

Calcium Oxide with Low Reactivity Towards Water

According to a process described by *G. De Clippeleir and A. Vanderlinden; U.S. Patent 4,086,170; April 25, 1978; assigned to Labofina SA, Belgium* overbased calcium sulfonates and concentrated oily solutions are prepared by reacting a solution of alkylbenzene sulfonic acids with an excess of a calcium oxide having a medium or low activity towards water and with carbon dioxide. Oily solutions of overbased calcium sulfonate obtained from such a calcium oxide are perfectly limpid and are easily filtrable.

Thus, it has been found that calcium oxide having a medium reactivity towards water, e.g., as measured according to the ASTM-C-110 method, that is to say a lower reactivity than that of quicklimes or commercial calcium oxides, allows preparation of calcium sulfonates having a high degree of alkalinity, for instance, higher than 300 mg KOH/g, the oily solutions of which are perfectly limpid and easily filtrable.

Such a calcium oxide is easily obtained by extending the roasting time of the limestone or by carrying out this roasting at a temperature which is much higher than that which is usually required for the complete conversion of the limestone into calcium oxide. When an oily solution of overbased calcium sulfonates is prepared, preferably the solvent which is used to prepare the starting solution of alkyl benzene sulfonic acids further comprises oxygen-containing solvents, preferably lower mono- or divalent alcohols, e.g., methanol, ethanol or the like and/or aromatic solvent, e.g., benzene or toluene.

It has been found that a small amount of water in methanol optimizes the reaction between calcium oxide and carbon dioxide, but if the calcium oxide is highly reactive towards water, it tends to agglomerate, thus reducing its surface area which is available for the contact with carbon dioxide. Whereas when a calcium oxide having a medium reactivity towards water is used under the same conditions it remains finely suspended in the mixture and it absorbs the carbon dioxide easily. Further to the abovementioned solvents, alkaline additives which are removable by evaporation and/or do not form insoluble products such as ammonia or organic amines may be added during the reaction. The volatile solvents and other additives are removed by distillation after the reaction is completed. After evaporation of the solvents and additives, the concentrated oily solution of overbased sulfonates which is obtained from the calcium oxide having a medium reactivity to water is easily filtrable and has a high degree of alkalinity whereas oily solutions of calcium sulfonates which are obtained from a calcium oxide highly reactive to water have a lower degree of alkalinity, and their filtration is difficult or even impossible.

The calcium oxide which is preferably used, fulfills the following specifications: CaO, minimum 92%; CO_2, maximum 1%; SiO_2, maximum 1.7%, MgO, maximum 1.4%; S, maximum 0.02%; Fe_2O_3, maximum 1.5%; and ignition loss, maximum 1.5%.

A purified alkylbenzene sulfonic acid, which is generally diluted with a mineral oil, is reacted with the calcium oxide of medium reactivity such as above defined and carbon dioxide is then bubbled through the mixture, to confer to the final product its potential of alkalinity or its overalkalization. It has been found that this overalkalization as well as the filtration of the overbased sulfonate solution are made easy when, first, the bubbling through of the carbon dioxide and, thereafter, the evaporation of the solvents are carried out in an alkaline medium. In order to obtain this alkaline medium, any basic compound which does not form insoluble products and/or is removable by evaporation, may be added, such as, for instance, ammonia or organic bases, typically mono- or diamines such as ethylenediamine. The selection of this basic compound principally depends on economic conditions, and therefore, ammonia is advantageously used. The abovementioned solvents may be used during the over-alkalization.

Single Stage Low Temperature Addition of Carbon Dioxide

A.R. Sabol; U.S. Patent 4,137,186; January 30, 1979; assigned to Standard Oil Company (Indiana) describes a process for producing highly basic gel-free, overbased sulfonates by single stage low temperature addition of acidic material, preferably carbon dioxide.

For the purpose of this process, the amount of overbasing produced is reported as the Total Base Number (TBN) which is the number of milligrams of KOH equivalent to the amount of acid required to neutralize the alkaline constituents present in one gram of the composition. A standard procedure for measuring Total Base Number is ASTM D-2896. The metal ratio is the ratio of molar equivalents of an alkaline earth, for example, magnesium, to molar equivalents of organic acid in the composition.

The objects of this process can be attained by forming a composition comprising an oil-soluble organic sulfonic acid containing at least 0.1 wt % neutral ammonium sulfonate, a stoichiometric excess of basically reacting magnesium oxide based on the total equivalent of sulfonic acid compound, about 0.1 to 8 mols water per mol of magnesium compound, about 0.1 to 5 mols of alkanol per mol magnesium compound, and at least one substantially inert organic liquid diluent; hydrating the magnesium oxide at an elevated temperature (preferably at reflux), stripping the methanol from the reaction mixture and then adding an acidic material to the hydrated reaction mixture while maintaining the hydrated reaction mixture at a temperature of 80° to 155°F. Significantly, it has been found that overbased magnesium sulfonates produced in this manner are gel-free and have reproducible TBNs of over 400 even using sulfonic acids formed from soft alkylate detergent bottoms.

Studies have shown that if the carbonation step is carried out at reflux, a crystalline form of overbased magnesium sulfonate is formed, instead of the amorphous type of overbased magnesium sulfonate which is necessary to obtain a haze-free product having a TBN over 400. These studies have also shown that amorphous products can only be produced if the carbonation step is at no more than 155°F. Above 155°F crystallization of magnesium monohydrate salt tends to be induced. The higher the temperature above 155°F, the greater the crystallization. However, if methanol is present gelation occurs. Accordingly, the temperature range of 80° to 155°F is critical in this process.

Metal Salts of Substituted Ethylsulfonic Acids

A process described by *L. de Vries; U.S. Patent 4,116,873; September 26, 1978; assigned to Chevron Research Company* relates to oil-soluble Group I or Group II metal or lead hydrocarbyl ethylsulfonates.

The oil-soluble Group I and Group II metal or lead salts of substantially saturated aliphatic hydrocarbyl ethylsulfonic acids, the substantially saturated hydrocarbyl substituent containing at least 25 aliphatic carbon atoms, are prepared by reacting an aryl ester of the formula described below with a Group I or Group II metal oxide or hydroxide:

(1)
$$\overset{R^1}{\underset{|}{H_2C-CH_2-SO_2-O-Aryl}}$$

or

(2)
$$\overset{R^1}{\underset{|}{H_3C-CH-SO_2-O-Aryl}}$$

where R^1 is a substantially saturated aliphatic hydrocarbyl substituent containing enough carbon atoms to make the sulfonate oil-soluble.

The aryl sulfonates of formulas (1) and (2) are prepared by adducting an aryl vinylsulfonate to a hydrocarbon. This adduction is carried out using conventional techniques such as those used to adduct maleic anhydride to hydrocarbon substituents in preparing hydrocarbyl succinic anhydrides. The aryl substituent is not displaced during the adduction reaction.

The metal sulfonates are prepared using any Group I and Group II metals or lead compound which forms a salt with the sulfonic acid moiety and which yields a salt useful as a detergent in lubricating oil compositions. Preferably, the Group I metal compounds are lithium, sodium and potassium compounds and the Group II metal compounds are magnesium, calcium, strontium, barium and zinc. The lead compound must be in the +2 valence state, i.e., Pb^{++}. More preferably, the Group I metal compounds are sodium and potassium compounds and the Group II metal compounds are magnesium, calcium and barium compounds.

Example 1: Preparation of Sodium Sulfonate — A 1-liter, 3-neck flask is charged with 210 g (ca 0.1 mol) of o-chlorophenyl polyisobutenylethylsulfonate in which the polyisobutenyl group has a number average molecular weight of about 950. 200 ml of an inert hydrocarbon thinner are added and the sulfonate and the solvent are stirred to mix them. With stirring an aqueous solution of 11 g (0.27 mol) of sodium hydroxide in 15 ml of water is added and the reaction mass is heated at 120°C for 1.5 hours. The reaction mass is allowed to cool and to settle overnight and then is filtered through Hyflosupercel (diatomaceous earth).

The filtrate is transferred to a 2-liter separatory funnel, using 200 ml of the thinner as a rinse to insure complete transfer. 300 ml of 2-butanol and 300 ml of water are added and the contents of the funnel are mixed thoroughly. The contents of the funnel are allowed to settle for one-half hour and the bottom layer is drained off (ca 260 ml). The nonaqueous phase remaining in the funnel is washed four times with approximately 600 ml of aqueous sodium chloride solution. The sodium chloride solution is used because water alone did not phase-separate well enough. The hydrocarbon phase is stripped of the butanol and any entrained water to an end point of 190°C at about 200 mm Hg.

The stripped product is cooled and 700 ml of hexane is added. The resulting mixture is filtered through Hyflosupercel. The filtrate is stripped free of solvent to 200°C at 5 mm Hg. 194 g of product having a slight haze are recovered. 80 g of a neutral solvent-refined lubricating oil having a viscosity of 100 SUS at 100°F (38°C) is added. This mixture is heated to 150°C with stirring to homogenize and then filtered through Hyflosupercel to yield a clear product. The product is analyzed with the following results: Na, 0.66 wt %; S, 0.96 wt %; and Cl, 0.02 wt %.

Example 2: Preparation of Calcium Sulfonate by Metathesis — To a 5-liter flask, 1,060 g of o-chlorophenyl polyisobutenylethylsulfonate having a number average molecular weight of 950 and 1,000 ml of an inert hydrocarbon thinner are added. Thereafter, 45 g of sodium hydroxide in 70 ml of water are added. With stirring the temperature is raised to 100°C and maintained there for 2 hours. After cooling to 95°C, a solution of 156 g of calcium chloride in 1,000 ml of water is added. The reaction mass is stirred for 1 hour at 85°C.

The reaction mass is transferred to two 4-liter separatory funnels and to each is added 750 ml of 2-butanol. The aqueous phase is drained off the bottom, the hydrocarbon phases are transferred to two 5-liter, 3-neck flasks and to each is added 80 g calcium chloride in 500 ml water. The mixtures are stirred for 1 hour at 85°C and then transferred to two separatory funnels as before. The aqueous phase is drained off the bottom and the hydrocarbon phase is washed again with a mixture of 80 g calcium chloride in 500 ml water and four times with water. The hydrocarbon phase is stripped free of solvent to 175°C at 5 mm Hg to yield 1,018 g product. To the product is added 509 g of a neutral solvent-refined lubricating oil having a viscosity of 100 SUS at 100°F. The mixture is stirred at 150°C to homogenize it and is then filtered through Hyflosupercel, after which it was analyzed with the following results: Ca, 0.56 wt %; S, 0.84 wt %; and Cl, 0.01 wt %.

Example 3: Preparation of Calcium Sulfonate by Metathesis — To a 5-liter, 3-neck flask, equipped as in Example 1, is added 1,350 g of phenyl polyisobutenyl-ethylsulfonate, in which the polyisobutenyl group has a number average molecular weight of 950, dissolved in 2,025 ml of an inert hydrocarbon solvent. A solution of 99 g (1.5 mols) of 85% potassium hydroxide dissolved in 300 ml of methanol is added. With stirring, the methanol is distilled off and the reaction mixture is maintained at 100°C for 2 hours. 800 ml of 2-butanol is added and the mixture is stirred at 79°C for 4 hours.

The reaction mass is divided into two equal parts which are charged to 5-liter, 3-neck flasks, each of which is equipped with a stirrer, thermometer and reflux condenser. To each flask is added 400 ml of 2-butanol, 800 ml of water, and 500 ml of the inert hydrocarbon solvent. The reaction masses are stirred at 80°C for one-half hour and then transferred to separatory funnels where they are allowed to settle and the water layers are drawn off. Each of the remaining hydrocarbon phases is transferred to separate 5-liter, 3-neck flasks equipped as above. Each is stirred three times with a solution of 147 g of calcium chloride dihydrate in 800 ml of water for one hour at 80°C and is then water-washed three times with 800 ml of water for 0.75 hour at 80°C.

After the last water wash has been separated from the hydrocarbon layer, the supernatant liquid is filtered through Hyflosupercel. The filtrate is stripped free of the hydrocarbon solvent to 170°C bottoms temperature of 5 mm Hg to yield 1,142 g of combined product. The quantity of product was low because some product was lost during the workup. 613 g of a neutral solvent-refined lubricating oil having a viscosity of 100 SUS at 100°F is added to the product to yield 1,755 g of concentrate. The concentrate had a slight haze, which was removed by filtering through Hyflosupercel. The concentrate is analyzed and found to contain: Ca, 0.65 wt %; S, 1.07 wt %; and Cl, less than 0.01 wt %.

Example 4: The lubricating oil compositions of this process are tested in the well-known 1-G Caterpillar test. The base oil used in these tests is a mid-continent base stock SAE 30 oil containing a conventional succinimide dispersant, a calcium phenate, and a zinc dithiophosphate. To this base oil is added 10 mmol of the calcium sulfonate to be tested. For comparison, commercially available sulfonates are tested. The sulfonate designated as A in the table below is a commercially available calcium sulfonate prepared by acid-treating a neutral solvent-refined lubricating oil having a viscosity of between 300 and 480 SUS. The sulfonate designated as B in the table is a commercially available mixture of calcium sulfonates derived from various sources such as acid-treated neutral solvent-refined lubricating oils as well as certain of the hard alkylates produced as by-products in detergent manufacture. The results of testing the lubricating oils of this process as well as lubricating oils containing the commercially available sulfonates are set forth in the table.

120 Hr 1-G Caterpillar Test

Sulfonate	Hours	Grooves	Lands	Underhead
A	60	58-4-0.6-0.5	160-310-215	4.8
B	60	50-6-0.6-0.7	390-120-515	4.8
Example 2	60	22-4-1.0-0.9	105-20-25	6.6
Example 3	60	27-8-1.0-0.6	165-30-25	5.7
	120	38-7-2-0.6	168-39-48	4.5

From the data in the table above, it can be seen that the calcium sulfonates of this process are good detergents in lubricating oil compositions. It should be noted that the lubricating oil compositions of this process provide a significant improvement in the rating of the lower lands of the pistons compared to the lubricating oils containing the commercially available sulfonate detergents.

Alkaline Earth Metal Sulfonates and Dispersants

L. de Vries; U.S. Patent 4,159,958; July 3, 1979; assigned to Chevron Research Company has found that the combination of a hydrocarbon-substituted succinic acid ester with an alkaline earth metal substantially saturated aliphatic sulfonate yields a lubricating oil additive of superior detergency as compared to the combination of hydrocarbon-substituted succinic acid ester with conventional aromatic sulfonate.

The lubricating oil compositions of the process are useful for lubricating internal combustion engines, automatic transmissions and as industrial oils such as hydraulic oils, heat-transfer oils, torque fluids, etc. The lubricating oils can not only lubricate the engines but, because of their dispersancy properties, help maintain a high degree of cleanliness of the lubricated parts.

Example 1: Preparation of Calcium Polyisobutenyl Sulfonate — To a 10-gallon glass-lined reactor are added 14,430 g of polyisobutylene having a number average molecular weight of 330 and an approximate average carbon number of 24, and 20,600 g of 1,2-dichloroethane. To this mixture is slowly added over a period of 1¼ hours 7,650 g chlorosulfonic acid. The reaction mixture is cooled continuously during the chlorosulfonic acid addition to maintain the temperature at 60°F. After the addition is completed, the reaction mixture is heated to 140°F. After maintaining the temperature of the reaction mixture at 140°F for 5½ hours, there is added slowly over a period of 1 hour a solution of 3,200 g NaOH in 6,400 ml methanol.

The reaction mixture is then stripped to 196°F at atmospheric pressure, and 1 gallon of hydrocarbon thinner and 130 g NaOH in 260 ml methanol are added and the stripping operation continued to 248°F at atmospheric pressure. The contents of the reactor are cooled and transferred to a larger reactor and sec-butyl alcohol and a solution of 6,300 g CaCl$_2$ in 32 liters of water is then added. This mixture is stirred at 100° to 120°F for 45 minutes. After settling, the water layer is drained off and the metathesis repeated twice with 3,900 g CaCl$_2$ in 18 liters of water.

The reaction mixture is then washed 3 times with approximately 4 gallons of water. 1 kg Ca(OH)$_2$ is added after the first water wash. After the water from the last wash is drained off, the supernatant product solution is filtered through diatomaceous earth. 3,000 g of diluent oil is added to the filtrate and the mixture is stripped to 280°F and 60 mm Hg pressure to yield 17,070 g of calcium sulfonate concentrate containing 1.85% Ca, 4.57% S; and 0.30% Cl. Neutral calcium as sulfonate, determined by Hyamine titration, a procedure published in *Analytical Chemistry*, Vol. 26, September 1954, pp 1492-1497, authors R. House and J.L. Darragh, is 1.81%.

Example 2: Preparation of Sodium Polyisobutenyl Sulfonate — To a 10-gallon glass-lined reactor are added 12,000 g of polyisobutylene having a number average molecular weight of 950 and an approximate average carbon number of 68, and 6,000 g of 1,2-dichloroethane. To this mixture is added slowly over a period of 1½ hours a solution of 2,100 g chlorosulfonic acid in 6,000 g butyl ether. The reaction mixture is cooled continuously to maintain the temperature at 40°F. After the addition is completed, the reaction mixture is warmed to 104°F. After maintaining the temperature of the reaction mixture at about 100°F for about 5 hours, there is added slowly over a period of 2 hours 3,810 ml of a 25% aqueous sodium hydroxide solution (approximately 1,150 g NaOH). 1,000 ml of hydrocarbon thinner is added and the reaction mixture is stripped to 195°F at atmospheric pressure. An additional 10,000 ml of hydrocarbon thinner is then added to yield 32,090 g of product.

Example 3: Preparation of Sodium Polyisobutenyl Sulfonate — The procedure of Example 2 is repeated with the exception that the reaction mixture is neutralized with a methanolic solution of sodium hydroxide prepared from 1,020 g NaOH and 4,300 ml of methanol. The product is 26,780 g of sodium polyisobutenyl sulfonate solution.

Example 4: Preparation of Calcium Polyisobutenyl Sulfonate — To the product solutions of Examples 2 and 3 are added half a volume of hydrocarbon thinner and half a volume of isobutyl alcohol, which are mixed thoroughly. This is the feed used in the continuous metathesis process.

The apparatus consists of a metathesis column 100 x 5 cm and a water-wash column 100 x 11.5 cm, both packed with one-quarter inch Penn State packing and maintained at 40°C with heating tape.

The metathesis column is filled with 20% aqueous $CaCl_2$ solution. $CaCl_2$ solution and water are fed into the columns 20 cm from the top at 40 and 80 ml/min, respectively. The outlets are at the very bottom of the columns. The height of the $CaCl_2$ solution and the water level in the columns is controlled by raising or lowering the outlet of $^5/_{16}$ inch tubing connected to the bottom outlet of the columns and usually maintained 15 cm from the top.

The product feed solution is pumped into the metathesis column 20 cm from the bottom at 20 ml/min and taken off 2 cm from the top. Residence time of the product in the metathesis column is 20 minutes. The metathesized product is then pumped into the water-wash column 20 cm from the bottom at 20 ml/min and taken off 2 cm from the top.

To the water-washed product is then added enough $Ca(OH)_2$ to neutralize any acid product that may have formed and enough diluent oil to give a 70% concentrate after stripping off of the solvent. The stripped and filtered product contains by x-ray fluorescence analysis 1.31% calcium, 1.97% sulfur, 0.07% chlorine and 1.10% neutral calcium as sulfonate by Hyamine titration.

Example 5: The compositions of this process were tested in a Caterpillar 1-G test in which a single-cylinder diesel engine having a 5⅛" bore by 6½" stroke is operated under the following conditions: timing, degrees BTDC 8; brake mean effective pressure, 141 psi; brake horsepower 42; Btu/min 5,850; speed 1,800 rpm; air boost, 53" Hg absolute; air temperature in, 255°F; water temperature out, 190°F; and sulfur in fuel, 0.4 wt %. At the end of each 12 hours of operation, sufficient oil is drained from the crankcase to allow addition of one quart of new oil. In the test on the lubricating oil compositions of this process, the 1-G test is run for 60 hours. At the end of the 60-hour period, the engine is dismantled and rated for cleanliness. The ring lands are rated on a scale of 0 to 800, with 0 representing clean and 800 representing black deposits. The ring grooves are rated on a scale of 0 to 100 groove fill, with 0 representing clean. The underhead of the piston is rated on a scale of 0 to 10, with 0 representing dirty and 10 representing clean.

The base oil used in these tests is a mid-continent base stock SAE 30 oil containing 15 mmol/kg of a zinc dihydrocarbyl dithiophosphate, 31 mmol/kg of a calcium phenate, and the amount noted in the following table of sulfonate and 1.5 wt % of Lz936, a hydrocarbyl succinate prepared from pentaerythritol.

Test Results: 1-G Caterpillar Test

Sulfonate (10 mmol/kg)	Grooves	Lands	Underhead
Commercial calcium petroleum sulfonate	24-3.0-0.7-0.8	245-170-105	2.0
Product of Example 4	33-10-0.5-0.5	200-20-20	1.5

In related work, *L. de Vries; U.S. Patent 4,159,956; July 3, 1979; assigned to Chevron Research Company* has found that the combination of a monosuccinimide with an alkaline earth metal substantially saturated aliphatic sulfonate yields a lubricating oil additive of superior detergency as compared to the combination of succinimide and conventional sulfonate.

L. de Vries; U.S. Patent 4,159,959; July 3, 1979; assigned to Chevron Research Company has found that the combination of a monohydroxyalkyl-substituted hydrocarbyl thiophosphonate with an alkaline earth metal substantially saturated aliphatic sulfonate yields a lubricating oil additive of superior detergency as compared to the combination of monohydroxyalkyl hydrocarbyl thiophosphonate with conventional aromatic sulfonate.

L. de Vries; U.S. Patent 4,159,957; July 3, 1979; assigned to Chevron Research Company also describes a lubricating oil of good detergency which contains an alkaline earth metal aliphatic sulfonate and a Mannich base condensation product.

Magnesium-Containing Complexes

A process described by *J.W. Forsberg; U.S. Patent 4,094,801; June 13, 1978; assigned to The Lubrizol Corporation* relates to noncarbonated magnesium-containing complexes which are prepared by heating, at a temperature above about 30°C, a mixture comprising: (a) at least one of magnesium hydroxide, magnesium oxide, hydrated magnesium oxide or a magnesium alkoxide; (b) at least one oleophilic organic reagent comprising a carboxylic acid, a sulfonic acid, a pentavalent phosphorus acid, or an ester or alkali metal or alkaline earth metal salt of any of these; (c) water; and (d) at least one organic solubilizing agent for component (b); the ratio of equivalents of magnesium to component (b), calculated as the free carboxylic or sulfonic acid or as the phosphoric acid ester,

being at least about 5:1, and the amount of water present being at least sufficient to hydrate a substantial proportion of component (a) calculated as magnesium oxide. The preparation of the magnesium complexes of this process is
illustrated by the following examples. All parts are by weight.

Example 1: A blend is prepared of 135 parts of magnesium oxide and 600 parts
of an alkylbenzenesulfonic acid having an equivalent weight of about 385, and
containing about 24% unsulfonated alkylbenzene. During blending, an exothermic reaction takes place which causes the temperature to rise to 57°C. The
mixture is stirred for one-half hour and then 50 parts of water is added. Upon
heating at 95°C for one hour, the desired magnesium oxide-sulfonate complex is
obtained as a firm gel containing 9.07% magnesium.

Example 2: A blend of 600 parts of the alkylbenzenesulfonic acid of Example
1 and 225 parts of magnesium oxide is prepared and heated for 2 hours at 60°
to 65°C. There is then added, over one hour, a solution of 10 parts of 30% ammonium hydroxide and 75 parts of water. The mixture is heated for 3 hours at
60° to 65°C, and then an additional 10 parts of 30% ammonium hydroxide solution is added over 5 minutes. Upon heating for 2 more hours at 60° to 65°C
and cooling, the desired magnesium oxide-sulfonate complex is obtained as a dark
brown gel.

Example 3: Following the procedure of Example 2, a blend is made of 600 parts
of the alkylbenzenesulfonic acid of Example 1 and 225 parts of magnesium oxide, and a solution of 30 parts of 30% ammonium hydroxide in 75 parts of water is added. After heating for 4 hours at 60° to 65°C, the mixture is cooled and
900 parts of hexane is added. The hexane-diluted mixture is centrifuged and the
hexane is removed by vacuum stripping at 150°C. The residue is cooled to 130°C
and 18 parts of triethanolamine is added. The product is the desired magnesium
oxide-sulfonate complex, a soft brown gel which contains 11.3% magnesium.

Example 4: Magnesium hydroxide, 233 parts, is added to 600 parts of the
alkylbenzenesulfonic acid of Example 1. There is then added 1,250 parts of water and the mixture is heated gradually to about 80°C over about 2 hours, whereupon a gel forms. The mixture is allowed to stand and a water layer of 830
parts is decanted; 570 parts of toluene is then added and an additional 300 parts
of water is removed by azeotropic distillation.

A 602-part portion of the resulting gel is diluted with 200 parts of toluene. The
solution is centrifuged and the toluene removed by blowing with nitrogen at
160° to 170°C to yield the desired magnesium oxide-sulfonate complex as a soft
gel.

The magnesium complexes of this process may be homogeneously incorporated
into lubricants, in which they function primarily as ash-producing detergents.
The products of Examples 1 through 4 are particularly useful for this purpose.
They can be employed in a variety of lubricants based on diverse oils of lubricating viscosity, including natural and synthetic lubricating oils and mixtures thereof.

Overbased Magnesium Phenates

A process described by *V.C.E. Burnop; U.S. Patent 4,104,180; August 1, 1978;
assigned to Exxon Research & Engineering Co.* involves producing overbased car-

bonates by hydrolyzing a metal alkoxyalkoxide in the presence of a phenolic or sulfonic surfactant and carbonating the product of hydrolysis.

A problem associated with the production of overbased metal compounds is that of the viscosity of both the reaction mixture and the final product itself. The overbased materials consist of an alkaline earth metal compound, generally a carbonate, dispersed in the alkaline earth metal salt of the dispersing agent; the amount of dispersed alkaline earth metal being known as the overbased amount. Generally these overbased materials are used as detergents in lubricating oils to react with acid residues formed in the oil, thus, the greater the basicity of the material the better since this allows smaller amounts of the materials to be used for a given effect in a certain lubricating oil.

However, to increase basicity it is necessary to increase the dispersed alkaline earth metal content which tends to increase the viscosity of the reaction mixture leading to processing problems. This problem is particularly marked if the alkaline earth metal is introduced in the form of the carbonated complex previously described and in order to overcome this problem it may be necessary to increase the amount of solvent used thus reducing reactor capacity and requiring solvent recovery.

These investigators have found that if the metal is introduced as its alkoxide and the carbonate formed after the alkoxide has been mixed with the surfactant and hydrolyzed it is possible to obtain highly basic materials consistently without viscosity problems and in many instances without the need for an additional solvent.

Thus, according to this process a colloidal suspension in oil of a metal carbonate is prepared as follows:

(1) forming a reaction mixture comprising:
  (a) an alkaline earth metal alkoxyalkoxide together with the alkoxyalcohol,
  (b) a surfactant being a sulfonic acid or metal, preferably Group II-A metal, sulfonate or one or more hydrocarbyl substituted phenols or metal phenates wherein the or each hydrocarbyl group contains up to 60 carbon atoms, or one or one or more sulfurized phenols having one or more hydrocarbyl group substituents containing up to 60 carbon atoms, or mixtures of the surfactants,
  (c) a nonvolatile diluent oil, and
  (d) at least one mol of water for every gram atom of the alkaline earth metal present in excess of the amount of the metal required to neutralize the surfactant.
(2) hydrolyzing the alkaline earth metal alkoxyalkoxide,
(3) introducing carbon dioxide into the reaction mixture under conditions at which the volatiles do not distill from the reaction mixture, and
(4) removing volatiles from the reaction mixture.

The alkaline earth metal alcoholate may be prepared in situ by reacting a metal oxide or hydroxide with an alkoxy alcohol such as ethoxy ethanol. Alternatively, the alcoholate itself may be used as starting material in which case it is preferred to use a solution of the alcoholate in the ether alcohol which is preferably ethoxyethanol. The alkaline earth metal may be calcium, barium, strontium or magnesium, although the techniques of the process are particularly useful in the production of overbased magnesium dispersants especially magnesium phenates. The alcoholate may conveniently be prepared by dissolving the metal in the alkoxyalcohol and when this technique is used with the preferred metal magnesium it is preferred to use ethoxyethanol since the magnesium ethoxyethoxide is readily soluble in ethoxyethanol.

In Examples 1 through 5 sulfurized nonyl phenol A was a product made from nonyl phenol containing 35 wt % of dinonyl phenol diluted with a nonvolatile oil to 70 wt % active ingredient. Sulfurized nonyl phenol A contained 7.4 wt % of sulfur and had a hydroxyl number of 131.

Sulfurized nonyl phenol B was prepared by adding 200 g of sulfur dichloride to 660 g of monononyl phenol dissolved in hexane followed by vacuum stripping. The product contained 8.7 wt % of sulfur and had a molecular weight of 484.

The magnesium alkoxyalkoxide was prepared by dissolving magnesium metal in about ten times its weight of ethoxyethanol to give a solution containing from 8.8 to 9.0 wt % of magnesium.

Example 1: The following mixture was made: 374 g magnesium alkoxy alkoxide (9.0% Mg), 250 g sulfurized nonyl phenol A, and 250 g of naphthenic hydrocarbon diluent oil of viscosity 5 cs at 210°F, and stirred at 50°C over a period of 2 hours while a second mixture containing 27 g of water and 27 g ethoxyethanol was run in. Carbon dioxide was then passed into the resulting product which was maintained at 50°C until 29 g of carbon dioxide has been absorbed. The addition of carbon was then continued while the temperature was slowly raised to 150°C. The pressure in the reaction vessel was then slowly reduced to about 25 cm of mercury and the volatile materials stripped off. The resulting product was found to contain only 0.04% by volume of sediment which was removed by filtration.

The final product had a TBN of 250 and a viscosity of 529 cs at 210°F and performed well as an additive in lubricating oils.

Example 2: The process of Example 1 was repeated with 220 g of sulfurized nonyl phenol B and 280 g of the diluent oil. The mixture was hydrolyzed with a mixture of 36 g of water and 36 g of ethoxyethanol and 43 g of carbon dioxide were absorbed before stripping. The final product weighed 609 g before filtering, had a TBN of 254 and a viscosity of 427 cs at 210°F.

Example 3: Example 1 was repeated except that 300 g of sulfurized nonyl phenol A were used, the amount of diluent was reduced to 200 g and the carbon dioxide charge increased to 44 g. The total base number of the product was 249 and the viscosity at 210°F was 753 cs.

Example 4: The process of Example 3 was repeated using only 18 g of water. The product obtained had a TBN of 248 but formed a skin upon exposure to air and could not be poured.

Example 5: An overbased sulfurized magnesium phenate was prepared in a 40-gallon reactor by charging 60 kg of the magnesium alkoxyalkoxide, 32.3 kg of the naphthenic hydrocarbon oil and 48.4 kg of the sulfurized nonyl phenol A into the reactor. The mixture was stirred at 60°C and hydrolyzed by adding 8.6 kg of a 50 wt % mixture of water and ethoxyethanol evenly over a period of 2 hours. The mixture was then stirred for a further 2 hours and the reactor heater then turned off. Carbon dioxide was passed into the bottom of the reactor at the rate of 30 liters per minute and the amount of unabsorbed carbon dioxide was measured by weighing a caustic soda trap attached to the exit from the reactor.

The reaction temperature and CO_2 absorbed by the reaction mixture was as follows:

Hours of CO_2 Introduction	CO_2 Absorbed (kg)	Reaction Temperature (°C)
0	0	60
0.5	1.8	69
1	3.6	75
1.5	4.9	79
2	6.0	80
3	6.1	79
4	6.4	77

The rate of introduction of carbon dioxide was then reduced to 10 liters per minute, the temperature raised to 150°C and the pressure in the vessel reduced to 60 mm of mercury to remove volatile materials.

The product was found to contain 0.04% by volume of sediment which was completely removed by adding 250 g of Dicalite Special speed flow filter aid and then filtering through a plate and frame filter.

The yield of filtered material was 94.6 kg of colloidal dispersion in oil having a sediment less than 0.01% with a TBN of 236 and a viscosity of 321 cs at 210°C.

The products of all the above examples were found to perform well as dispersants and neutralizing agents in lubricating oils.

Example 6: The following ingredients were charged to a 40-gallon reaction vessel: (1) 65 kg of a solution of magnesium ethoxyethoxide in ethoxyethanol containing 8.75 wt % magnesium; (2) 43.6 kg of sulfurized monononyl phenol having a hydroxyl number of 178 and containing 7.7 wt % sulfur; (3) 3.3 kg of C_{24} alkyl benzene sulfonic acid; and (4) 32.4 kg of a paraffinic diluent oil.

The mixture was heated to 60°C and 11.4 kg of a 50 wt % water and ethoxyethanol mixture were then run into the reactor over a period of 1 hour.

Carbon dioxide was next passed into the reactor for a period of 4 hours at a rate of 20 liters per minute while the temperature was allowed to increase by 10°C per hour. Absorption of carbon dioxide was complete after 4 hours and the rate of introduction was then reduced to 10 liters per minute and the temperature raised to 150°C and the pressure reduced to 30 mm of mercury. 59.4 kg of ethoxyethanol containing 1 wt % of water were distilled off.

After centrifugation for 30 minutes in white spirit the product had a sediment of 0.08 wt % and after addition of 250 g of filter aid and then filtering and washing the reactor with 5.0 kg of diluent oil the combined weight of filtrate and washing was 99 kg. The product had zero sediment, a TBN of 252 and a viscosity at 210°F of 240 cs. The product contained 5.5 wt % magnesium, 3.44 wt % sulfur, and 0.26 wt % chlorine.

Example 7: 1,500 g of nonyl phenol (65 wt % mononyl phenol and 35 wt % dinonyl phenol) having a hydroxy number of 216 were mixed in a reaction vessel with 606 g of a hydrofined paraffinic oil of viscosity 4 cs at 210°F at a temperature between 80° and 90°C. 430 g of a mixture containing 60% SCl_2 and 40% S_2Cl_2 were then added over a period of 4 hours after which the product was blown with nitrogen at 100°C over a period of 8 hours to remove hydrochloric acid.

2,300 g of a product were obtained which had a hydroxyl number of 138 and contained 6.87 wt % sulfur and 0.25 wt % chlorine.

2,100 g of this product were mixed with 1,260 g of a paraffinic diluent oil having a viscosity of 4 cs at 210°C, 140 g of an 88% active ingredient oil solution of a C_{24} alkylbenzene sulfonic acid and a magnesium alkoxy alkoxide prepared by dissolving 240 g of magnesium in 2,400 g of ethoxyethanol. This mixture was stirred vigorously at 60°C while 380 g of a 50 wt % water-ethoxyethanol mixture was run in over a period of 2 hours. The product was then saturated with carbon dioxide at a temperature between 60° and 80°C, 258 g of carbon dioxide being absorbed. After stripping, 4,350 g of material of TBN 251 and having a viscosity at 210°F of 502 cs was obtained. When the product was diluted with more diluent oil to have a TBN of 241, the viscosity at 210°F reduced to 333 cs.

Example 8: 52 kg of the sulfurized nonyl phenol A containing 35 wt % dinonyl phenol and 65 wt % mononyl phenol and 6.7 wt % sulfur having a hydroxyl number of 121 were charged to a reaction vessel together with 62 kg of a solution of magnesium ethoxyethoxide containing 8.7 wt % magnesium and 31.3 kg of a diluent oil viscosity of 4 cs at 210°F. The mixture was stirred at 60°C and 8.4 kg of a 50 wt % mixture of water and ethoxyethanol added over a period of 2 hours.

Carbon dioxide was then passed into the reaction vessel at the rate of 20 liters per minute for a period of 4 hours during which the temperature was allowed to rise to 80°C and 6.2 kg of carbon dioxide were absorbed. When there was no further absorption of carbon dioxide, the rate at which it was introduced was reduced to 10 liters per minute and the temperature slowly raised to 150°C; when the temperature reached 120°C the pressure was gradually reduced until ethoxyethanol started to distill and this was allowed to continue until there was no further distillation of ethoxyethanol.

The product which contained 0.05% sediment had a TBN of 242 and a viscosity of 372 cs at 210°F.

Example 9: 986 parts of 70% active ingredient sulfurized mononyl phenol of hydroxyl number 156 containing from 6.6 to 7.0 wt % sulfur, 664 parts of diluent oil and 58 parts of a C_{24} alkyl benzene sulfonic acid were added to a solution of 115 parts of magnesium in 1,150 parts of ethoxyethanol held at 60°C. 120 parts of water in 260 parts of ethoxyethanol were then added slowly to this mixture which was held at 65°C. The mixture was then carbonated with carbon dioxide and when 140 parts of carbon dioxide had been absorbed the temperature was raised to 150°C and the pressure reduced to 50 mm of mercury so that the ethoxyethanol distilled off.

Upon filtration a product was obtained that had a TBN of 259 and a viscosity of 252 cs at 210°F.

Example 10: 1,225 parts of 70% active ingredient dodecyl phenol sulfide of hydroxyl number 133 containing 7.2 wt % sulfur, 728 parts of diluent oil and 65 parts of C_{24} alkyl benzene sulfonic acid were added to 144 parts of magnesium dissolved in 1,444 parts of ethoxyethanol held at 60°C. 144 parts of water mixed with 432 parts of ethoxyethanol were added slowly to the original mixture which was then at 65°C.

After the addition of the water/ethoxyethanol mixture carbon dioxide was passed into the reaction vessel whose temperature was held at 65°C until 137 parts of carbon dioxide had been absorbed. The introduction of carbon dioxide was continued while the product was stripped to 150°C and 50 mm mercury pressure. Upon filtration a product was obtained that had a TBN of 242 and a viscosity of 391 cs at 210°F. Analysis of the product showed 5.3% magnesium and 3.8% sulfur.

Example 11: The products of Examples 8, 9 and 10 were included in lubricating oils based on a MIL-C lubricating oil and containing 3.7 wt % of an ashless polyamine dispersant and 1.2 wt % of neutral calcium phenate, 1.2 wt % of a zinc dialkyldithiophosphate and 1.2 wt % of magnesium sulfonate.

These oils were subject to the following tests to determine their suitability as lubricating oils:

(1)   The MS IIC antirust test (ASTM STP 315 F);
(2)   The L38 bearing corrosion test (SAE publication 680538);
(3)   The Modified Catalyzed Oxidation Test (Mod COT) in which an oil sample containing 0.1 wt % iron naphthenate is stirred with a paddle made from a Petter W-1 engine bearing with air blowing at 171°C and the % increase in the viscosity of the oil after 30 and 48 hours is measured.
(4)   Differential Scanning Calorimeter (DSC) in which an oil sample is heated in an aluminum pan in air at 210°C and 100 psi pressure. After a certain length of time (the induction time) the oil oxidizes and the heat of oxidation is measured. The induction time and the heat of oxidation are then combined according to the following formula: $\angle = 1.5188 \log_{10} G - 0.3167 \log_{10} F$, where G = induction period in minutes and F = heat of oxidation in calories per gram to give the "$\angle$" value which gives an indication of the oxidation stability of the oil. The lower the $\angle$ value the greater the oxidation stability.

The results of these tests are shown in the table on the following page.

Magnesium Phenate	Weight Percent	MS IIC Value	BWL (mg)	Mod COT Viscosity Increase, % 30 hr	48 hr	DSC L Value
Example 7	2.5	–	26.0	–	–	–
Example 8	1.5	–	65.5	110	390	0.6
Example 8	2.5	8.1	59.5	120	550	1.0
Example 9	1.5	–	28.0	140	–	1.0
Example 9	2.5	7.6	25.0	230	–	1.0
Example 10	1.5	–	44.3	170	780	0.8
Example 10	2.5	8.5	45.5	170	670	0.6

Overbased Metal Phenates

T. Hori and S. Hayashida; U.S. Patent 4,123,371; October 31, 1978; assigned to Maruzen Oil Co., Ltd., Japan have found that overbased sulfurized alkaline earth metal phenates having a high TBN (total basic number) can be prepared more effectively by use of a blending ratio between the alkaline earth metal agent and the phenol of smaller than the ratio between the metal and the phenol in the normal phenate, i.e., of about 0.99 equivalent or less of alkaline earth metal agent per equivalent of phenolic hydroxyl group of the blending phenol.

The process for preparing the overbased sulfurized alkaline earth metal phenate comprises reacting a phenol, sulfur, a dihydric alcohol and an alkaline earth metal oxide or hydroxide with the gram equivalent ratio of the alkaline earth metal agent per phenolic hydroxyl group being maintained at about 0.99 to about 0.001 and then reacting the reaction product with carbon dioxide at a temperature of about 50° to about 230°C.

Example 1: A 2-liter, 4-necked flask equipped with a stirrer, a condenser tube, a nitrogen gas inlet tube and a thermometer was charged with 1,233.7 g (5.6 mols) of nonyl phenol, 10.8 g of sulfur and 32.0 g (0.56 mol) of calcium oxide having a purity of 98.3%, and the starting materials were stirred. Ethylene glycol (118.2 g) was added to the resulting suspension in a stream of nitrogen at 132°C under atmospheric pressure. The mixture was stirred at 135°C for about 5 hours. Then, while the pressure in the reaction system was gradually reduced, the water generated in the reaction, most of the unreacted ethylene glycol and a small amount of the nonyl phenol were distilled off, whereupon 1,276.4 g of a dark yellowish green liquid distillation residue was obtained. The temperature of the final distillate was 87°C (6 mm Hg).

Then, 1,266.2 g of the distillation residue obtained as described above was placed in an autoclave, and caused to absorb carbon dioxide under an elevated pressure (not more than 11 kg/cm²) at a temperature of 123° to 126°C. The reaction system was then maintained at 155°C for 2 hours under an elevated pressure (not more than 8 kg/cm²) to produce 1,289.5 g of a dark yellowish green reaction product solution.

A 2-liter pear-shaped two-necked flask was charged with 1,278.3 g of the reaction product solution obtained after the carbon dioxide treatment described above and 133.7 g of a 150 neutral oil (a paraffinic lubricating oil having a viscosity of 4,386 cs at 210°F). A small amount of ethylene glycol, most of the unreacted nonyl phenol and a small amount of a lubricating oil fraction were distilled off from the mixture under reduced pressure to obtain 262.1 g of a distillation residue. The temperature of the final distillate was 167°C (3 mm Hg).

After the extremely small amounts of insoluble materials present in the distillation residue were removed by, e.g., filtration or centrifugal separation, 261.4 g of a very dark yellow, clear, viscous liquid product was obtained.

Material balance calculations showed that the product contained 281%, based on the theoretical amount, of calcium per phenolic hydroxyl group equivalent of the nonyl phenol reacted. Analysis of the final product gave the following results: viscosity (cs at 210°F), 318.7; TBN (total basic number) (JIS K 2500; KOH mg/g), 232; calcium, 8.40 wt %; and sulfur, 2.31 wt %.

Example 2: The same experimental device as described in Example 1 was charged with 771.1 g (3.5 mols) of nonyl phenol, 22.5 g of sulfur and 78.5 g (1.4 mols) of calcium oxide having a purity of 99.9% in a stream of nitrogen under atmospheric pressure, and the starting materials were stirred. Ethylene glycol (313.0 g) was added to the resulting suspension in a stream of nitrogen at 130°C under atmospheric pressure. The mixture was stirred for about 5 hours at 135°C. While the pressure in the reaction system was gradually reduced, the water generated in the reaction and 99.3 g of ethylene glycol (a part of the unreacted ethylene glycol) were distilled off, whereupon 1,053.7 g of a dark yellowish green liquid distillation residue was obtained. The temperature of the final distillate was 105°C (15 mm Hg).

Then, 1,045.6 g of the distillation residue produced as described above was placed in an autoclave, and reacted with carbon dioxide under an elevated pressure (not more than 11 kg/cm²) at a temperature of 127°C. Then, the system was maintained at 155°C and under an elevated pressure of not more than 8.7 kg/cm² for 2 hours to produce 1,119.7 g of a crimson reaction product solution.

A 2-liter pear-shaped two-necked flask was charged with 1,108.9 g of the reaction product solution obtained after the carbon dioxide treatment as described above and 226.7 g of a 150 neutral oil (described in Example 1), and in a stream of nitrogen under reduced pressure, the unreacted ethylene glycol, most of the unreacted nonyl phenol and a small amount of an oil fraction were distilled off from the mixture to obtain 536.5 g of a distillation residue. The temperature of the final distillate was 179°C (3 mm Hg).

When the small amounts of insoluble materials present in the distillation residue were removed by, e.g., filtration or centrifugal separation, 532.7 g of a very dark yellow, clear, viscous liquid product was obtained. The final product had the following characteristics: viscosity (cs at 210°F), 879.4; TBN (JIS K 2500; KOH mg/g), 275; calcium content (% based on the theoretical amount), 361; calcium, 10.1 wt %; and sulfur, 3.67 wt %.

Improved Filtration Properties for Overbased Metal Naphthenates

A process described by *W.J. Powers III; U.S. Patent 4,100,084; July 11, 1978; assigned to Texaco Inc.* comprises forming a clarified overbased metal naphthenate lubricating oil composition having a metal ratio greater than 1 and up to 10 and a total base number as defined by ASTM D 2896 of at least about 50 and up to 500 or higher, desirably between about 280 and 450. The method comprises first forming an initial reaction mixture, preferably having a water content less than 1 wt %, composed of the following ingredients:

(1) An oil-soluble metal naphthenate reactant having a metal ratio from 1 to 2. The acids from which the naphthenate reactants are derived are advantageously of a molecular weight of between about 230 and 600;
(2) A metal hydroxide;
(3) An alcohol selected from the group consisting of alkanol and alkoxylated alcohol having a carbon number from 1 to 5;
(4) A hydrocarbon lubricating oil having an SUS viscosity at 100°F of between about 50 and 300; and
(5) Optionally and preferably a volatile inert liquid hydrocarbon diluent having a boiling point between about 150° and 300°F.

In examples of the practice of the process, the naphthenic acid, diluent oil, hydrocarbon diluent and one equivalent of slaked lime were charged to a nitrogen blanketed 3-liter, 3-neck flask fitted with an air driven stainless steel stirrer and a reflux condenser equipped with a water separator. The stirrer was started and the flask was heated until overhead water formulation ceased. The crude soap mixture was then transferred to a 2,000 ml Parr stirred autoclave fitted with two turbine impellers. The remaining lime and the methanol were added. The mixture was heated to 140° to 155°F.

CO_2 was added through a sparger. During runs at greater than 6:1 overbasing ratios, the reactor system was bled, as required, to hold reactor pressure at a maximum of 10 psig. This resulted in a substoichiometric charge of CO_2, basis metal hydroxide. The previously recognized phenomenon referred to as "overcarbonation" was found not to occur when overbasing ratios of less than 6 mols $M(OH)_2$/mol metal naphthenate were used, but does appear to occur at higher overbasing ratios. Overcarbonation must be avoided to assure that the product is filterable. After the CO_2 charge was complete, the temperature was held at 140° to 155°F for 30 minutes. 5 wt % of a filter aid was added and the product was filtered through blotter paper in a pressure bomb filter. The product was then stripped on a rotary vacuum stripper using a bath temperature of 250°F.

Promoter System for Alkyl Phenol Sulfide Reactions

A. Peditto, F. Fossati and V. Petrillo; U.S. Patent 4,100,085; July 11, 1978; assigned to Liquichimica Robassomero SpA, Italy describe a process of the type in which an alkylphenolsulfide, having the formula

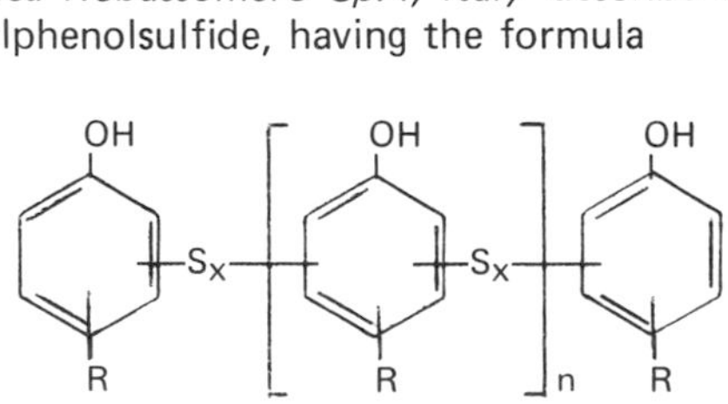

where R is a C_{8-20} alkyl radical, x has an average value of between 1 and 2, preferably of between 1 and 1.5, and n is an integer of between 0 and 3, is reacted with an oxygen bearing compound of an alkaline earth metal and carbonated with carbon dioxide. The process is characterized in that the reaction between the alkylphenolsulfide, the oxygen bearing compound of the alkaline earth metal and the carbon dioxide, is carried out in the presence of a promoter consisting of either anhydrous NH_3 or NH_4OH, and of a copromoter selected from the alcohols having a short chain, anyhow not higher than C_5, the lower homologues being preferred.

According to the preferred case, the addition of the promoter takes place at or immediately before the neutralization step, whereas the addition of the copromoter takes place at or immediately before the carbonating step, the two steps being thus distinct, the first step occurring at a temperature of between 70° and 130°C and the second step taking place at a temperature of between 55° and 130°C, depending on the boiling point of the copromoter, the neutralization step being completed in a preferred time of between 30 and 180 minutes and the carbonating step being completed preferably in a time of between 90 and 360 minutes.

Example: A reaction vessel was charged with 315 g of n-nonane, 155 g of a sulfurized alkylphenol having the above formula (R being a propylene oligomer, x being on the average 1.25 and n being 1), and 5 g of a 32% solution of NH_4OH. The mixture, under stirring, was brought to 80±2°C and then 112 g of $Ca(OH)_2$; were added; thereafter, by maintaining the stirring action and the temperature of 80±2°C, the neutralization was carried out for 60 minutes. Then, the temperature was raised to 125±2°C, the water and ammonia excess being removed. The mixture was thereafter cooled to 64±2°C and 168 g of methanol were added; the mixture was then carbonated, under stirring, at 64±2°C by 49 g of CO_2. At the end of the carbonating step, the temperature was slowly raised to 125°C, the methanol being removed and 196 g of a lubricating oil, having viscosity of 150 SSU at 100°F (37.8°C), being added. The product was then filtered and evaporated for the removal of the reaction solvent. A product was obtained containing 9.54% Ca and 3.23% sulfur, the viscosity being 152 cs at 210°F (about 97°C).

Phenoxide-Halocarboxylic Acid Condensates

T.F. Steckel; U.S. Patents 4,128,488; December 5, 1978 and 4,131,554; Dec. 26, 1978; both assigned to The Lubrizol Corporation has found that compositions made by reacting (1) a metal phenoxide substituted with at least one hydrocarbon-based group of at least about 30 carbon atoms with (2) a carboxylic acid reagent containing from 1 to 3 carboxyl-based groups and a halogen-substituted hydrocarbon-based aliphatic or alicyclic group containing a halogen atom are useful as additives for lubricants and normally liquid fuels. Analogous thiophenoxide-based compositions are similarly useful.

In the following examples, all percentages and parts are by weight (unless otherwise stated expressly to the contrary) and the molecular weights are number average molecular weights ($\overline{M}n$) as determined by gel permeation chromatography (GPC) or vapor phase osmometry (VPO).

Example 1: A mixture of 2,240 parts of a poly(isobutene)-substituted phenol ($\overline{M}n$ = 885 VPO), 800 parts xylene and 83.4 parts of sodium hydroxide is heated to reflux and dried by azeotropic distillation. The resulting phenoxide-containing mixture is cooled to 100°C and 500 parts of a commercial mixture of alcohols containing approximately 61% isobutyl alcohol and 39% amyl alcohol is added as solvent. At 65°C, 233 parts of sodium chloroacetate is then added.

The mixture is held at reflux (122° to 123°C) for five hours and stripped to about 173°C under nitrogen. After cooling the mixture to about 95°C, 200 parts of toluene and 208 parts of aqueous hydrochloric acid are added. This mixture is held at about 90° to 95°C for two hours and then stripped to 150°C under vacuum. Diluent oil (600 parts) is added and the mixture is filtered to yield 2,708 parts of an oil solution of the desired product.

Example 2: A mixture of 784 parts (0.5 equivalent) of the product solution described in Example 1, 135 parts of toluene, 150 parts of diluent oil and 36.3 parts (0.875 equivalent) of a commercial ethylene polyamine mixture corresponding in empirical formula to pentaethylene hexamine is heated at 155° to 165°C for 8 hours while water is removed by the use of a Dean-Stark trap. The mixture is stripped to 160°C under vacuum and filtered to yield an oil solution of the desired product in an oil solution containing 1.23% nitrogen.

Example 3: At 85°C, a mixture is prepared by adding 33 parts of paraformaldehyde to 783 parts of the solution described in Example 1, 312 parts of diluent oil, 145 parts toluene and 78 parts of a commercial ethylene polyamine mixture, wherein the amines have an average of 3 to 10 nitrogen atoms per molecule, containing about 34% nitrogen. The resulting mixture is heated at 105°C for 2 hours; then water is removed by azeotropic distillation. The mixture is stripped at 165°C under vacuum and filtered. The filtrate (1,128 parts) is an oil solution of the desired product containing 2.04% nitrogen.

Example 4: A mixture of 627 parts of the product solution of Example 1, 219 parts of diluent oil and 27.2 parts of pentaerythritol is heated to 222°C in 3 hours and held at 222° to 230°C for 4.5 hours. The mixture is stripped at 230°C under vacuum and filtered to yield an oil solution of the desired product.

Example 5: The procedure for Example 4 is repeated except the pentaerythritol is replaced on an equivalent basis by glycerol.

The compositions of this process are useful in and of themselves as antirust and anticorrosion agents for fuels and lubricants, particularly when they are free acids, esters of the higher alcohols, carboxamides or ammonium carboxylates of the polyamines. These esters, carboxamides and carboxylates can also function in fuels and lubricants as detergents and dispersants for sludge and varnish formed in internal combustion engines.

Metallic Compounds of Polyarylamine Sulfides

J.C. Nnadi; U.S. Patent 4,083,792; April 11, 1978; assigned to Mobil Oil Corporation describes oil-soluble compounds identified as overbased metallic compounds of polyarylamine sulfides and polyarylaminephenol sulfides. The metallic compounds of polyarylamine sulfides and polyarylaminephenol sulfides, prior to overbasing, have the general chemical formula

In the above formula, n is an integer of from 1 to about 10, preferably from 1 to about 5; A is an aromatic moiety, preferably phenyl or naphthyl; M is a polyvalent metal, such as, e.g., Be, Mg, Ca, Ba, Mn, Co, Ni, Pd, Cu, Zn and Cd; X is a radical selected from the group consisting of organophosphoro, organocarboxyl, organoamino, organosulfonyl, organothio, organooxy, nitrate, phosphate, sulfate, sulfonate, oxide, hydroxide, carbonate, sulfite, fluoride, chloride, bromide and iodide; R_1 and R_2 are alkyl of from 1 to about 10 carbon atoms, aryl, hydrogen,

$$\begin{matrix} O \\ \| \\ -C-R' \end{matrix}$$

or a combination thereof; R' is alkyl of from 1 to about 10 carbon atoms, aryl or hydrogen; and R_3, R_4, R_5 and R_6 are hydrogen, alkyl of from 1 to about 200 carbon atoms, aryl, alkyl-substituted aryl where the alkyl substituent is comprised of from 1 to about 200 carbon atoms, carboxyaryl, carbonylaryl, aminoaryl, mercaptoaryl, halogenoaryl or combinations thereof.

The oil-soluble overbased compounds may be obtained by reacting one or more of the above compounds with one or more basic metallic compounds such as, e.g., $Ca(OH)_2$ or $Ba(OH)_2$, in the presence of a promoter such as, e.g., methanol.

Example 1: 1 mol (262 g) of dodecylaniline was dissolved in 262 g of Promor No. 5 process oil and then neutralized with hydrogen chloride gas until the theoretical amount of HCl (36.5 g) was taken up. To the above dodecylaniline hydrochloride was added 76.5 g of sulfur dichloride (about 0.75 mol) at such a rate as to keep the reaction temperature at 30° to 70°C. The temperature was slowly raised to 150°C during a period of 6 hours and kept at 150°C for 4 hours under nitrogen purge. The reaction was cooled, mixed with 500 cc of toluene and 100 cc of tetrahydrofuran and then washed once with 100 cc of 10% NaOH solution and twice with 250 cc of H_2O.

The washed product was distilled to remove the solvents and unreacted dodecylaniline. The final stage of distillation was carried out to a pot temperature of 180°C and 3 mm Hg, where it was held for 2 hours. The yield of product was 509 g and it contained 2.7 wt % N and 2.94 wt % S and had a total base number of 81. This product consisted mostly of trimer, tetramer and pentamer.

A 100 g sample of the above product was diluted with 40 g of process oil and then reacted with 5 g of $Ca(OH)_2$ in 100 g of methanol with CO_2 gas bubbling therethrough for 2 hours at 40° to 50°C. The reaction mixture was then heated to remove the methanol and held at 150°C for 1 hour. The filtered overbased compound product contained 1.48 wt % S.

Example 2: A 100 g sample of polydodecylaniline sulfide (50% active ingredient and 50% process oil), 25 cc of water and 3 g of zinc oxide (ZnO) were mixed together and heated to gradually distill off the water. The mixture was held at about 150°C for 1 hour under atmospheric pressure and at 150°C for 1½ hours under house vacuum. The remaining residue was filtered through a funnel packed with Hyflo filter aid on filter paper. The product filtrate weighed 80 g. Chemical analysis of the product gave the following results in weight percent: N, 2.51; S, 2.16; and Zn, 0.81.

A 50 g quantity of the product filtrate of this example was then diluted with 20 g of process oil and reacted with 2 g of Ca(OH)$_2$ as in Example 1 to yield the oil-soluble overbased compound of this process.

Example 3: A 210 g sample of the polydodecylaniline sulfide of Example 2, 5 g of zinc methane sulfonate and 40 g of water were mixed together and heated to gradually distill off the water. The mixture was heated and filtered as in Example 2 and 185 g of product filtrate was obtained. Chemical analysis of the product gave the following results in weight percent: N, 2.44; S, 3.23; and Zn, 0.68.

A 100 g sample of the product filtrate above prepared was then diluted with 40 g of process oil and reacted with 5 g of Ba(OH)$_2$ as in Example 1 to yield the oil-soluble overbased compound of this process.

The antioxidant properties of the overbased compounds of this process were measured by adding these compounds to a suitable oil and subjecting the oil to oxidation at high temperatures. The test was a bulk oil catalytic oxidation process in which a stream of dry air was passed through a heated sample of the lubricant composition for a time at various elevated temperatures in the presence of iron, copper, aluminum and lead as catalysts. The metal samples consisted of 15.6 in^2 of sand-blasted iron wire, 0.78 in^2 of polished copper wire, 0.87 in^2 of polished aluminum wire, and 0.167 in^2 of polished lead surface. The antioxidant activity was evaluated as the ability of the additive to control the acid number (NN) and viscosity (KV) of the oil and to prevent them from rising at an unduly rapid rate. The sludge formation during the oxidation was estimated visually.

The base stocks used in this evaluation were a synthetic ester oil lubricant (made by reacting pentaerythritol with an equimolar mixture of C$_5$ and C$_9$ monocarboxylic acids) and a solvent refined paraffinic neutral oil stock having the following properties: Pour, °F, 20; SSU at 100°F, 130; SSU at 210°F, 42; and V.I. min, 115. Results of this evaluation for sample compounds of this process are tabulated below in the table.

Compound	Temperature (°F)	ΔNN	Percent ΔKV	Sludge
Base oil*	425	4.6	300	moderate
Base oil* + 1 wt % Ex. 1 product	425	3.1	100	trace
Base oil**	375	12.1	400	heavy
Base oil** + 1 wt % Ex. 1 product	375	5.7	150	moderate

*Synthetic ester oil lubricant.
**Solvent refined paraffinic neutral oil stock.

J.C. Nnadi; U.S. Patent 4,076,636; February 28, 1978; assigned to Mobil Oil Corporation also describes oil-soluble overbased polyarylamine-arylhydroxy (alkoxy) sulfides and overbased organic and inorganic metallic salts, complexes, mixed salt complexes of polyarylamine-arylhydroxy (alkoxy) sulfides. These overbased compounds function as high temperature detergents and dispersants.

OTHER COMPOSITIONS

Solid Graphite Particles

A process described by *D.L. DeVries and J.M. DeJovine; U.S. Patent 4,132,656;*

January 2, 1979; assigned to Atlantic Richfield Company relates to lubricating oil compositions which include solid materials to enhance the properties of such compositions.

These compositions comprise a major amount by weight of oil of lubricating viscosity; a minor amount by weight of solid particles effective to improve the lubricating properties of the composition; and a minor amount by weight of at least one of certain specific nitrogen-containing polymers.

One preferred nitrogen-containing polymer is a graft polymer having a dialkylaminoalkylmethacrylate, or mixtures thereof, grafted to the polymer backbone.

A second preferred nitrogen-containing polymer is an oil-soluble interpolymer prepared from a long chain n-alkyl methacrylate and a dialkylaminoalkylmethacrylate or a N-(alkanone) acrylamide.

The compositions include a minor amount by weight of solid particles effective to improve the lubricating properties of the compositions. Preferably, a major portion, by weight, and more preferably substantially all, of such solid particles, have a maximum transverse dimension in the range of about one millimicron to about 2 microns, and most preferably in the range of about 1 millimicron to about 1 micron. Suitable solid particles for use in the process include those materials known to provide improved lubricating properties to lubricating oil compositions. Such solid particles include, e.g., graphite, molybdenum disulfide, zinc oxide, tungsten disulfide, mica, boron nitrate, borax silver sulfate, cadmium iodide, lead iodide, barium fluoride and tin sulfide. The solid particles useful in the compositions are preferably selected from the group consisting of graphite, molybdenum disulfide, zinc oxide, and mixtures thereof; more preferably, from the group consisting of graphite, molybdenum disulfide and mixtures thereof; and most preferably, graphite.

Examples 1 through 4: A series of four lubricating oil compositions were prepared by blending together individual components, noted below, at a slightly elevated temperature, i.e., about 100° to about 130°F, to insure proper mixing. The final compositions were as follows.

The Sequence VC Test described for Examples 1 and 2 illustrates the problem in formulating a suitable solid-particle-containing lubricating composition. Examples 3 and 4 are lubricating oil compositions in accordance with the process. These lubricating compositions containing graphite and nitrogen-containing graft polymer and nitrogen-containing interpolymers of the process provide reduced sludge and varnish deposition relative to lubricating compositions not containing these nitrogen-containing polymers, and are illustrative of the improved solids containing lubricating compositions of the process.

Component	Example 1	Example 2	Example 3	Example 4
	(wt %)			
Mineral oil, 125 SUS at 100°F	84.0	74.7	75.0	75.0
Conventional additive mixture	7.4	7.5	7.5	7.5
Methacrylate polymer	8.6	7.8	—	—
Nitrogen-containing graft polymer	—	—	7.5	—
Nitrogen-containing interpolymer	—	—	—	7.5
Graphite dispersion	—	10.0	10.0	10.0

This additive mixture is a commercially available combination of materials each of which is conventionally used in lubricating oil compositions. This mixture includes alkyl zinc dithiophosphate, both overbased and neutral calcium sulfonates, calcium phosphonate-phenate and both an ashless dispersant and an ashless rust inhibitor. This mixture also included about 50 wt % of a light mineral oil as solvent for the active ingredients.

The methacrylate polymer is conventionally used to improve the viscosity index of lubricating oil polymers. Such polymer includes essentially no N-vinyl pyrrolidone. The material as used includes about 50 wt % of a mineral oil as solvent for the polymer. The polymer is believed to have an average molecular weight of about 800,000 and to be derived from a methacrylic ester containing about 16 carbon atoms per molecule.

The nitrogen-containing graft polymer is prepared in accordance with U.S. Patent 3,923,930 and has a terpolymer backbone of ethylene, propylene and 4,4-hexadiene present in a weight ratio of about 50:46:4. This terpolymer has an intrinsic viscosity of about 1.1, about one mol of unsaturation per 2,000 g of terpolymer and about 1,500 carbon atoms per terpolymer molecule. N,N-dimethyaminoethyl methacrylate is grafted on the terpolymer backbone. About 2.67 g of the methacrylate is employed. The graft polymer contains about 0.04 wt % of nitrogen.

A series of nitrogen-containing interpolymers having a molecular weight between 30,000 and 120,000 were prepared by reacting a mixture of C_{12-16} alkylmethacrylates (Neodol 25L) and C_{16-20} alkylmethacrylates (Alfor 1620 SP) with either dimethyaminoethyl methacrylate or N-(1,1-dimethylbutan-3-one) acrylamide in the following proportions:

Monomers				(wt %)				
Dimethylaminoethyl methacrylate	4	–	–	5	–	8	10	–
N-(1,1-dimethylbutan-3-one) acrylamide	–	4	5	–	8	–	–	10
C_{12-16} alkyl methacrylate (Neodol 25L)	71	71	70	70	67	67	70	70
C_{16-20} alkyl methacrylate (Alfol 1620 SP)	25	25	25	25	25	25	20	20

The monomers are combined with 100 g of a hydrofined paraffin base oil having a viscosity of about 145 SUS at 100°F and are charged to a one liter resin kettle and purged with purified nitrogen for 40 minutes. The reaction mixture is then heated to 80° to 83°C and 0.5 g of azobisisobutyronitrile and 0.3 g of dodecyl mercaptan are added and the polymerization allowed to proceed to completion over a period of four hours. The temperature is then raised to 100°C and held for one hour at this temperature at which point 300 g of a hydrofined dewaxed paraffin base oil having a viscosity of about 100 SUS at 100°F are added and the temperature is held at 100°C for an additional hour. The reactor contents are nitrogen-containing interpolymers dissolved in an oil diluent. The weight amount employed is on a diluent-free basis.

The graphite dispersion is a mineral oil-based dispersion containing about 10 wt % of solid graphite particles which have an average (by weight) particle size of about 200 millimicrons. The dispersion also includes about 6% of a nitrogen-and methacrylate-containing dispersant to aid in maintaining dispersion stability. The dispersant is believed to be derived from a methacrylic ester containing about 16 carbon atoms per molecule.

Each of the lubricating oil compositions identified as Examples 1 and 2 was used to lubricate an internal combustion engine which, in turn, was operated through a Reference Sequence VC Test. This test, in which the engine is operated for 192 hours, is described in "Multicylinder Test Sequences for Evaluating Automotive Engine Oils—ASTM Special Technical Publication 315F," American Society for Testing and Materials (1973). This procedure is known to produce data which can be used to make valid comparisons of the effects various lubricating oil compositions have on engine sludge and varnish ratings under normal operating conditions.

Sludge and varnish ratings in the Reference Sequence VC Test are based upon visual inspection of various engine components and comparison with a series of CRC reference standards.

Results of this test using each of the abovedescribed lubricating compositions are summarized below. For comparison purposes, minimum SE standard lubricating oil qualification ratings are also presented.

	Composition		SE
	1	2	Minimum Ratings
Average overall sludge rating	8.7	8.1	8.5
Average overall varnish rating	8.3	7.8	8.0
Piston skirt	8.3	8.0	7.9

The above data indicate that compositions which include solid particles cause a substantial decrease in sludge and varnish ratings (increase in sludge and varnish formation). This conclusion is apparent by comparing the results from Composition 2 with those from the nongraphite containing Composition 1. Thus, the inclusion of the conventional viscosity index improver, which provides adequate sludge and varnish formation protection (see Example 1) when included in a composition without solid particles, fails to meet the SE qualification standards when such solid particles are added.

D.L. DeVries and J.M. DeJovine; U.S. Patents 4,134,844; January 16, 1979; 4,136,040; January 23, 1979; and 4,094,799; June 13, 1978; all assigned to Atlantic Richfield Company also describe a number of other compositions employing solid graphite particles and polymeric additives. These include oxidized olefin polymer-polyamine reaction products (U.S. Patent 4,134,844), nitrogen-containing mixed ester of a carboxy-containing interpolymer (U.S. Patent 4,136,040), and N-vinyl-pyrrolidone-acrylic ester copolymers (U.S. Patent 4,094,799).

Overbased Calcium Sulfonates in Diester Lubricating Oils

According to a process described by *W.C. Crawford; U.S. Patent 4,138,347; February 6, 1979; assigned to Texaco Inc.* small amounts of adducts of nonylphenol and ethylene oxide defined by the formula shown on the following page:

$$C_9H_{19}-\left\langle\bigcirc\right\rangle-O(CH_2CH_2O)_nH$$

wherein n ranges from 1 to 9.5 are found to have a dispersing and/or solubilizing action on overbased calcium sulfonates in 100% synthetic diester base lubricating oils. Best results are obtained where an adduct in which n is 6 is used in an oil comprising essentially di(2-ethylhexyl) azelate.

Preferred among these is the adduct marketed by Jefferson Chemical Company under the name "Surfonic N-60," wherein n is 6. The total adduct concentration basis weight of oil ranges from 0.5 to 1.50 wt %. A dosage lower than 0.5 wt % is ineffective.

In the process, calcium sulfonate is blended with the synthetic oils and the adduct is then blended therein at 150°F using a laboratory mixer.

The process is effective particularly when the calcium sulfonates have a relatively high overbasing ratio, as expressed by their Total Base Number (TBN) where a TBN of 0 is assigned to the neutral salt, ranging from 30 to 400.

Column A of the table shows the insolubility of calcium sulfonates in the synthetic ester at a treating level typically used to formulate 1% ash oils of SE quality. The additive is insoluble. Columns B through E provide data establishing that the minimum amount of Surfonic N-60 to solubilize 0.25 wt % Ca from sulfonates is 0.5 wt %. This is also the optimum concentration of Surfonic N-60 to maintain calcium in solution at low temperatures. Blends with higher concentrations of Surfonic N-60 develop a definite haziness on storage at low temperature. The blend with 0.25% Surfonic N-60 developed a barely perceptible haze on storage. The high lumetron turbidity for this blend (Col. E, 26.0 lumetron) also indicates that 0.25% Surfonic N-60 is too low a dosage to solubilize calcium sulfonates.

Blend F shows that even 0.12% calcium sulfonate will not dissolve in di(2-ethylhexyl) adipate.

Using 0.11% Ca from a calcium sulfonate with a lower overbasing ratio than the other and, hence, less CaCO$_3$, shows that the solubility is greater, even in the absence of Surfonic N-60, in this case (Blend F vs Blend J). Higher overbasicity would lead to solubility problems, basis high lumetron number for Blend I (0.16% Ca).

Using 6.2 overbased calcium sulfonates shows that 0.23% Ca from this additive is not soluble (Blend G), and that lower calcium than 0.23% concentrations could have solubility problems, e.g., Blend H.

Surfonic N-60 can also be used to solubilize calcium sulfonates in a synthetic base stock composed of esters of naphthenic acids which are not dimer acids (see Blends L, M). Surprisingly, calcium sulfonate (420 TBN) was found to be soluble in 2-ethylhexyl esters of dimer acids but the 7.0 lumetron number (Blend N) indicates that solubility is not as great as desired and thus Surfonic N-60 can be used to improve the solubility in this case too.

Use of Adducts to Solubilize Highly Overbased Calcium Sulfonates in Esters

Blend	A	B	C	D	E	F	G
Di(2-ethylhexyl) adipate	98.50	97.00	97.50	98.00	98.25	99.25	90.00
Esters of naphthenic acids	–	–	–	–	–	–	–
2-Ethylhexyl esters of dimer acids	–	–	–	–	–	–	–
Calcium sulfonates*							
I	1.50	1.50	1.50	1.50	1.50	0.75	–
II	–	–	–	–	–	–	10.0
III	–	–	–	–	–	–	–
Surfonic N-60	–	1.50	1.00	0.50	0.25	–	–
% Ca	0.25	0.25	0.25	0.25	0.25	0.12	0.20**
Appearance	two phases	clear	clear	clear	slight haze	two phases	two phases
Lumetron turbidity, %	***	1.0	3.0	3.0	26.0	***	***

Blend	H	I	J	K	L	M	N
Di(2-ethylhexyl) adipate	98.50	98.50	99.0	–	–	–	–
Esters of naphthenic acids	–	–	–	98.50	97.50	98.00	–
2-Ethylhexyl esters of dimer acids	–	–	–	–	–	–	98.50
Calcium sulfonates*							
I	–	–	–	1.50	1.50	1.50	1.50
II	1.50	–	–	–	–	–	–
III	–	1.50	1.0	–	–	–	–
Surfonic N-60	–	–	–	–	1.00	0.50	–
% Ca	0.026	0.16	0.11	0.25**	0.21	0.22	0.23
Appearance	clear	clear	clear	two phases	clear	clear	clear
Lumetron turbidity, %	7.0	9.0	4.5	***	4.0	26.0	7.0

*I–TBN is 420; II–TBN is 6.2; III–TBN is 300.
**Calculated.
***Unsuitable.

VISCOSITY INDEX IMPROVERS
AND OTHER ADDITIVES

Mineral lubricating oils have been modified by a vast array of additives for purposes of improving viscosity index, thermal stability, oxidation stability, detergency, and other properties. The viscosity index is highly important especially
in multigrade oils in order to provide lubricating oil compositions having much
flatter viscosity-temperature curves than the unmodified oils.

It is especially vital that the lubricating oil compositions exhibit a specified range
of viscosities at relatively low temperatures. A multigrade lubricant designates
those lubricants which meet a 0°F viscosity specification and a 210°F viscosity
specification, such as is shown for motor oils by the following table derived from
SAE, J300a taken from the SAE Handbook for 1969.

Specification Oil Grade (SAE)	Viscosity at 0°F (poises)	Specification Oil Grade (SAE)	Viscosity at 210°F (SUS)
—	—	20	45-58
5W	12*	30	58-70
10W	12-24	40	70-85
20W	24-96	50	85-110

*Maximum

According to the table, for example, an SAE 10W/50 oil must have a viscosity
at 0°F between 12 and 24 poises and a viscosity at 210°F of between 85 and
110 SUS.

The pour point of an oil is a characteristic which determines the oil's usefulness
and serviceability in the colder climates. An oil's pour point is an approximate
indication of the lowest temperature at which the oil can be poured or removed
from containers or can be caused to flow through tubing and piping. Service
oils including lubricating oils, automatic transmission fluids and the like, must
be capable of flowing through automotive systems, their associated transmission
lines, or the like, at the lowest temperature at which they are used. In a similar
manner, fuel oils must be capable of flowing through conveying and transmission

lines at the lowest temperature conditions to which they are subjected.

Although the pour point of a mineral oil is only a general indicator of its low temperature flow properties, it provides a useful function and is commonly found in an oil's specifications. Most service oils, and particularly the paraffinic base oils, require a pour point depressant as an additive in order to meet established specifications.

Many of these requirements are increasingly being met by the development of multipurpose additives, which combine viscosity index, pour point and dispersancy properties.

VISCOSITY INDEX IMPROVERS

Styrene and tert-Butylstyrene Block Copolymers

T.L. Staples; U.S. Patent 4,136,048; January 23, 1979; assigned to The Dow Chemical Company describes a composition having an improved or even constant viscosity/temperature relationship. The composition comprises a fluid and one or more block copolymers having an AB configuration wherein the A portion is a polymeric structure that is insoluble in the fluid below a characteristic temperature and soluble above that temperature and the B portion is a polymeric structure that is soluble in the fluid over the complete range of temperature for which the liquid composition will be used and wherein the chain length of the B portion is small relative to that of the A portion.

The polymers are unusual in exhibiting a reversible solution/emulsion phase transition that is temperature dependent. The temperature at which the composition passes from a solution to a colloidal dispersion shall be referred to as the transition temperature.

The useful polymers are those block copolymers particularly of a monoalkenyl aromatic structure. The polymers are also characterized by an AB structure.

The A portion is a polymeric structure or block that is insoluble in the fluid below a characteristic temperature and soluble above that temperature. Typical of such blocks are linear structures of polystyrene or of copolymers of styrene and nuclear alkylated styrene where the alkyl group contains from about 3 to 8 carbon atoms. The A blocks may also include significant amounts of comonomers such as α-methylstyrene, chlorostyrene or like materials.

The B portion of the block polymer is a polymeric structure that by itself is soluble in the oil over the complete range of temperatures for which it is desired to maintain the constant viscosity/temperature relationship. This B portion particularly may be a hydrocarbon such as a polyalkylated styrene including, for example, poly-tert-butylstyrene. The molecular weight of the B portion should be small relative to the A portion, advantageously being from about 10,000 to 50,000.

Example 1: A number of block copolymers were prepared by anionic polymerization techniques. Those copolymers were evaluated for transition temperature in a light hydrocarbon oil (Calumet 3800). The oil compositions were heated and cooled to determine the state of the copolymer below the transition temperature. The results are shown in the table.

In all cases, the copolymer/oil compositions of this process became stable emulsions below the transition temperature, while in all cases the copolymers lacking a B portion precipitated and settled out of solution.

The emulsions of the process remained stable for several months, as shown by no visible sediment even after storage in a freezer below 0°C.

| | | .... A Block ..... | B Block | Total | T_{trans} |
| | S/TBS | Calculated | Calculated | 10^{-5} $\bar{M}$ | C-3800 |
Sample	(wt %)	10^{-5} $\bar{M}$	10^{-5} $\bar{M}_n$	Calculated	(°C)
Process					
A	67/33	4.9	0.36	5.3	35–40
B	75/25	3.0	0.15	3.15	~55
C	80/20	3.0	0.15	3.15	~65
D	20/80	3.55	0.14	3.7	<25
E	40/60	3.55	0.14	3.7	<25
F	33/67	3.55	0.14	3.7	<25
G	47/53	3.55	0.14	3.7	<25
H	67/33	3.4	0.22	3.6	~35
I	80/20	3.4	0.22	3.6	~65
Contrast					
J	70/30	5.4	—	5.4	—
K	67/33	4.9	—	4.9	35–40
L	83/17	2.7	—	2.7	92–94

Example 2: The viscosity/temperature relationship was determined on a sample of Calumet 3800 oil by heating the sample and periodically taking viscosity measurements using a Ubbelohde tube in a bath controlled to within 0.5°C.

An amount of the block copolymer identified in Example 1 as H was dissolved in the oil at 120°C to give a 2% solution. The viscosity determinations were repeated. The results are shown graphically in Figure 2.1a.

Example 3: A hydrocarbon oil composition was prepared by dissolving at 120°C 1% of copolymer H and 1.65% of copolymer I. Viscosity determinations were made as in Example 2.

In contrast, a similar composition was made with 2% of copolymer H and another with 3.3% of copolymer I. Viscosity determinations were made.

The results are shown graphically in Figure 2.1b where the curve resulting from the blend of two copolymers indicates a relatively constant viscosity over the range from 20° to 80°C, while the compositions with only one copolymer show some fluctuation in viscosity and viscosity control over only a part of the temperature range.

All compositions show improvement over the base oil.

Figure 2.1: Viscosity Index Improvers

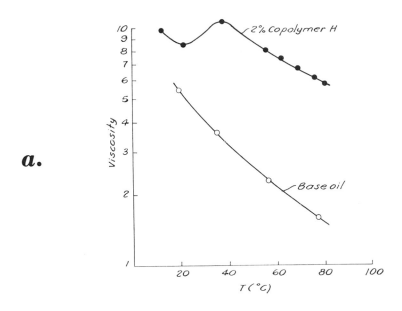

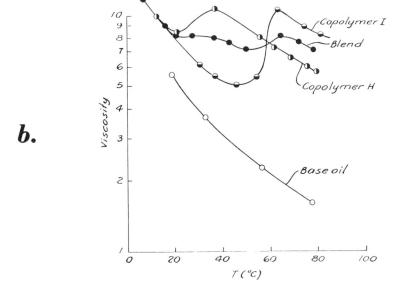

Source: U.S. Patent 4,136,048

Styrene-Isoprene Block Copolymers and Polybutene

A. Richardson; U.S. Patent 4,081,390; March 28, 1978; assigned to Orobis Limited, England describes a composition suitable for use as a lubricant additive which comprises from 1 to 15% of a vinyl aromatic/isoprene sequential block copolymer having a number average molecular weight in the range of 25,000 to 125,000 and containing from 10 to 40% by weight of the vinyl aromatic component, from 5 to 45% of a polybutene having a number average molecular weight in the range of 5,000 to 60,000, and the remainder of the composition comprising a solvent neutral base oil. The percentages of vinyl aromatic/isoprene copolymer, polybutene and base oil are expressed as weight percentages based on the total weight of the composition.

Vinyl aromatic/isoprene sequential block copolymers may be prepared by techniques known in the art. The most common technique is that of anionic polymerization, sometimes known as living polymerization, wherein a predetermined amount of a polymerization initiator such as an organolithium compound, e.g., n- or sec-butyl lithium, dissolved in a hydrocarbon solvent is added to a predetermined quantity of the vinyl aromatic monomer, preferably in the presence of a diluent, which diluent may be a hydrocarbon solvent, e.g., toluene. After the vinyl aromatic monomer is completely polymerized, pure isoprene monomer is added. The nonterminated vinyl aromatic polymer chains initiate polymerization of the isoprene monomer which adds thereto until the isoprene monomer is consumed.

Preferably the vinyl aromatic/isoprene copolymer is Shellvis 50 VI-Improver, a hydrogenated styrene/isoprene sequential block copolymer of number average molecular weight in the range of 50,000 to 100,000 and containing about 75% isoprene and 25% styrene, greater than 95% of the isoprene component being present in the 1,4-form in which greater than 95% of the olefinic double bonds are hydrogenated and the styrene component having less than 5% of the aromatic nucleus double bonds hydrogenated.

Preferably the polybutene is Hyvis 7000/45, which is a solution of a polybutene having a number average molecular weight in the range of 30,000 to 42,000 in 150 solvent neutral base oil, the polybutene forming 45% by weight of the solution.

The solvent neutral base oil is suitably a 100 to 150, preferably a 130 to 150 solvent neutral base oil.

The process is illustrated with reference to the following examples.

Comparison Test 1: Hyvis 7000/45 was used to dissolve a number of conventional VI improver additives at a temperature above 100°C and the solutions cooled to ambient temperature with the following results:

   (1)   25% Hyvis 7000/45:75% Shell Slurry (Lubad 125 solution), a styrene/butadiene copolymer; polymer separates out on cooling;

   (2)   45% Hyvis 7000/45: 55% Shell Slurry; polymer separates out on cooling and the whole mixture almost solidifies;

   (3)   45% Hyvis 7000/45: 55% Viscoplex 6-50 (a dispersant type methacrylate ester polymer); polymer separates out on cooling;

(4)   55% Hyvis 7000/45:45% ECA5792 (olefin copolymer); polymer separates out on cooling;

(5)   80% Hyvis 05 (polybutene having a number average MW of 360): 20% solid Lubad 125; polymer separates out of solution on cooling; and

(6)   Hyvis 7000/45:solid Lubad 125 (insoluble at all levels).

All percentages are weight percentages based on the total weight of the composition.

Example: Shellvis 50 was dissolved in Hyvis 7000/45 by mixing the two at a temperature above 100°C with stirring. The solution was cooled to room temperature without any polymer separating and its viscosity at 210°F was measured with the following results:

(1)   95% Hyvis 7000/45:5% Shellvis 50; viscosity at 210°F = 2,075 cs, and

(2)   90% Hyvis 7000/45:10% Shellvis 50; viscosity at 210°F = 4,600 cs.

The viscosity of Hyvis 7000/45 in the absence of any polymeric additive was 900 cs.

Comparison Test 2:   Shellvis 50 was dissolved in 150SN oil by mixing the two at a temperature above 100°C with stirring and the solutions cooled to ambient temperature with the following results:

(1)   5% Shellvis 50:95% 150SN oil, a highly viscous, but nevertheless mobile solution; and

(2)   10% Shellvis 50:90% 150SN oil, an immobile jelly.

All percentages are by weight based on the total weight of the composition.

Block Copolymers of Styrene and Methacrylate Components

J.B. Rogan and S.A. White; U.S. Patent 4,136,047; January 23, 1979; assigned to Standard Oil Company (Indiana) describe block copolymers having molecular weights ranging from about 10,000 to 500,000 and having a weight average molecular weight to number average molecular weight ($\overline{M}w/\overline{M}n$) ratio less than 2, which comprises:

(a)   from about 5 to 50 wt % in blocks of styrene (S) or α-methylstyrene (AMS); and

(b)   from about 50 to 95 wt % in blocks of 3,4-dimethyl-α-methylstyrene (DMAMS) or lauryl methacrylate (LMA), which can be incorporated into a lubricating composition in effective amounts to improve its viscosity index.

Examples: In all operations where the absence of water and air was important, the glassware was dried in a 170°C oven for at least 30 minutes and then allowed to cool in a stream of argon. The use of "dried" will be understood to mean this procedure.

Tetrahydrofuran (THF) was distilled from excess sodium naphthalene under an argon atmosphere into a dried 3,000 ml flask or bottle. Subsequently, this bottle was fitted with a dried dispenser assembly which permitted the THF to be forced under argon pressure directly into the reaction flask.

Sodium naphthalene was prepared in THF solution as described by Sorenson and Campbell, *Preparative Methods of Polymer Chemistry,* p. 197, Interscience Publishers, Inc., New York (1961), and was transferred under argon pressure through 1 to 2 mm i.d. polyethylene tubing to a dried flask fitted with a stopcock. Samples of solution were withdrawn conveniently through the stopcock with a syringe. The concentration of sodium naphthalene was determined by titration of aliquots with standard HCl, using methyl red as the indicator. Shortly before use, all monomers were passed through a 4" bed of activated silica gel and stored under argon.

Polymerizations were carried out in three-neck flasks fitted with a stirrer in a ground joint bearing an opening that was just large enough for a serum cap. Argon gas inlet and outlets were through syringe needles that pierced the serum cap. Although the dried equipment was assembled while hot, the THF solvent was added after cooling in an argon stream. All subsequent additions of catalyst or monomer were made with dried syringes and needles. Impurities in the THF were titrated away with sodium naphthalene and then the calculated quantity of sodium naphthalene was added to give the desired molecular weight. The reaction flask was cooled to approximately –40°C internal flask temperature in a dry ice bath and the monomer then added as rapidly as possible with a syringe. While styrene polymerized immediately, AMS and DMAMS required approximately one hour while the reaction mixture was cooled to –78°C.

After initial polymerization either a second monomer was added or the reaction was quenched with methanol. With LMA as the second monomer enough 1,1-diphenylethylene was added to cap all anionic ends before adding the LMA. Polymers were isolated by precipitation in methanol or isopropyl alcohol, filtration and resuspension in fresh methanol in a Waring blender, filtering again and drying. Yields were usually quantitative.

A series of block copolymers of styrene, DMAMS and LMA were prepared using standard experimental procedures with nearly constant total molecular weight of about 200,000, but with varying composition. The $\overline{M}w/\overline{M}n$ ratio for the block copolymers was less than 1.5 as determined by gas phase chromatography. For these polymers, the inside block was formed first by polymerization of the appropriate monomer and then the monomer for the outside block added. As was true of all α-substituted styrenes, it was necessary to polymerize the DMAMS at low temperatures (–78° to –40°C) due to its low ceiling temperature.

To prepare a block copolymer with styrene inside, it is possible to use a mixture of the monomers due to the large difference in reactivity ratios. In contrast, block copolymers of methacrylates and styrene can be prepared only with styrene inside, because the polystyryl anion will initiate methacrylate polymerization but the methacrylate anion will not initiate a styrene polymerization. The widely different acidities of the α-hydrogens of esters and ethylbenzene point to the same conclusion.

The relative acidity of two compounds is a measure of the relative stabilities of their conjugate bases. Since esters are approximately ten powers of ten more acidic than ethylbenzene, it follows that the methacrylate ion is more stable than the polystyryl ion and, thus, the ability of one anion to initiate polymerization of the other monomer is understandable.

When lauryl methacrylate was added to polystyryl dianion directly, there was only a low yield of block copolymer produced. This was attributed to attack of the anion at the carbonyl carbon of LMA, thus terminating the chain rather than at the carbon as desired. This reaction was avoided by first adding 1 mol of 1,1-diphenylethylene for each anion equivalent of the polystyryl dianion. The resulting anion is less reactive and more discriminating so that it attacks LMA only at the β-carbon.

A series of copolymers were tested as viscosity index improvers and the results are shown in Table 1 and Figure 2.2. The data show that a S-DMAMS-S block copolymer containing 12% styrene in SX-10 fuel oil is superior to Paratone and is comparable to Acryloid. Sonic shear stability data given in Table 2 and Figure 2.2 show that the copolymer is superior to either Acryloid or Paratone.

Table 1

Polymer (V.I. Improver)	Styrene in Polymer (wt %)	Base Oil Type	Concentration of Polymer in Base Oil (wt %)	100°F Viscosity SUS	210°F Viscosity SUS	VI$_E$
—	—	SX-5	—	89.7	38.4	92
—	—	SX-10	—	175.0	44.6	95
Acryloid	—	SX-5	0.6	126.0	44.5	169
Acryloid	—	SX-5	1.0	153.0	49.1	198
Acryloid	—	SX-5	1.4	186.0	55.3	225
Acryloid	—	SX-10	0.2	190.0	47.6	123
Acryloid	—	SX-10	0.6	238.0	53.6	152
Acryloid	—	SX-10	1.0	287.0	60.7	179
Paratone	—	SX-5	1.0	140.0	44.7	144
Paratone	—	SX-5	1.4	163.0	47.5	155
Paratone	—	SX-5	2.0	212.0	52.7	168
Paratone	—	SX-10	1.0	250.0	52.0	124
Paratone	—	SX 10	1.6	300.0	56.9	137
Paratone	—	SX-10	2.0	380.0	64.9	146
LMA	—	SX-10	1.0	234.1	51.9	132
LMA	—	SX-10	2.0	268.1	55.1	144
LMA	—	SX-10	4.0	388.	68.8	160
DMAMS	—	SX-5	0.5	98.2	39.5	106
DMAMS	—	SX-5	1.0	108.2	40.9	128
DMAMS	—	SX-5	2.0	138.2	44.1	135
S-DMAMS-S	12	SX-5	0.5	97.0	39.8	119
S-DMAMS-S	12	SX-5	1.0	110.7	41.7	136
S-DMAMS-S	12	SX-5	2.0	144.4	47.7	190
S-DMAMS-S	12	SX-10	0.5	195.3	47.2	112
S-DMAMS-S	12	SX-10	1.0	219.7	50.0	126
S-DMAMS-S	12	SX-10	2.0	285.2	57.9	152
LMA-S-LMA	20	SX-10	0.5	234.9	50.7	121
LMA-S-LMA	20	SX-10	1.0	330.5	61.4	147
LMA-S-LMA	20	SX-10	2.0	615.0	101.0	190
LMA-S-LMA	20	SX-10	4.0	2336.	371.5	261
LMA-S-LMA	30	SX-10	0.5	202.6	48.6	125
LMA-S-LMA	30	SX-10	1.0	218.	50.9	138
LMA-S-LMA	30	SX-10	2.0	309.	60.5	154
LMA-S-LMA*	40	SX-10	0.5	203.	47.7	112

(continued)

Table 1: (continued)

Polymer (V.I. Improver)	Styrene in Polymer (wt %)	Base Oil Type	Concentration of Polymer in Base Oil (wt %)	100°F Viscosity SUS	210°F	VI$_E$
LMA-S-LMA*	40	SX-10	1.0	240.	51.6	127
LMA-S-LMA*	40	SX-10	2.0	349.	64.6	157

*Oil solution hazy.

Figure 2.2: Block Copolymers—Viscosity Index Improvers

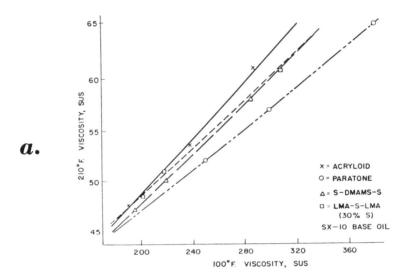

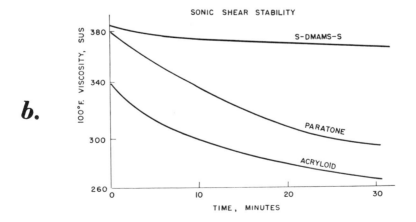

Source: U.S. Patent 4,136,047

Table 2:

| Time | . . . . Viscosity, SUS* . . . . | | | | | |
| (min) | . . Paratone. . . | | . . Acryloid. . . | | . S-DMAMS-S . | |
	100°F	210°F	100°F	210°F	100°F	210°F
0	382	68.6	343	66.7	286	66.4
₊10	340	62.8	299	59.8	376	65.7
+20	320	60.4	281	57.6	373	65.6
+30	306	58.8	270	56.2	371	64.9
Percent disintegration after 30 min	19%	14%	21%	16%	4%	2%

*Lubricant composition = SX-10 base oil 89%; VI improver 2%; and other additives 9%.

Styrene, Butylstyrene and Vinyl Benzyl Ether Terpolymers

H.A. Smith; U.S. Patent 4,080,304; March 21, 1978; assigned to The Dow Chemical Company describes a liquid composition which exhibits a reversible phase transition between a solution and a colloidal dispersion that is temperature-dependent resulting in a relatively constant viscosity over a range of temperatures. The composition consists essentially of a hydrocarbon oil and an interpolymer of at least one hydrocarbon monomer, an oil-solubilizing monomer, and at least one dispersant monomer which has been interpolymerized.

The temperature at which the composition passes from a solution to a colloidal dispersion is referred to as the transition temperature.

The useful interpolymers are those composed of an ethylenically unsaturated hydrocarbon monomer including preferably a monoalkenyl aromatic monomer. Typical of such monomers are styrene and α-methylstyrene.

The interpolymer must also contain an oil-solubilizing monomer. Included within the term "oil-solubilizing monomer" are the nuclear alkylated styrenes where the alkyl group contains up to about eight carbon atoms. Representative of the latter monomers are vinyl toluene, tertiary butyl styrene and tertiary octyl styrene. The long chain alkyl and oxyalkyl methacrylates having from about 12 to 18 carbon atoms are also useful.

Those compounds include lauryl methacrylate and octadecyl methacrylate. The dispersant monomer is a bisoluble compound which is interpolymerizable with the hydrocarbon monomer and which, in addition, has a functional moiety that, when interpolymerized with that hydrocarbon and oil-solubilizing monomers in minor amount, will cause formation of a stable colloidal dispersion below the transition temperature.

The dispersant monomer may be selected from a wide variety of such compounds. A preferred class of such dispersant monomers are certain vinyl benzyl ethers of the formula:

$$CH_2=C \underset{R}{\overset{}{\mid}} \text{—} \underset{}{\boxed{}} \text{—} CH_2(OCHCH_2)_m \underset{R}{\overset{}{\mid}} \text{—Y}$$

where R is hydrogen or methyl, m is about 10 to 100, and Y is $-OR_1$, $-SR_1$,

$$-NR_2, \qquad \overset{O}{\underset{R_3}{\overset{\|}{-OCR_1}}} \quad or \quad \overset{O}{\underset{R_3}{\overset{\|}{-NCR_2}}}$$

where R_1 is an alkyl, aralkyl or alkaryl hydrophobic group of 10 to about 22 carbon atoms, R_2 is an alkyl group of 1 to about 22 carbon atoms and R_3 is hydrogen or an alkyl group of 1 to about 22 carbon atoms, provided R_2 and R_3 in combination have at least 10 carbon atoms.

The composition of the interpolymer will vary with a number of factors including the particular hydrocarbon oil to be stabilized; and the temperature range over which control is desired.

The interpolymers may be prepared by known polymerization techniques including bulk, suspension, solution or emulsion polymerization. The advantages and disadvantages of each such technique are well-known.

The fluid/polymer compositions are readily prepared by adding the polymer to the fluid, while being stirred and held at a temperature above that at which the polymer is insoluble. In the examples, parts and percentages are by weight.

Example 1: Polymer Preparation — A mixture of 30 g styrene, 87.6 g t-butyl-styrene, 2.4 g of dispersant monomer, and 0.12 g benzoyl peroxide was made up in a citrate bottle, flushed with nitrogen, and capped. The capped bottles were then placed in a tumbler at 80°C for 4 days, cooled, and removed. The polymer so-obtained was removed from the bottles and ground.

Such polymers contain 25% styrene and 2% dispersant monomer and have a weight average molecular weight of $4.5\text{-}5.5 \times 10^5$.

In a similar manner, polymers containing 30% styrene and 0, 5 and 10% dispersant monomer were prepared.

Example 2: 2.5 and 5% solutions of the polymers made in Example 1 above were made up in transmission oils (Amoco SAE 40 oil and Calumet 3500 oil) and the behavior of viscosity of these systems with temperature determined as well as the polymer precipitation temperature and the nature of the precipitate (phase separated or dispersed). This data is given in the following table.

From the data it can be seen that solutions of the polymers containing internal surfactants have viscosities at 210°F that are comparable to those without surfactant and comparable viscosity indexes.

However, on precipitation of the polymer at temperatures below T_{ps}, the resulting mixture remains homogeneously dispersed in the cases of the emulsified systems with one exception, whereas the unemulsified systems allow the polymer to separate out which could lead to plugging of filters and small passages in equipment through which these mixtures might pass.

Polymer*	Emulsifier	Emulsifier (%)	MW x10⁵	Tps** (°C)	η210°F	η100°F	Room Temperature Pourability	Room Temperature Dispersion
. . . . . 2.5% Solutions								
C-3500 oil 75/25	vinyl silicone	0	3.5	55	7.4	12.0	yes	fluid, polymer precipitated
		2	4.9	62	11.6	21.4	yes	well dispersed
		5	6.95	63	18.5	16.4	yes	well dispersed
SAE-40 oil 25/75	vinyl silicone	0	4.1	54	52.0	282	none	polymer precipitated
		2	3.7	54	48.2	275	yes	very fluid if agitated, well dispersed
		5	4.1	60	50.8	520	yes	very fluid if agitated, well dispersed
		10	3.75	65	49.0	332	yes	very fluid if agitated, well dispersed
SAE-40 oil 25/75	VBE***	0	4.1	54	52.0	282	none	polymer precipitated
		2	5.2	62	59.4	628	yes	fluid if agitated, well dispersed
		5	5.45	87	56.8	646	yes	very fluid if agitated, well dispersed
		10	2.1	100	38.4	~270	yes	polymer precipitated
. . . . . 5% Solutions								
SAE-40 oil 25/75	vinyl silicone	0	4.5	62	125	281	no	polymer precipitated
		2	3.7	62	108	354	yes	especially if agitated, well dispersed
	VBE***	2	5.3	75	138	267	yes	especially if agitated, well dispersed
	OD-E1800 VBE†	2	5.6	74	148	279	some	well dispersed
SAE-40 oil 30/70	vinyl silicone	0	4.1	79	117	357	no	polymer precipitated
		2	3.5	80	105	331	yes	especially if agitated
	VBE***	2	6.0	97	158	272	yes	especially if agitated
	OD-E1800 VBE†	2	5.6	92	147	264	some	well dispersed

*Styrene/tert-butylstyrene.

**Temperature of polymer precipitation.

***Vinyl benzyl ether of condensation product of nonylphenol and 40 mols ethylene oxide.

†Vinyl benzyl ether of polyoxyethylene (MW 1,800) monooctadecyl ether.

Hydrogenated Block Copolymers of Butadiene and Isoprene

R.L. Elliott; U.S. Patent 4,073,737; February 14, 1978; Exxon Research & Engineering Co. has found that there is a type of copolymer which is highly useful in a viscosity modifier in lubricating oils containing a polymeric pour point depressant since it provides excellent low temperature viscometric performance to the blended oil, while retaining the desired levels of shear and oxidative stability. This type of copolymer is an oil-soluble copolymer of the following formula: $(A)_x(B)_y(C)_z$, wherein A is a conjugated diene of the formula:

$$CH_2{=}CH-\overset{\overset{\displaystyle R}{|}}{C}{=}CH_2$$

where R is a C_{1-8} alkyl group, preferably CH_3, i.e., isoprene, and present in mol % proportion as indicated by x which may vary from 45 to 99 mol %; B is butadiene and present in mol % proportion as indicated by y which may vary from 1 to 10, preferably 2 to 9; and C is a C_{8-20} monovinyl aromatic compound and/or aromatic-substituted diene and present in mol % proportion as indicated by z, which may vary from 0 to 45 mol %, preferably 5 to 40, whereby the most useful composite properties of oxidative stability and $-18°C$ viscosity of the lubricating oil blend is realized.

A block copolymer according to this process is a copolymer obtained by anionically copolymerizing in hydrocarbon solution the monomers of butadiene and at least one other conjugated diene in the presence of an alkali metal or an alkali metal compound as a catalyst until at least 99% of the monomers have been incorporated into the copolymer and thereafter, if desired, incorporating a polymer block of a monovinyl aromatic monomer.

Example 1: Into a clean, dry polymerization vessel, under a nitrogen atmosphere, was charged toluene (150 ml) and n-butyllithium (0.55 meq). Butadiene (1.9 g, 0.036 mol) and isoprene (25.0 g, 0.368 mol) were simultaneously added to the vessel cooled to $-5°C$. Polymerization was initiated by slowly heating the solution to $50°C$. The solution was allowed to continue reacting for an additional 30 minutes whereupon there was added, at ambient temperature, styrene (2.0 g, 0.019 mol) in 50 ml of toluene.

The resulting solution was slowly heated to $50°C$ and allowed to continue reacting for an additional 30 minutes. The reaction was terminated with 3 ml of methanol. The copolymer was isolated by precipitation into 2 liters of methanol containing 0.1 wt % antioxidant and drying in a vacuum oven at $100°C$ for 24 hours. This yielded a clear block copolymer (26.5 g, 88.4% of theoretical yield). This copolymer had an ($\overline{M}n$) of 59,000 (membrane osmometry) with approximately 85% of the diene monomer units in a 1,4 configuration (determined by nuclear magnetic resonance).

Example 2: Following the procedure of Example 1, isoprene (17.7 g, 0.26 mol), butadiene (9.1 g, 0.168 mol) and styrene (3.0 g, 0.029 mol) were copolymerized with n-butyllithium (0.35 meq) in 250 ml toluene. An isolation procedure as in Example 1 yielded a clear block terpolymer (27.9 g, 93% of theoretical yield) having an (Mn) of 103,000 (membrane osmometry) of about 85% 1,4 configuration.

Example 3: The copolymers of Examples 1 and 2 were, in turn, subjected to chemical hydrogenation as described in *Die Makromoleculare Chemie,* 163, 1 and 13 (1973). This hydrogenation is illustrated with the copolymer of Example 1 as follows.

Into a dry flask, under a nitrogen atmosphere, was carefully placed a sample of 5 g of the copolymer, xylene (250 ml) and p-toluenesulfonylhydrazide (TSH) (35 g) and the resulting solution maintained at reflux (135° to 140°C) under mild agitation for 2 hours. After this time, the solution was hot filtered (about 90°C), then cooled to ambient temperature and isolated by precipitation from 1,500 ml methanol. The resulting polymer was washed with methanol (500 ml), then dried in a vacuum oven at 100°C for 24 hours to yield 4.7 g of a hydrogenated butadiene (8 mol %), isoprene (85 mol %), styrene (7 mol %) block terpolymer.

Example 4: For the hydrogenation of the copolymer of Example 2, the procedure of Example 3 was followed, except for the following changes. 7 g of copolymer, 500 ml of xylenes and 28 g TSH were used, and mildly agitated for 4 hours. The resulting copolymer yield was 6.4 g of a hydrogenated butadiene (35 mol %), isoprene (54 mol %), styrene (11 mol %) block terpolymer.

Example 5: To demonstrate their viscosity index improving characteristics, the resulting hydrogenated polymers of Examples 3 and 4 were blended to a viscosity of about 12.4 cs in ENJ-102, a mineral lubricating oil. This oil was a blend of two basic oils which contained 0.5 wt % of a commercial polymeric pour point depressant.

Both oils were paraffinic, solvent refined neutral oils. The first had a viscosity of about 150 SUS at 100°F and constituted 25.75 wt % of the blend. The second oil had a viscosity of about 300 SUS at 100°F and constituted 73.75 wt % of the blend. Polyisobutylene (Paratone N) was also blended with the test oil for comparison. The stability of the several lubricating oil test compositions was examined by determining the extent of viscosity loss in a sonic breakdown test. The sonic breakdown test is a measure of shear stability and is conducted according to the procedure described in *ASTM Standards, Vol 1, Test for Shear Stability of Polymer-Containing Oils,* p. 1160 (1961). The results of these tests are summarized in the table.

Hydrogenated Polymer of Example*	KV at 210°F** (cs)	0°F Viscosity*** (p)	Pour Point† (°F)	Sonic Shear†† (%)
3	12.71	25.5	-35	22.8
4	12.54	24.3	0	9.10
Paratone N	12.4	28.9	-37	22

*Additive is blended in an amount to provide a viscosity of composition approximating 12.4 cs at 210°F.

**Determined in accordance with ASTM D-445.

***Determined in accordance with ASTM D-2602.

†Determined in accordance with ASTM D-97.

††Run at 0.75 amps and 40°C for 15 min according to ASTM standards, Vol. 1 (1961) p 1160.

The block copolymer (Example 3) of the process has viscosity improving properties comparable to a commercial viscosity index improver with comparable shear stability and pour point effect. The low temperature viscometrics provided by a blend of the copolymer of the process to the lubricating oil containing a pour point depressant are somewhat better than available from a copolymer containing large amounts of butadiene (the copolymer of Example 4) as evidenced by the fact that the $0°F$ viscosity of each for the latter is 1.2 poises higher.

Hydrogenated and Sulfonated Block Copolymers of Styrene and Isoprene

D.L. Wood, R.J. Moore and R.B. Rhodes; U.S. Patent 4,086,171; April 25, 1978; assigned to Shell Oil Company describe a lubricating composition comprising a major proportion of a mineral lubricating oil and between about 0.1 and 15.0 wt % of a viscosity index improver having the general configuration A-B, where each A is a sulfonated monoalkenyl arene polymer block having a number average molecular weight of between about 25,000 and 55,000; and B is a hydrogenated polymer block. Block B, prior to hydrogenation, is a conjugated diene selected from butadiene, isoprene, and butadiene/isoprene blocks, at least 90% of the olefinic unsaturation of block B being reduced by hydrogenation of the block copolymer. Block B has a number average molecular weight between about 20,000 and 150,000. The weight ratio of block A to block B is between about 0.45:1 and 0.8:1.

The sulfur content of the block copolymer is more than about 0.1% by weight and less than 1.0% by weight. Preferably, the viscosity index improver is a selectively hydrogenated, selectively sulfonated styrene-isoprene two block copolymer.

The viscosity index improver of the process results in lubricating oil compositions having improved high temperature (above $270°F$) viscosity performance characteristics.

Hydrogenated Copolymers of Butadiene and Styrene

V. Ladenberger, K. Bronstert, F. Hovemann and P. Simak; U.S. Patent 4,073,738; February 14, 1978; assigned to BASF Aktiengesellschaft describe mineral lubricating oil compositions which contain a predominant proportion of a mineral lubricating oil, together with a pour point depressant based on alkyl acrylate or alkyl methacrylate polymers and a viscosity index improver comprising a special selectively hydrogenated copolymer of styrene and a conjugated diene. The viscosity index improver employed is a selectively hydrogenated, random butadiene-styrene copolymer containing from 35 to 45% by weight of styrene and possessing a particular structure and distribution of the monomers. The compositions exhibit high stability to shear and may, therefore, in particular be used for lubricating IC engines or gearboxes.

Hydrogenated Star-Shaped Polymer

A process described by *R.J.A. Eckert; U.S. Patent 4,116,917; September 26, 1978; assigned to Shell Oil Company* is directed to a particular hydrogenated star-shaped polymer that possesses some unique properties when employed as a viscosity index improver in lubricating oil compositions.

In addition, this polymer may also be employed in fuel oils, such as middle distillate fuels, as well as with mineral lubricating oils and synthetic lubricating oils. In particular, the hydrogenated star-shaped polymer has a poly(polyalkenyl coupling agent) nucleus and at least 4, preferably 7 to 15, polymeric arms linked to the nucleus. These arms are selected from the group consisting of: (a) hydrogenated homopolymers and hydrogenated copolymers of conjugated dienes; (b) hydrogenated copolymers of conjugated dienes and monoalkenyl arenes; and (c) mixtures thereof.

This hydrogenated star polymer, when employed as a viscosity index improver in a lubricating oil composition, is present in an amount of between 0.15 and 10.0% by weight.

The lubricating oils containing viscosity index improvers possess excellent thickening efficiency at high temperature, while also possessing very good low temperature viscosity characteristics. Most importantly, by employing these viscosity index improvers, as opposed to the prior art viscosity improvers, a lower amount of viscosity index improver is required in order to obtain the required thickening performance.

The polymers not only possess much superior oxidative shear stability and permanent shear stability, they also possess significantly improved temporary shear loss. Temporary shear loss refers to the temporary viscosity loss at high shear stress conditions resulting from the non-Newtonian character of the polymeric viscosity index improvers.

High Viscosity Index Gear Oil Formulations

According to a process described by *B. Mitacek; U.S. Patent 4,082,680; April 4, 1978; assigned to Phillips Petroleum Company* shear-stable, high viscosity index gear oil formulations are formed by the inclusion into such formulations of a small amount of a hydrogenated butadiene-styrene copolymer having a butadiene content of 30 to 44 wt % and a weight average molecular weight in the range of about 12,000 to 20,000.

Example: A shear-stable gear oil was prepared with the following formulation:

> 51.1 wt % neutral oil 10H, viscosity at 100°F = 95 SUS, viscosity index 100 (KC-10);
> 36.0 wt % bright stock oil 250H, viscosity at 210°F = 205 SUS, viscosity index 96 (KC-250);
> 7.0 wt % gear oil additive containing phosphorus and sulfur (Anglamol 99LS);
> 0.2 wt % polymethacrylate pour point depressant (Acryloid 152); and
> 5.7 wt % butadiene-styrene hydrogenated copolymer.

The butadiene-styrene copolymers of this process were made by first copolymerizing styrene and butadiene using butyllithium catalyst and this polymer was then hydrogenated as described in Example 1 of U.S. Patent 3,554,911 as follows.

The hydrogenated polymer for this example was prepared using the following recipe and conditions.

Butadiene	35 parts by weight
Styrene	65 parts by weight
Cyclohexane	800 parts by weight
Tetrahydrofuran	1.5 parts by weight
Sec-butyllithium	0.154 parts by weight
Initiation temperature	122 °F (50°C)
Initiation pressure	20 psig (138 kPa)

Charge order was cyclohexane, reactor purged with nitrogen, butadiene, styrene, tetrahydrofuran, and sec-butyllithium. Essentially quantitative conversion was obtained in 3 hours. At that time the unterminated product was transferred to a hydrogenation reactor, 0.13 g of nickel (as nickel octoate) and 1.05 g of triethylaluminum in cyclohexane were added, the reactor was pressured to 50 psig (345 kPa) with hydrogen, the temperature was increased slowly to 350°F (177°C), the hydrogen pressure was increased to 400 psig (2,760 kPa) and the temperature rose quickly to 395°F (201.5°C) and fell to 350°F (177°C) in about 30 minutes.

Reaction was continued for 1.5 hours at 350°F and 400 psig, the reactor was cooled to 170°F (77°C), and the essentially completely hydrogenated polymer was recovered.

With respect to the polymer in the above example, essentially completely hydrogenated means that 95 wt % or more of the olefinic groups are hydrogenated and 5 wt % or less of the phenyl groups (when present) are hydrogenated.

The unhydrogenated polymer had the following properties: 16.1% trans unsaturation, 9.6% vinyl unsaturation, 17,500 $\overline{M}w$, 14,100 $\overline{M}n$, and 63.5 wt % total styrene.

The hydrogenated polymer had the following properties: 1.5% trans unsaturation, nil percent vinyl unsaturation, and 16,400 $\overline{M}w$, 13,100 $\overline{M}n$.

The test gear oil was charged to the differential of two 1972 Buick Skylarks running on mileage accumulators. Viscosities were measured at intervals. The results are given below.

Oil Miles	Viscosity SUS/210°F	Oil Miles	Viscosity SUS/210°F
0	94.76	0	94.84
8,492	87.74	13,841	85.39
29,689	83.21	33,676	81.48
46,871	80.93	54,776	78.80

From the tests it was concluded that the gear oil would have a viscosity SUS 210°F above 74 after 50,000 miles operation as required by the specification for this type of gear oil.

Addition of Oil-Soluble Acid to Prevent Hazing

J.B. Gardiner, M.W. Hill and J. Ryer; U.S. Patent 4,069,162; January 17, 1978; assigned to Exxon Research & Engineering Co. have found that a wide variety of unwanted catalysts, metal weak acid salts which result from the by-products of the polymerization, finishing process or other steps in the manufacture of ethylene-containing copolymers of oil concentrates can cause haze in and create filtration problems of lubricating oil compositions prepared from the ethylene copolymers.

These haze and/or filtration problems can be overcome by treating the hydrocarbon polymer or its oil concentrate which comprises a hydrocarbon solvent and from 0.1 to 50, preferably 5 to 30 wt %, based on the solution, of a soluble hydrocarbon polymeric material having viscosity index improving characteristics plus an oil-soluble strong acid.

Useful strong acids which eliminate the hazing property of the hazing substance are represented by oil-soluble derivatives of maleic acid, malonic acid, phosphoric acid, thiophosphoric acids, phosphonic acid, thiophosphonic acids, phosphinic acid, thiophosphinic acids, sulfonic acid, sulfuric acid, and α-substituted halo- or nitro- or nitrilo-carboxylic acids.

POUR POINT DEPRESSANTS

α-Olefin Copolymers

A process described by *W.J. Heilman and T.J. Lynch; U.S. Patent 4,132,663; January 2, 1979; assigned to Gulf Research & Development Company* relates to a mineral oil composition containing a minor amount of an α-olefin copolymer of about 60 to 95 mol percent 1-hexene and about 5 to 40 mol percent 1-octadecene, the oil composition having a significantly lower pour point than the mineral oil without the copolymer additive.

The following examples illustrate the process. In preparing the copolymers all reactants, solvents and catalysts were of ultra-high purity for the reaction. The pour point determinations were made by the method described in ASTM D97, which results in an accuracy of ±5°F. In all examples the copolymer was completely soluble in the oil at the temperature of the pour point determination and evidenced no crystallinity at this temperature.

The 1-octadecene used was approximately 90 wt % of the specified 1-olefin, about 8 wt % of other olefins of the same carbon number as the specified 1-olefin, primarily a vinylidene isomer, and about 1 wt % of the next lower and next higher 1-olefin. The 1-hexene used herein was greater than 96 wt % of the specified 1-olefin.

Examples 1 through 24: A high molecular weight copolymer of 1-hexene and 1-octadecene was made in a 250 ml Erlenmeyer flask in a shaker oil bath. 1 g of $(TiCl_3)_3 \cdot AlCl_3$ [grade AAX (Stauffer Chemical Company)], 2 ml of triethyl aluminum (Al/Ti atomic ratio of 4.0) and 100 ml of n-heptane were placed in the flask which was then purged with hydrogen for 30 minutes at a bath temperature of 49°C.

A mixture of 26.9 g (88 mol percent) 1-hexene and 11.02 g (12 mol percent) 1-octadecene was then injected into the flask. After reacting for 2 hours at a constant bath temperature of 49°C, 20 cc of a 50/50 wt % mixture of n-butanol and 2,4-pentadione was added to solubilize and quench the catalyst. After the mixture was treated with 10% aqueous sodium hydroxide, the aqueous layer was separated from the organic layer. The copolymer was precipitated from the organic layer by isopropyl alcohol, dried and dissolved as a 33 wt % concentrate in a dewaxed, hydrofined oil having a 210°F viscosity of 3.2 cs.

The resulting copolymer comprising 12% of its repeating units from 1-octadecene and 88% of its repeating units from 1-hexene had a weight average molecular weight of 170,000 and a distribution factor Mw/Mn of 9.2 as determined by gel permeation chromatography using an instrument that was calibrated using known standard poly(1-hexene) fractions.

The average number of repeating units, n, was determined to be about 1,600 using the average molecular weight of 104.2 for the copolymerization mixture. This copolymer and a series of related copolymers made in an equivalent manner were mixed with a commercially available No. 2 fuel oil having an unaided pour point of 0°F to determine the pour point depression effected by the copolymer additives.

Various homopolymers as well as a copolymer of higher α-olefins were also tested in this No. 2 fuel oil for possible pour point depression. The results of these experiments are set forth in the table in which only the mol percent of the higher olefin is indicated for various amounts (weight percent) of polymer additive.

Example No.	Lower Olefin	Higher Olefin	Mol %	. . . . Pour Point (°F).		
				0.05%	0.1%	0.2%
1	C_6	C_{16}	7.5	–	–5	–5
2	C_6	C_{16}	10.0	–	–5	–5
3	C_6	C_{18}	7.0	–15	–20	–25
4	C_6	C_{18}	8.0	–10	–25	–
5	C_6	C_{18}	9.0	–15	–25	–
6	C_6	C_{18}	10.0	–	–25	–40
7	C_6	C_{18}	11.0	–20	–30	–
8	C_6	C_{18}	12.0	–15	–30	–
9	C_6	C_{18}	15.0	–15	–25	–
10	C_6	C_{18}	18.0	–20	–30	–
11	C_6	C_{18}	24.0	–25	–30	–
12	C_6	C_{18}	33.0	–25	–30	–
13	C_6	C_{18}	39.0	–5	–25	–
14	C_6	C_{20}	7.0	–10	–	–
15	C_6	C_{20}	10.0	–10	–20	–
16	C_8	C_{16}	7.5	–	–	–5
17	C_8	C_{18}	10.0	–	–	–5
18	C_{10}	C_{16}	7.5	–	–	–5
19	C_{10}	C_{18}	7.5	–	–	–5
20	none	C_{12}	100	–5	–5	–
21	none	C_{14}	100	–	–10	–
22	none	C_{16}	100	–	–5	–
23	none	C_{18}	100	–	0	–
24	none	C_{20}	100	–	*	–

*Insoluble.

Polyalkylacrylate and Ethylene-Propylene Copolymers

C.M. Cusano and R.E. Jones; U.S. Patent 4,146,492; March 27, 1979; assigned to Texaco Inc. describe a special interpolymer of (A) one or more C_{1-15} alkylacrylates and (B) one or more of C_{16-22} alkylacrylates having a weight ratio of A:B of between about 90:10 and 50:50, an average molecular weight of 1,000 to 25,000 and an average alkyl carbon side chain length of between about 11 and 16 carbons produced under neat conditions which has improved pour depressing effects and an improved compatibility with ethylene-propylene copolymer VI improvers in concentrate and finished lubricating oil formulations.

Lubricating oil compositions have further been discovered that contain at least about 50 wt % lubricating oil, between about 0.005 and 10 wt % interpolymer and between about 0.5 and 30 wt % ethylene-propylene copolymer of a molecular weight between 10,000 and 150,000 having a propylene content of between 20 and 70 mol % and a polydispersity index of less than 5, of superior pour depression and polyalkylacrylate-ethylene-propylene copolymer compatibility.

Example 1: This example illustrates the preparation of the polyalkylacrylate under neat conditions. The equipment employed consists of a 1 liter resin kettle equipped with external heaters, thermocouple, nitrogen inlet, stirrer, thermometer and condenser.

In a typical reaction 450 g of Neodol 25L ($\sim C_{12-15}$)alkylmethacrylate mixture having an average alkyl carbon chain length of 13.6 and 150 g of Alfol 1620SP methacrylate ($\sim C_{16-20}$ n-alkylmethacrylate) of an average alkyl carbon chain length of 17.1 carbons and 6 g of n-dodecylmercaptan were heated under a nitrogen blanket with stirring. When the pot temperature reached 95°C, 0.45 g of azobisisobutyronitrile (AIBN) polymerization initiator was added. Polymerization proceeded (95° to 100°C) as monitored by refractive index increase.

After two hours of reaction, 0.2 g of additional AIBN was added and heating at 95°C was continued for an additional 1.5 hours. At the end of the 1.5 hour period, an additional 0.1 g AIBN was added and heating continued for a further 2.5 hours giving a total reaction time of 6 hours. At the end of the 6 hour time, the formed polyalkylmethacrylate gave the following analysis:

Description	Value
Molecular weight (VP Osmometry)	3,300
Kinetic viscosity (2 wt % polymer in 53 SUS 210°F oil) 210°F	9.07 cs
Furol viscosity at 210°F	440 sec
Alkyl side chain carbon length, average	14.4
Residual monomer content	0.3 wt %

Example 2: This example still further illustrates the preparation of the polyalkylacrylates. The procedure of Example 1 was repeated in five separate runs with the exception that the methacrylate monomers were replaced as follows.

Run A — 50:50 weight ratio isodecylmethacrylate (IDMA):Alfol 1620SP methacrylate (AMA);

Run B — 10:40:50 weight ratio methylmethacrylate (MMA):IDMA: AMA;

Run C — 10:40:50 weight ratio butylmethacrylate (BMA):IDMA:AMA;
Run D — 21:59:20 weight ratio BMA:MMA:AMA.

The neat polyalkyl methacrylates prepared in the above four runs were analyzed and the results are as follows:

Description	. . . . Runs			
	A	B	C	D
MW (VP osmometry)	3,500	3,500	3,500	3,500
Kinetic viscosity (210°F), cs*	9.32	9.11	9.29	9.33
Furol viscosity (210°F), sec	~700	~450	~550	~700
Average alkyl side chain carbon length	13.6	12.7	13.0	12.3
Residual monomer content, wt %	0.7	0.1	0.1	0.3

*2 wt % in 53 SUS 210°F oil.

Example 3: This example illustrates the preparation of a comparative polyalkylacrylate pour depressor. The equipment employed was that used in Example 1. In a typical reaction, 150 g of Neodol 25L methacrylate, 50 g of Alfol 1620SP methacrylate, 100 g of mineral oil of an SUS viscosity at 100°F of approximately 145 and 0.2 g of n-dodecylmercaptan were heated under mixing and nitrogen atmosphere conditions.

When the reaction mixture reached 83°C, 0.3 g of AIBN was added and the reaction was maintained at 83°C for approximately 4 hours. Polymerization as in Example 1 was monitored by measuring refractive index incrementally. At the end of the 4 hour period, an additional 0.15 g of AIBN was added and kept at 83°C for 1.5 hours. Still another 0.15 g AIBN was added at the end of the 1.5 hour period with the reaction maintained at 85°C for an additional 1.5 hours.

At the end of the second 1.5 hour period, 306 g of 100 SUS (100°F) mineral oil were additionally added and the mixture was heated at 100°C for 1 hour to finish the preparation. Analysis of the lube oil solution product found it to contain 33.0 wt % polyalkylmethacrylate of an average molecular weight of 180,000 having an average alkyl side chain carbon length of 14.4.

Example 4: This example illustrates the effectiveness of the polyalkylacrylate of the process in pour depressing lubricating oil compositions containing the ethylene-propylene copolymer and the pour depressing superiority in many instances of the polyalkylacrylates over conventional polyalkymethacrylates.

To a mineral lubricating oil of an SUS viscosity of about 130° and 100°F and a pour of 0°F, there was mixed an ethylene-propylene copolymer of an average molecular weight of 100,000 having a propylene content of 32.8 mol %, and a representative polyalkylacrylate prepared in Example 1 to give a final composition containing 1.0 wt % ethylene-propylene copolymer and 0.2 wt % of polyalkylmethacrylate, the remainder being the base oil. The resultant formulation was designated as Formulation 1.

Comparative Formulation 2 was prepared using the same ingredients as Formulation 1, except that the comparative polyalkylmethacrylate of Example 3 was substituted for the polyalkylmethacrylate of representative Example 1. The pour point data for the above test formulations and the base oil is set forth below.

Formulation	E-P Copolymer	. . . . (wt %)* . . . . Polyalkylmethacrylate Polymer	Pour Point (°F)
Base oil	0	0	0
Formulation 1	1.0	0.20	−25
Formulation 2	1.0	0.20	+5

*Neat basis.

Example 5: This example still further illustrates the pour depressancy effect of the polyalkylacrylate in lubricant compositions of the process.

To a paraffinic base oil having an SUS viscosity of about 130 at 100°F and pour of 0°F, there was mixed an oil solution of 13 wt % ethylene-propylene viscosity index improver of about 40,000 average molecular weight and 45 mol % propylene, 87 wt % mineral oil (100 SUS at 100°F) and the polyalkylmethacrylate of Example 1. Additional test formulations were formed which were identical in all aspects to the described formulation except containing varying amounts of the poly-n-alkylmethacrylate of Example 1. This series was designated as the Formulation E series.

Another formulation series was also prepared for test which was identical to the Formulation E series with the exception that comparative polyalkylmethacrylate of Example 3 was substituted for representative polyalkylmethacrylate of Example 1. This comparative series formulation was designated as the Formulation F series. The pour point test data is reported below.

Description	E-P Copolymer	. . . . (wt %)* . . . . Polymethacrylate	Pour Point (°F)
Base oil	0	0	+5
Formulation			
E-1	1.5	0.01	−30
E-2	1.5	0.02	−30
E-3	1.5	0.03	−35
E-4	1.5	0.08	−30
E-5	1.5	0.17	−35
F-1	1.5	0.01	−10
F-2	1.5	0.02	−35
F-3	1.5	0.04	−30
F-4	1.5	0.08	−35
F-5	1.5	0.17	−40

*Neat.

Example 6: This example illustrates the superiority of the polyacrylates of the process in respect to compatibility with ethylene-propylene (EP) copolymer under concentrate conditions as measured by Lumetron Turbidity (LT). The lower the turbidity, the greater the term of compatibility.

Ten formulations were prepared. The polymethacrylates employed were prepared utilizing the reactants, reactant quantity ratios and general procedure of Example 1, except that polymerization initiator quantities, temperatures and end stopper were varied to vary the molecular weight. The ethylene-propylene copolymers were in the form of filtered and unfiltered lubricating oil of the polymer.

The neat polymethacrylate and ethylene-propylene polymer lube oil solutions were blended and the turbidities of the resultant blends were measured.

The test data and results are reported in the table below where (A) is unfiltered 13 wt % ethylene-propylene copolymer lube oil solution having LT of 14.0; EP copolymer has an ethylene content of 55 mol % and a molecular weight of about 40,000; (B) is filtered 13 wt % ethylene-propylene copolymer lube oil solution having LT of 5.0; EP copolymer has an ethylene content of 55 mol % and a molecular weight of about 40,000; and (C) is unfiltered 13 wt % ethylene-propylene copolymer lube oil solution having LT of 13.5; EP copolymer has an ethylene content of 67.2 mol % and a molecular weight of about 40,000.

Run No.*	Polymethacrylate (MW)	EP Copolymer	LT
1	3,160	A	12.0
2	79,100	A	17.5
3	3,160	B	4.5
4	128,000	B	13.0
5	78,100	B	13.5
6	71,100	B	13.0
7	3,160	C	12.5
8	128,000	C	20.0
9	78,100	C	17.0
10	71,100	C	17.0

*All blends contained 3.0 wt % polymethacrylate, 12.6 wt % ethylene-propylene copolymer and 8.4 wt % lubricating oil (~100 SUS at 100°F).

In the above table, Runs 1, 3 and 7 are the representative concentrate compositions of the process, the remainder are comparative. As can be seen from the above, the polymethacrylates of the process reduce the turbidity of the resultant ethylene-propylene-polymethacrylate-lube oil blend, whereas the comparative polymethacrylates increase turbidity.

Vinyl Acetate, Ethylene and Vinyl Chloride Terpolymer

According to a process described by *W.M. Sweeney; U.S. Patent 4,127,138; November 28, 1978; assigned to Texaco Inc.* a low pour point fuel oil composition is prepared from a major amount of a high pour point, low sulfur, waxy, residual fuel and a minor amount of low wax, low pour, residual fuel oil by adding from 0.01 to 0.5% by weight of an oil-soluble terpolymer such as vinyl acetate-ethylene-vinyl chloride or allyl chloride having a number average molecular weight of about 4,000 to 70,000. The copolymer may be added either in a water-glycol emulsion or in a hydrocarbon to one of the blend components which has been heated to between about 25° and 150°C.

Example: Ethylene, vinyl acetate, vinyl chloride and benzene are fed continuously at rates of 10.01, 4.49, 0.01 and 2.70 lb/hr into a 2 liter stirred reactor maintained at a temperature of 80° to 110°C at 4,000 psig. Azobisisobutyronitrile is employed as the catalyst and is introduced into the reactor as a benzene solution at the rate of 0.8 lb/1,000 lb of polymer. The residence time in the reactor is 15 minutes. After the reaction mixture is removed from the reactor it is stripped of solvent and unreacted materials yielding the terpolymer product.

The composition of the terpolymer is 28 wt % vinyl acetate, 0.7 wt % vinyl chloride and 71.3 wt % ethylene with a number average molecular weight of about 20,000 as determined by vapor pressure osmometry.

A fuel oil composition is prepared by mixing at 60°C for 1 hour 60% by volume of F/18 residual fuel, about 40% by volume of Louisiana No. 6 fuel oil and a sufficient amount of the aboveprepared terpolymer so that the concentration of the terpolymer is 0.125 wt %. The pour point of this composition is determined by the method of ASTM D-97 and found to be substantially below that of the same fuel oil mixture without terpolymer which exhibits a pour point of 80°F. The pour point of the F/18 residual fuel alone is 95°F, while the pour point of the Louisiana No. 6 residual fuel alone is 30°F.

Alkylated Polybenzyl Polymers

W.P. Broeckx and L.G. Dulog; U.S. Patent 4,142,865; March 6, 1979; assigned to S.A. Texaco Belgium NV, Belgium describe low pour hydrocarbon oil compositions which comprise a blend of a hydrocarbon oil or oils and about 0.01 to 0.50 wt % of an oil-soluble, alkylated polybenzyl polymer or copolymer with styrene, propylene or 1-hexene.

Preparation of the abovedescribed polymers can be conveniently carried out according to the process set forth in U.S. Patent 3,418,259. For example, polymerization of p-dodecylbenzyl chloride proceeds as follows:

$$nC_{12}H_{25}\!-\!\!\langle\bigcirc\rangle\!\!-\!CH_2Cl \xrightarrow[\;MX_3\;]{-55°C} \left[-CH_2\!-\!\langle\bigcirc\rangle_{nC_{12}H_{25}}\right]_y + yHCl$$

where MX_3 represents a Friedel-Craft catalyst such as $AlCl_3$, etc.

The polybenzyl polymers and copolymers are useful as pour depressants in a wide variety of hydrocarbon oils such as middle distillates, gas oils, residual fuels, crude oils, etc.

VISCOSITY INDEX IMPROVERS AND DISPERSANTS

Epoxidized Terpolymers

A process described by *J.J. Pappas, N. Jacobson and E.N. Kresge; U.S. Patent 4,156,061; May 22, 1979; assigned to Exxon Research & Engineering Co.* is directed to epoxidized terpolymers of ethylene, C_{3-8} α-olefins and nonconjugated dienes in which the chain of carbon atoms forming the terpolymer backbone is essentially saturated and any substantial unsaturation in the terpolymer, prior to epoxidation, is in an alkylene radical pendant to the backbone or is pendant to or part of a cyclic structure attached to the backbone. The epoxidized terpolymer may be further reacted with a functional reagent having a replaceable hydrogen which will react with an oxirane in the presence of an acidic or basic catalyst to form a functional adduct. The oil-soluble epoxidized terpolymer or its functional adducts are useful as additives for lubricants, such as sludge dispersants and viscosity index improvers, and as sludge dispersants in mineral oil fuels, e.g., distillate fuel oil and gasoline.

Example 1: 30 g of an ethylene-propylene-5-ethylidene-2-norbornene terpolymer containing 49% by weight of ethylene, 47.5% by weight of propylene, 3.5% by weight of 5-ethylidene-2-norbornene and of a viscosity average molecular weight ($\overline{M}v$) of 150,000 were dissolved in 800 ml of chloroform contained in a 1 liter flask fitted with a stirrer, dropping funnel and thermometer. A solution of 1.96 g of m-chloroperbenzoic acid (85% purity) in 40 ml of chloroform was added dropwise to the stirred polymer solution over the course of one-half hour.

During the addition, the temperature of the reaction mixture rose from 25° to 28°C indicating reaction. The homogeneous solution was allowed to stand overnight (16 hours) and was then poured in a slow stream into 4 liters of methanol with rapid stirring. The precipitated product was filtered on a Buchner funnel, washed with a further quantity of methanol and dried under vacuum.

The dried product weighted 29.2 g (some mechanical loss) and exhibited a strong absorption band at 8.0 μ and two moderate absorption bands at 11.2 and 12 μ when examined by infrared spectroscopy, showing the presence of oxirane functionality. Analysis showed an oxygen content of 0.55 wt %.

Example 2: The experimental procedure of Example 1 was repeated with 30 g of a terpolymer containing 45% by weight of ethylene, 46% by weight of propylene, 9.0% by weight of 5-ethylidene-2-norbornene and an $\overline{M}v$ of 165,000. 5.02 g of the m-chloroperbenzoic acid dissolved in 50 ml of chloroform was added to the terpolymer dissolved in 800 ml of chloroform. The epoxidized terpolymer was worked up and isolated in the same manner as the product in Example 1.

The product weighed 27.6 g, exhibited absorption at 8.0, 11.2 and 12 μ by infrared analysis and had an oxygen content of 1.15% by weight.

Example 3: 3 g of the epoxidized product of Example 1 were mixed with 0.50 g of N,N-dimethyl-1,3-propanediamine, 0.004 g of p-toluene sulfonic acid and 50 ml of 1,2,4-trichlorobenzene in a 200 ml flask fitted with a reflux condenser and nitrogen purge line. The mixture was heated under nitrogen for 7 hours at a temperature of 170° to 175°C, cooled to room temperature and worked up in methanol. Purification was attained by resolution in toluene and reprecipitation in methanol. Analysis of the dried product, which weighed 2.68 g showed a nitrogen content of 0.16 wt %.

Example 4: The procedure of Example 3 was repeated with the product of Example 2. 3 g of the epoxidized product of Example 2, 1.0 g of N,N-dimethyl-1,3-propanediamine, 0.004 g of p-toluene sulfonic acid and 50 ml of 1,2,4-trichlorobenzene were heated for 7 hours at 170° to 175°C under an atmosphere of nitrogen. Reprecipitated dry product weighed 2.69 g and showed on analysis a nitrogen content of 0.20 wt %.

Example 5: The epoxidized terpolymer products of Examples 1 and 2 and the functional adducts of Examples 3 and 4 were made up as 10 wt % concentrates in solvent 150 neutral oil, which is a solvent extracted, neutral, paraffinic oil having a viscosity of 150 SUS at 100°F as determined by ASTM D-567.

Aliquots of the concentrates were dissolved in a used automobile crankcase mineral lubricating oil and tested for ability to maintain sludge in a dispersed state by means of a sludge inhibition bench test.

The concentrations used are shown in Table 1. After dissolving the epoxidized polymers, the samples were heated at 280°±2°F for 16 hours. After this length of time, the amount of sludge that is not dispersed by the additive was measured by centrifuging the samples at 2,000 rpm for 30 minutes. The oil was carefully decanted from the sludge and the sludge was then washed twice with 25 cc of n-pentane. After washing, the sludge was dried at room temperature to constant weight (about 1 hour). Blanks were also run on used oil containing no additive to determine total sludge. The percentage of sludge dispersed by the additives was then calculated as set forth in Table 1.

Table 1

Polymer	Concentration (g polymer/10 g used oil)	Sludge Dispersed (%)
Example 1	0.02	38
Example 1	0.04	39
Example 1	0.06	53
Example 1	0.10	59
Example 2	0.02	69
Example 2	0.04	75
Example 2	0.06	72
Example 2	0.10	77
Example 3	0.01	60
Example 3	0.02	85
Example 4	0.01	89
Example 4	0.01	93
Example 4	0.02	88
Example 4	0.025	92
Example 4	0.05	83

These data show the effectiveness of the process polymers in dispersing sludge as no sludge is dispersed without the additive.

The functional adducts of the epoxidized polymers, prepared in Examples 3 and 4 were tested as viscosity index improvers in solvent 150 neutral oil, which is a conventional mineral lubricating oil. The results are shown in Table 2, where thickening efficiency is defined as the ratio of the weight percent of Paratone N (a polyisobutylene of $\overline{Mv}$ 125,000 as determined in diisobutylene at 20°C according to the relationship $[\eta] = 36 \times 10^{-5} \times \overline{Mv}^{0.64}$) required to thicken solvent 150 neutral oil to a viscosity of 12.4 cs at 210°F to the weight of the experimental polymer required to thicken the same oil to the same viscosity at the same temperature.

Table 2: Thickening Efficiency of Functional Adducts

	Thickening Efficiency
Product of Example 3	1.9
Product of Example 4	2.7

As seen by the above example, oil-soluble epoxidized terpolymers, or their functional adducts, of the process can be made as automotive crankcase oil dispersants and can also be effective in neutralizing acidity in lubricants.

Oxidized Copolymer and Acrylonitrile

R.L. Elliot and B. Gardiner, Jr.; U.S. Patent 4,098,710; July 4, 1978; assigned to Exxon Research & Engineering Co. describe additives which are nitrile-containing copolymers prepared by a condensation reaction of reactants such as acrylonitrile onto oxidized ethylene-propylene copolymers induced by a thermal means such as heat or in the presence of a strong base catalyst such as sodium hydroxide.

Thus, it has been found that multifunctional viscosity improvers of enhanced dispersancy can be obtained by condensing an oxidized copolymer of ethylene and one or more C_{3-50}, preferably C_{3-18}, α-monoolefins with a C_{3-50} anionically polymerizable monomer by effecting the condensation through physical means, i.e., heat, or catalytic means, i.e., the presence of a strong base.

Example 1: To 10 g of an oxidized/masticated ethylene-propylene copolymer (containing about 44 wt % ethylene and about 56 wt % propylene) with $\overline{M}n$ 23,000 in 200 ml of freshly distilled tetrahydrofuran (THF) maintained at ambient temperature and under a nitrogen atmosphere was rapidly added 1 g of a 50% NaOH solution. The solution was stirred for 15 minutes while heating to 35°C. To the stirring solution was slowly added over 15 minutes, a solution of acrylonitrile (1g, 18.5 mmol) in THF (5 ml). The solution was heated to 40° to 50°C for 3 hours, cooled to ambient temperature, then stirred an additional 18 hours.

The condensed polymeric product was recovered by precipitation from a large volume of methanol, then washed with methanol and finally dried in a vacuum oven at about 100°C for 18 hours, after which 7.6 g of product were recovered. The polymeric additive product obtained contained 0.085 wt % nitrogen (Kjeldahl).

Example 2: Following the procedure as in Example 1, 10 g of an oxidized/-masticated ethylene-propylene copolymer (containing about 44 wt % ethylene and about 56 wt % propylene) of $\overline{M}n$ 34,000 was condensed. In this example, however, the solution was heated at 45° to 60°C for 3 hours.

Precipitation yielded 9.6 g of polymer with a nitrogen content of 0.12 wt % (Kjeldahl).

Example 3: Following the procedure of Example 1 with the copolymer of Example 2 yielded a functionalized polymer. Precipitation yielded 8.4 g of a polymer with a nitrogen content of 0.065 wt % (Kjeldahl).

Example 4: Following the procedure as in Example 1 with the copolymer of Example 2, the solution was heated at reflux, about 66°C for 4 hours.

Precipitation followed by a second precipitation yielded 9.4 g of a polymeric additive product with a nitrogen content of 0.09 wt % (Kjeldahl).

Example 5: Following the procedure as in Example 4, an oil-oxidized/masticated ethylene-propylene copolymer (containing about 44 wt % ethylene and about 56 wt % propylene) with an $\overline{M}n$ 42,000 (membrane osmometry) was condensed with acrylonitrile. Precipitation yielded 9.82 g of a product with nitrogen content of 0.075 wt % (Kjeldahl).

Example 6: Following the general procedure of Example 1, an oil-oxidized/-masticated copolymer (containing acrylonitrile) was prepared by refluxing in THF (~66°C) for 5.5 hours. Precipitation yielded 9.03 g of product with a nitrogen content of 0.18 wt % (Kjeldahl).

Example 7: 5 g of an ethylene-propylene copolymer (containing 46 wt % ethylene and 54 wt % propylene) which was air-oxidized from a number average molecular weight of about 72,000 to 23,000 (membrane osmometry) was dissolved in 45 g of solvent 150N mineral oil and placed in a reaction vessel on an electric heater so that the temperature of the reactants could be controlled.

5.0 g (0.094 mol) of acrylonitrile was introduced into the reaction vessel after which the reaction vessel was flushed with nitrogen and subjected to a pressure of about 2" of mercury, which elevated pressure was maintained during the reaction period. The reaction was carried out by heating the ingredients with agitation at a temperature ranging from 128° to 140°C for 7 hours. The reaction vessel was cooled to room temperature and the contents subjected to dialysis whereby 4.0 g of product was obtained which contained 2.25 wt % nitrogen (Kjeldahl). The product exhibited strong absorption at 2,270 cm^{-1} by infrared analysis.

Example 8: In this example the efficacy of the polymeric additive products of this process, particularly with regard to their unusual dispersancy properties in lubricating oil applications, is illustrated by comparison with a commercially available multifunctional viscosity improver [Lz 3702 (Lubrizol Corporation)] in a Sludge Inhibition Bench Test (SIB). The SIB has been found, after a large number of evaluations, to be an excellent test for assessing the dispersing power of lubricating oil dispersant additives.

Using this test, the dispersant action of the several functionalized polymers prepared in accordance with this process were compared with the dispersing power of a dialyzed product obtained from dialysis of a commercial dispersant (Lz 3702). Sufficient dialyzed residue which analyzed about 0.4 wt % nitrogen was dissolved in solvent 150N mineral oil to provide a 10% active ingredient concentrate.

The dialyzed residue and polymer products of the process were appropriately diluted in mineral oil to furnish the 0.025, 0.05 and 0.1 wt % of added additive to the used oil.

The test results are given in the table below, the results of which can be summarized as showing the nitrogen-containing functionalized polymers of the process to have comparable or superior dispersancy at 1 and 0.5 wt % additive levels to that shown by a commercially available multifunctional viscosity index improver.

Example	Product of Ex. No.	Concentration g Product/10 g Used Oil	Percent Sludge Dispersed
1	1	0.1	77.9
		0.05	65.2
		0.025	41.4
2	2	0.1	84.8
		0.05	53.6
		0.025	29.0
3	3	0.1	76.8
		0.05	32.6
		0.025	21.5
4	4	0.1	90
		0.05	60.2
5	5	0.1	88.0
		0.05	65.8
6	6	0.1	90.7
		0.05	65.8
7	7	0.1	80
		0.05	64
		0.025	22
8	Dialyzed Lz 3702 commercial dispersant	0.1	89
		0.05	74
		0.025	31

Aminated Oxy-Degraded Copolymers

L.J. Engel and J.B. Gardiner; U.S. Patent 4,113,636; September 12, 1978; assigned to Exxon Research & Engineering Co. describe a process in which copolymers, having a degree of crystallinity of 3 up to 25 wt %, comprising about 68 to 80 mol % ethylene, and one or more C_{3-8} α-olefins are mechanically degraded at elevated temperatures, in the presence of air or oxygen-containing gas, to form an oxygenated-degraded polymer, which is reacted with an amine compound. Preferably, the polymer is only partially reacted with the amine. The resulting aminated polymers are useful as sludge dispersants for fuels and lubricants. When the aminated polymers have a high molecular weight, they are also useful as viscosity index improvers.

Example 1: A typical laboratory synthesis of a high ethylene content copolymer useful in the process is as follows. Ethylene and propylene were continuously polymerized in the presence of n-heptane solvent, with a Ziegler-Natta catalyst consisting of vanadium oxytrichloride ($VOCl_3$) and ethyl aluminum sesquichloride ($Et_3Al_2Cl_3$), while hydrogen was fed continuously to the reactor. The reaction was carried out in a 2 liter glass reactor equipped with a stirrer, catalyst and cocatalyst inlet tubes, monomer inlet tubes, solvent inlet tube, product overflow, and a jacket for circulation of chilled water.

Monomers were purified by contact with hot (150°C) copper oxide and molecular sieves. Solvent was purified and dried by percolation through molecular sieves and silica gel. Hydrogen was similarly dried by passage through a bed of silica gel. The reaction pressure was maintained at one atmosphere by controlling the rate of monomer addition and product removal. Principal catalyst ($VOCl_3$) was fed as a 0.03 molar solution in purified n-heptane, while the cocatalyst ($Et_3Al_2Cl_3$) was fed as a 0.15 molar solution in similarly purified heptane. The reaction conditions are shown in Table 1 under steady state conditions.

Table 1

Volume of reactor to overflow	1,500 ml
Heptane feed rate	75 ml/min
Ethylene feed rate	0.75 l/min
Propylene feed rate	2.25 l/min
$VOCl_3$ catalyst solution feed rate	1.0 ml/min
$Et_3Al_2Cl_3$ cocatalyst solution feed rate	1.0 ml/min
Hydrogen feed rate at $25°C$	30.0 cc/min
Reaction temperature	55 $°C$

A sample of the product was recovered after inactivation of the catalyst with isopropanol, the solvent removed by stripping with steam, catalyst residues extracted from the polymer by slurrying in a Waring blender with dilute hydrochloric acid, and after thorough water-washing, was filtered and dried under vacuum at $50°C$. Analysis of the product is shown in Table 2.

Table 2

Ethylene in product*	68.5 wt %
Ethylene in product	77.0 mol %
I.V. at $135°C$ in Decalin	2.01 dl/g
Thickening efficiency (T.E.)	2.7
Soluble in n-decane at $45°C$	100.0 wt %
$\overline{M}w/\overline{M}n$	2.2
Crystallinity**	11.5 wt %

*Determined by the method of Gardner, Cozewith & VerStrate:
 Rubber Chem. & Tech. 44, 1015 (1971).
**Determined by the method of VerStrate & Wilchinsky:
 J. Polymer Sci. A-2, 9, 127 (1971).

Example 2: A copolymer of ethylene and propylene, prepared on a commercial scale in essentially the same manner as the copolymer of Example 1, was used in this example. In brief, the copolymer was synthesized at $55°C$ and 65 psig with a Ziegler-Natta catalyst system comprising $VOCl_3$ and $Et_3Al_2Cl_3$, and a mol ratio of ethylene to propylene of 1:2.5, respectively, in the feed, which also contained hydrogen. The product had an ethylene content of 68 wt % (76 mol %); a $\overline{M}w$ 245,000; a $\overline{M}n$ of 81,000; a $\overline{M}w/\overline{M}n$ ratio of 3.0; a T.E. of 2.8; and a crystallinity of 7.5 wt %.

A 2½ gallon Bramley Beken Blade Mixer, fitted with a 5 hp Reeves Vari-Speed Moto-Drive geared to provide a speed at the mixer of from about 17 to 85 rpm was used for the following: (a) 6 lb of the above commercial copolymer, which has been granulated from a slab in a Rietz chopper, was fed to the mixer which was set at a speed of 35 rpm at the input shaft, while at the same time 120 psig steam was turned on at the jacket. Temperature, measured with a thermocouple, varied between 300° and 365°F over a period of 5 hours during which time the

T.E. of the masticated copolymer decreased from an initial 2.8 to 1.69, while the oxygen content determined by infrared absorption increased.

(b) 6 lb of the above commercial polymer was masticated in the mixer set at an input speed of 66 rpm and 120 psig steam on the jacket. The temperature varied from 325° to 370°F, while the T.E. of the copolymer during mastication dropped from an initial 2.80 to 1.09.

(c) 6 lb of the above commercial polymer was masticated in the mixer set at an input speed of 85 rpm and 120 psig steam on the jacket. The T.E. dropped from an initial 2.8 to 0.80 over the course of 5 hours, while the oxygen content of the masticated polymer rose to 0.933 wt %. Temperature during the 5 hour mastication period varied between 336° to 392°F.

Example 3: (a) 6 lb of a commercially available copolymer of ethylene and propylene having an ethylene content of about 68 wt % (76 mol %); a $\overline{M}w$ of 433,000; a $\overline{M}n$ of 94,000; a $\overline{M}w/\overline{M}n$ ratio of 4.6; and an initial T.E. of 4.2 was masticated in the Bramley Beken Mixer at an input shaft speed of 35 rpm and a steam pressure on the jacket of 120 psig for 5 hours. During this time, the temperature of the masticated, e.g., oxy-degraded polymer varied between 248° and 350°F. At the end of the run, the T.E. had been reduced to 1.22.

(b) 6 lb of the same polymer as was used in Example 3(a) was masticated in the same equipment at an input shaft speed of 85 rpm and a steam pressure of 120 psig for a period of 5 hours. During that time, the temperature varied between 324° and 394°F. The T.E. was reduced at the end of 2 hours from an initial 4.2 to 1.32 with an oxygen uptake of 0.375 wt %; at the end of 3 hours to a T.E. of 1.10 and oxygen content of 0.590 wt % and at the end of the run (5 hours) to a T.E. of 0.97 and oxygen content of 0.886 wt %.

Example 4: A 50 gal Beken Blade Mixer, fitted with a 30 hp motor geared to provide input shaft speeds to the mixer of 32, 48 and 72 rpm, and a steam jacket supplied with 75 psig steam was used for the following run. The input shaft speeds noted above refer to the rpm of the fast rotor, the slow rotor revolving at one-half the rpm of the fast rotor.

Three runs were made at the above input speeds with a charge to the mixer for each run of 190 lb of the copolymer of Example 2, i.e., copolymer of ethylene and propylene having an ethylene content of about 68 wt % (76 mol %), a T.E. of 2.8 and a crystallinity of 7.5 wt %, in order to determine the time required to reduce the T.E. of the masticated polymer to a T.E. of 1.4. The results obtained are shown in Table 3.

Table 3

Mastication Time	32 rpm	48 rpm	72 rpm
1 hr T.E.	2.7	2.5	2.4
O_2, wt %	0.15	0.17	0.19
Temp, °F	350	370	394
2 hr T.E.	2.2	2.1	1.5
O_2, wt %	0.21	0.20	0.34
Temp, °F	373	410	424
Time (hr) to T.E. of 1.4	3.0	2.8	2.2
O_2, wt %	0.32	0.35	0.36
Temp, °F	402	414	455

At the termination of each of the runs of Examples 2 through 4, a quantity of a solvent extracted paraffinic base neutral, low pour lubricating oil, having a viscosity of about 100 or 150 SUS at 100°F, equal in weight to the weight of copolymer charged to the mixer, was added to the masticated polymer in the mixer. The mixer was then run for about 2 to 5 minutes, with no additional heat input, and then drained to a blending tank. Dilution of the masticated polymer in this manner facilitates drainage from the mixer and subsequent dilution.

Example 5: 100 g of a 7 wt % solution of a masticated copolymer of ethylene and propylene in a solvent extracted midcontinent base oil having a viscosity of 105 SUS at 100°F was reacted with stirring and a gentle sparge of nitrogen with 0.2 g of a technical grade of tetraethylenepentamine for 4 hours at a temperature of 120°C. The masticated copolymer was that of Table 3, and had a T.E. of 1.4, an oxygen content of 0.36 wt % and was prepared by oxy-degradation of a copolymer of ethylene and propylene which had an ethylene content of 67 wt %, a T.E. of 2.8 and a crystallinity of 7.5 wt %. On cooling, a clear product, free of haze, was obtained which passed the Haze Dilution Test.

This Haze Dilution Test is based on the finding that free amine will precipitate zinc C_{4-5} dialkyl dithiophosphate from an oil solution. This test is carried out by making up a mixture of 10 wt % of the reaction mixture of the amine and the oxidized polymer in oil, 1.0 wt % of an oil concentrate containing about 25 wt % of a light mineral oil and dissolved therein, about 75 wt % of a zinc dialkyl dithiophosphate prepared by reacting a mixture of about 65 wt % of isobutyl alcohol and about 35 wt % of mixed primary amyl alcohols with P_2S_5 and then neutralizing with zinc oxide, and about 89 wt % of a white mineral oil. After 4 to 20 hours heating at 180°F, if there is unreacted, i.e., free, amine present, then a visually observable haze will form due to precipitation of the zinc dialkyl dithiophosphate.

Example 6: (a) 8,040 g of an oil solution containing 12.7 wt % of the oxy-degraded copolymer of ethylene and propylene, used in Example 5, in the oil of Example 5, were heated with stirring and nitrogen sparge with 22.5 g of a commercial polyethylene polyamine (Polyamine 400) in a reaction vessel at 140°C for 3½ hours. On cooling, the product was clear and free of haze, passed the Haze Dilution Test noted above, and when compared to Paratone N for thickening efficiency, gave a result of 1.38 for the concentrate.

The above aminated oxy-degraded polymer was tested for sonic breakdown. This is a standard measurement for determining the shear stability of polymer-oil compositions, the lower percentage reflecting which compositions have the greatest resistance to shear breakdown and, hence, which are the most stable under automotive lubricating conditions. In this method, the sample under test is blended with an approved base stock to a viscosity at 210°F of 15.0±0.5 cs.

A portion of the blend is subjected to sonic shearing forces at a specific power input and a constant temperature for 15 minutes. Viscosities are determined on the blend both before and after the treatment; the decrease in viscosity after the treatment is a measure of the molecular breakdown of the polymer under test.

It is customary to examine the blend of a standard sample of known behavior each time a test is made, and to use this as a reference to establish the correct value of the sample under test. The corrected value is reported as a percent sonic breakdown which is calculated from the formula:

$$\frac{\text{viscosity of blend before test } - \text{ viscosity of blend after test}}{\text{viscosity of blend before test } - \text{ viscosity of base oil}} \times 100$$

The aminated polymer gave a sonic shear of 4.82%, which is low, indicating good shear stability.

(b) A 10W-30 SAE crankcase oil was made up using 8.0 wt % of the oil concentrate of 6(a), 3 wt % of an ashless dispersant additive, 1.2 wt % of a magnesium-containing detergent additive and 1.6 wt % of a zinc dialkyl dithiophosphate additive in 86.2 wt % of a mineral lubricating oil. For comparison purposes, the formulation was made up to the same viscosity without the oil concentrate of 6(a) and instead using a commercial VI-dispersant nitrogen-containing acrylate polymer.

The above formulations were tested in the Sequence V-C Engine Test, which is described in "Multicylinder Test Sequences for Evaluating Automotive Engine Oils", *ASTM Special Technical Publication 315F,* p. 133ff (1973). The V-C test evaluates the ability of an oil to keep sludge in suspension and prevent the deposition of varnish deposits on pistons, valves and other engine parts. The results given below clearly show the superior properties of the oil in which the aminated oxy-degraded copolymer of this process has been incorporated over the competitive product.

| | . . . . . MS/V-C Test Results. . . . . | | |
	Sludge	Piston Skirt Varnish	Total Varnish
Oil with aminated copolymer of 6(a)	8.5	8.2	8.1
Competitive oil	7.7	8.1	8.0
Passing criteria for test	8.5	8.0	8.0

In the above test, the ratings are on a scale of 0 to 10, with 0 being an excessive amount of sludge and varnish, while 10 is a completely clean engine.

Aminated Interpolymers

J.B. Gardiner and I. Kuntz; U.S. Patent 4,139,480; February 13, 1979; assigned to Exxon Research & Engineering Co. describe multipurpose lubricating oil additives having utility as viscosity index improvers, antiwear agents, sludge dispersants and pour point depressants. The process involves the reaction products of (1) nitrogen compounds having one or more amino groups and/or (2) oxygen compounds having one or more hydroxyl, epoxide or ether groups and/or (3) sulfur compounds and/or (4) hydrogen with alternating interpolymers of monomers comprising (A) one or more polar monomers, (B) one or more olefinic monomers, and (C) a monomer similar to (A) or (B), but containing, in addition, a reactive group that reacts with (1), (2), (3) or (4).

Example 1: Preparation of an Interpolymer of Dodecyl Acrylate, Isobutylene and VBC (Polymer A) — The polymerization was carried out in a pressure vessel fabricated from a solid cylinder of polypropylene which has been bored so as to create a cylindrical cavity of 800 ml. The vessel was sealed by means of a threaded cap and an oil-resistant "O" ring fashioned from an NBR rubber.

The polymerization vessel, contained in a dry-box from which air and moisture were excluded by means of a positive internal pressure of oxygen- and moisture-free nitrogen, was charged with 300 ml of toluene which had been purified by percolation through a column of Linde 3A molecular sieves, 120 g (0.5 mol) of a commercial grade of dodecyl acrylate containing 15 ppm of 4-methoxy-phenol as an antioxidant and 15.3 g (0.1 mol) of a commercial grade of chloro-methylstyrene having an isomer distribution of about 60% of the 3-isomer and about 40% of the 4-isomer, referred to as VBC (commercially designated as vinyl benzyl chloride).

The pressure vessel was then immersed in a trichlorofluoromethane bath maintained at $-20°C$, located in the dry-box and the vessel and contents cooled to $-15°C$. There was then added to the vessel in succession 7.5 ml of a 1.0 molar solution of ethyl aluminum sesquichloride ($Et_{1.5}AlCl_{1.5}$) in purified n-heptane, 56 g (1.0 mol) of liquefied isobutylene and 1 mmol of lauroyl peroxide dissolved in 20 ml of purified toluene. The reaction vessel was sealed, removed from the bath and allowed to come to room temperature over about 1 hour.

The reaction vessel was then placed in a tumbling water bath maintained at $32°C$ and tumbled for a period of 30 hours. The pressure vessel contents were then transferred to a flask and the reaction terminated by the addition of 15 ml of isopropyl alcohol and 10 ml of methanol.

An aliquot of the solution was removed for analysis and added to boiling water whereupon the interpolymer precipitated as a slurry. The polymer was filtered from the water and dried in a vacuum oven for 4 hours at $60°C$ at a pressure of 20 torrs. The polymer had an inherent viscosity of 0.55 when determined in toluene at a concentration of 0.1 g/dl at $25°C$, a chlorine content of 0.93 wt % and when examined by NMR in CCl_4 solution at 60 MH_z showed the monomer distribution in the polymer to be: 55 mol % dodecyl acrylate, 40 mol % iso-butylene, and 5 mol % vinyl benzyl chloride. The structure of the interpolymer was determined by using the chemical shifts at 7.0 ppm as a measure of aromatic protons, the chemical shift at 4.5 ppm for the $-CH_2Cl$ group and the $-OCH_2-$ signal at 3.95 ppm for the ester.

The remainder of the inactivated toluene-heptane solution of the interpolymer was added to a quantity of a low pour solvent extracted midcontinent neutral oil having a viscosity of 100 SUS at $37.8°C$ to give a 30 wt % solution after the solvents were removed by steam distillation and the oil solution dried by blotter pressing. Reaction of this solution with an amine compound to yield an example of the products of this process is described in Example 4.

Example 2: Polymer B — Using the same equipment and procedure as was used in Example 1, an interpolymer was prepared from dodecyl acrylate, iso-butylene and VBC. The interpolymer had an inherent viscosity of 0.60 when determined in toluene at a concentration of 0.1 g/dl at $25°C$ and a chlorine content of 0.81 wt %. The deashed polymer was dissolved in the same neutral oil as was used in Example 1 to give 25 wt % solution. Reaction of this solution with several amine compounds is described in Examples 5 through 7.

Example 3: Polymer C — Using the same equipment and procedure as was used in Example 1, an interpolymer was prepared from dodecyl acrylate, isobutylene and VBC. The interpolymer had an inherent viscosity of 0.89 when determined in toluene at a concentration of 0.1 g/dl at 25°C and a chlorine content of 0.81 wt %. The deashed polymer was dissolved in the same neutral oil as was used in Example 1 to give 25 wt % solution. Reaction of this solution with an amine compound is described in Example 8.

Example 4: 10 g of the interpolymer oil solution of Example 1 was mixed in a beaker with 0.200 g of N,N-dimethyl-1,3-diaminopropane under an atmosphere of nitrogen at 150°C for 3 hours. The mol ratio of amine compound to the gram atoms of chlorine present in the oil solution was 0.027. A clear, gel-free solution was obtained.

Example 5: 10 g of the oil solution of Example 2 was mixed with 0.205 wt % of a poly(ethylenediamine) available commercially as PA-400 with a molecular weight of 210 and heated with stirring under an atmosphere of nitrogen at 150°C for 3 hours. The mol ratio of amine to available chlorine was 0.17. The resultant aminated interpolymer, when subjected to the Zinc Precipitation Haze Test, had a transmittance of 92% compared to the blank oil.

Zinc Haze Precipitation Test (ZHPT)— The ZHPT is based on the observation that a zinc dialkyl dithiophosphate reagent, when heated with an oil solution containing free amine, will interact with the amine and precipitate, giving rise to a visual haze. The reagent is prepared by reacting a mixture of 65 wt % of isobutyl alcohol and 35 wt % of mixed primary amyl alcohols with P_2S_5 and neutralizing the reaction mixture with zinc oxide. A 75 wt % solution of the zinc dialkyl dithiophosphate is prepared in a light mineral oil as a stock solution.

The test is carried out by mixing 10 parts by weight of the aminated interpolymer oil solution with 1 part by weight of the stock solution and 89 parts by weight of a highly refined, acid-treated USP white mineral oil and heating the mixture for 3 to 24 hours at 82°C. During the heating period the percentage of the incident light transmitted through the sample is measured by means of a photocell nephelometer. Values of more than 35% are considered hazy and do not pass.

Example 6: 10 g of the oil solution of Example 2 was mixed with 0.22 wt % of N,N-dimethyl-1,3-diaminopropane and heated with stirring under an atmosphere of nitrogen at 150°C for 3 hours. The mol ratio of amine to available chlorine was 0.03. The resultant aminated interpolymer, when subjected to the ZHPT, was perfectly clear and free of any haze.

Example 7: 10 g of the oil solution of Example 2 was mixed with 0.102 wt % of PAM-400 and heated with stirring under an atmosphere of nitrogen at 150°C for 3 hours. The mol ratio of amine to available chlorine was 0.102. The resultant aminated interpolymer, when subjected to the ZHPT, showed a transmittance of 67% at the end of 20 hours.

Example 8: 10 g of the oil solution of Example 3 was mixed with 0.21 wt % of N,N-dimethyl-1,3-diaminopropane and heated with stirring under an atmosphere of nitrogen at 150°C for 3 hours. The mol ratio of amine to available chlorine was 1.54. The resultant aminated interpolymer, when subjected to the ZHPT gave a perfectly clear, haze-free solution.

Example 9: The aminated product of Example 8 was tested in a low-ash formulation for varnish inhibition by the following comparative procedure. Two 1 g samples of lubricating oil (one sample containing about 0.2 g of the product of Example 8 and the other only the oil) were each admixed with 9 g of a commercial lubricating oil obtained from a taxi after 2,000 miles of driving with the lubricating oil. Each 10 g sample was heat-soaked overnight at about 140°C and thereafter centrifuged to remove the sludge.

The supernatant fluid of each sample was subjected to heat cycling from about 115°C to room temperature over a period of 3½ hours at a frequency of about 2 cycles per minute. During the heating phase a gas containing a mixture of about 0.7 vol % SO_2, 1.4 vol % NO and the balance air was bubbled through the test samples and during the cooling phase water vapor was bubbled through the test samples. At the end of the test period, which testing cycle can be repeated as necessary to determine the inhibiting effect of any additive, the wall surfaces of the test flasks in which the samples were contained are visually evaluated as to the varnish inhibition.

The results showed a significant reduction in varnish deposition in the wall of the flask containing the product of Example 8 relative to the flask containing the same oil without the aminated interpolymer.

Polyamine-Hydroperoxidized Ethylene-Propylene Copolymer Reaction Products

J.O. Waldbillig and C.M. Cusano; U.S. Patent 4,132,661; January 2, 1979; assigned to Texaco Inc. describe additives which consist of the reaction product of a polyamine containing both tertiary and primary amino groups and a hydroperoxidized ethylene-propylene copolymer or a terpolymer (termonomer consisting of an unsaturated hydrocarbon such as a diene, cycloalkene or bicycloalkene) having a molecular weight in the range of 5,000 to 500,000. Examples of suitable types of polyamines include di(lower)alkylamino(lower)alkylamines such as dimethylaminopropylamine, diethylaminopropylamine or dimethylaminoethylamine and those in which the tertiary amino group exists in a ring system such as 2-aminopyridine, 2-piperidinoethylamine, 5-aminoquinoline, N-(3-aminopropyl)morpholine or 2-aminopyrimidine.

The additives of this process have a 210°F thickening power in the range of 1.0 to 100 SUS/1.0 wt % of the polymer in lubricating oil.

The additives are extremely shear stable. They show a characteristic infrared spectrum with absorbance peaks at frequencies in the range of 1,550 to 1,750 cm⁻¹.

The additives are prepared by dissolving the ethylene-propylene polymer in an inert solvent at a temperature of around 70°C using agitation. A free radical initiator such as azobisisobutyronitrile is added and air is bubbled through the reaction medium for 2 to 48 hours. A solvent such as the lubricating oil whose properties are to be improved is then added and the inert solvent is removed by vacuum distillation. Next the amine is introduced into the reaction mass and the mixture is heated for 0.5 to 20 hours at 80 to 250°C under an atmosphere of nitrogen at a pressure of 0 to 1,000 psig.

The reaction mass again is distilled under vacuum at 80° to 300°C to remove excess amine and catalyst residue. The additive can be precipitated by boiling in isopropyl alcohol and filtered off.

The dispersant additives were tested for their effectiveness in mineral lubricating oil compositions in the dispersancy and sequence V-C tests. The dispersancy test is conducted by heating the test oil, mixed with a synthetic hydrocarbon blowby and a diluent oil at a fixed temperature for a fixed time period. After heating, the turbidity of the resultant mixture is measured. A low percent turbidity (0 to 10) is indicative of good dispersancy while high results (20 to 100) indicate increasingly poor dispersancy.

The sequence V-C test is detailed in the ASTM Special Technical Publication, 310-F. The test is used to evaluate crankcase motor oils with respect to sludge and varnish deposits as well as their ability to keep the positive crankcase ventilation (PVC) valve clean and functioning properly. Ratings of 0 to 10 are given, 10 representing absolutely clean and 0 representing heavy sludge and varnish deposits.

The rust inhibiting properties of the lubricants were determined in the SE required standard MS-IIC Rust Test. This test was developed and is effective for evaluating crankcase oils with respect to low temperature rusting.

Components used to formulate lubricating compositions containing the additives of the process are identified as follows:

A   Dinonyldiphenylamine (mostly 4,4' substituted),

B   18:1 overbased calcium sulfonate in oil BB,

C   Ethylene-propylene copolymer of 20,000 to 50,000 molecular weight (13.0 wt % in oil CC),

D   Polyester type methacrylate copolymer,

E   Reaction product of sulfurized polybutene and tetraethylenepentamine,

F   50% reaction of polybutenyl succinimide (molecular weight 1,300) in oil BB,

G   10 to 20% hydroperoxidized dimethylaminopropylene aminated ethylene-propylene copolymer in 90 to 80% of oil CC,

H   Zinc dialkyldithiophosphate.

Typical base oils used in the practice of the process had the following inspection values:

Oil No.	Viscosity (SUS) 100°F	Viscosity (SUS) 210°F	Specific Gravity	Pour Point (°F)	S (%)
AA	127	41.5	0.8644	−5	0.16
BB	98	38.8	0.8844	+10	0.12
CC	99	39.1	0.8639	0	0.25
DD	332	53.2	0.8822	−5	0.40
EE	846	78.1	0.8927	+15	0.34
FF	333	53.3	0.8838	+10	0.29

The process is further illustrated by the following example.

Example: The materials used in this example are as follows: 200 g amorphous copolymer of ethylene and propylene (EPSyn 5006), 8.0 g azobisisobutyronitrile (AIBN), 2,000 ml benzene, 4.0 g dimethylaminopropylamine (DMAPA), and 1,800 g oil AA.

The procedure used in this example is as follows. The ethylene-propylene copolymer used [EPSyn 5006 (Copolymer Rubber and Chemical Corporation)] has a mol % ethylene content of the polymer of between 60 and 63 and its molecular weight, as expressed by a Mooney viscosity, is nominally 50±5 ML 1 + 8 at 250°F with a moisture content of less than 0.5%.

The EPSyn 5006 was dissolved in the benzene at 70°C with stirring. AIBN was added at 70°C and air was bubbled through the rapidly stirred solution at 400 ml/min for 18 hours. After adding oil AA, the solution was stripped to 158°C (0.13 mm). The DMAPA was charged and the solution was heated 10 hours at 160°C. Stripping the product to 146°C (0.07 mm) yielded 2,007 g of product (Z).

The sample was filtered through Super-Cel. The polymer, isolated by precipitation in boiling isopropyl alcohol, analyzed 0.10% N.

Table 1: Test Data in Oil AA

Concentration of Z, wt % in oil	5.0	10.0	15.0
Dispersancy test	15.0	9.5	5.5
210°F thickening power, SUS	–	8.8	–

The composition made above was tested in an engine test under similar conditions as a hydroperoxidized ethylene-propylene copolymer aminated with tetraethylene-pentamine and described in U.S. Patent 3,785,980 as Y and as a dispersant, X, consisting of a 41% tetrapolymer of butyl, dodecyl, octadecyl and N,N-dimethyl-aminoalkyl methacrylates in a molar ratio of 21:50:25:4.

The data is given below in Table 2.

Table 2

	Blends		
	1	2	3
Components, wt %			
Z	9.0	–	–
Y	–	9.0	–
X	–	–	9.0
Oil AA	74.1	74.1	79.10
Oil DD	10.0	10.0	10.0
Zinc dialkyldithiophosphate	0.65	0.65	0.65
Overbased calcium sulfonate	1.00	1.00	1.00
Ethyl mono- and dinonyl-diphenylamine	0.25	0.25	0.25
Copolymer ethylene/propylene (20,000–50,000 MW, 13%; diluent oil CC, 87%)	5.0	5.0	–
Methyl silicone fluid, ppm	150	150	–
Engine Rust Rating IIC Test	7.3	5.4	5.4

The explanatory data given below in Table 3 show that the lubricating composition (1) containing the additive of the process is superior to one containing a typical additive prepared by the method of U.S. Patent 3,785,980 (2) or a composition containing a methacrylate dispersant-VI improver (3).

Table 3 gives the results of dispersancy VC engine tests which have been conducted with the dimethylaminopropylamine aminated ethylene-propylene polymer (G).

As shown in the table, G just barely failed (average varnish 7.9 vs 8.0 for SE oil) the VC engine test with 2.40 wt % component E present as a supplementary dispersant (5). In a formulation where 1.40 wt % of component F was added as a supplementary dispersant, a passing SE quality oil was obtained (6).

Table 3

	. . . . Blends			
	4 Certified (SwRI)	5	6	7	8
		. . . . .Screener			
Composition, wt %					
Oil					
AA	75.81	73.37	80.39	79.19	EE:FF*
H	1.36	1.38	1.36	1.36	1.36
A	0.25	0.25	0.25	0.25	0.25
B	1.48	1.50	1.50	1.50	1.51
C	6.00	6.00	–	6.20	–
D	0.10	0.10	0.10	0.10	0.05
E	–	2.40	–	–	–
F	–	–	1.40	4.70	7.01
G**	1.50	1.50	2.25	1.00	0.30
Dispersancy tests	2.5	2.5***	3.5	4.0	3.5
Reference oils†					
FREO 126	2.5	2.5	2.5	2.5	2.5
FREO 127	21.5	20.0	14.5	12.5	23.5
FREO 179	69.0	64.0	48.5	31.5	64.0
VC Test					
Average sludge	9.5	9.7	9.7	9.7	9.5
Average varnish	8.2	7.9	8.0	7.1	7.5
Piston skirt varnish	7.0	7.9	8.0	7.2	8.1
Oil ring clogging	0	0	0	0	0
Oil screen clogging	0	0	0	0	0

*22.82:65.00.
**Concentration expressed on a "neat" polymer basis.
***The dispersancy test was determined with a formulation containing 1.10% of component E.
†The FREO oils are Ford Motor oils used for referencing the Seq. VC Engine Test.

Ethylene, Propene and Cyclic Imide Terpolymers

G. Marie, A. Lang and G. Chapelet; U.S. Patent 4,139,417; February 13, 1979; assigned to Societe Nationale Elf Aquitaine, France describe amorphous copolymers of monoolefins or of monoolefins and nonconjugated dienes with unsaturated derivatives of imides. These copolymers containing from 99.9 to 80% by weight of nonpolar units derived from at least two monoolefins containing 2 to 18 carbon atoms, particularly ethylene and propene or ethylene and butene-1, and possibly one or more nonconjugated dienes, and from 0.1 to 20% units derived from an amide having the formula:

$$
\begin{array}{c}
\diagup \!\! C{=}O \\[-2pt]
A \qquad\qquad N{-}Z \\[-2pt]
\diagdown \!\! C{=}O
\end{array}
$$

wherein Z is an alkenyl radical containing 2 to 16 carbon atoms, and A designates a saturated or unsaturated divalent hydrocarbon radical which contains 2 to 12 carbon atoms and which may possibly carry amino, halogen or carboxyl groups.

The copolymers may be used as polymer additives in lubricating compounds to improve their viscosity index and to disperse the slurry which they may contain.

Examples of unsaturated cyclic imides which may produce the polar units of the copolymers are those N-alkenylated derivatives of the imides which are selected from the group comprising the succinimide, glutarimide, maleimide, citraconic imide, phthalimide, the imide of the himic anhydride, the hexahydrophthalimide and the imides of the tricarboxylic butane acid and of the esters of the acid, and particularly the N-vinylsuccinimide, N-allylsuccinimide, N-butenylsuccinimide, N-vinylglutarimide, N-vinylmaleimide, N-vinylphthalimide, N-vinylcitraconimide, N-vinylhexahydrophthalimide, the N-vinyl imide of the himic anhydride and the N-vinyl imide of the tricarboxylic butane acid or of the esters of that acid.

Ethylene, Propene and N-Vinylimidazole Terpolymers

G. Chapelet, H. Knoche and G. Marie; U.S. Patent 4,092,255; May 30, 1978; assigned to Entreprise de Recherches et d'Activites Petrolieres (E.R.A.P.), France describe polymeric additives which are olefinic copolymers containing by weight, x% units derived from ethylene; y% units derived from a monoolefin having 3 to 6 carbon atoms or from a monoolefin having 3 to 6 carbon atoms and a nonconjugated diene, the proportion of units resulting from the diene being equal to or lower than 20%; and z% units derived from one or more nitrogenous monomers selected from the group comprising the vinylimidazoles, the vinylimidazolines and their derivatives.

Examples of nitrogenous monomers are N-vinylimidazole (vinyl-1-imidazole), N-vinyl methyl-2-imidazole, N-vinyl ethyl-2-imidazole, N-vinyl phenyl-2-imidazole, N-vinyl dimethyl-2,4-imidazole, N-vinylbenzimidazole, N-vinyl methyl-2-benzimidazole, N-vinylimidazoline (vinyl-1-imidazoline), N-vinyl methyl-2-imidazoline, N-vinyl phenyl-2-imidazoline and vinyl-2-imidazoline.

Preferred polymer additives are constituted by the terpolymers of ethylene and propene or butene-1 with one of the abovementioned nitrogenous monomers, mainly N-vinylimidazole, N-vinylimidazoline, N-vinylpyrrolidone, N-vinylcarbazole, N-allylphenothiazine and allylthiobenzothiazole. The proportions (by weight) x, y and z of the units derived respectively from ethylene, propene or butene-1, and of the nitrogenous monomer, are comprised within the defined limits such that $20 \leqslant x \leqslant 75$, $20 \leqslant y \leqslant 75$ and $0.15 \leqslant z \leqslant 15$ with $(x + y + z) = 100$.

The statistic copolymers may be prepared by means of a coordination catalysis method wherein the ethylene, the olefin having 3 to 6 carbon atoms or the olefin having 3 to 6 carbon atoms and the nonconjugated diene, and the nitrogenous monomer or monomers are contacted, in a suitable solvent, with a catalyst of the Ziegler-Natta type, while, if required, the nitrogenous monomer is complexed by a Lewis acid.

The statistic copolymer may be prepared by coordination catalysis with complexation of the nitrogenous monomer described in Luxemburg Patent 69,835, where the monomer is of the vinylimadazole or vinylimidazoline type and in Luxemburg Patent 69,836, where the monomer is derived from a lactam or thiolactam.

Example: Following an operating procedure similar to that described in Example 1 of the Luxemburg Patent 69,835, a certain number of amorphous statistic terpolymers of ethylene, propene and N-vinylimidazole were prepared by coordination catalysis (tests 1 through 10).

The physico-chemical properties of these terpolymers, which are adapted to be used as polymer additives in the lubricating compositions according to the process, are given in Table 1, hereinafter as compared to those of a reference sample, namely a copolymer of ethylene and propene (test 11) commercially available as an additive to improve the viscosity index of lubricating oils.

The polydispersity of the terpolymers is lower than that of the copolymer, which is an advantageous feature as far as their stability when submitted to mechanical shearing effects (or shearing stability) is concerned.

Table 1

Test No.	N-Vinyl Imidazole Content (wt %)	Ethylene Content (wt %)	Reduced Viscosity (Decalin at 135°C)	Poly- dispersity
1	0.8	65.2	0.81	2.2
2	1.71	52.7	0.96	2.7
3	1.30	58.7	0.98	—
4	0.47	64.1	1.09	2.6
5	0.32	61.8	1.11	—
6	0.65	58.6	1.11	—
7	0.64	69.9	1.24	—
8	0.55	54.9	1.36	—
9	0.29	57.5	1.40	—
10	0.49	57.7	1.59	—
11	0	54.4	1.10	3

To study the effect of the abovementioned terpolymers on viscosity index of the lubricating oils, the viscosity index determination according to the ASTM-D 2270 Standard was made for lubricating compositions prepared by adding variable amounts of the terpolymers or of the reference copolymer to a reference oil, commercially designated "200 Neutral", which is neutral paraffinic oil extracted by means of a solvent, having a viscosity of 44.1 cs at 37.8°C, and of 6.3 cs at 98.9°C, and a viscosity index of 100.

The results obtained have been listed in Table 2. It should be recalled that the viscosity index of an oil is a number characterizing, on a conventional scale, the viscosity variation of that oil as a function of temperature; the slighter the variation, the higher the index.

Table 2

Test No.	200 Neutral Oil +					
	. .0.9% by wt Terpolymer . .			. .1.2% by wt Terpolymer . .		
	. Viscosity, cs at. .			. Viscosity, cs at. .		
	37.8°C	98.9°C	V.I.	37.8°C	98.9°C	V.I.
1	74.88	9.74	120	–	–	–
2	79.13	10.33	124	–	–	–
3	78.31	10.21	123	–	–	–
4	90.14	11.51	127	–	–	–
5	88.4	11.7	134	–	–	–
6	75.8	10.2	128	88.94	11.94	137
7	77.94	10.53	131	90.45	12.17	139
8	93.27	12.01	131	–	–	–
9	84.0	10.89	126	104.83	13.55	138
10	83.33	11.36	137	102.74	13.41	139
11	–	–	–	–	–	–

Test No.	200 Neutral Oil +		
	. . . . .1.5% by wt Terpolymer		
	. . . . Viscosity, cs at.		
	37.8°C	98.9°C	V.I.
1	86.91	11.82	139
2	–	–	–
3	96.56	12.67	137
4	–	–	–
5	–	–	–
6	108.04	14.01	140
7	–	–	–
8	–	–	–
9	139.19	17.27	145
10	–	–	–
11	100.3	12.99	136.5

By incorporating the terpolymers according to the process to the reference oil, the viscosity index of that oil is substantially improved.

In addition, when comparing the results obtained with the terpolymer No. 6 and the reference copolymer No. 11, which have the same reduced viscosity, it can be seen that the terpolymer has a greater effect on the viscosity index than the reference copolymer.

Another advantage of these terpolymers resides in the fact that they have only a very slight thickening effect at low temperatures. Thus, the viscosity of the lubricating composition containing 1.5% of terpolymer No. 6, as determined according to the ASTM-D 2602 Standard, was only 26.5 poises.

Copolymer, Amine Compound and Oxygen Reaction Products

According to a process described by *L.J. Engel and J.B. Gardiner; U.S. Patent 4,068,056; January 10, 1978; assigned to Exxon Research and Engineering Company* aminated polymers, resulting from the reaction of a hydrocarbon polymer with an oxygen-containing gas and an amine compound at elevated temperatures of from 130° to 300°C provide a multifunctional additive for lubricants and hydrocarbon fuels. This reaction can be carried out if desired in an oil solution.

The resulting aminated polymers are useful as sludge dispersants for fuel lubricants. When the aminated polymers have a high molecular weight, they are also useful as viscosity-index improvers with dispersant and/or pour point depressant activity. The following examples illustrate the process.

Example 1: 6 lb of a copolymer of ethylene and propylene containing about 55 wt % ethylene and having a thickening efficiency (TE) of 2.86 were put into a masticator which was 2½ gal Bramley Beken Blade Mixer, fitted with a 5 hp Reeves vari-speed motodrive geared to provide a speed at the mixer of from 13 to 150 rpm. The masticator was heated with a steam jacket from 345° to 368°F and 8.5 g (0.305 wt %) of triethylenetetramine was added. The fast blade rotated at 52 rpm while the slow blade rotated at 26 rpm for 4.5 hours under ambient air atmosphere.

The final material showed a thickening efficiency of 1.46. The polymer was shown by analysis to have 0.071 wt % nitrogen incorporated into it. The product was blended into solvent 100 Neutral, low pour at 12.5 wt %, to form a concentrate.

Examples 2 through 5: Additional aminated polymeric additives according to the process were prepared in accordance with the procedure and apparatus of Example 1. The results are set forth in the following table.

| | .Reaction Conditions. | | | | . . . Product . . . | |
| Ex. | Time | Temperature | | | | Nitrogen |
No.	(hr)	(°C)	. . .Amine Compound (wt %) . . .		TE	(wt %)
2	4.4	171–191	diethylenetriamine	0.305	1.35	0.05
3	4.7	171–188	PA-400*	0.268	1.36	0.12
4	5.0	188	PA-400*	0.154	1.43	0.05
5	4.8	174–191	tetraethylenepentamine	0.382	1.43	0.09

*Polyamine 400 (PA-400).

The utility of the aminated polymeric additives of the process is demonstrated by the following formulations subjected to engine test. A 10W-30SAE crankcase oil was made up using 9.1 wt % of the oil concentrate of Example 1, 4 vol % of an ashless dispersant additive, 1 vol % of a magnesium containing detergent additive, 0.9 vol % of a zinc dialkyldithiophosphate additive, 0.3 vol % of an overbased phenated dispersant and 0.5 vol % of an antioxidant, in a mineral lubricating oil.

For comparison purposes, a formulation was made up to the same viscosity replacing the oil concentrate of Example 1 with a comparable concentrate of non-nitrogen containing ethylene propylene copolymer sold commercially as a viscosity index improver.

The above formulations were tested in the Sequence V-C engine test, which is described in *Multicylinder Test Sequences for Evaluating Automotive Engine Oils*, ASTM Special Technical Publication 315F, page 133ff (1973). The V-C test evaluates the ability of an oil to keep sludge in suspension and prevent the deposition of varnish deposits on pistons, valves, and other engine parts. The test results given below clearly show the superior properties of the oil in which the aminated oxy-degraded copolymer has been incorporated.

| | | . . . . . . MS-VC Test Results. . . . . . | |
	Sludge	Piston Skirt Varnish	Total Varnish
Oil with aminated copolymer of Ex. 1	8.3	8.1	8.0
Oil with commercial additive	6.4	7.0	7.3
Passing criteria for test	8.5	8.0	8.0

In the above test the ratings are on a scale of 0 to 10, with 0 being an excessive amount of sludge and varnish, while 10 being a completely clean engine. The amination/oxidation can be conducted at pressures ranging from 0.1 to 20 atm.

For the aminated polymeric additives of higher molecular weight, $\overline{M}n$, of from 10,000 to 500,000 (useful for pour point depressant and/or V.I. improving-dispersant applications) the nitrogen content ranges from 0.001 to 5, preferably 0.01 to 0.5 wt %; for additives of lower molecular weights $\overline{M}n$ of less than 10,000 (useful in dispersant or pour point depressant applications) the nitrogen content ranges broadly from 0.001 to 25, preferably from 0.01 to 8 wt %.

In a related process described by *L.J. Engel and J.B. Gardiner; U.S. Patent 4,068,058; January 10, 1978; assigned to Exxon Research and Engineering Company* hydrocarbon polymers, optimally, copolymers having a degree of crystallinity of up to 25 wt %, comprising 2 to 98 mol % ethylene, and one or more C_3 to C_{28} alpha-olefins are mechanically grafted at elevated temperatures, under an inert atmosphere, to form an aminated polymer which is thereafter mechanically-oxidatively degraded at an elevated temperature in the presence of air or oxygen-containing gas to form an aminated polymeric reaction product.

The resulting aminated polymers are useful as sludge dispersants for fuels and lubricants. When the aminated polymers have a high molecular weight, they are also useful as viscosity-index improvers with dispersant and/or pour point depressant activity.

L.J. Engel and J.B. Gardiner; U.S. Patent 4,068,057; January 10, 1978; assigned to Exxon Research and Engineering Company also describe a process in which hydrocarbon polymers, preferably polymers having a degree of crystallinity of less than 25 wt % and comprising 2 to 98 wt % ethylene, and one or more C_3 to C_{28} alpha-olefins, are mechanically grafted under an inert atmosphere and at elevated temperatures, in the presence of an amine compound, to form an amino-grafted polymer.

The resulting aminated polymers are useful as sludge dispersants for fuels and lubricants. When the aminated polymers have a higher molecular weight, they are also useful as viscosity-index improvers with dispersant and/or pour point depressant activity.

Nitrogen-Containing Copolymers from Lithiated Copolymers

According to a process described by *W.J. Trepka; U.S. Patent 4,145,298; March 20, 1979; assigned to Phillips Petroleum Company* nitrogen-containing copolymers are prepared by the reaction of lithiated hydrogenated conjugated dienemonovinylarene copolymers with nitrogen-containing organic compounds. The nitrogen-containing copolymers are oil additives which combine the aspects of viscosity index improvers with dispersant properties. The following example illustrates the process.

Example: This example illustrates the use of 4-dimethylamino-3-methyl-2-butanone in preparing a V.I. improver having dispersant properties. A 41/59 hydrogenated butadiene-styrene block copolymer having a block styrene content of 20 wt % and having properties as shown in Table 1, was metalated and reacted with 4-dimethylamino-3-methyl-2-butanone according to the following recipe and conditions.

Step 1

Hydrogenated butadiene-styrene copolymer, pbw	100
Cyclohexane, pbw	1520
n-Butyllithium mhp*	12
Tetramethylethylenediamine, mhp	12
Temperature, °C	70
Time, hours	1.5

Step 2

4-Dimethylamino-3-methyl-2-butanone, mhp	24
Temperature, °C	22
Time, minutes	1

*mhp = millimols per 100 grams of polymer.

Metalation and grafting were carried out employing essentially anhydrous reagents and conditions under an inert nitrogen atmosphere. The hydrogenated butadiene-styrene copolymer was dissolved in cyclohexane and the resulting polymer-cement given a 5 min nitrogen purge to assure absence of dissolved oxygen. After addition of n-butyllithium and tetramethylethylenediamine, the solution was tumbled in a constant temperature bath at 70°C for 1.5 hours. After cooling to room temperature, 4-dimethylamino-3-methyl-2-butanone was added, and the mixture then shaken vigorously whereupon the red-orange color of the mixture discharged immediately.

The resulting mixture was coagulated in isopropyl alcohol, filtered, and purified by three successive dissolutions in cyclohexane and coagulations in isopropyl alcohol. The modified polymer was dried for about 15 hours at 60°C under reduced pressure.

Properties of the modified polymer and the starting, unmodified polymer are given in Table 1. Molecular weights were determined from gel permeation chromatography curves by a procedure described by G. Kraus and C.J. Stacy, "Symposium 43," *J. Poly. Sci.*, 329-343 (1973). Heterogeneity index is the quotient of the weight average molecular weight ($\overline{M}w$), divided by the number average molecular weight ($\overline{M}n$). Inherent viscosity was determined according to the procedure given in U.S. Patent 3,278,508, column 20, Note (a) with the modification that the solution was not filtered through a sulfur absorption tube but rather a sample of the solution was filtered through a fritted glass filter stick of grade C porosity and pressured directly into the viscometer. Nitrogen was determined using the Dohrmann Micro Coulometric Method.

Table 1

	4-Dimethylamino-3-Methyl-2-Butanone-Modified Polymer	Unmodified Polymer
$\overline{Mn}$	43,000	58,000
Heterogeneity index	1.32	1.25
Inherent viscosity	0.80	0.80
Nitrogen, wt %	0.087	None

Viscosity index, pour point, and sonic shear, as shown in Table 2, were determined on oil solutions of polymer blended to 74±2 SUS viscosity at 210°F (99°C) using a premium motor oil as shown in the following formulation.

Mid-Continent SAE 10 stock	66.76
Mid-Continent SAE 20 stock	19.89
Phil-Ad 100 (overbased calcium petroleum sulfonate)	8.15
Lubrizol 934 (alkyl succinic ester)	4.07
Lubrizol 1395 (zinc dialkyldithiophosphate)	0.73
Paraflow 46 (pour depressant-flow improver)	0.40

For the sonic shear test, the SUS viscosity at 210°F (99°C) of a 2.0 wt % solution of polymer in a base oil is determined before and after irradiation for 6.5 min at 100°F (38°C) jacket temperature in a Ratheon Model DF-101 sonic oscillator operated at 10 kc/sec. The SUS viscosity is determined according to ASTM D455-74.

Table 2

	4-Dimethylamino-3-Methyl-2-Butanone-Modified Polymer	Unmodified Polymer
Concentration, wt %[*]	2.2	2.0
Viscosity index[**]	159	151
Pour point, °F[***]	−40	−30
Sonic shear, SUS viscosity loss	0.6	0.2

[*]As required to give 74±2 SUS viscosity at 210°F in motor oil formulation.
[**]ASTM D2270-75.
[***]ASTM D97-66.

These data illustrate the modification of the copolymer does not appreciably alter the viscosity index, the pour point, or the sonic shear of oil formulations containing the copolymers.

Data presented illustrate that the 4-dimethylamino-3-methyl-2-butanone-modified copolymer provides dispersancy equivalent or better than that provided by the commercial dispersants.

Grafted Copolymers

J.-P. Durand, P. Gateau, F. Dawans and B. Chauvel; U.S. Patent 4,085,055; April 18, 1978; assigned to Institut Francais du Petrole and Rhone Poulenc Industries, France have developed graft copolymers which are obtained by reacting, in the presence of a compound generating free radicals, a hydrogenated polymer or copolymer of conjugated diolefin with at least one polymerizable vinyl compound, either as monomer or in a prepolymerized form.

More particularly, the polymeric products used as substrates for grafting consist of hydrogenated homopolymers of conjugated dienes having from 4 to 6 carbon atoms, hydrogenated copolymers of at least two conjugated dienes having from 4 to 6 carbon atoms, or hydrogenated copolymers of conjugated dienes having from 4 to 6 carbon atoms with styrene, which have an average molecular weight by weight of from 10,000 to 300,000, preferably 30,000 to 100,000, and which have been hydrogenated up to a residual olefinic unsaturation lower than 15% and, preferably, from 0 to 5% of the units contained therein.

Examples of grafting substrates are homopolymers of 1,3-butadiene, isoprene, 1,3-pentadiene, 1,3-dimethylbutadiene, copolymers formed with at least two of these conjugated dienes and copolymers of the latter with styrene, these homopolymers and copolymers having been hydrogenated up to the abovementioned residual unsaturation degree.

The process will be further illustrated by the following examples. Examples 1 through 6, and 8 and 9 form no part of the process and are given only for comparison purposes.

Example 1: To a solution of 50 g of 1,3-butadiene in 500 ml of n-heptane, there is added 0.17 g of tetrahydrofuran and 0.8 mmol of normal-butyl-lithium. The reaction mixture is stirred at 30°C for 4 hours. There is thus obtained a solution containing 48 g of polybutadiene. The microstructure of the polymer, determined by IR spectrophotometry consists of 66% of 1,2 units and 34% of 1,4 units; the average molecular weight by weight is 90,000 and the ratio of the average molecular weights by weight and in number $\overline{M}w/\overline{M}n$ is 1.35. These features are reported in Table 1 below.

40 g of this polymer is dissolved in benzene and the resulting solution is added to 360 g of a lubricating oil of the 200 Neutral type. Benzene is then evaporated under reduced pressure, up to a constant weight.

Example 2: Example 1 is repeated. At the end of the polymerization, there is added a suspension resulting from the reaction of 11.6 mg of cobalt, in the form of cobalt octoate, with 67 mg of triethyl aluminum. The reaction vessel is stirred at 90°C for 2 hours under a hydrogen pressure close to 25 bars. The polymer is then separated by precipitation of the reaction solution in an excess of isopropyl alcohol and dried under reduced pressure up to constant weight. There is thus obtained 49 g of hydrogenated polybutadiene having a residual unsaturation content lower than 3 mol %.

40 g of this polymer is dissolved in benzene and the resulting solution is added to 360 g of a lubricating oil of the 200 Neutral type. Benzene is then evaporated, under reduced pressure, up to constant weight.

The 10% by wt solution of polymer in oil thus obtained is then added to a 200 Neutral oil in a sufficient proportion to raise the viscosity at 98.9°C of oil to 15 cs (20% by wt of solution has been added, i.e., 2% by wt of polymeric additive). The resulting composition exhibits properties which are given by way of comparison in Table 2 and show a good behavior to shearing and oxidation, but no dispersing power.

Example 3: 1.27 g of tetrahydrofuran and 6 mmol of n-butyllithium are added to a solution of 50 g of 1,3-butadiene in 500 ml of n-heptane. The reaction mixture is stirred at 30°C for 3 hours. There is thus obtained a solution containing 48 g of a polybutadiene whose microstructure, determined by IR spectrophotometry, consists of 65% of 1,2 units and 35% of 1,4 units. The average molecular weight by weight is 11,000 and the ratio of the average molecular weights by weight and in number $\overline{Mw}/\overline{Mn}$ is 1.35 (Table 1). The polymer is then hydrogenated and dissolved in a proportion of 10% by wt into a lubricating oil of the 200 Neutral type, as described in Example 2.

Examples 4 through 6: Butadiene-styrene, butadiene-isoprene and isoprene-styrene copolymers have been prepared according to the anionic method in the presence of butyllithium, and then hydrogenated in a manner similar to that described in Example 2. The characteristics of these copolymers are reported in Table 1.

Table 1

Additive of Example	. . . . . . Composition, mol %			$\overline{Mw}$	$\overline{Mw}/\overline{Mn}$
	Butadiene	Isoprene	Styrene		
2	100	—	—	90.000	1.35
3	100	—	—	11.000	1.35
4	50	—	50	80.000	1.30
5	—	50	50	85.000	1.25
6	65	35	—	90.000	1.30

The residual olefinic unsaturation of each of these hydrogenated copolymers is lower than 3%. These different copolymers are added to a 200 Neutral lubricating oil as in Example 2. The characteristics of the resulting compositions, reported in Table 2, show a good resistance to shearing, but the absence of dispersing power.

Example 7: To 90 g of the solution of the polymer of Example 2 in a 200 Neutral oil, there is added 0.34 g of N-vinyl imidazole, 0.66 g of methyl methacrylate and 0.2 g of tert-butyl perbenzoate. The mixture is sitrred at 130°C; there is added two times 0.04 g of tert-butyl perbenzoate (first after 1 hour and then after 3 hours of reaction), the total reaction time being 8 hours.

Certain characteristics of the 200 Neutral oil containing 20% by wt of the resulting solution are summarized in Table 2. It can be observed that the dispersing power is satisfactory, as well at 200°C as after cooling to 20°C, and that the efficiency is improved, without, however, resulting in a significant decrease of the shearing strength and of the resistance to oxidation.

Example 8: When Example 7 is repeated, except that the solution of polymer of Example 2 is replaced by that of Example 1, there is obtained a lubricating composition whose characteristics, summarized in Table 2, show, in addition to a decrease of efficiency, a very poor resistance to oxidation due to the unsaturation of the nonhydrogenated polybutadiene. This defect excludes such a composition from the scope of the process.

Example 9: When Example 7 is repeated, except that the solution of polymer of Example 2 is replaced by that of Example 3, there is obtained a composition which, on the one hand, requires the use of a too large proportion of polymeric additive (more than 10% by wt with respect to the lubricating oil) in order to obtain the desired viscosity (15 cs) at 98.9°C, and on the other hand, has a too low viscosity index (V.I. = 120) for including this formulation in the scope of the process.

Example 10: When Example 7 is repeated except that N-vinyl imidazole is replaced by N-vinyl pyrrolidone, there is obtained a lubricating composition whose characteristics, summarized in Table 2, are close to those of Example 7.

Table 2

Ex. No.	Conc. Additive* (% by wt)	V.I.**	Effi- ciency***	Partial Shear- ing† (%)	. . . Stability to Oxidation. . . . Induction Period (min)	Absorbed O₂ After 7 hr (mol/l)	Viscosity Variation (%)	Dispersing . .Power . . A	B
1	0	100	–	–	260	0.9	+35	–	–
2	2	140	0.70	6	300	0.3	−6	0.40	0.33
4	2	136	0.74	5	290	0.4	−8	0.39	0.34
5	2	134	0.72	5.5	310	0.3	−7	0.40	0.36
6	2	142	0.68	8	310	0.3	−6	0.41	0.35
7	2	151	0.96	7	250	0.4	−10	0.70	0.70
8	2	140	0.70	9	35	1††	−39††	0.68	0.68
9	>10	120	0.60	5	200	0.5	−12	0.74	0.73
10	2	145	0.94	8	350	0.2	−5	0.70	0.70

*In the lubricating oil to obtain the viscosity of 15 cs at 98.9°C.
**Calculated according to ASTM D-2270 standard.
***Ratio of specific viscosities determined by capillarity at 100°C and 0°C.
†Viscosity drop due to ORBAHN shearing of the polymer, after 30 cycles, determined by DIN 51 382 standard.
††Test discontinued after 2 hr.

Stability to oxidation is the measurement, in relation to time, of the absorbed oxygen at 150°C and of the viscosity of an oil containing the polymeric additive and 0.4% of 4,4-methylenebis(2,6-di-tert-butylphenol) in the presence of 24 ppm Cu and 24 ppm Pb in the form of naphthenates.

Dispersing power is defined as the ratio of diameters after 24 hours (diameter of the carbon black spot/diameter of the oil spot) of the two coaxial spots formed when depositing a drop of oil containing the additive and carbon black on a sheet of filter paper, according to the spot method described by Gates, V.A., et al. in *SAE Preprint 572* (1955) or by A. Schilling in "Les huiles pour moteurs et le graissage des moteurs" *Eds Technip*, Tome I, p. 89 (1962). The results (A) are determined at 200°C and (B) after cooling to 20°C.

Stabilized Imide Graft of Ethylene Copolymers

A process described by *J.B. Gardiner, J. Zielinski, R.L. Elliott and S.J. Brois; U.S. Patent 4,137,185; January 30, 1979; assigned to Exxon Research & En-*

gineering Co., relates to the viscosity stable solutions of substantially saturated polymers comprising ethylene and one or more C_3 to C_{28} alpha-olefins, preferably propylene, which have been grafted in the presence of a free radical initiator with an ethylenically-unsaturated dicarboxylic acid material preferably at an elevated temperature and in an inert atmosphere.

These intermediates are then reacted first with a polyamine, preferably an alkylene polyamine, having at least two primary amino groups, such as diethylene triamine, and then with an anhydride of an organic acid, to form multifunctional polymeric reaction products characterized by viscosity stabilizing activity in mineral oil solutions.

Thus, it has been discovered that the reaction of the imidated products/by-products of the graft reaction with organic acid anhydrides, e.g., acetic anhydride, has been found to provide amide derivatization of any primary amino groups of the imidated ethylene copolymer whereby viscosity stabilizing activity is provided to the copolymers.

It is believed that the viscosity stabilization involves conversion of the primary amino groups of the imide reaction product to an amide preventing chain extension of the derivatized ethylene copolymeric multifunction V.I. improver.

The reaction appears to be an acylation of pendant primary amine groups by their reactions with the organic acid anhydride which can be represented as follows:

alkylene aminoimide of a grafted ethylene copolymer

hydrocarbyl amide of an alkylene imide grafted ethylene copolymer

This acylation of the free primary amino group with the anhydride produces an amide structure which limits the multifunctionalized copolymer's property of solution chain extension thereby inhibiting viscosity increase of oil solutions containing the class of additives of the process.

To enhance the freedom from haze of the mineral oil solutions, the mineral oil compositions can be further reacted with an oil-soluble hydrocarbyl substituted acid having from 10 to 70 carbon atoms having a pK of less than 2.5, preferably a polymethylene substituent benzene sulfonic acid, said polymethylene substituent having from 18 to 40, optimally 24 to 32 carbons, in an amount of from 0.01 wt % to 8 wt % at a temperature within the range of 150°C to 200°C for a period from 0.1 to 20 hours, e.g., for 1 hour at 190°C. This further step results in an additive oil composition of improved viscosity stability which has no visually perceptible haze.

The following examples illustrate the process. In these examples, all parts are by wt unless specifically indicated otherwise and all nitrogen was determined by Kjeldahl analysis.

Example 1: Preparation of Imide Graft Ethylene Copolymer — 5314 kg of a 20.2 wt % solution of an ethylene-propylene copolymer concentrate (made by the Zeigler-Natta process using H_2 moderated $VOCl_3$/aluminum sesquichloride catalyst having a crystallinity less than 25%; containing about 45 wt % ethylene and 55 wt % propylene; and having a thickening efficiency (TE) of 1.4 ($\overline{M}n$ = 27,000) in S130N (Solvent 130 Neutral Mineral Oil) was heated to 250°F under N_2 sparge and stirring, taking 1 hour and 5 min. Under N_2 blanket 31 kg of maleic anhydride were added over 10 min. The solution was heated to 310°F, taking 2 hours and 15 min.

Lupersol 130, 6 kg, (2,5-dimethyl hex-3-yne-2,5-bis-tert-butyl peroxide) was added in three equal changes over a 2 hour and 50 min period. Excess maleic anhydride was stripped out with N_2 over 2 hours and 20 min. 20 kg of diethylene triamine (DETA) was charged and allowed to react for 1.5 hrs. Excess DETA was stripped with vacuum and N_2 for 6 hours. The resulting material was diluted with S130N to 14 wt % polymer, cooled and drummed. The final material had 0.262 wt % DETA incorporated.

Example 2: Preparation of Acylamidate of Maleimide Graft of Ethylene Copolymer — 2528 g (0.065 mol of DETA) of the product of Example 1 was heated to 120°C under a nitrogen sparge. To this heated solution, 16.9 g (0.669 wt %, 0.166 mol, an excess) of acetic anhydride was slowly added over a period of 30 min with stirring. The mixture was allowed to soak at a temperature of 120°C under the nitrogen blanket for 1.5 hours after which the reaction by-products including the acetic anhydride were sparged off for 2 hours at a temperature of 120°C with nitrogen.

The resulting product showed under differential IR a substantial absorption peak at 1650 cm^{-1} and a lack of acetic acid, since there is substantially no absorption at 1720 cm^{-1}. The resulting copolymer solution had a color of 5 with a haze reading of 108, nephelos unchanged from the starting material, as measured on a nephelometer (Model 9, Kohlmann Industries).

This material had a viscosity of 1543 cs at 210°F, active ingredient of 15.42 wt % by dialysis, N wt % of 0.12% (0.49 wt % N on polymer), flash point of 420°F and TE of 1.43. On blends with a test oil of 6.2 cs, 9.5 wt % gave a 12.8 cs, 210°F viscosity, 13% sonic shear breakdown, a pour point (with 0.4 wt % of a vinyl acetate fumerate pour depressant) of less than −35°F and a 0°F viscosity of 25.3 poises in a Cold Cranking Simulator (ASTM method).

Example 3: Acetic Acid Salt of Maleimide of Graft Ethylene Copolymer — 3000 g (0.077 mol of DETA) of the product of Example 1 were charged to a flask and heated to 118°C with N_2 sparge. 10 g (0.167 mol, an excess) of glacial acetic acid were injected. The resulting admixture was reacted at 118°C with stirring and under a nitrogen blanket and maintained at 118°C for 1 hour. The solution was then heated to 155°C and maintained there for 2 hours with nitrogen blanket. The mixture was then sparged at 155°C for 2 hours.

The differential IR showed presence of acetic acid during reaction, which was almost completely lost after sparging. Storage stability tests showed no improvement in stability over the starting material. It is concluded that there is no significant amidation occurring when the acid itself is used rather than the anhydride and literature information indicate that excessive heat and pressure (if the acid is volatile) is necessary to convert the acid salt to the amide.

Example 4: The utility of the additives were measured by subjecting the products of Examples 1 and 2 to a standard engine test of blended formulations containing these additives. A 15W50 SAE crankcase oil formulation was made up using 12.5 wt % of the soil concentrate of Example 2, 2 vol % of an ashless dispersant additive, 1.1 vol % of an overbased magnesium sulfonate, 0.8 vol % of overbased calcium phenate, 0.5 vol % of an antioxidant, and 1.43 vol % of a zinc dialkyldithiophosphate and a mineral lubricating oil blend of base stocks. For comparison purposes, a formulation was made up in the same manner replacing the oil concentrate of Example 2 with the same wt % of the oil concentrate of Example 1.

The above formulations were tested in the Sequence V-C Engine Test, which is described in "Multicylinder Test Sequences for Evaluating Automotive Engine Oils," *ASTM Special Technical Publication 315F*, page 133 ff (1973). The V-C tests evaluate the ability of an oil to keep sludge in suspension and prevent the deposition of varnish deposits on pistons, valves, and other engine parts. The test results are given below and show that the two blends are not statistically different in performance.

| | | . . . . MS-VC Test Results . . . . | |
	Sludge	Piston Skirt Varnish	Total Varnish
Oil with product of Ex. 2	9.28	7.86	8.27
Oil with product of Ex. 1	9.31	7.98	7.99
Passing criteria for test	8.5	7.9	8.0

In the above tests, the ratings are on a scale of 0 to 10, with 0 being an excessive amount of sludge and varnish while 10 being a completely clean engine.

Example 5: In order to show the surprising viscosity stability provided to the maleimide amide graft products of ethylene copolymeric V.I. improvers, the resulting products of Examples 1, 2 and 4 were subjected to a test whereby the change in viscosity of the products was measured over a period of 2 hours. while maintaining the solutions at 99°C. The results are shown as follows.

| Test No. | Product of Example | . . . . Viscosity, cs . . . . . | | Viscosity Increase (%/hr) |
		Initial at 99°C	After 2 hr at 99°C	
1	2	1,543	1,547	0.13
2	1	1,224	1,251	1.10
3	4	1,273	1,368	1.07

The above results show that there is a surprising enhancement of the viscosity stabilization activity of solutions of polymer in oil when the products of the process are used. This 2 hour test has been found to correlate with the long-term storage stability results when the solutions containing polymer are stored at temperatures of 180°F for periods of up to 2 months.

Polyolefin Copolymers with Maleic Anhydride Grafting Reactions

R.L. Stambaugh and R.A. Galluccio; U.S. Patents 4,160,739; July 10, 1979; and 4,161,452; July 17, 1979; both assigned to Rohm and Haas Company describe graft copolymers which combine the efficient thickening properties of polyolefinic viscosity index improvers and the dispersancy provided by nitrogen-containing materials.

The process involves the grafting of a monomer system comprising maleic acid or anhydride and at least one other (different) monomer which is addition-copolymerizable therewith, the grafted monomer system then being postreacted with a polyamine. The copolymerizable monomers are selected for their reactivity with maleic acid or anhydride so that more maleic acid or anhydride may be incorporated into the polymer than would occur in the absence of the comonomers.

In another aspect of the process, it has been found that the graft copolymers are efficiently produced with little or no wasteful by-product by forming an intimate mixture of backbone polymer, copolymerizable monomer system, and a free radical initiator, wherein the temperature of the mixture, at least during the time that the initiator is being uniformly dispersed therein, is maintained below the decomposition temperature of the initiator. Thereafter, the temperature is increased to or above the temperature at which the initiator decomposes, preferably while continuing agitation of the reaction mixture, to form a graft copolymer, followed by postreaction with a polyamine.

The detergency test data are based on the following test procedures and all parts and percentages are by weight, unless otherwise stated.

A method for determining the dispersing activity of any given polymer is based on the capacity of the polymer to disperse asphaltenes in a typical mineral oil. The asphaltenes are obtained by oxidizing a naphthenic oil with air under the influence of a trace of iron salt as catalyst, such as ferric naphthenate. The oxidation is desirably accomplished at 175°C for approximately 72 hours by passing a stream of air through a naphthenic oil to form a sludge which may be separated by centrifuging. The sludge is freed from oil (extracting it with pentane). It is then taken up with chloroform and the resulting solution is adjusted to a solids content of about 2% (weight by volume).

When a polymer is to be examined for its dispersing activity, it is dissolved in a standard oil, such as a solvent-extracted 100 neutral oil. Blends may be prepared to contain percentages varying from 2% to 0.01% or even lower of polymer in oil.

A 10 ml sample of a blend is treated with 2 ml of the standard solution of asphaltenes in chloroform. The sample and reagent are thoroughly mixed in a test tube and the tube is placed in a forced draft oven at either 90° or 150°C for 2 hours to drive off volatile material. The tube is then allowed to cool and the appearance of the sample is noted.

If the polymer has dispersing activity, the oil will appear clear although colored. Experience has demonstrated that, unless a polymer exhibits dispersing activity, at concentrations below 2% in the above test, it will fail to improve the cleanliness of engine parts in actual engine tests.

Example 1: (A) Preparation of Graft Copolymer — A 350 g sample of a 50/50 ethylene/propylene copolymer (Epcar 506) is charged to a 5 liter flask containing 1050 g o-dichlorobenzene. After sparging with nitrogen, the contents of the flask are heated to approximately 150°C and mixed using a metal C type stirrer to effect a homogeneous solution. During this step and throughout the entire reaction cycle a nitrogen blanket is maintained. Within 2 hours a homogeneous solution is obtained.

The polymer solution is then cooled to 80°C at which temperature 35.0 g methyl methacrylate (MMA) and 17.5 g maleic anhydride (MAH) are added. These monomers are thoroughly blended into solution over 35 minutes. A mixture of 1.75 g tert-butylperbenzoate initiator in 25 g o-dichlorobenzene is then added. The temperature is then increased to bring the reaction mixture to 140°C in about 45 min. At 120°C the grafting reaction begins, as indicated by an increase in solution viscosity. The solution temperature is maintained at 140°C for 1 hour to complete the graft reaction step before diluting with 1500 g 100 neutral viscosity oil.

The nitrogen flow is then discontinued and vacuum stripping apparatus consisting of a condenser and receiver are attached. The solution is stripped at 1.0 mm Hg and 140°C. Total stripping time is about 3 hours. Infrared assay of isolated polymer confirms the presence of both MMA and MAH in the final polymer product. Titration of the polymer solution indicates that 89.1% of the MAH charged is grafted. Assay of the distillate recovered after stripping indicates that 88.5% of the MMA charged is grafted. The grafted olefin polymer contains 8.9% MMA and 4.5% MAH.

(B) Preparation of Postreacted Product — 3500 g of an 8.4% polymer solution prepared in Part A is charged to a 5 liter flask equipped with thermometer, stirrer and nitrogen inlet. With nitrogen flowing, heat is applied. Upon reaching 140°C, 21.6 g N-(3-aminopropyl)morpholine (NAPM) is charged to the polymer solution. Within 15 min the solution viscosity increases significantly. The solution viscosity returns to approximately its original level within about 2 hours of the NAPM addition, presumably as the amic acid is converted to imide. The solution is further heated with stirring for 16 hours before the nitrogen inlet is replaced by a condenser and collection vessel to facilitate vacuum stripping of unreacted NAPM. After approximately 1 hour at 150°C and 2.0 mm Hg, all unreacted NAPM is removed.

After stripping, the solution is homogenized at 16,000 psi in a Kobe Size 3, 30,000 psi Triplex Pump to reduce the polymer molecular weight. Analytical assay of the final product gives a Kjeldahl nitrogen content of 0.76%, indicating a 6.1% NAPM content, expressed as the maleimide. Infrared assay of isolated polymer confirms imide formation with no apparent residual anhydride bands.

Utilizing the standard asphaltenes dispersancy test described above, 0.0625 gm of polymeric product disperses 0.4% asphaltenes at 150°C.

An SAE 10W40 oil is prepared by blending 12.4 wt % of the concentrate of the final homogenized imide product, 1.5 wt % of a commercial zinc dialkyl-dithiophosphate antiwear additive and 2.0% of an overbased magnesium sulfonate, in a Mid-Continent solvent refined lubricating oil base stock. This oil thus contains 1.04% of the product copolymer. When evaluated in the Sequence V-C engine test, a merit rating of 9.0 sludge, 8.3 varnish and 7.2 piston skirt varnish is achieved after 200 test hours.

Example 2: The procedures of Example 1 are repeated in all essential respects except for use of N,N-dimethylaminopropylamine (N,N-DMAPA, 22.5 g) in place of the N-(3-aminopropyl)morpholine and the polymer product of Part A is used as a 9.8% polymer solution. Analytical assay of the final product gives a Kjeldahl nitrogen content of 0.94, indicating a 6.1% N,N-DMAPA content, expressed as maleimide. Infrared assay of isolated polymer indicates imide formation with no residual anhydride bands.

Using the standard asphaltenes dispersancy test described above, 0.0625 g of polymeric product disperses 0.4% asphaltenes at 150°C.

An SAE 10W40 oil is prepared by blending 14.6 wt % of the final homogenized product (1.43% polymer in the final blend), 1.5 wt % of a commercial zinc di-alkyldithiophosphate antiwear additive and 2.0% of an overbased calcium sulfonate, in a Mid-Continent, solvent-refined lubricating oil base stock. When evaluated in the Sequence V-C engine test, a merit rating of 9.6 sludge, 8.4 varnish and 7.9 piston skirt varnish is achieved after 192 hours.

Star-Shaped Isoprene-Divinylbenzene Polymers

According to a process described by *T.E. Kiovsky; U.S. Patent 4,077,893; March 7, 1978; assigned to Shell Oil Company* ashless, oil-soluble additives having both dispersant and viscosity index (V.I.) improving properties are prepared by the process comprising:

  (a) reacting a selectively hydrogenated star-shaped polymer with an alpha-beta unsaturated carboxylic acid, anhydride or ester at a temperature of between 150° and 300°C, for between 1 hour and 20 hours. The star-shaped polymer comprises a poly(polyalkenyl coupling agent) nucleus, and at least four polymeric arms linked to the nucleus. The polymeric arms are selected from the group consisting of:

    hydrogenated homopolymers and hydrogenated co-polymers of conjugated dienes;
    hydrogenated copolymers of conjugated dienes and monoalkenyl arenes; and
    mixtures thereof;

  and wherein at least 80% of the aliphatic unsaturation of the star-shaped polymer has been reduced by hydrogenation while less than 20% of the aromatic unsaturation has been reduced; and

  (b) the product of step (a) is reacted with an alkane polyol having at least two hydroxy groups at a temperature of between 150° and 250°C.

The dispersant V.I. improvers possess excellent viscosity improving properties, oxidative stability, mechanical shear stability, and dispersancy. In particular, the lubricating oils containing the V.I. improver/dispersants possess excellent thickening efficiency at high temperature while also possessing very good low temperature viscosity characteristics.

Preparation of the Base Polymer: The base polymer employed in making the dispersant V.I. improver is a star polymer. These polymers are generally produced by the process comprising the following reaction steps:

(a) polymerizing one or more conjugated dienes and, optionally, one or more monoalkenyl arene compounds, in solution, in the presence of an ionic initiator to form a living polymer;

(b) reacting the living polymer with a polyalkenyl coupling agent to form a star-shaped polymer; and

(c) hydrogenating the star-shaped polymer to form a hydrogenated star-shaped polymer.

The living polymers produced in reaction step (a) of the process are the precursors of the hydrogenated polymer chains which extend outwardly from the poly(polyalkenyl coupling agent) nucleus.

Example: A hydrogenated star-shaped polymer made from isoprene and a divinylbenzene coupling agent was reacted with maleic anhydride and pentaerythritol to form a dispersant/V.I. improver according to the process.

The star-shaped polymer was prepared by first polymerizing isoprene in a cyclohexane solvent with a secondary butyllithium initiator. The polymer branch A-Li had a molecular weight of 45,700. The living polymer was then coupled with commercial divinylbenzene (55 wt %, Dow Chemical) in a molar ratio of divinylbenzene to lithium of 3:1. The coupled polymer had a total molecular weight of 577,000 on a polystyrene equivalent basis.

Then the polymer was hydrogenated with an aluminum triethyl/nickel octoate catalyst. The final molecular weight was 609,000, the coupling yield was 96%, the saturation index was 10%, and residual unsaturation by ozone titration was 0.11 meq/g.

Then 10 g of the hydrogenated star-shaped polymer prepared above was dissolved in 190 g 1,2,4-trichlorobenzene. To the polymer solution was added maleic anhydride (0.80 g, 7.0 mmol) and the mixture heated to 205°C for 4 hours. Excess maleic anhydride was removed by distillation under vacuum. A small part of the solvent also came over. A nitrogen atmosphere was maintained up until vacuum was applied.

Pentaerythritol (0.95 g, 7.0 mmol) was added to the maleated polymer solution and the mixture heated to 205°C for 4 hours under nitrogen. Unreacted pentaerythritol was removed by distillation under vacuum.

An amount of the lube base stock was added which was equal in volume to the reaction mixture and the trichlorobenzene solvent distilled under vacuum. The oil solution was then diluted with heptane, filtered, washed with methanol and stripped of volatiles.

Dispersant power of the oil-soluble product was readily apparent from the formation of a stable emulsion during the washing step. In addition, the dispersancy of the product was assessed by a Spot Dispersancy Test. In this test, 1 part of a 2% wt solution of the additive to be tested in 100 N oil is mixed with 2 parts used, sludge-containing oil and heated overnight at 150°C. Blotter spots are then made on filter paper and the ratio of sludge spot diameter to oil spot diameter is measured after 24 hours. A poor value is under 50%. The additive prepared above yielded a value of 56%. Unmodified star polymer gave a value of 27%.

A 2% by wt concentration of the aboveprepared additive in a common mineral lubricating oil base stock increased the 210°F KV from 4 cs for the lube stock alone to 15 cs for the lube oil plus additive. This viscosity increase demonstrates the usefulness of the additive as a V.I. improver.

Multistep Polymerization for Polymer-in-Oil Solution

F. Wenzel, U. Schoedel, H. Jost and H. Pilz; U.S. Patent 4,149,984; April 17, 1979; assigned to Rohm GmbH, Germany describe a method for making a polymer-in-oil solution, useful for improving the viscosity-temperature relationship and low-temperature properties of lubricating oils.

The method comprises a first step of polymerizing a methacrylic acid ester of an alcohol having 8 to 18 carbon atoms in a solution, in a lubricating oil, of a polyolefin polymer of an olefinic hydrocarbon monomer having 2 to 4 carbon atoms, the oil solution of the polyolefin having a viscosity of less than 15,000 cs at 100°C, and then, in a second step, adding further polyolefin polymer of the type defined herein until the total polymer content in the oil solution is from 20 to 55% by wt of the solution and the methacrylate ester comprises from 50 to 80% of the polymer content.

The process permits the preparation of products with outstanding dispersing and detergent effect, in that the already-mentioned polymerizable heterocyclic compounds can be incorporated into the polymers in question. This can take place by a common polymerization, with the methacrylate, of these compounds, particularly vinylpyridine, vinylpyrrolidine, vinylpyrrolidone, and/or vinylimidazol. Nevertheless, it is particularly advantageous to graft polymerize the heterocyclic monomers in a second polymerization step after extensive polymerization of the methacrylate or after conclusion of the methacrylate polymerization. The following examples illustrate the process.

Example 1: A mixture comprising 90% of long-chain methacrylic acid esters (C_{10-18}-alcohols) and 10% of methyl methacrylate is polymerized in a known manner using a free-radical polymerization catalyst at a temperature of 90°C in a 10% by wt solution in 150-neutral oil mineral oil, of a degraded polyolefin copolymer having an average molecular weight of 90,000 which copolymer contains 72% of ethylene and 28% of propylene. The polyolefin-oil solution has a viscosity of 820 cs at 210°F. The ester mixture is polymerized in such an amount that the total polymer content at the conclusion of the polymerization is 40%. The viscosity of this product at 210°F is 530 cs.

At 120°C, a degraded polyolefin copolymer having an average molecular weight of 90,000 and comprising 54% of ethylene and 46% of propylene is dissolved in the product (a) at a temperature of 120°C in an amount such that the total polymer content reaches 45%.

If, in step (a), a degraded polyolefin copolymer is used which also contains 72% of ethylene and 28% of propylene, but which has an average molecular weight of 120,000 and a 10% solution of which in mineral oil has a viscosity of 3,200 cs, a product is obtained at the end of the polymerization whose viscosity at 210°F is 1,200 cs. This is further treated as in (b) above at a temperature of 150°C.

Example 2: (a) A mixture comprising 85% of long-chain methacrylic acid esters (C_{10-18}-alcohols) and 10% of butyl methacrylate, and 5% of butyl acrylate is polymerized in a known manner using a free-radical polymerization catalyst at a temperature of 90°C in a 10% solution, in 150-neutral oil mineral oil, of a degraded polyolefin copolymer having an average molecular weight of 90,000, which copolymer contains 72% of ethylene and 28% of propylene and which has a viscosity, in this solution, of 820 cs at 210°F. The mixture is polymerized in such an amount that the total polymer content is 40% at the end of polymerization. The viscosity of this product is 420 cs at 210°F.

(b) 5% of vinylpyrrolidone, calculated on the polymer content, was polymerized in the polymer product prepared in this fashion, said polymerization occurring at a temperature of 130°C in the presence of a free-radical polymerization catalyst. Essentially, the polymerization can be carried out according to German Patent 1,520,696.

At a temperature of 140°C, a degraded polyolefin copolymer having an average molecular weight of 90,000 and which contains 54% of ethylene and 46% of propylene is dissolved in the product so prepared such that a total polymer content of 47% is reached.

A mixture of 2% of vinylpyrrolidone with 2% of vinylimidazole can also be employed instead of 5% of vinylpyrrolidone.

MULTIFUNCTIONAL VISCOSITY INDEX IMPROVERS

Sulfone Copolymers

H.N. Miller; U.S. Patent 4,070,295; January 24, 1978; assigned to Exxon Research and Engineering Company describes sulfone copolymeric oil additives having utility as a lubricating oil pour point depressant, viscosity index improver, dispersant, load carrying agent, rust inhibitor and/or antioxidant, or as a cold flow improver for heavy distillates and residual fuels.

These products comprise the hydrocarbon soluble copolymers of an ethylenically unsaturated polar monomer and sulfur dioxide. This sulfone copolymer may contain one or more additional monomers including: C_{2-50}, preferably at least C_6, substantially linear alpha olefins; C_{2-12}, preferably C_{2-9}, cyclic olefins; and C_{4-6} conjugated diolefins.

The process can be more fully understood by reference to the following examples. Examples 1 through 4 and 6 through 8, all containing a polar monomer, represent the process. Example 5, without the polar monomer, represents prior art.

Example 1: Sulfone Copolymer of SO$_2$ and Allyl Alcohol — 29 g (0.5 mol) of allyl alcohol was dissolved in 100 ml of ethyl acetate contained in a 1 liter reaction flask. The solution was saturated with sulfur dioxide and maintained with a sulfur dioxide atmosphere during polymerization under a pressure slightly in excess of atmospheric. 0.5 g of tert-butyl hydroperoxide dissolved in 40 ml of ethyl acetate was periodically added in two 10 ml and four 5 ml aliquot portions to the reactive solution over a time period of about 2 hours.

The reaction vessel was maintained in a water bath whereby the temperature of the reaction was maintained between 5° and 19°C. Approximately 31 g (0.5 mol) of SO$_2$ was consumed in the reaction to produce a sulfone copolymer which precipitated from the solution. After the polymerization reaction was completed, the system was sparged with nitrogen and the polymer precipitated in normal hexane. The yield was 58.8 g of polymer (approximately 96.4% yield) having a grayish white appearance.

Example 2: Sulfone Copolymer of SO$_2$ and Allyl Acetate, Tetradecene-1 and Hexadecene-1 — The general procedure of Example 1 was used; however, the amount of monomers, introduction and nature of catalyst solution, and temperature were changed as follows.

19.6 g (0.1 mol) of tetradecene-1 (>90% purity), 22.4 g (0.1 mol) of hexadecene-1 (>90% purity), and 2.1 g (0.021 mol) of allyl acetate was dissolved in 92 ml of benzene. 11.7 g (0.18 mol) of SO$_2$ were consumed during polymerization. 0.9 g of tert-butyl hydroperoxide (dissolved in 80 ml of benzene) was periodically introduced in 3 aliquot portions of 10 ml and 2 aliquot portions of 20 ml over the 85 min period of the reaction; and the temperature was maintained between 4° and 9°C.

The sulfone copolymer was precipitated in methyl alcohol and reprecipitated in a mixture of toluene and methyl alcohol and finally vacuum dried. The yield was 48.3 g (82.9% of theoretical). The sulfone copolymer had a $\overline{M}n$ of 7,918.

Example 3: Sulfone Copolymer of SO$_2$, Allyl Alcohol and Hexene-1 — The general procedure of Example 2 was followed with variations in the monomers and process noted hereafter.

42 g (0.5 mol) hexene-1 (>90% purity) and 1.1 g (0.02 mol) of allyl alcohol were dissolved in 50 ml of benzene. 23.5 g (0.37 mol) of SO$_2$ were consumed. 0.5 g of tert-butyl hydroperoxide was dissolved in 50 ml of benzene, and the reaction was carried on for 88 min and maintained at a temperature between 5° and 10°C with the catalyst solution being added in 10 ml aliquot portions at intervals of approximately 15 to 20 min. The resulting sulfone copolymer, precipitated in methyl alcohol, provided, after drying, 54 g (70.7% of theoretical) of an off-white to light amber colored, amorphous product.

Example 4: Sulfone Copolymer of SO$_2$, Allyl Acetate and C$_{16-32}$ Alpha Olefins Mixture — The general procedure of Example 2 was followed with variations in the monomers and process noted hereafter.

23.4 g (0.08 mol) of a mixture of C_{16-32} alpha olefins (90.3 wt % were C_{18-28} alpha olefins), and 0.4 g (0.004 mol) of allyl acetate were dissolved in 125 ml of cyclohexane. An excess of 19 g SO_2 was present during the polymerization. 0.5 g of 5-butyl hydroperoxide was dissolved in 50 ml of cyclohexane. The polymerization was conducted for 30 min during which 10 ml aliquot portions of the solution of the free-radical catalyst was added at approximately 5 min intervals and the temperature maintained at 10°C. A nitrogen sparge was carried out thereafter and the resulting polymer subsequently precipitated in methyl alcohol and vacuum desiccated. The polymer yield was 19.8 g.

Example 5 (for Comparison): Sulfone Copolymer of SO_2, Dodecene-1 and Octadecene-1 — The procedure of Example 2 was substantially followed but with variations in the monomers and process noted hereafter.

18 g (0.106 mol) of dodecene-1 (~97% pure) and 36.2 g (0.14 mol) of octadecene-1 (~97% pure) were dissolved in 54.2 g of benzene. 45 g (0.7 mol) of SO_2 was present during the polymerization. 0.5 g of tert-butyl hydroperoxide dissolved in 50 ml of benzene was the catalyst solution. The reaction was carried on for approximately 1 hour at 10° to 21°C. The copolymer was precipitated in methyl alcohol and yielded 61.6 g of sulfone copolymer having a $\overline{M}n$ of 10,673.

Example 6: Sulfone Copolymer of SO_2, Allyl Alcohol, Dodecene-1 and Octadecene-1 — The procedure of Example 5 was followed but with variations in the monomers and temperature as follows.

17.1 g (0.10 mol) of dodecene-1 (~99% pure), 34.3 g (0.133 mol) of octadecene-1 (~99% pure) and 1.25 g (0.022 mol) of allyl alcohol were dissolved in 50 ml of benzene. 16.3 g (0.26 mol) of SO_2 were consumed in the polymerization, and the temperature was maintained at from 6° to 20°C. The yield was approximately 100% of said sulfone copolymer.

Example 7: Sulfone Copolymer of SO_2, Allyl Acetate, Dodecene-1 and Octadecene-1 — The procedure of Example 6 was followed with a change in the monomers as follows.

18.0 g (0.107 mol) of dodecene-1 (~99% pure), 36.2 g (0.143 mol) of octadecene-1 (~99% pure) and 2.5 g (0.025 mol) of allyl acetate were dissolved in 58 ml of benzene. 15.8 g (0.25 mol) of SO_2 were consumed in the polymerization. The yield was 66.9 g (90% of theoretical) of a copolymer having a $\overline{M}n$ of 10,329.

Example 8: Sulfone Copolymer of SO_2 and the Allyl Esters of Dodecanoic and Octadecanoic Acids — The C_{12} aliphatic and C_{18} aliphatic esters of allyl alcohol were prepared as follows.

A 20% molar excess of allyl alcohol was reacted in separate reactions with the respective aliphatic acid in cyclohexane. The esterifications were each catalyzed by paratoluene sulfonic acid. The temperature of each esterification was between 70° and 80°C and was so maintained for a period of 3 to 4 hours during which the water of esterification was distilled off by maintaining a reduced pressure over each reaction.

Each resultant product solution was neutralized with sodium bicarbonate, water washed 3 times, after which 25 ml of cyclohexane was added and the system left standing overnight in the presence of magnesium to produce the respective product esters. Thereafter, each system was rotovacuated to recover the respective allyl ester.

The copolymer was produced by the process of Example 3 with variations in the monomers and process as follows.

23.6 g (0.10 mol) of dodecanoate ester of allyl alcohol and 22.2 g (0.068 mol) of octadecanoate ester of allyl alcohol were dissolved in 50 ml of benzene. 3.5 g (0.06 mol) of SO_2 were consumed during polymerization, and polymerization was conducted for 50 min. The yield was 18.6 g (32.9% of theoretical) of a sulfone copolymer having a $\overline{Mn}$ of 1,129.

The utility of the additives of the process is demonstrated, in part, by data hereafter presented which information was derived from tests indicative of lubricating oil pour point depressant activity, residual fuel flow activity, and extreme pressure lubricity. Illustrative and comparative lubricating oil pour depressant activity is shown in various solvent neutral oils.

Table 1: Pour Points

. .Additive, wt % . .		. . . . .Pour Points, $^\circ$C			
		SN 75	SN 150	SN 330	SN 450
	0	−18	−15	−9	−9
Example 5	0.1	−34	−34	−26	−26
Example 7	0.1	−29	−32	−29	−32
Example 8	0.1	−23	−23	−26	−26

Illustrative residual fuel flow activity is shown in Table 2 wherein the results of adding 0.15 wt % of the additive to a residual fuel known as 343°C FVT (final vapor temperature) Brega North African Residuum are set forth.

Table 2: Flow Points

	Flow Points at Reheat Temperature, $^\circ$C			
	38	46	54	66
Residuum neat	41	41	41	41
Residuum + copolymer of Example 4	21	21	21	13

The additive property of extreme pressure lubricity provided by the sulfone copolymers according to this process is illustrated in the data of Table 3. This data was obtained by testing lubricants modified by the addition of sulfone copolymers in a Falex lubricant testing machine. This machine provides for rotation of a steel pin (lubricated by the test lubricant) in a chuck provided by 2 cooperating aluminum members pressing against a portion of said pin.

The test is discontinued at the moment when the pin breaks. The test conditions were 2 min at 250 rpm, followed by 500 rpm until breakage occurs. The test oil was mineral oil with the additive added in an amount of 0.5 wt %, based on the weight of the oil.

Table 3: Falex Testing

Additive	Minutes to break at 500 rpm
−	1.5
Ex. 2	13
Ex. 5*	0.5
Ex. 6	14

*Comparison example.

This data illustrates the enhanced extreme pressure lubricity of lubricating oils treated according to this process relative to nontreated or treated with a copolymer of SO_2 and alpha olefins (Example 5).

Carboxylic Acid Grafting

A process described by *R.L. Elliott and J.B. Gardiner; U.S. Patents 4,144,181; March 13, 1979; and 4,089,794; May 16, 1978; both assigned to Exxon Research and Engineering Company* relates to haze-free polymeric dispersant additives for lubricating oils which may also be useful as viscosity index improvers for lubricating oils.

The process provides haze-free solutions of substantially saturated polymers comprising ethylene and one or more C_{3-28} alpha-olefins, preferably, propylene, which have been grafted in the presence of a free-radical initiator with an ethylenically-unsaturated carboxylic acid material preferably at an elevated temperature and in an inert atmosphere, and thereafter reacted first with a polyfunctional material reactive with carboxy groups, such as a polyamine, a polyol, or a hydroxy amine, or mixtures thereof, to form multifunctional polymeric reaction products and thereafter with at least a haze preventing amount of an oil-soluble acid.

The following examples illustrate the process. In these examples, all parts are by weight unless specifically indicated otherwise and all nitrogen values were determined by Kjeldahl analysis.

Example 1: 23.4 gal (180 lb) of a 15 wt % solution of an ethylene-propylene copolymer concentrate [made by the Ziegler-Natta process using H_2-moderated $VOCl_3$/aluminum sesquichloride catalyst, having a crystallinity of less than 25%, containing about 45 wt % ethylene and 56 wt % propylene, and having a TE of 2.11 ($\overline{M}w$ = 56,000)] in S100N (Solvent 100 Neutral Mineral Oil) was heated to 154°C under a nitrogen blanket.

To this was added with stirring 0.87 lb (0.33 wt %) of 2,5-dimethyl hex-3-yne-2,5-bis-tert-butyl peroxide and 3.42 lb (1.29 wt %, 0.0132 mol %) of maleic anhydride. After about 2.5 hours at 154°C, the system was sparged with nitrogen for 1.5 hours to remove all of the untreated maleic anhydride.

To this system at 154°C, 3.6 lb (1.36 wt %, 0.0096 mol %) of N-aminopropyl-morpholine was added. All wt % values are based on the weight of the initial oil solution. The reaction was conducted for 1 hour at 154°C, followed by vacuum stripping (84 kPa) at 154°C for 5 hours.

To the final product was added 77 lb of S100 N. This resulted in a nitrogen-containing grafted copolymer which contains 0.46 wt % nitrogen and a final oil concentrate of 10.5 wt % graft copolymer whose haze reading was >230 nephelos (off-scale) as measured on a nephelometer purchased from Coleman Industries, Model 9.

Example 2: The same general procedure of Example 1 was followed modified by change to the charges below.

1.34 lb (0.496 wt %) 2,5-dimethyl hex-3-yne-2,5,-bis-tert-butyl peroxide;
5.48 lb (2.03 wt %, 0.2 mol %) maleic anhydride; and
5.8 lb (2.15 wt %, 0.015 mol %) N-aminopropylmorpholine.

The final resulting polymers contained 1.27 wt % nitrogen and the haze reading of the final oil concentrate was again >230 nephelos.

Example 3: The same general procedure of Example 1 was followed modified by change to the charges below given as wt % of total charge.

Maleic anhydride	0.995
2,5-dimethyl-hex-3-yne- 2,5-bis-tert-butyl peroxide	0.198
N-aminopropylmorpholine	1.003
Ethylene-propylene polymer concentrate of Ex. 1	97.9

The resulting grafted copolymer contained 0.21 wt % nitrogen and the haze reading of the final oil concentrate was again >230 nephelos.

Example 4: The same general procedure of Example 1 was followed modified by change to the charges below given as wt % of the total charge. Maleic anhydride 1.52; 2,5-dimethyl hex-3-yne-2,5-bis-tert-butyl peroxide 0.392; N,N-dimethylaminopropylamine 1.150; and the ethylene-propylene polymer concentrate of Example 1, 97.3.

The dialysate of the resulting grafted copolymer contained 0.36 wt % nitrogen and the haze reading of the oil concentrate was again >230 nephelos.

Example 5: 300 g of the oil concentrate of Example 3 were placed into a 4-necked 1 liter flask under a nitrogen blanket. The solution was heated to 205°C with good agitation and thereafter treated with 3 g of SA-119, which is a commercial alkaryl sulfonic acid sold by Esso SA France. SA-119 is a 90% active oil concentrate of primarily C_{24} (average) alkylbenzene sulfonic acid having a $\overline{Mn}$ of ~500. The reaction was continued for 15 min at 205°C and then the clear solution cooled and bottled. Haze reading of the final oil concentrate was 48 nephelos as measured on the nephelometer.

Example 6: 600 g of the oil concentrate of Example 4 were placed into a flask under a nitrogen atmosphere. The solution was heated to 205°C with good agitation and then treated with 6 g of SA-119 at 205°C for 4 hours after which the clear solution was cooled and bottled. Haze reading of the final oil concentrate was + 59 nephelos as measured on the nephelometer.

Example 7: 93 lb of the oil concentrate of Example 1 were treated with 3.72 lb of SA-119 (a 1:1 wt ratio of SA-119 to S100 N oil was utilized) at 190°C for 4 hours. The final clear oil concentrate had a haze reading of 80 nephelos.

Example 8: 148 lb of the oil concentrate of Example 2 were treated with 8.8 lb of SA-119 (a 1:1 wt ratio of SA-119 to S100 N was used) at 190°C for 4 hours. The final oil concentrate had a haze reading of 144 nephelos.

In summary, the preceding examples have demonstrated that haze reduction of ethylene copolymer dispersant viscosity index improving oil compositions is readily realized when such compositions are treated according to the process. Not only is the haze reduced but these compositions remain visually improved in haze reduction for periods of time usually met in the shelf life required for such oil compositions.

The treatment of the oil compositions with the antihazing acid reagent also has the further advantage of improving the sludge dispersancy of the lubricating oil to which the compositions of the process are added as seen from the following example.

Example 9: In this example, the efficacy of the derivatized copolymers of this process as dispersants in lubricating oil is illustrated by comparison with the untreated materials of the prior art in a Sludge Inhibition Bench Test (designated SIB). The SIB test has been found, after a large number of evaluations, to be an excellent test for assessing the dispersing power of lubricating oil dispersant additives.

The medium chosen for the SIB test was a used crankcase mineral lubricating oil composition having an original viscosity of 325 SUS at 38°C that had been used in a taxicab that was driven generally for short trips only, thereby causing a buildup of a high concentration of sludge precursors. The oil that was used contained only a refined base mineral lubricating oil, a viscosity index improver, a pour point depressant and zinc dialkyldithiophosphate antiwear additive. The oil contained no sludge dispersant. A quantity of such used oil was acquired by draining and refilling the taxicab crankcase at 1,000 to 2,000 mile intervals.

The SIB test is conducted in the following manner. The abovedescribed used crankcase oil, which is milky brown in color, is freed of sludge by centrifuging for 1 hour at 39,000 g. The resulting clear bright red supernatant oil is then decanted from the insoluble sludge particles thereby separated out. However, the supernatant oil still contains oil-soluble sludge precursors which on heating under the conditions employed by this test will tend to form additional oil-insoluble deposits of sludge.

The sludge inhibiting properties of the additives being tested are determined by adding to portions of the supernatant used oil, a small amount, such as 0.5, 1 or 2 wt %, on an active ingredient basis, of the particular additive being tested. 10 g of each blend being tested is placed in a stainless steel centrifuge tube and is heated at 138°C for 16 hours in the presence of air. Following the heating, the tube containing the oil being tested is placed in a stainless steel centrifuge tube and is heated at 138°C for 16 hours in the presence of air. Following this heating, the tube containing the oil being tested is cooled and then centrifuged for 30 min at 39,000 g. Any deposits of new sludge that form in this step are separated from the oil by decanting the supernatant oil and then carefully washing the sludge deposits with 25 ml of pentane to remove all remaining oil from the sludge. Then the weight of the new solid sludge that has been formed in the test, in mg, is determined by drying the residue and weighing it.

The results are reported as % of sludge dispersed by comparison with a blank not containing any additional additive. The less new sludge formed, the larger the value of % sludge dispersant, and the more effective is the additive as a sludge dispersant. In other words, if the additive is effective, it will hold at least a portion of the new sludge that forms on heating and oxidation stably suspended in the oil so it does not precipitate down during the centrifuging.

Using the above test, the enhanced sludge dispersant activity of the sulfonic acid treated materials over the untreated materials is shown below.

Test Sample	.....Polymeric Additive.....		Percent Sludge Dispersed
	Ex. No.	g/10 g Oil	
1	1	0.1	63
2	1	0.05	74
3	7	0.1	80
4	7	0.05	88
5	2	0.1	8
6	2	0.05	26
7	8	0.1	8
8	8	0.05	48
9	3	0.1	0
10	3	0.05	0
11	5	0.1	41
12	5	0.05	0
13	4	0.1	32.2
14	4	0.05	0
15	6	0.1	80.1
16	6	0.05	14.5

The data shown above indicate that at a concentration of 1% sulfonium amine salt of the morpholine derivative of a maleic acid graft copolymer had superior dispersant activity to the morpholine derivative (compare Test Sample 1 with 3, 9 with 11, and 13 with 15. Comparable results are shown for a concentration level of 0.5 wt %.

Electrophilically Terminated Ethylene-Alpha Olefin Copolymers

R.L. Elliott and J.B. Gardiner; U.S. Patent 4,138,370; February 6, 1979; assigned to Exxon Research and Engineering Company have found that polymeric viscosity improvers can be multifunctionalized to provide enhanced dispersancy by electrophilically terminating the anion of an oxidized copolymer of ethylene and one or more C_{3-50}, preferably C_{3-18}, alpha monoolefins with a C_{3-50} reactant containing an imine structure.

It has also been found that other electron seeking groups of electrophilic terminating compounds can be utilized to react with the copolymer anion to introduce reactive sites. These sites can in turn be derivatized to provide multifunctionality, e.g., aminated in order to introduce nitrogen into the copolymer for enhanced dispersancy. The electrophilic terminating compound is characterized by a structural group which contains a carbon atom doubly bonded to an oxygen, sulfur, or nitrogen heteroatom, e.g., an inorganic compound such as carbon monoxide or an organic compound such as acetone.

This finding has made possible the realization of a new class of multifunctionalized polymeric products containing oxygen, nitrogen, sulfur, boron and/or phosphorus characterized by an electrophilically terminated, oxidized ethylene copolymer portion which have utility as additives for lubricating oil compositions.

In their preferred form, this class of products can be characterized as oil-soluble nitrogen-containing polymers formed either directly (as with a ketimine) or indirectly (from subsequent amination) from an electrophilic termination of an anion of an oxidized copolymer of ethylene and at least one C_{3-50} alpha olefin monomer.

In their optimum form, the products of the process are oil-soluble, nitrogen-containing copolymers of ethylene and propylene having a number average molecular weight ($\overline{Mn}$) of from 1,000 to 500,000 and containing from 0.005 to 4%, preferably 0.05 to 2%, optimally 0.1 to 1.0% by weight of nitrogen, and from 0.005 to 6% by weight of oxygen which demonstrate outstanding dispersancy and have utility as ashless sludge dispersants.

The following examples illustrate the process.

Example 1: To a stirring solution of 10 g oxidized (air masticated) ethylene-propylene copolymer [44 wt % (54 mol %) ethylene and 56 wt % propylene] of 23,000 $\overline{Mn}$ in 200 ml dry tetrahydrofuran maintained at ambient temperature and under a nitrogen atmosphere was rapidly added (ca 10 sec) 1.5 ml of a 1.6 molar solution of n-butyllithium in hexane.

The mixture was allowed to stir under the same conditions for 1 hour after which time it was treated with 1 ml of N-diethylene triamine-2-propylidene-imine. The solution was slowly heated to 50°C and additional stirring continued for 2 hours. The reaction was terminated with 2 ml of methanol, and the polymer isolated by precipitation with 1.5 liters of methanol. The resulting electrophilically terminated copolymer was washed with 500 ml additional methanol, then dried in a vacuum oven at 100°C for 15 hours, after which time 9.4 g of polymeric product was recovered. The nitrogen level of the product was 0.04 wt % (Kjeldahl).

Example 2: To a stirring solution of 10 g of oxidized (air masticated) ethylene-propylene copolymer [44 wt % (54 mol %) ethylene and 56 wt % propylene] of 23,000 $\overline{Mn}$ in 200 ml dry tetrahydrofuran maintained at ambient temperature and under a nitrogen atmosphere was rapidly added (ca 10 sec) 1.0 ml of a 1.6 molar solution of n-butyllithium in hexane.

The mixture was allowed to stir under the same conditions for 1 hour after which time it was treated with 1.2 g of methylchloroformate in 25 ml of tetrahydrofuran at ambient temperature. The solution was stirred for an additional 1.5 hours at ambient temperature. To the solution was added 2 ml of methanol, and the electrophilically terminated copolymer isolated by precipitation with 2 liters of methanol. The resulting copolymer was washed twice with two 500 ml portions of additional methanol, then dried in a vacuum oven at 100°C for 53 hours, after which time 9.0 g of polymeric product was recovered (yield of

90%). Infrared spectra show a strong ester band at 1755 cm^{-1}.

Example 3: 5 g of the functionalized copolymer of Example 2 was dissolved in 100 ml toluene, then carefully refluxed at 110°C under a nitrogen atmosphere with a solution of diethylene triamine (DETA), 1 g in 10 ml toluene, for 6 hours. The solution was cooled to ambient temperature and the polymer product recovered by precipitation from 1.5 liters of methanol.

The resulting product was washed with 500 ml additional methanol and then dried in a vacuum oven at 100°C for 15 hours, after which time 4.72 g of polymeric product was recovered (yield of 94%). The nitrogen level of the resulting polymeric product was 0.28 wt % (Kjeldahl).

Example 4: To a stirring solution of 50 g of oxidized (air masticated) ethylenepropylene copolymer [44 wt % (54 mol %) ethylene and 56 wt % propylene] of 23,000 $\overline{M}w$ in 500 ml dry tetrahydrofuran maintained at 40°C temperature and under a nitrogen atmosphere was rapidly added (ca 10 sec) 6 ml of a 1.6 molar solution of n-butyllithium in hexane. The mixture was stirred under the same conditions for 1.5 hours after which time it was treated with a solution of 4.9 g of methylchloroformate in 35 ml tetrahydrofuran. The solution was slowly heated to 60°C and stirred while cooling for 1 hour.

To the solution was added 6 ml methanol, and the electrophilically terminated copolymer isolated by precipitation with 1.5 liters isopropanol. The resulting polymeric product was washed with 750 ml acetone and then dried in a vacuum oven at 100°C for 15 hours, after which time 47.2 g of polymeric product was recovered (yield of 94%). Infrared spectra show a strong ester band at 1755 cm^{-1}.

Example 5: 5 g of the functionalized polymer of Example 4 was dissolved in 100 ml xylenes and then carefully refluxed at 140°C under a nitrogen atmosphere with a solution of 1 g diethylene triamine in 5 ml xylene for 16 hours. The solution was cooled to ambient temperature and the polymeric product recovered by precipitation from 2.1 liters of methanol.

The resulting polymeric product was washed with additional 500 ml methanol then dried in a vacuum oven at 100°C for ca 15 hours, after which time 4.8 g of polymeric product was recovered (yield of 96%). The nitrogen level of the resulting polymer was 0.20 wt % (Kjeldahl).

Example 6: In this example the efficacy of the derivatized electrophilically terminated copolymers of this process as dispersants in lubricating oil applications, is illustrated by comparison with a commercially available multifunctional V.I. improver, sold as Lz3702 by Lubrizol Corporation, in a Sludge Inhibition Bench Test (designated SIB). The SIB test has been found, after a large number of evaluations, to be an excellent test for assessing the dispersing power of lubricating oil dispersant additives.

Using the above test, the dispersant activity of derivatized electrophilically terminated copolymers prepared in accordance with this process were compared with the dispersing power of a dialyzed product obtained from dialysis of a commercial dispersant previously referred to as Lz3702. Sufficient dialyzed residue, which analyzed about 0.4 wt % nitrogen, was dissolved in S-150 N mineral oil to provide a 10% active ingredient concentrate. The dialyzed residue and poly-mer products of the process were appropriately diluted in mineral oil to furnish the 0.05 and 0.1 wt % of added additive to the used oil. The test results are given below.

. . . .Polymeric Additive. . . .		Percent Sludge Dispersed
Example	g/10 g Oil	
1	0.1	50
	0.05	10
3	0.1	51
	0.05	34
5	0.1	84
	0.05	68
Lz3702	0.1	89
	0.05	73

The results of the above table can be summarized as showing that the nitrogen-containing polymeric products of Examples 1 and 3 provide dispersancy at the 1 wt % and 0.5 wt % additive levels. The polymeric product of Example 5 pro-vides dispersancy at the 1 and 0.5 wt % additive levels comparable to that shown by a commercially available multifunctional V.I. improver.

Polymers Containing Postreacted Phenol Antioxidant Functionality

R.H. Hanauer and G.L. Willette; U.S. Patent 4,098,709; July 4, 1978; assigned to Rohm and Haas Company have discovered V.I. improvers which provide a significant improvement in high temperature diesel engine operation. This ad-vance has been achieved by incorporating hindered phenol antioxidant function-alities directly into suitable polymeric V.I. improvers. As such, the antioxidant moiety prevents extensive oxidative decomposition of the polymer and other fluid additives, thus reducing the tendency to lacquer, TGF formation and land deposits. Incorporation of the polymers of this process in diesel lubricants greatly enhances the possibility of the adoption of multigraded oils for both mild and severe diesel operation.

The antioxidant polymers are formed by postreacting a carboxylic acid con-taining V.I. improving polymer with a hindered phenol containing compound to obtain a polymer of the structure:

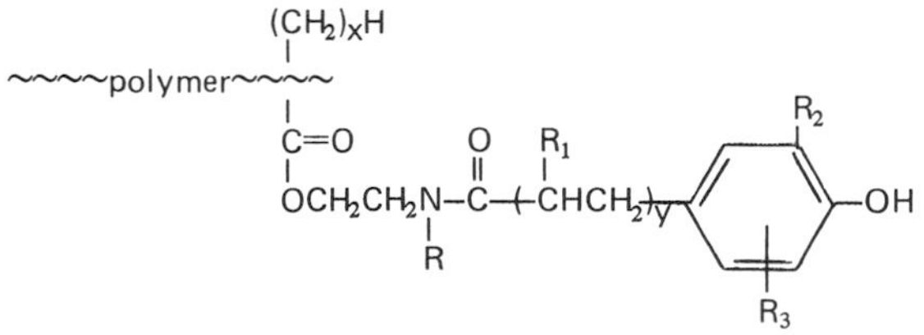

where x is 0 or 1, R is H or C_{1-12} alkyl, y is 0 or 1, R_1 is hydrogen or methyl, R_2 and R_3 independently are C_{1-12} alkyl, and the polymer backbone is the residue of the V.I. improving starting polymer.

Dithiophosphorylated Copolymers of Aziridineethyl Acrylates

According to a process described by *J.P. Pellegrini, Jr. and H.I. Thayer; U.S. Patent 4,136,042; January 23, 1979; assigned to Gulf Research & Development Company* lubricating oils comprising mineral oils and synthetic oils having high pour points are provided with one or more enhanced characteristics such as improved pour point, viscosity, viscosity index, metal-free and antiwear properties by the addition of a dialkyl, diaryl, dialkaryl, diarylalkyl, or diaryloxyalkyl dithiophosphate copolymer of a monomeric aziridineethyl acrylate or methacrylate and a monomeric alkyl acrylate or alkyl methacrylate.

The following examples illustrate the process.

Example 1: A light neutral mineral oil having enhanced pour point, viscosity, viscosity index, metal-free and antiwear properties is prepared by blending a 2.5:97.5 mol ratio of dibutyl dithiophosphate, aziridineethyl methacrylate/isodecyl methacrylate copolymer with the mineral oil.

The copolymer was prepared by charging a 3 liter flask, equipped with a reflux condenser, thermometer and mechanical stirrer, with 450 g of light neutral mineral oil, 220.7 g of isodecyl methacrylate and 4.22 g of aziridineethyl methacrylate and 1.6 g of α,α'-azodiisobutyronitrile. The system was flushed with nitrogen for ½ hour and the mixture was heated at 65°C for 10 hours.

The copolymer was dithiophosphorylated by charging 45 g of the copolymer, 90 g of light neutral mineral oil and 2.610 g of dibutyl dithiophosphoric acid to a 500 ml flask equipped with mechanical stirrer, thermometer and a standard reflux condenser. The system was next flushed with nitrogen and the mixture heated with stirring for 5 hours at 176°F (80°C).

The resulting dithiophosphorylated copolymer was blended with a light neutral mineral oil at a concentration of 3% by wt. The light neutral mineral oil was analyzed before and after addition of the dithiophosphorylated copolymer. The results are shown below.

Component	...Viscosity.... 100°F	210°F	Viscosity Index	Pour Point (°F)
Light neutral mineral oil	22.47	4.25	101	-5
Light neutral mineral oil + 3% dithiophosphorylated copolymer	49.93	9.67	196	-10

The dibutyl dithiophosphate, aziridineethyl methacrylate/isodecyl methacrylate copolymer (2.5:97.5 mol ratio) above is especially suited for use as metal-free, antiwear agents when formulated with engine lubricating compositions. As can be determined from the above, the viscosity, viscosity index, and pour point are additionally enhanced by the addition of the copolymer herein.

It is to be noted that substantially the same results are obtained when the dialkyl, diaryl, dialkaryl, diarylalkyl, or diaryloxyalkyl dithiophosphorylated aziridineethyl acrylate or methacrylate/alkyl acrylate or alkyl methacrylate copolymers, substantially as described herein are substituted for the dibutyl dithiophosphate, aziridineethyl methacrylate/isodecyl methacrylate above.

Example 2: A lubricating oil having improved antiwear and metal-free characteristics, as well as improved viscosity, viscosity index and pour point properties was formulated by preparing a 1:19 mol ratio of dibutyl dithiophosphoric acid, aziridineethyl methacrylate/lauryl methacrylate copolymer by combining 10 g of dibutyl dithiophosphate and 3.38 g of aziridineethyl methacrylate and 100 ml of dry benzene.

The procedure of Example 1 was followed with the following exceptions. The mixture was heated and stirred at 80°C for 3 hours under a nitrogen atmosphere. The solvent (benzene) was removed by distillation and the resulting product was added to 96.5 g of lauryl methacrylate in 200 g of light neutral mineral oil which contained 0.66 g of α,α'-azodiisobutyronitrile. This solution was heated at 65°C for 12 hours.

A lubricating oil blend was prepared by mixing 3% by wt of the resulting copolymer with a light neutral mineral oil. An analysis of the oil before and after addition of the copolymer gave the results shown below.

Component	... Viscosity. ... 100°F	210°F	Viscosity Index	Pour Point (°F)
Light neutral mineral oil	22.47	4.25	101	–5
Light neutral mineral oil + 3% dithiophosphorylated copolymer	43.37	8.28	182	–55
Light neutral mineral oil + 3% dithiophosphorylated copolymer	49.95	9.84	201	–50

The order of addition in forming the dithiophosphorylated copolymer above was varied by reacting the aziridineethyl methacrylate with the lauryl methacrylate to form the 1:19 mol ratio copolymer and then reacting the appropriate weight of dibutyl dithiophosphoric acid with the resulting copolymer. In both instances, the viscosity, viscosity index and pour point of the lubrication oil was enhanced. Additionally, the antiwear and metal-free properties of the oil composition are improved.

When the other dithiophosphorylated copolymers described herein are substituted for the dibutyl dithiophosphate, aziridineethyl methacrylate, lauryl methacrylate copolymer above, substantially the same results are obtained.

The light neutral mineral oil used above had the following properties.

Gravity, API	32.9
Specific gravity, 60/60°F	0.8607
Viscosity, SUS, 100°F	102.5
Viscosity, SUS, 210°F	39.5
Viscosity index	96
Pour point, °F	+10
Iodine number	4.6

Polyoxyalkylene Phenothiazines for Ester Lubricants

D.A. Williams and R. Carswell; U.S. Patent 4,072,619; February 7, 1978; assigned to The Dow Chemical Company have found that from 1.0 to 90 wt % of N-polyoxyalkylene phenothiazines having a weight average MW range from 300 to 5,000 can be blended with synthetic ester lubricants to provide lubricant compositions that have superior viscosity and pour point characteristics over a wide range of temperatures and also have superior oxidative and thermal stability. A preferred range of MW for the polyoxyalkylene phenothiazines is from 375 to 1,300. A preferred range of the amount of the polyoxyalkylene phenothiazines is from 5 to 50 wt % of the blend.

The polyoxyalkylene group of the phenothiazines is derived from ethylene oxide, propylene oxide, 1,2-butylene oxide, 2,3-butylene oxide, and styrene oxide. The only limitation is that when ethylene oxide is used for the polyoxyalkylene phenothiazine compounds the amount used must be such that less than 85% by wt of the compounds is made up from ethylene oxide.

The polyoxyalkylene phenothiazines are prepared by the general methods set forth in U.S. Patent 2,815,343 wherein pure or mixed alkylene oxides are reacted with phenothiazine in the presence of an alkali metal hydroxide or alkoxide to form the adducts.

A mixture of alkylene oxides can be reacted with the phenothiazine to give random copolymer adducts or the alkylene oxides can be reacted in sequence to give block copolymer adducts. Specific examples of useful random copolymer adducts are:

> N-polyoxypropylene-polyoxybutylene (9:1 wt ratio) phenothiazine
> of 1,000 to 1,200 MW;
> N-polyoxypropylene polyoxyethylene (1:1 wt ratio) phenothiazine
> of 1,000 to 1,200 MW;
> N-polyoxyethylene-polyoxybutylene (1:1 wt ratio) phenothiazine
> of 1,200 to 1,400 MW; and
> N-polyoxypropylene-polyoxyethylene (1.3:1 wt ratio) phenothiazine
> of 2,900 to 3,100 MW.

Specific examples of useful homopolymer adducts prepared from the reaction of a pure alkylene oxide and phenothiazine are:

> N-polyoxypropylene phenothiazine of 300 to 400 MW;
> N-polyoxypropylene phenothiazine of 1,000 to 1,200 MW;
> N-polyoxybutylene phenothiazine of 300 to 400 MW; and
> N-polyoxybutylene phenothiazine of 1,000 to 1,200 MW.

LOAD-CARRYING ADDITIVES

It is well known that various additives can be added to lubricating oils in order to improve various oil properties and to make a more satisfactory lubricant. Antiwear agents are intended to decrease wear of the machine parts. Wear inhibitors for incorporation in motor oils and industrial oils are finding greater use as a result of the greater stress placed on moving parts in high performance engines. Numerous additives have been developed for use in such oil compositions to improve the lubricating characteristics and lessen the wear of the moving parts. Zinc dialkyl dithiophosphates (ZOP) have been long used as antiwear additives and antioxidants in hydraulic oils, motor oils, and aromatic transmission fluids.

In particular, there is a need for additives which are intended to protect the devices to be lubricated from wear due to friction. The demands made upon such extreme-pressure additives for lubricants are that they are adequately soluble, that they increase the load-bearing capacity and that they do not have a corrosive action on the particular metal parts. This chapter provides synthesis and formulation details for the use of effective additives for lubricants, which meet these requirements.

MINERAL OILS

Ammonium Salts of Oxa- and Thiaphosphetanes

A process described by *A. Schmidt and P.C. Hamblin; U.S. Patent 4,080,307; March 21, 1978; assigned to Ciba-Geigy AG, Switzerland* relates to certain ammonium salts of oxa- and thiaphosphetanes and their use as additives for lubricants. The compounds correspond to the general formula:

$$\left[(X_1)_n \overset{CH_2}{\underset{CH_2}{\diagdown}} \overset{X_2}{\underset{}{\overset{\uparrow}{P}}} - X_3 \right]_m^{-} (ZH_m)^{m+}$$

in which X_1, X_2 and X_3 independently of one another denote O or S and, if X_1 is O, n denotes 1 and, if X_1 is S, n denotes 1 to 6, and Z is a monoacidic or diacidic, nitrogen-containing, oil-soluble organic base and, if Z is a monoacidic base, m denotes 1 and, if Z is a diacidic base, m denotes 2.

The formula comprises the following basic types of ammonium salts of 1,3-oxa- or 1,3-thiaphosphetanes:

Ammonium salts of 1,3-oxaphosphetanes of the formula are those in which X_1, X_2 and X_3 are O; those in which X_1 is O, one of X_2 and X_3 is O and the other is S; and those in which X_1 is O and X_2 and X_3 are S, n always being 1 in the oxa compounds.

Ammonium salts of 1,3-thiaphosphetanes of the formula are those in which X_1, X_2 and X_3 are S; those in which X_1 is S, one of X_2 and X_3 is S and the other is O; and those in which X_1 is S and X_2 and X_3 are O, n having a value of 1 to 6, and especially of 1, in the thia compounds.

Preferred ammonium salts of phosphetanes of the formula are those in which X_1 is S and X_2 and X_3 are O and n is 1 to 6 and preferably 1.

Even in very small amounts, the compounds of the formula act as high-pressure additives in lubricants. Thus, mineral and synthetic lubricating oils, and also mixtures thereof, which are provided with 0.001 to 5% by weight, and preferably 0.02 to 3%, relative to the lubricant of a compound of the formula display excellent high-pressure lubricating properties which manifest themselves in greatly reduced wear phenomena of the parts which rub against one another and are to be lubricated.

Example 1:

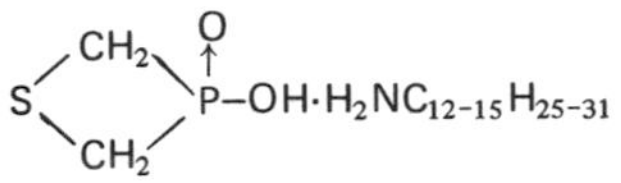

$$S \underset{CH_2}{\overset{CH_2}{\diagdown}} \overset{\overset{O}{\uparrow}}{P} - OH \cdot H_2 N C_{12-15} H_{25-31}$$

26.7 g (0.215 mol) of 3-hydroxy-3-oxo-1,3-thiaphosphetane are suspended in 200 ml of toluene and 37.5 g (0.196 mol) of Primene 81-R (mixture of primary C_{12-15} tert-alkylamines) are added dropwise at room temperature, while stirring. The reaction is exothermic. The reaction mixture is then heated at the reflux temperature for 1 hour.

The solution is filtered and concentrated completely under reduced pressure. The residual resin is dried for 1 hour in vacuo at 100°C and 0.1 mm Hg. This gives a yellowish, transparent resin which is readily soluble in hexane and mineral oil. (Additive No. 1).

Example 2: 12.4 g (0.1 mol) of 3-hydroxy-3-oxo-1,3-thiaphosphetane are suspended in 100 ml of toluene and 19.1 g (0.1 mol) of Primene 81-R (mixture of primary C_{12-15} tert-alkylamines) are added dropwise at room temperature, while stirring. Working up is carried out as in Example 1. (Additive No. 2).

If, in Example 2, the Primene 81-R is replaced by the amines listed in the table on the following page, an otherwise identical procedure gives the corresponding ammonium salts.

Additive No.	Amine	Melting point	Ammonium salt
3	n-dodecylamine	93° C	OH . $H_2NC_{12}H_{25}$-n
4	n-octylamine	78° C	OH . $H_2NC_8H_{17}$-n
5	Primene J.M.T.	resin	OH . $H_2NC_{18}H_{37}$-tert.
6	n-dibutylamine	99° C	OH . $NH(C_4H_9$-n$)_2$
7	n-tributylamine	resin	OH . N-$(C_4H_9$-n$)_3$
8	N,N-diethylaniline	resin	OH . $N(C_2H_5)_2$-C_6H_5
9	N-methylaniline	resin	OH . $HN(CH_3)$-C_6H_5
10	2,6-diethylaniline	94–97° C	OH . NH_2-$C_6H_3(C_2H_5)_2$

Example 3: The exceptional load-bearing properties of the additives, according to the process, for lubricants are shown in the test in the FZG gear wheel distortion test rig.

For this purpose, mixtures of the additives according to the process in a nondoped mineral lubricating oil (viscosity: 20 cs/50°C) were prepared and tested using the FZG machine according to DIN 51,354 (standard test A/8.3/90). For comparison, the nondoped mineral lubricating oil without an additive, and also mixtures of this lubricating oil with commercially available high-pressure additives, were also tested using the FZG machine. The results of these tests are summarized in the table below.

The distinct superiority of the additives according to the process, for lubricants over commercially available high-pressure additives can be seen, at a markedly lower use concentration, from the considerable reduction in the specific weight change of the gear wheels and also from a simultaneous improvement in the load-bearing capacity by 1 to 2 power stages.

Additive No.	Test No.	Concentration (wt %)	SWC* (mg/kWh)	Power Stage at Which Damage Occurs
—	1	—	0.61	7
1	2	0.03	0.1	12
2	3	0.05	0.1	>12
2	4	0.08	0.1	>12
ZDTP**	5	0.5	0.29	11
Triphenyl phosphate	6	1.0	0.37	10
5	7	0.1		>12

*Specific weight change of the gear wheels.
**Commercially available zinc dialkyldithiophosphate (about 80% active substance).

Amine Salt of Dialkyldithiophosphate

H. Shaub; U.S. Patent 4,101,427; July 18, 1978; assigned to Exxon Research & Engineering Co., has found that the antiwear and antifriction properties of a fully formulated lubricating composition containing a zinc dialkyldithiophosphate or a similar metal dialkyldithiophosphate is improved by adding an amine salt of a dialkyldithiophosphate. The amine salt is prepared with a tert-alkyl primary amine. The amine salt may be simply added to a formulated composition or used as a substitute for the metal dialkyldithiophosphate which is often included in lubricating compositions.

Example 1: In this example, four formulations were prepared in a 10-W-40 SE quality automotive engine oil and the relative wear and friction were then determined with a ball on cylinder test. The apparatus used in the ball on cylinder test is described in the journal of the American Society of Lubrication Engineers, entitled *ASLE Transactions*, vol. 4, pages 1-11, 1961.

In this example, the standard was a 10-W-40 SE quality automotive engine oil containing a dispersant, a rust inhibitor, a detergent, an oxidation inhibitor and a VI improver and a zinc dialkyldithiophosphate in which the alkyl groups were a mixture of such groups having between about 4 and 5 carbon atoms. The wear and friction of this composition were then determined and assigned relative values of 1.00.

In a second composition, the same formulated 10-W-40 SE quality automotive engine oil was used, but 0.1 wt % of an amine salt prepared by neutralizing a di-n-propyl dithiophosphate with a mixture of C_{12}, C_{13} and C_{14} tert-alkyl primary amines were added. This mixture is designated Primene 81-R. The composition of Primene 81-R is described in the Rohm and Haas brochure. After formulation, the wear and friction were determined and, on a relative basis were 1.00 and 0.82 respectively.

In a third composition, the same formulated 10-W-40 SE quality automotive engine oil used in the first two compositions was used and 0.5 wt % of the amine salt used in the second composition was used. The wear and friction exhibited by this composition, on a relative basis, were 1.0 and 0.72, respectively.

The fourth composition tested in this example was identical to the second and third compositions except that 1.0 wt % of the amine salt was used. After formulation, the wear and friction were again determined in the same manner as that previously described and were found to be 0.90 and 0.62, respectively.

From the foregoing, it should be readily apparent that the addition of a tert-alkyl primary amine salt of a dialkyldithiophosphate to a fully formulated 10-W-40 SE quality automotive engine oil containing a zinc dialkyldithiophosphate significantly reduces friction.

Example 2: In this example, the friction of two lubricating compositions was determined with a ball on cylinder test with a load of 4 kg, a temperature of 220°F, 0.26 rpm and a period of 70 minutes. In each formulation, a solvent 150 neutral, low pour base oil was used and the base oil without additives was assigned a relative friction of 1.

In the first composition, 0.1 wt % of an amine salt identical to that used in Example 1 was added to the base oil. The relative friction was then found to be 0.8.

In a second formulation, 0.5 wt % of an amine salt identical to that used in Example 1 was added to the base oil. The relative friction of this composition was 0.51.

From the foregoing, it is clear that the friction of an unformulated base oil is significantly reduced when a tert-alkyl primary amine salt of a dialkyldithiophosphate is added thereto.

Imidazoline-Phosphonate Reaction Products

M. Braid; U.S. Patent 4,125,472; November 14, 1978; assigned to Mobil Oil Corporation has found that an improvement in the antiwear and friction modifying properties of a lubricant can be obtained by incorporating a product formed by reacting: (a) a dialkylalkane phosphonate having the formula:

$$R'-\overset{\displaystyle O}{\overset{\displaystyle \|}{P}}-(OR)_2$$

where R' is a substantially unbranched paraffinic alkyl group containing from about 10 to 36 carbon atoms, but preferably is a substantially unbranched paraffinic alkyl group containing from about 10 to 20 carbon atoms, and most preferably is octadecyl; and R is a hydrocarbyl group containing from 1 to 4 carbon atoms with at least 1 hydrogen atom present on the carbon atom which is bonded to oxygen, but R preferably is methyl or ethyl and most preferably is methyl; with (b) a substituted imidazoline of the formula:

$$\begin{array}{c} H_2C \text{————} N-R^3 \\ | \qquad\qquad | \\ H_2C \diagdown \quad /C-R^2 \\ N \end{array}$$

where one of the R^2 and R^3 substituents must be a substantially unbranched paraffinic or olefinic hydrocarbyl group containing from about 12 to 35 carbon atoms; and the other R^2 or R^3 substituent is selected from the group consisting of: paraffinic alkyl containing from 1 to about 35 carbon atoms; alkenyl, containing from 1 to about 35 carbon atoms; and hydroxyalkoxy-, alkoxymethoxy-, and oxo-substituted alkyl and alkenyl containing from 1 to about 20 carbon atoms.

Example 1: A solution of 35 g of dimethyloctadecyl phosphonate and 35 g of 1-(2-hydroxyethyl)-2-heptadecenylimidazoline in 400 ml of petroleum ether, BP 30° to 60°C was chemically dried with anhydrous calcium sulfate. The petroleum ether was removed by distillation and the temperature of the homogeneous residue was raised to 150°C and maintained for 1.5 hours under reduced pressure of about 200 mm of Hg while a slow stream of nitrogen was passed through subsurface. The reaction product was obtained as an amber, very viscous semioil.

Example 2: As in Example 1, a mixture of 5 g of dimethyloctadecyl phosphonate and 5 g of 1-(2-hydroxyethyl)-2-heptadecenylimidazoline was heated for 1.5 hours at 150°C at a reduced pressure of less than 0.1 mm of Hg. The product was an amber, very soft semisolid.

Example 3: Under an atmosphere of nitrogen, 2.9 of dimethyloctadecyl phosphonate and 2.9 g of 1-(2-hydroxymethyl)-2-heptadecenylimidazoline was heated together for 2 hours at 130° to 140°C. The resulting product was an amber, soft solid.

Example 4: As in Example 3, a mixture of 2 g of 1-(2-hydroxyethyl)-2-heptadecenylimidazoline and 2 g of dimethyloctadecyl phosphonate was heated at atmospheric pressure under nitrogen for 2 hours. The product was an amber, soft semisolid.

Example 5: A mixture of 25 g of dimethyloctadecyl phosphonate and 25 g of 1-(2-hydroxyethyl)-2-heptadecenylimidazoline was heated at 170°C for about 2 hours under a pressure of less than 1 mm of mercury while a very slow stream of nitrogen was passed through the mixture. The residue after cooling was an amber, very soft, waxy solid, 47.8 g.

Example 6: An aliquot portion of the product of Example 1 was heated in an atmosphere of nitrogen under a reduced pressure of less than 1 mm of Hg at a temperature of 200°C for 0.5 hour. The product produced by this treatment was slightly darker in color and slightly more viscous than the product of Example 1.

Example 7: A mixture of 35 g of dimethyloctadecyl phosphonate and 40 g of the imidazoline prepared from the reaction of N-β-hydroxyethylethylenediamine and behenic acid, 1-(2-hydroxyethyl)-2-eicosanylimidazoline, is stirred under nitrogen for 2.5 hours while heating at 195°C. The reaction product is obtained as a yellowish waxy solid.

Compositions containing 0.5 wt % of the products of Examples 1 through 6, and 99.5 wt % of a base oil were tested for water tolerance, antiwear and chatter characteristics.

The test results are reported in the table on the following page.

It is seen from the data presented in the table that the products represented by Examples 5 and 6 impart effective antiwear and water tolerance properties to the base oil. It is noted that the products made from reactions at temperatures up to and including 150°C fail in the same regard. Thus, the criticality of the greater than 150°C temperature requirement is shown.

Formulation*	International Harvester . . . Water Tolerance Test . . . . Haze Observed Duplicate Runs	Rating	Shell Four Ball Wear Test Wear Scar Diameter (mm)	John Deere . . Chatter Test. . Chatter Index	Rating	Timken EP Test Wt Loss (mg)
Base oil	–		–	–	–	–
Example 1	heavy/heavy	unsatisfactory	–	–	–	–
Example 2	heavy/heavy	unsatisfactory	–	–	–	–
Example 3	very heavy/heavy	unsatisfactory	–	–	–	–
Example 4	heavy/heavy	unsatisfactory	–	–	–	–
Example 5	trace/medium	satisfactory	0.425	170	pass	16
Example 6	light/medium	satisfactory	–	–	–	–

*99.5 wt % base oil and 0.5 wt % of product from example indicated.

(N,N-Diorganothiocarbamyl) Phosphorothioites

P.S. Landis and A.O.M. Okorodudu; U.S. Patent 4,104,181; August 1, 1978; assigned to Mobil Oil Corporation have found that (N,N-diorganothiocarbamyl) phosphorothioites having the following general structures:

$$[R_2N-\overset{\overset{\displaystyle S}{\parallel}}{C}-S\tfrac{}{3}P \qquad R'-OP\tfrac{}{}[S-\overset{\overset{\displaystyle S}{\parallel}}{C}-NR_2]_2$$

and mixtures thereof impart good antiwear and extreme pressure properties to lubricant compositions containing them. In the formulas, R and R' are C_{1-32} alkyl, C_{6-32} aryl or C_{7-32} alkaryl. R and R' may also be (C_nH_{2n-1}) where n is 3 to 20. R may also be $(CH_2)_n$ in heterocyclic systems with n being 2 to 10.

The additives may be conveniently prepared by reacting a suitable phosphorus amide with carbon disulfide. The resultant reaction product is thereafter incorporated into a mineral or synthetic lubricating oil.

Example 1: The following compound was prepared as described below.

$$[(C_4H_9)_2N-\overset{\overset{\displaystyle S}{\parallel}}{C}-S\tfrac{}{3}P$$

Benzene (150 ml) and hexabutylphosphorous triamide (120 g, 0.29 mol) were charged into a reaction flask and carbon disulfide (132 g, 1.7 mols) added dropwise over a 30 minute period. The reaction mixture was then refluxed for 2 hours and cooled. Upon standing, solids separated which were collected and washed with petroleum ether. Product—MP 117° to 120°, 6.9% nitrogen (calculated 6.5%), and 5.3% phosphorus (calculated 4.8%).

Example 2: The following compound was prepared as described below.

$$[(C_2H_5)_2N\overset{\overset{\displaystyle S}{\parallel}}{C}-S\tfrac{}{3}P$$

Hexaethylphosphorous triamide (100 g, 0.4 mol) and 200 ml of benzene were charged into a reaction flask and stirred. To this, carbon disulfide (185 g, 2.4 mols) was added dropwise over a 30 minute period controlling the exothermic reaction temperature at around 40°C. Following the addition, the mixture was refluxed for 30 minutes and cooled. Solids which separated were collected, washed with benzene and air dried. The product contained 6.7% phosphorus (calculated 6.4%) and 8.9% nitrogen (calculated 8.7%).

Example 3: The following compound was prepared as in Example 2, except that tristetramethylenephosphorous triamide was used.

$$\left[\overbrace{}^{} N{-}\underset{\underset{S}{\|}}{C}{-}S{-}\!\!\!\right]_3 P$$

The product contained 7.7% phosphorus (calculated 6.5%).

Example 4: The following compound was prepared as in Example 2, except that tetraoctylphosphorous triamide was used.

$$[(C_8H_{17})_2N{-}\underset{\underset{S}{\|}}{C}{-}S{\tfrac{}{}}]_3 P$$

The liquid product obtained after stripping the reaction mixture contained 5.4% nitrogen (calculated 4.2% nitrogen).

Example 5: The following compound was prepared as in Example 2.

$$[(C_2H_5)_2N{-}\underset{\underset{S}{\|}}{C}{-}S]_2 P{-}O{-}\!\!\!\bigcirc\!\!\!{-}C_9H_{19}$$

0.4 mol of nonylphenyl tetraethylphosphorous diamide and 0.8 mol of carbon disulfide were reacted. The product obtained after stripping the reaction mixture under vacuum contained 6% phosphorus (calculated 5.7%) and 4.9% nitrogen (calculated 5.1%).

The additive compounds were evaluated in a 4-Ball Wear Test using ½" 52100 steel balls at a load of 60 kg for 30 minutes under the conditions set forth in the table below. The oil used was an 80/20 mixture of a solvent refined mid-Continent paraffinic 150/160 second bright mineral oil and a 200/210 second refined mid-Continent neutral mineral oil.

4-Ball Wear Test

Additive	Weight Percent	Temperature (°F)	500	1,000	1,500	2,000
			. . . . Speed (rpm)			
			. . . . . . Scar Diameter (mm)			
Base stock	—	room	0.50	0.60	0.88	2.34
(control)		200	0.60	1.06	1.86	2.23
		390	1.0	1.31	2.06	—
Example 1	1	room	0.40	0.40	0.50	0.60
		200	0.40	0.55	0.60	0.75
		390	0.95	0.90	1.05	1.20
Example 2	1	room	0.40	0.40	0.50	0.70
		200	0.40	0.50	0.50	0.45
		390	0.45	0.70	1.10	1.30

(continued)

Additive	Weight Percent	Temperature (°F)	Speed (rpm) Scar Diameter (mm)			
			500	1,000	1,500	2,000
Example 3	1	room	0.40	0.60	0.65	1.40
		200	0.55	0.60	0.70	0.70
		90	0.60	0.80	1.05	1.90
Example 4	1	room	0.45	0.60	1.70	0.90
		200	0.60	0.70	0.80	1.80
		390	0.90	1.0	2.35	2.0
Example 5	1	room	0.40	0.50	0.50	0.70
		200	0.50	0.50	0.55	0.50
		390	0.50	0.70	1.85	1.70

Phosphorus Acid Compounds-Acrolein-Ketone Reaction Products

D.E. Ripple; U.S. Patent 4,081,387; March 28, 1978; assigned to The Lubrizol Corporation describes additives which are made by the reaction of certain phosphorus acid compounds with certain aldehydes or ketones and certain thiols or dithiols.

More specifically the compounds of the process correspond to the formula:

$$Y^a-S-Y^b$$

where Y^a corresponds to

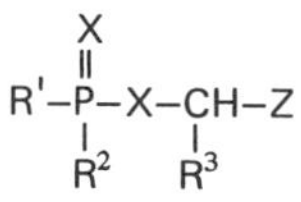

$$R^1-\overset{\overset{\displaystyle X}{\|}}{P}-X-\underset{\underset{\displaystyle R^3}{|}}{\overset{\overset{\displaystyle}{|}}{C}}H-Z$$
$$R^2$$

and Z is

$$-CH+\underset{R^6}{\overset{R^5}{\underset{|}{\overset{|}{C}}}}\underset{}{\overset{}{}}_n\underset{R^7}{\overset{OH}{\underset{|}{\overset{|}{C}}}}-,\ -CH=\underset{R^7}{\overset{R^4}{\underset{|}{\overset{|}{C}}}}-,\ -CH+\underset{R^6}{\overset{R^5}{\underset{|}{\overset{|}{C}}}}\underset{}{\overset{}{}}_n CH=\underset{R^7}{\overset{}{\underset{|}{\overset{|}{C}}}}-$$

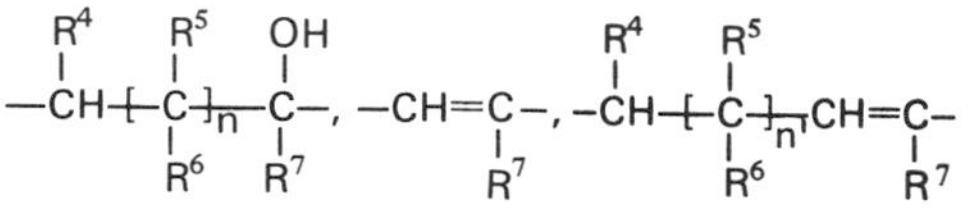

where each R^1 and R^2 is independently a member selected from the group consisting of hydrocarbyl, hydrocarbyloxy and hydrocarbyl mercapto of 1 to about 30 carbon atoms; R^3 is hydrogen or hydrocarbyl of up to about 10 carbon atoms; R^4 is hydrogen or lower alkyl; R^5 and R^6 are each independently selected from hydrogen, hydrocarbyl, hydrocarbyloxy and hydrocarbyl mercapto of 1 to about 10 carbon atoms; R^7 is hydrogen or hydrocarbyl of 1 to about 30 carbon atoms; each X is independently oxygen or divalent sulfur; n is zero or an integer of 1 to about 10; n' is zero or an integer of 1 to about 9; Y^b is

$$-R^8-H$$

or

$$-R^8 S-R^9$$

where R^8 is a divalent hydrocarbyl group of 1 to about 30 carbon atoms and R^9 is hydrogen or Y^a.

A particularly preferred type of phosphorus acid compound is prepared by the reaction of phosphorus pentasulfide or homologs thereof (e.g., P_4S_{10}) with one or more hydroxy compounds which contain the organic groups R^1 and R^2 as defined above; that is, R^1OH and R^2OH or mixtures of two or more of any of these. An example of this type of reaction is the reaction of phosphorus pentasulfide with ethyl alcohol to produce O,O-diethyl phosphorodithioic acid.

Examples of the preferred aldehydes which can be used include: acrolein, croton-aldehyde, 2-pentenal, 3-octenal, 5-decenal, 2-ethyl-2-hexenal, 10-hexadecenal, methacrolein and cinnamaldehyde. Because of their commercial availability, acrolein and crotonaldehyde are especially applicable.

Examples of the preferred ketones which can be used include: methyl vinyl ketone, methyl isopropenyl ketone, methyl propenyl ketone, mesityl oxide, 5-hexen-2-one, 5-methyl-5-hexen-2-one, 5-hepten-2-one and 3-hepten-2-one.

The following examples illustrate the process. All references to percentages, parts, etc., refer to percentages, parts, etc., by weight unless expressly stated otherwise.

Example 1: A phosphorodithioic dialkylaryl acid is prepared by reacting at a temperature of about 149° to 154°C, 1 mol P_2S_5 with 4 mols of a propylene tetramer alkylated phenol. The resulting acid is characterized by a phosphorus content of 4.70%, a sulfur content of 9.63% and an acid neutralization number to bromophenol blue [i.e., NNA (bpb)], as determined by ASTM Procedure D-974, of 80.

Example 2: Freshly distilled acrolein (57 g, 1.01 equivalents) is added to the phosphorodithioic dialkylaryl acid (643 g, 0.92 equivalent) as prepared in Example 1 under a nitrogen purge over a 1-hour period at about 60° to 70°C. The material is then held at about 70°C for 3 hours, cooled to room temperature and is characterized by a NNA (bpb) of 6. The material is then stripped to about 80°C under a vacuum of 20 torrs. The stripped material is then maintained at about 60°C while propylene oxide (7 g, 0.12 mol) is added. The material is filtered through diatomaceous earth. The filtrate is characterized by a phosphorus content of 4.37%, a carbonyl content of 2.81% and a NNA (bpb) of 6.

Example 3: Textile spirits (100 ml), an aliphatic petroleum naphtha having a distillation range of 63° to 79°C at 760 torrs, n-dodecyl mercaptan (12 g, 0.06 equivalent), the product of Example 2 (60 g, 0.06 equivalent), 0.1 g p-toluene sulfonic acid and 11 g Linde 3A molecular sieve pellets are held at about 25° to 30°C for 20 hours. The material is filtered through paper to remove the sieves, stripped to about 80°C under a vacuum of 30 torrs and then filtered through dia-tomaceous earth. The product is characterized by a sulfur content of 10.25%, a phosphorus content of 3.49% and a NNA (bpb) of 3.

Example 4: A phosphorodithioic dialkyl acid is prepared by reacting at a tem-perature of about 110° to 118°C, 1 mol P_2S_5 with 4 mols of decyl alcohol:iso-octyl alcohol in a 30:70 weight ratio. The resulting acid is characterized by a phosphorus content of 7.9%, a sulfur content of 16.2% and a NNA (bpb) of 131.

Example 5: Freshly distilled acrolein (130 g, 2.33 equivalents) is added to the phosphorodithioic dialkyl acid (905 g, 2.12 equivalents) as prepared in Example 4 under a nitrogen purge over a 1 hour period at about 60° to 70°C. The material is held at about 60° to 70°C for 3.5 hours, cooled to room temperature and is characterized by a NNA (bpb) of 5. The material is then stripped to about 48°C under a vacuum of 24 torrs. At about 40°C, propylene oxide (10 g, 0.17 equivalent) is added and the material is filtered through diatomaceous earth. The filtrate is characterized by a phosphorus content of 5.80%, a carbonyl content of 5.43% and a NNA (bpb) of 1.2.

Although the phosphorus-containing compositions of this process as described above are, in themselves, useful as extreme pressure, antiwear and load-carrying agents, they are nevertheless susceptible to improvement by the addition of 1 or more chemical additives to supplement their action to give the properties desired when incorporated into the lubricating and hydraulic compositions of this process.

Dimethyl Octadecylphosphonate

A process described by *A.G. Papay; U.S. Patent 4,158,633; June 19, 1979; assigned to Edwin Cooper, Inc.* provides an internal combustion engine crankcase lubricating oil composition having a lubricating viscosity up to SAE 40, comprising a major amount of a lubricating oil and a minor friction-reducing amount of a phosphonate having the formula

$$R^1-\overset{\displaystyle O}{\overset{\displaystyle \|}{P}}-(OCH_3)_2$$

wherein R^1 is an alkyl or alkenyl group containing about 12 to 30 carbon atoms.

Examples of these phosphonates are dimethyl triacontylphosphonate, dimethyl triacontenylphosphonate, dimethyl eicosylphosphonate, dimethyl hexadecylphosphonate, dimethyl hexadecenylphosphonate, dimethyl tetracontenylphosphonate, and dimethyl hexacontylphosphonate.

The most preferred additive is dimethyl octadecylphosphonate.

Tests have been carried out which demonstrate the ability of the oil composition to significantly improve fuel economy. Initially, friction tests were conducted. These tests were made using a bench apparatus in which a steel annulus and a steel plate were pressed against each other under 229 psi load. The steel annulus was rotated at 40 lineal ft/min and the torque required to start (static friction) and to maintain rotation (kinetic friction) was measured. The rubbing interface of the annulus and steel plate was lubricated with the test lubricating oil.

The base motor oil used in the test was formulated using neutral mineral oil. The base formulation included a commercial ashless dispersant (i.e., polyisobutyl-succinimide of polyethylene polyamine), a zinc dialkyldithiophosphate, an overbased calcium alkylbenzene sulfonate (300 base number), a phenolic antioxidant and a commercial polyacrylate V.I. improver. Both static and kinetic coefficients of friction were measured for the base oil and the base oil containing various concentrations of dimethyl octadecylphosphonate. The results are given in the table on the following page.

Concentration (wt %)	Coefficient of Friction		Percent Decrease in Friction	
	Static	Kinetic	Static	Kinetic
0	0.0620	0.0536	—	—
0.5	0.0539	0.0490	13.1	8.6
0.75	0.0521	0.0485	16.0	9.5
1.0	0.0501	0.0444	19.2	17.2
1.5	0.0478	0.0444	22.9	17.2

Similar, but slightly lower friction reductions were obtained using a different base oil which contained a different phenolic antioxidant.

Since it has been found that some additives which reduce friction in a bench test do not improve fuel economy in actual use, further tests were carried out in a 1977 U.S. production automobile having a V6 engine. The car was operated on a chassis dynamometer under controlled temperature and humidity conditions. Each test sequence consisted of four consecutive EPA city/highway cycles plus a 50 mph steady state cycle. The first cycle started with a cold engine (32°F). The subsequent three cycles started with the warmed-up engine. Gasoline consumption in mpg during the test was measured at frequent intervals by weight and also by volume.

The base oil used in the automobile test was the same base oil used in the bench test except it was presheared by operating for the equivalent of 1,000 miles in a dynamometer engine to eliminate the effect of oil thinning during the actual test. Viscosity index improvers tend to shear during their initial use causing a decrease in the viscosity of the oil. Improved mileage due to such thinning can mask the effect of the test additive.

The mpg data obtained during each cycle of the test sequence was then analyzed by computer regression analysis to eliminate variances due to barometric change and inherent engine trend during the test. This gave the true mpg at the 95 to 99% confidence level. The test sequence was conducted using the base oil and again using the same base oil plus 1 wt % of dimethyl octadecylphosphonate. The results are reported in the following table in terms of percent improvement over base line.

Test Cycle	Improvement in mpg
Cold start transient*	3.4
Cold start city cycle	2.8
Hot start city cycle	2.0
Hot start highway cycle	1.1
50 mph steady	1.3

*Measured during first 3.6 miles of the 11.1 mile city cycle.

Mixture of Zinc Salts of Dialkyldithiophosphoric Acid

A process described by *J. Crawford; U.S. Patent 4,101,428; July 18, 1978; assigned to Chevron Research Company* provides a mixture of the zinc salts of O,O'-dialkyldithiophosphoric acids having the structural formula shown on the following page where R^1 and R^2 are the same or different and are either C_1 to C_{30} primary alkyl groups or C_4 to C_{20} secondary alkyl groups.

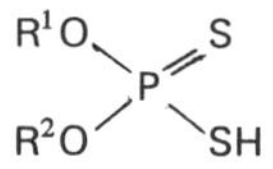

$$\begin{array}{c} R^1O \\ \diagdown \\ R^2O \diagup \end{array} P \begin{array}{c} \diagup\!\!\!=\!\! S \\ \diagdown SH \end{array}$$

The mixture is prepared by contacting a feedstock comprising a mixture of one or more C_{1-30} primary alcohol(s) and one or more C_{4-20} secondary alcohol(s) with phosphorus pentasulfide to form a mixture of O,O'-dialkyldithiophosphoric acids of the above structure, and thereafter neutralizing the mixture of acids by contact with up to 60% molar excess of an alkaline zinc compound, the secondary alcohol(s) forming not less than 25 mol % of the feedstock mixture.

Preferably the primary alcohol is a C_{4-18}, even more preferably a C_{6-12}, primary alcohol.

Preferably the secondary alcohol is a C_{4-12}, even more preferably a C_{4-10}, secondary alcohol.

Preferably the alkaline inorganic zinc compound is zinc oxide.

The mixture of the zinc salts of O,O'-dialkyldithiophosphoric acids dissolved in solvent neutral base oil, as prepared, conveniently forms a concentrate which may be blended with oils of lubricating viscosity to form a finished lubricant composition. Other additives conventionally incorporated into finished lubricant compositions may be blended into the concentrate or incorporated directly into the finished lubricant composition.

Example 1: 222 g (1 mol) of phosphorus pentasulfide was added to a mixture of primary and secondary alcohols made up as follows: 286 g (2.2 mols) of a C_8 primary alcohol (alphanol); 112.2 g (1.1 mols) of methylisobutyl carbinol (secondary alcohol); and 81.4 g (1.1 mols) of sec-butanol (secondary alcohol).

The temperature of the mixture was increased gradually to 80°C over a period of 2 hours. The mixture was heated for a further 6 hours to facilitate the removal of hydrogen sulfide, aided by the introduction of a nitrogen purge.

The resulting liquid was allowed to cool and then poured onto a slurry of 100 g of zinc oxide and 100 g of 100 solvent neutral oil. Vacuum was applied and the temperature raised to 100°C, water of reaction being removed in the conventional manner by a condenser receiver system. After 2 hours the mixture was filtered to give 699 g of a mixture of zinc salts of dialkyldithiophosphoric acids dissolved in solvent neutral base oil.

The product mixture was analyzed for phosphorus, zinc and sulfur. The results of the analysis are given in Table 1.

Example 2: 222 g (1 mol) of phosphorus pentasulfide was added to a mixture of: 286 g (2.2 mols) of a primary C_8 alcohol (alphanol) and 162.8 g (2.2 mols) of sec-butanol.

Thereafter the same procedure was followed as that described in Example 1. The results of the analysis of the product are given in Table 1 on the following page.

Table 1: Chemical Analysis of Products

Analysis	. . . . .Example		. Commercial Products . .	
	1	2	X	Y
	. . . . (%).			
P	7.33	7.63	7.0–7.8	6.8–7.6
Zn	8.06 (110 P)	8.60 (112.7 P)	(105–115 P)	7.14–8.74
S	13.61 (186 P)	14.37 (188.3 P)	(190 P min)	(190 P min)

Note: The numbers in parentheses indicate the amounts of the elements in percentage relationships to phosphorus.

X is a commercial product consisting of the zinc dialkyldithiophosphate manufactured from the primary C_8 alcohol, alphanol. Y is a commercial product consisting of the zinc dialkyldithiophosphate manufactured from a mixture of 70 wt % sec-butanol and 30 wt % methylisobutyl carbinol.

Example 3: The products obtained in Examples 1 and 2 and the commercial products were evaluated in a cloud point test in which a 5% w/w solution of the additive in white oil (liquid paraffin) is heated at a constant rate (5°C/min) in a test tube. The cloud point is the temperature at which decomposition occurs and is identified by the mixture turning white. The observed values of the cloud point are given in the following Table 2.

Table 2: Cloud-Points of Products

Product	Cloud Point, °C
X	240
Z*	227
Y	208
Example 1	232
Example 2	229
50/50 X/Y	212

*A commercial zinc dialkyldithiophosphate manufactured from 1.5 mols isobutanol and 1 mol n-hexanol.

Examination of Table 2 shows that the product prepared from a secondary alcohol (Y) has the lowest thermal stability. The product prepared from a C_8 primary alcohol (X) has the highest thermal stability. A 50/50 mixture of the most stable and the least stable has a thermal stability very much closer to the least stable. However, the products of Examples 1 and 2 have thermal stabilities comparable to or better than Z and approaching X in the case of the product of Example 1.

Example 4: The products of Examples 1 and 2 and the commercial products were evaluated in a Petter W1 single cylinder petrol engine run to Def 2101D, but extended beyond the normal 36 hours. The results of these engine tests are given in Table 3 on the following page.

The results confirm that the antiwear properties compare very well with zinc dialkyldithiophosphates prepared exclusively from primary or secondary alcohols and compare reasonably well with the blend.

Table 3: Engine Test Results

	Y	Z	50/50 X/Y Blend	Example 1	Example 2
Bearing wt loss, mg					
36 hr	9	54	4	11	9
48 hr	20	66	10	12	15
60 hr	22	70	22	15	17
% Viscosity increase					
36 hr	—	20	34.7	28	23
48 hr	44	45	51.3	35	43
60 hr	53	—	62.4	46	46
Undercrown (10 clean)	9.6	9.8	8.2	9.9	9.6

Metal Dithiophosphates

G. Caspari; U.S. Patent 4,085,053; April 18, 1978; assigned to Standard Oil Company (Indiana) describes a process for manufacturing metal dithiophosphates, and metal dithiophosphate compositions. The process for manufacturing zinc, barium, cadmium, magnesium or nickel dithiophosphates comprises reacting phosphorus pentasulfide with one or more alcohols to form a dithiophosphoric acid; neutralizing the dithiophosphoric acid with zinc, barium, cadmium, magnesium or nickel base in the presence of an acidic promoter; and reacting a substantial portion of the excess acidic promoter with a weak base.

Example 1: 317 g (1 mol) of a dithiophosphoric acid, prepared from a mixture of isobutanol, isoamyl alcohol, isooctanol and P_2S_5 in 15 wt % 5W oil, were neutralized by adding dropwise dithiophosphoric acid to a slurry of 47 g (0.57 mol) ZnO and 0.32 ml (0.005 mol) nitric acid (70%) in 20 g 5W oil at 175°F.

After addition of the acid, sweetness was checked (absence of H_2S) with lead acetate paper. If a base is to be added, it would be added at this point in the procedure. In this case, none is added.

The reaction mixture was heated to 200°F. Simultaneously, nitrogen was blown through the mixture to remove the water formed during the neutralization. The reaction mixture was then filtered through diatomaceous earth. The recovered filtrate contained 9.3% zinc, 8.45% phosphorus, and 18.2% sulfur.

Example 2: To the product of Example 1, 1.0 ml (0.015 mol) NH_4OH (28%) was added as base to the stirred reaction mixture. After 15 minutes, nitrogen was blown through the mixture and water was stripped off. At the same time, the temperature was raised to 200°F. To the water-free reaction mixture, 10 g of filter cell was added. Filtration through filter cell yielded a clear, light brown oil.

Example 3: The procedure of Example 1 was used except the neutralization promoter was 0.38 ml (0.005 mol) HCl (38%).

Example 4: To the product of Example 3, 1.0 ml (0.015 mol) NH_4OH (28%) was added as base following the procedure in Example 2.

Example 5: The procedure of Example 1 was followed except 2.5 g (0.01 mol) zinc sulfoxylate dissolved in 2 ml water, was added as a promoter after two-thirds of the dithiophosphoric acid had been dropped to the ZnO slurry.

Example 6: To the product of Example 5, 2.0 ml (0.03 mol) NH_4OH was added as base after complete neutralization of the dithiophosphoric acid following the procedure in Example 2.

Example 7: The procedure of Example 1 was followed except 3.8 g (0.01 mol) zinc acetate was used as neutralization promoter.

Example 8: To the product of Example 7, 2.0 ml (0.03 mol) NH_4OH was added as base following the procedure of Example 2.

Example 9: The procedure of Example 1 was followed except 0.32 ml (0.005 mol) HNO_3 (70%) was used as promoter.

Example 10: To the product of Example 9, 0.6 g (0.01 mol) urea dissolved in 1.5 ml H_2O was added as base following the procedure of Example 2.

Example 11: The procedure of Example 1 was followed except 0.32 ml (0.005 mol) HNO_3 (70%) was used as promoter.

Example 12: To the product of Example 11, 1.5 ml (0.15 mol) triethylamine was added as base following the procedure of Example 2.

Example 13: The procedure of Example 1 was followed except the dithiophosphoric acid was prepared from a mixture of isopropyl, isobutyl and isooctyl alcohols, and 0.32 ml (0.005 mol) HNO_3 (70%) was used as promoter.

Example 14: To the product of Example 13, 0.6 g (0.01 mol) urea was dissolved in 1.5 ml of water, and added as a base following the procedure of Example 2.

The effect of weak base treatment on acid promoted neutralized zinc dithiophosphates in regard to hydrolytic stability and deposit formation was demonstrated by heating the sample to 160°F for 48 hours and then allowing the sample to stand at room temperature for a period of time.

Effect of Base Treatment on Hydrolytic Stability and Deposit Formation

Example	Catalyst	Base	Appearance*
. . . . .Aliphatic ZOP from i-C_4H_9OH, i-$C_5H_{11}OH$, i-$C_8H_{17}OH$			
1	HNO_3	—	sweet, hazy, light deposits
2	HNO_3	NH_4OH	sweet, clear, no deposits
3	HCl	—	sour, deposits
4	HCl	NH_4OH	sweet, no deposits
5	Zn sulfoxylate	—	sour, light deposits
6	Zn sulfoxylate	NH_4OH	sweet, clear, no deposits
7	Zn acetate	—	sour, deposits
8	Zn acetate	NH_4OH	sweet, clear, no deposits
9	HNO_3	—	sweet, deposits
10	HNO_3	urea	sweet, clear, no deposits
11	HNO_3	—	sweet, deposits

(continued)

Example	Catalyst	Base	Appearance*
. . . . Aliphatic ZOP from i-C_4H_9OH, i-C_5H_{11}OH, i-C_8H_{17}OH			
12	HNO_3	triethylamine	sweet, clear, no deposits
. . . . Aliphatic ZOP from i-C_3H_7OH, i-C_4H_9OH, i-C_8H_{17}OH			
13	HNO_3	—	sour, heavy deposits
14	HNO_3	urea	sweet, clear, no deposits

*After heating to 160°F for 48 hr and standing at room temperature for 3 months.

Zinc Alkyl Dithiophosphate and Substituted Succinic Anhydride

T.M. Warne; U.S. Patent 4,094,800; June 13, 1978; assigned to Standard Oil Company (Indiana) describes lubricating oil compositions having improved antiwear properties which comprise a major portion of a lubricating oil and an effective amount of an oil-soluble additive combination comprising a basic zinc alkyl dithiophosphate having alkyl groups made from primary alcohols containing from about 6 to 20 carbon atoms and a nonacidic lubricating oil antirust compound comprising a succinic anhydride substituted with an alkenyl group which has about 8 to 50 carbon atoms reacted with an alcohol, an amine, or mixtures thereof.

The zinc dialkyldithiophosphate of this process, such as commercially available Oronite OLOA 269 N, OLOA 269 and ELCO 108 are generally made from dialkyldithiophosphoric acid having the formula:

$$\begin{array}{c} RO \\ \diagdown \\ \quad\; P \\ RO \diagup \quad \diagdown \end{array} \begin{array}{c} S \\ \diagup\!\!\diagup \\ \\ SH \end{array}$$

wherein R comprises an alkyl group containing about 7 to 12 carbon atoms, the alkyl groups being made from primary alcohols. Examples of suitable alcohols are normal alcohols such as n-heptyl, n-octyl, n-decyl, and n-dodecyl or branched chain alcohols such as methyl or ethyl branched isomers of the above.

Formation of sludge due to thermal degradation was determined when formulated oil was heated at 300°F in the presence of bubbling air and copper and iron catalyst for 96 hours. The percent sludge was calculated after filtration through a 0.5 micron Millipore filter. 100 g of the formulated oil is placed in a glass tube ca 12" long and 1¼" in diameter. 15" lengths of copper and iron wire are cleaned and coiled as described in ASTM D-943 and immersed in the oil. The tube is inserted to a depth of 9" in an aluminum block electrically heated to 300°F. Dry air is bubbled through the oil at a rate of 50 cc/min.

A water-cooled condenser is attached to the top of the tube. After 96 hours, the tube is removed from the heated block and allowed to cool to room temperature. The oil is decanted and the tube and metal catalyst washed with 100 ml ASTM isooctane. The oil and wash solvent are combined and filtered through a 0.45 micron Millipore filter.

Hydrolytic instability was demonstrated by heating at 100°C for 48 hours a sample of formulated oil in which has been dispersed 1% of distilled water. Evolution of H_2S and/or formation of solid deposits show poor water tolerance. 100 g of the formulated oil and 1.0 g distilled water are placed in an 8 oz bottle and heated to 210° to 215°F in an oven. When the oil reaches test temperature, the bottle is removed from the oven, capped tightly and shaken vigorously to mix oil and water. The capped bottle is returned to the oven for 24 hours. It is then removed, reshaken, and returned to the oven. After an additional 24 hours, the bottle is removed from the oven, reshaken and allowed to cool to room temperature in the dark for at least 24 hours. The oil is then observed for evidence of instability. H_2S formation is detected by odor and/or blackening of moistened lead acetate test paper. The oil is filtered through a 5 micron Millipore filter and the time required for filtration is noted. The weight and appearance of the residue are determined.

Example 1: Anglamol 75, a zinc dialkyldithiophosphate (ZOP) made from mixed secondary alkyl alcohols plus an alkenylsuccinic acid rust inhibitor, at 1.0 vol % was tested in a base oil made by blending SAE 10 and 20 weight solvent-extracted, hydrogenated base stocks to give an oil with a viscosity of 210 SUS at 100°F. The test oil additionally contained a polymethylacrylate pour depressant (Acryloid 703) at 0.2 vol % and a silicone antifoam (Dow Corning 200 fluid) at 2 ppm.

Example 2: 0.8 vol % Lubrizol 1360, a ZOP made from mixed primary alkyl alcohols, some of which may be branched, and 0.15 vol % of rust inhibitor Hitec E536, was tested in the same base oil as Example 1. The same pour depressant and antifoam were used. Hitec E536 is a nonacidic antirust condensation product of dodecenylsuccinic acid and a polyamine, having a total of about 2.6% nitrogen and an acid number of about 56.

Example 3: 1.0 vol % Oronite 973B, a ZOP made from primary alkyl alcohols, plus an alkenylsuccinic acid rust inhibitor, was tested in the same base oil as Example 1. The same pour depressant and antifoam were used.

Example 4: 0.8 vol % of OLOA 269N, a ZOP made from primary octyl alcohol, and 0.15 vol % Hitec E536 were tested in the same base oil as Example 1. The same pour depressant and antifoam were used.

Example 5: 1.0 wt % Lubrizol 1060, a ZOP prepared from secondary aliphatic alcohols, and 0.1 wt % of an acidic rust inhibitor, comprising a 50 vol % solution of dodecenylsuccinic acid in transformer oil, were tested in a solvent-extracted SAE 10 weight Midcontinent petroleum stock.

Example 6: 1.0 wt % of Lubrizol 1060, a ZOP prepared from secondary aliphatic alcohols and 0.1 wt % of Hitec E-536 were tested in the same base oil as Example 5.

Example 7: 1.0 wt % of Lubrizol 1360, a ZOP made from mixed primary alkyl alcohols, some of which may be branched, and 0.1 wt % of a 50 vol % solution of dodecenylsuccinic acid in transformer oil, were tested in the same base oil as Example 5.

Example 8: 1.0 wt % of Lubrizol 1360, a ZOP made from mixed primary alkyl alcohols, some of which may be branched, and 0.1 wt % of Hitec E-536 were tested in the same base oil as Example 5.

Example 9: 1.0 wt % OLOA 269N, an overbased ZOP made from primary alkyl alcohol, and 0.1 wt % of a 50 vol % solution of dodecenylsuccinic acid in transformer oil, were tested in the same base oil as Example 5.

Example 10: 1.0 wt % OLOA 269N, an overbased ZOP made from primary alkyl alcohol, and 0.1 wt % Hitec E-536, were tested in the same base oil as Example 5.

Example 11: 1.0 wt % OLOA 269R, an overbased zinc di-(2-ethyl-1-hexyl) dithiophosphate and 0.1 wt % of a 50 vol % dodecenylsuccinic acid in transformer oil, were tested in the same base oil as Example 5.

Example 12: 1.0 wt % OLOA 269R, an overbased zinc di-(2-ethyl-1-hexyl) dithiophosphate, and 0.1 wt % Hitec E-536, were tested in the same base oil as Example 5.

	Example			
	1	2	3	4
Sludge formation				
300°F, 96 hr, % sludge	0.40	0.033	0.021	0.033
Hydrolytic instability				
Insoluble residue, %	0.049	0.132	0.010	0.011
H₂S evolved	yes	yes	no	no
D2619 wet copper corrosion				
Copper loss, mg/cm²*	0.38, 0.52	0.38, 0.78	0.19, 0.28	0.17, 0.25
Copper appearance	2-C	4A-4B	2-D	1A-1B
H₂S evolved	no	no	no	no

*First number is after solvent cleaning only of copper strip. second number is for same strip after removal of chemically-bound copper with a 10% KCN solution.

	Example							
	5	6	7	8	9	10	11	12
Sludge formation, % 300°F, 72 hr**	0.45	0.43	0.0009	0.0011	0.0007	0.0032	0.0006	0.0006
Hydrolytic instability								
Insoluble residue, %	0.15	0.17	0.19	0.010	0.061	0.007	0.007	0.002
H₂S evolved	yes	yes	yes	no	yes	no	no	no
D2619 wet copper corrosion								
Copper loss, mg/cm²*	1.31	0.63	1.30	1.19	0.19	0.27	0.07	0.08
	1.61	0.87	1.92	1.52	0.43	0.29	0.14	0.12
Copper appearance	***	***	4-B	***	4-A	1-B	2-D	1-B
	4-C	4-C		4-C				
H₂S evolved	no	no	no	no	yes	no	no	no

*See note in above table.
**Type C test only 72 hr, rather than 96 hr as in previous table. There is no standard test length and the time before oxidation of the base oil begins to take place is somewhat less for this oil than for the oil used in the previous table.
***Brown.

Phosphorothionate Derivatives of Alkylphenol Sulfides

A.L.P. Lenack; U.S. Patent 4,136,041; January 23, 1979; assigned to Exxon Research & Engineering Co. describes phosphorothionate derivatives of alkylphenol sulfides and has found that certain of these materials which are ashless are effective antiwear and antioxidant additives in lubricating oils with sufficient oil solubility to allow them to be supplied as concentrates.

The process provides a compound of the general formula as shown on the following page.

In the above formula, Y is selected from phenoxy, alkylphenoxy, alkoxy, phenyl-thio, alkylphenylthio, thioalkyl, alkylamino and arylamino, and x is 1 or 2.

The aromatic rings of the compounds of the general formula, including Y, when Y contains an aromatic ring, may be substituted. For example, either or both rings may contain a substituent selected from alkyl, nitrogen-containing groups, oxygen, sulfur, halogen-containing groups or halogen itself, carboalkoxy or ether groups. Where the substituents are alkyl, it is preferred that they contain less than 30 carbon atoms, preferably less than 25.

Where the compounds are to be used as oil additives, the substituent must not of course inhibit the performance of the additive to an undesirable extent. If the compounds are to be used as oil additives, it is preferred that each aromatic ring of the alkylphenol sulfide nucleus carries an alkyl substituent containing from 2 to 24 carbon atoms, preferably from 6 to 15 carbon atoms, since those in which the alkyl substituent contains only 1 carbon atom have limited oil solubility. If Y is aromatic, it may also carry such an alkyl substituent.

The process for the production of compounds of the above general formula in-volves first reacting a phenol sulfide, preferably an alkylphenol sulfide, with a phosphorus halide, then reacting the product of the first stage with a compound of formula YH, where Y is selected from phenoxy, alkyl phenoxy, alkoxy, phenyl thio, alkylphenylthio, thioalkyl, alkylamino or arylamino to form the phosphite ester and reacting the phosphite ester with sulfur. The reaction may thus be depicted as follows:

This is, however, an oversimplification, since the final product obtained tends to be a mixture of materials. The exact composition of the mixture and the reason for its formation are not fully understood. Phenol sulfides, however, tend to be mixtures of compounds containing varying numbers of sulfur atoms in the bridge and also containing polymeric material.

Example 1: 563 g of phosphorous trichloride were added, drop by drop, to a solution of 1,300 g of a commercially available nonylphenol sulfide in 400 g of toluene. 2.7 g of water were added and the solution refluxed for 3 hours under a blanket of nitrogen after which the toluene and excess phosphorous trichloride were removed by vacuum distillation.

Elemental analysis of the product showed 65.0 wt % carbon, 8.4 wt % hydrogen, 7.5 wt % sulfur, 5.9 wt % phosphorus and 7.3 wt % chlorine with a molecular weight of 997. The theoretical content for the structure:

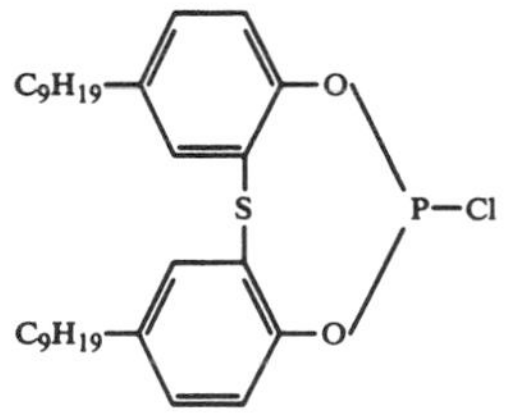

is 67.9 wt % carbon, 8.2 wt % hydrogen, 6.0 wt % sulfur, 5.8 wt % phosphorus and 6.6 wt % chlorine with a molecular weight of 534. This therefore indicates the probability of some molecules containing polysulfide linkages and some being polymeric.

66 g of nonylphenol were dissolved in 90 ml of toluene and 30.3 g of triethylamine added to this solution. A solution of 160.3 g of the bis(nonylphenoxy)-sulfide phosphorochloridite prepared above, dissolved in 250 ml of toluene was added to this nonylphenol solution held at 40°C and the resulting mixture stirred for 3 hours and the precipitated triethylamine chlorohydrate filtered off and the toluene removed by vacuum distillation.

Elemental analysis of this product showed it to contain 73.8 wt % carbon, 9.7 wt % hydrogen, 5.5 wt % sulfur, 4.4 wt % phosphorus and just a trace of chlorine. The molecular weight was 1,156.

The theoretical content for the structure:

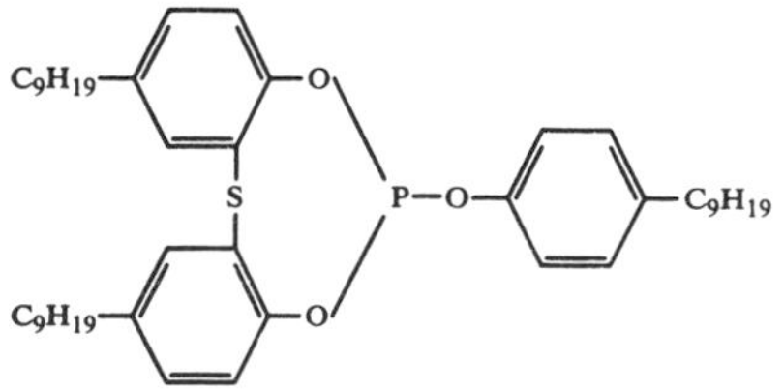

is 75.0 wt % carbon, 9.6 wt % hydrogen, 4.4 wt % sulfur and 4.3 wt % phosphorus

with a molecular weight of 718. This again indicates the probability of some polysulfide linkages and polymeric molecules.

200 g of the bis(nonylphenoxy)sulfide nonylphenoxyphosphite obtained as above were heated with 8.9 g of sulfur at 190°C for 30 minutes to yield bis(nonylphenoxy)sulfide nonylphenoxyphosphorothionate.

Example 2: The performance of the product of Example 1 as an oil additive was compared with the commercially available antiwear additive triphenyl phosphorothionate (TPPT), 2-nonylphenoxy-1,3,2-benzodioxaphosphate-2-sulfide, bis(nonylphenoxy)sulfide nonylphenoxyphosphite and a commercially available zinc dialkyldithiophosphate (ZDDP).

The solubility of the additive in oil was determined by tests conducted on blends of the neat additives in a paraffinic-type oil. The blends of additives in oil are made at various concentrations and are stored at room temperature for 1 week. The values given in the table below are the maximum additive concentrations observed above which a cloudy appearance and precipitate or layering tended to form.

The antiwear properties were assessed by the Hertz 4-ball test (ASTM-2266) at a concentration of 2.4 milliatoms of phosphorus per 100 g of oil. The hydrolytic stability of the material was tested according to ASTM D-2619. The results of these tests were as follows:

	TPPT	A*	B**	Example 1	ZDDP
Oil solubility, wt %	3.5	20	***	***	***
Hydrolytic stability					
Copper wt loss, mg/cm^2	0.1	10	1.5	0.1	0.3
NN on oil, mg KOH/g	0.1	0.5	—	0.1	0.3
Acid no. in water phase	2.7	196	147	16.7	4.0
Copper corrosion	1A	4C	1B	1B	2C
Antiwear properties, 4-ball data					
Scar, mm	—	0.5	0.5	0.4	0.4
Mean load, kg	—	—	34	46	52
Seizure load, kg	—	—	100	126	126
Weld load, kg	—	—	160	200	200

*A is 2-nonylphenoxy-1,3,2-benzodioxaphosphate 2-sulfide
**B is bis(nonylphenoxy)sulfide nonylphenoxy phosphite
***All proportions

Chlorinated Wax-Trihydrocarbyl Phosphite Reaction Products

G. Frangatos; U.S. Patent 4,098,707; July 4, 1978; assigned to Mobil Oil Corporation describes a lubricant composition comprising lubricant and an antiwear amount of a product prepared by reacting a chlorinated wax with a trihydrocarbyl phosphite.

Example 1: 150 g of a cracked wax containing 40% by weight of chlorine, a molecular weight of 610 and a viscosity at 210°F of 150 to 160 seconds and 50 g (0.2 mol) of tributyl phosphite were heated under nitrogen with stirring at 185° to 190°C for two hours. The clear, yellowish solution was cooled and kept at room temperature, under nitrogen, overnight. The following day, stirring under nitrogen at 185°C was resumed and continued for four hours. 18 g of phosphite was recovered when the contents were placed under house vacuum. 173 g of a clear, yellowish fluid was obtained. It should be noted that throughout the heating cycle, butyl chloride was distilling off.

The product had a chlorine content of 31.1% and a phosphorus content of 4.35%.

Example 2: 180 g (0.34 mol) of the cracked wax used in Example 1 and 50 g of triethyl phosphite (0.3 mol) were placed in a flask and heated at 170° to 175°C under nitrogen with continuous stirring for four hours. House vacuum was applied and 26 g of the phosphite was collected. This was returned to the reaction flask, the contents of which were then cooled to room temperature and kept under nitrogen overnight. Stirring at 170 to 175°C under nitrogen was begun and maintained for an additional three hours. Distillation gave 22.0 g of the phosphite and 193 g of a viscous, dark brown residue. The residue had a chlorine content of 37.8% by weight and a phosphorus content of 1.82% by weight.

To a 100 cc sample of a lubricating oil comprising an 80/20 mixture, respectively, a 150" solvent paraffinic bright mineral oil (210°F) and 200" solvent paraffinic neutral mineral oil (100°F) was added sufficient of the product of Example 1 (containing a 4.35% by weight of phosphorus).

The oils were tested in a modified 4-ball test at room temperature and at 200°F for 30 minutes under a load of 60 kg at varying rpm. At both temperatures, the average scar diameter, both horizontal and vertical was: 500 rpm, 0.40 mm; 1,100 rpm, 0.50 mm; 1,500 rpm, 0.60 mm and 2,000 rpm, 0.60 mm.

G. Frangatos; U.S. Patent 4,159,960; July 3, 1979; assigned to Mobil Oil Corporation has also found that the antiwear and load-carrying properties of lubricants are improved by incorporating a product prepared by reacting a trihydrocarbyl phosphate with a methallyl halide and a halogen.

Phosphorus Trihalide-Multifunctional Alcohol Reaction Products

A process described by *G. Frangatos; U.S. Patent 4,077,892; March 7, 1978; assigned to Mobil Oil Corporation* provides a lubricant composition having improved antiwear properties resulting from the addition thereto of a product made by reacting a partially esterified multifunctional alcohol with a phosphorus trihalide or a dihydrocarbyl phosphonate.

Example 1: A mixture of 20.4 g (9.15 mols) of pentaerythritol, 58.5 g (0.45 mol) of heptanoic acid and 150 ml of n-decane was stirred and refluxed for four hours. 8 ml of water was collected in a water trap. 30.3 g (0.3 mol) of triethylamine was added and the mixture was cooled. 13.7 g (0.1 mol) of phosphorus trichloride was added and the reaction mixture was heated at 70°C for one hour. It was cooled and then filtered, after which the solvent was removed, leaving a residue of 62 g having a phosphorus content of 3.29% by weight.

Example 2: A mixture of 360 g (0.6 mol) of pentaerythritol ester, 27.2 g (0.2 mol) of pentaerythritol and 0.5 g of calcium hydroxide was heated under nitrogen with stirring at 220° to 225°C for 3 hours. The reaction mixture was cooled and 81.0 g (0.8 mol) of triethylamine was added thereto. 34.3 g (0.25 mol) of phosphorus trichloride was slowly added to the mixture with stirring and cooling. Following the addition of phosphorus trichloride the reaction mixture was brought to 80°C and maintained there for one-half hour. The mixture was filtered and the filtrate was heated to 110°C under reduced pressure to remove any remaining phosphorus trichloride. The residue weighed 352 g and had a phosphorus content of 2.89% by weight.

Example 3: A mixture of 180.0 g (0.3 mol) of pentaerythritol ester, 27.2 g (9.2 mols) of pentearythritol and 1.0 g of calcium hydroxide was heated at 220° to 225°C for 3 hours. The reaction mixture was cooled to room temperature and 77.6 g (0.4 mol) of dibutyl phosphite, i.e.,

$$\overset{O}{\overset{\|}{(C_4H_9O)_2P-H}}$$

was added. This mixture was stirred at 190° to 195°C for 2 hours while stirring and bubbling nitrogen therethrough. Charcoal and a filter aid were added to the residue and it was filtered, leaving 247 g of product having a phosphorus content of 4.66% by weight.

The ester used in Examples 2 and 3 is obtained by reacting a mixture of C_{5-9} monocarboxylic acid and pentaerythritol. Its viscosity at 210°F was about 5 cs. In addition it had a hydroxyl number of about 3.4 and a saponification number of about 400. The products of the examples were tested in a modified 4-ball test.

One percent by weight of each product was placed in a blend of a 150" (210°F) solvent paraffinic bright mineral oil and a 200" (100°F) solvent paraffinic neutral mineral oil. These were blended in a ratio of 80/20, respectively. The samples were tested at various temperatures and speeds, but always at a load of 60 kg and for 30 minutes. The following table summarizes the test results.

Average Scar Diameter	. Room Temperature, rpm . .				. . . . .20°F, rpm				. . . . 390°F, rpm			
	500	1,000	1,500	2,000	500	1,000	1,500	2,000	500	1,000	1,500	2,000
	. . . . mm.											
Example 1												
Horizontal	0.40	0.40	0.40	0.50	0.40	0.50	0.80	0.50	0.50	0.60	0.80	2.30
Vertical	0.40	0.40	0.40	0.50	0.40	0.50	0.80	0.50	0.50	0.60	0.80	2.37
Example 2												
Horizontal	0.40	0.60	0.70	0.70	0.50	0.60	0.70	0.80	0.60	0.70	0.70	0.70
Vertical	0.40	0.60	0.70	0.70	0.50	0.60	0.70	0.80	0.60	0.70	0.70	0.70
Example 3												
Horizontal	0.40	0.40	0.40	0.40	0.40	0.40	0.50	0.50	0.50	0.50	0.70	0.70
Vertical	0.40	0.40	0.40	0.40	0.40	0.40	0.50	0.50	0.50	0.50	0.70	0.70
Untreated oil												
Final average	0.50	0.60	0.88	2.34	0.60	1.06	1.86	2.23	1.00	1.31	2.06	1.98

Monoaryl Phosphonates

In a process described by *R.F. Bridger and K.D. Schmitt; U.S. Patents 4,130,496; December 19, 1978; and 4,092,254; May 30, 1978; both assigned to Mobil Oil Corporation* monoaryl phosphonates are prepared by heating the corresponding aryl-di-t-alkyl phosphite with acid catalyst on an inert particulate matrix. Lubricant compositions of this process comprise oils and greases containing a minor amount of a monoaryl phosphonate to improve their antiwear properties.

Generally, the catalysts are heterogeneous acid catalysts on an inert matrix. Particularly suitable are macroreticular acid cation exchange resin catalysts, e.g., Amberlyst-15 which are characterized by substantial porosity, high surface area, low surface concentration, usually <0.5 meq hydrogen ion/m^2 surface area.

The monoaryl phosphonate additives are prepared as indicated below by thermal cleavage, preferably at a temperature of 80° to 130°C, of an aryl-di-t-alkyl phosphite, e.g., an aryl-di-t-butyl phosphite followed by acid-catalyzed elimination of a second molecule of, for example, isobutylene from the aryl-alkyl phosphonate.

Example: Preparation of 4-n-Nonylphenyl Phosphonate — A solution of 4-n-nonylphenol (0.04 mol, 8.8 g) and N,N-diethylaniline (0.04 mol, 5.96 g) in 75 ml hexane was added dropwise with stirring to excess phosphorous trichloride (0.4 mol, 55 g) in 100 ml hexane. The mixture was stirred 1 hour at room temperature and filtered. After removal of solvent and excess PCl$_3$ by vacuum distillation, 11.95 g (0.0372 mol) of 4-n-nonylphenyl phosphorodichlorodite was obtained. A hexane solution (100 ml) of t-butanol (0.0744 mol, 5.51 g) and N,N-diethylaniline (0.0744 mol, 11.09 g) was added dropwise to a solution of the phosphorodichlorodite in 50 ml of hexane.

Upon completion of the addition, the solution was refluxed 30 minutes, filtered and evaporated under vacuum to give a residue of 4-n-nonylphenyl-di-t-butyl phosphite (0.0332 mol, 13.14 g). The phosphite was heated at reflux in 50 ml toluene containing a suspension of Amberlyst-15 (2 g) resin for 90 minutes. The Amberlyst-15 was removed by filtration, and solvent was removed in vacuo to give 8.69 g (0.0306 mol) 4-n-nonylphenyl phosphonate as a colorless oil. Analyses confirmed the product structure.

The example illustrates that the method provides essentially pure monononylphenyl phosphonate, free of other phosphorus-containing impurities. Only a minute amount of di(nonylphenyl)phosphonate was present in the product due to an impurity in the nonylphenyl phosphorodichlorodite from which the nonylphenyl-di-t-butyl phosphite was prepared. No other phosphorus impurities were present.

The base stock oil employed in the 4-ball wear test results shown in the table below comprised a 150 SUS at 210°F solvent-refined paraffinic bright stock lubricating oil. The additive was tested at a concentration of 0.5 wt %; standard conditions of 40 kg load, 600 rpm, and 30 minutes test time were employed at 200°F.

As will be apparent from the data of the table, the lubricant composition of the process exhibits highly improved antiwear properties, as evidenced by the indicated comparative data with respect to wear scar diameter and wear rate.

Additive	Wear Scar Diameter (mm)	Wear Rate x 10^{12} (cc/cm-kg)
None	0.6858	4.60
Process	0.3885	0.282

Organophosphorus Derivatives of Hydroxycarboxylic Acids

A.O.M. Okorodudu; U.S. Patent 4,101,432; July 18, 1978; assigned to Mobil Oil Corporation describes compounds having the following general formula:

$$(RO)_m-\overset{\overset{(O)_n}{\|}}{P}-[O\overset{\overset{R''}{|}}{R'}(CH_2)_qCOOH]_{m'}$$

where R is alkyl of 1 to 32 carbon atoms, aryl or alkaryl and R' is alkylene or arylene and R'' is H, alkyl of 1 to 30 carbon atoms, aryl or alkaryl and where n is zero or 1 and q is zero or 1 to 30 and m and m' are 1 or 2 with the proviso that the sum of m and m' equals 3.

Although the compounds may be prepared in several ways, they are preferably prepared as described below by reacting an organophosphorus halide and a hydroxycarboxylic acid.

Suitable organophosphorus halides include compounds having the following general formula:

$$(RO)_m-\overset{\overset{(O)_n}{\|}}{P}-X_q$$

which includes such halides as $ROPX_2$, $(RO)_2PX$, $(RO)_2P(O)X$ and $ROP(O)X_2$, where R is alkyl of 1 to 32 carbon atoms, aryl or alkaryl, X is chloro-, bromo- or iodo-, m is 1 to 2, n is zero or 1, and q is 1 or 2, the sum of m and q being 3. Preferred are those halides having the structure $(RO)_2P(O)Cl$ or $(RO)_2PCl$.

Suitable hydroxycarboxylic acids include compounds having the following general structure:

$$R'-(CH_2)_q\overset{\overset{R''}{|}}{C}OOH$$

where R' is alkylene or arylene and R'' is H, alkyl of 1 to 30 carbon atoms and q is zero or 1 to 30. Preferred is lactic or 12-hydroxy stearic acid.

The organophosphorus halide, e.g., $ROPCl_2$, $(RO)_2PCl$ and $(RO)_2P(O)Cl$, may be conveniently reacted with hydroxycarboxylic acid in the following manner:

(1)
$$(RO)_2\overset{\overset{(O)_n}{\|}}{P}Cl + CH_3(CH_2)_5\overset{\overset{OH}{|}}{C}H-(CH_2)_{10}COOH \longrightarrow (RO)_2\overset{\overset{(O)_n}{\|}}{P}-O-\overset{\overset{(CH_2)_5CH_3}{|}}{C}H-(CH_2)_{10}COOH$$

(2)
$$(RO)-PCl_2 + 2HO\overset{\overset{CH_3}{|}}{C}H-COOH \longrightarrow (RO)-P-(O-\overset{\overset{CH_3}{|}}{C}H-COOH)_2$$

Various additives of the process were evaluated in the standard 4-Ball Wear Test using ½" 52100 steel balls at a load of 60 kg and for 30 minutes under the conditions set forth in Tables 1 and 2 below. The oils used were a 80/20 mixture of a solvent refined Mid-Continent paraffinic 150/160 second bright mineral oil and a 200/210 second refined Mid-Continent neutral mineral oil (Table 1), and a synthetic ester lubricant made by reacting pentaerythritol with an equimolar mixture of C_5 and C_9 monocarboxylic acids (Table 2).

Table 1

Ex.	Additive	Conc. Wt. %	Temp ° F	Scar Diameter (mm) (RPM)			
				500	1,000	1,500	2,000
	Mineral Oil Base Stock	100	Room	0.50	0.60	0.88	2.34
			200	0.60	1.06	1.86	2.23
			390	1.0	1.31	2.06	—
1	CH₃ | C₁₈H₃₇OP(OCHCOOH)₂	1	Room	0.50	0.40	0.50	0.60
			200	0.50	0.50	0.50	0.70
			390	0.50	0.60	0.60	0.70
2	C₉H₁₉—⟨ ⟩—OP(OCHCOOH)₂ with CH₃	1	Room	—	—	—	—
			200	0.50	0.60	0.60	0.60
			390	0.50	0.80	0.80	0.90
3	C₉H₁₉—⟨ ⟩—OP[OCH(CH₂)₁₀COOH]₂ with (CH₂)₅CH₃	1	Room	0.40	0.40	0.50	0.60
			200	0.40	0.50	0.60	0.60
			390	0.50	0.60	0.70	0.70
4	(C₉H₁₉—⟨ ⟩—O)₂P OCHCOOH with O, CH₃	1	Room	0.50	0.40	0.40	0.50
			200	0.50	0.40	0.50	0.60
			390	0.50	1.05	0.70	0.70
5	(C₉H₁₉—⟨ ⟩—O)₂P—OCH(CH₂)₁₀COOH with O, (CH₂)₅CH₃	1	Room	0.50	0.50	0.50	0.50
			200	0.50	0.50	0.50	0.60
			390	0.50	0.50	0.70	0.90
6	(C₉H₁₉—⟨ ⟩—O)₂POCHCOOH with CH₃	1	Room	0.50	0.50	0.50	0.60
			200	0.50	0.50	0.60	0.60
			390	0.50	0.70	0.80	0.70
7	(C₉H₁₉—⟨ ⟩—O)₂P OCH(CH₂)₁₀COOH with (CH₂)₅CH₃	1	Room	0.40	0.50	0.50	0.60
			200	0.40	0.50	0.50	0.50
			390	0.50	0.60	1.00	1.20
8	(C₄H₉O)₂POCHCOOH with O, CH₃	1	Room	0.50	0.60	0.80	0.90
			200	0.70	0.70	1.00	1.20
			390	0.70	0.80	0.90	1.00
9	(C₄H₉O)₂P—OCH—(CH₂)₁₀COOH with O, (CH₂)₅CH₃	1	Room	0.40	0.50	0.80	0.60
			200	0.50	0.50	0.60	0.60
			390	0.60	0.60	0.60	0.60
10	(C₉H₁₉—⟨ ⟩—O)₂P—O—⟨ ⟩—COOH	1	Room	0.40	0.50	0.50	0.53
			200	0.50	0.50	0.50	0.60
			390	0.50	0.60	0.60	0.70
11	(C₉H₁₉—⟨ ⟩—O)₂P—O—⟨ ⟩ with COOH	1	Room	0.40	0.43	0.40	0.65
			200	0.50	0.46	0.50	—
			390	0.50	0.60	0.65	0.70

Table 2

Example	Additive	Conc. Wt.%	Temp. ° F	Scar Diameter (mm) Speed (RPM)			
				500	1,000	1,500	2,000
1	Synthetic Base Stock	100	Room	0.70	0.90	0.90	1.95
			200	0.80	0.90	2.0	2.10
			390	0.90	1.30	1.50	2.40
2	C₉H₁₉—⟨ ⟩—OP[OCH(CH₂)₁₀COOH with (CH₂)₅CH₃]₂	1	Room	0.50	0.50	0.50	0.60
			200	0.40	0.50	0.60	0.70
			390	0.65	0.80	0.90	1.85
3	(C₉H₁₉—⟨ ⟩—O)₂POCHCOOH with CH₃	1	Room	0.50	0.60	0.60	0.58
			200	0.50	0.60	0.70	0.70
			390	0.60	0.80	0.65	2.3
4	(C₉H₁₉—⟨ ⟩—O)₂POCHCOOH with O, CH₃	1	Room	0.50	0.80	1.86	1.95
			200	0.60	0.80	1.80	1.95
			390	0.80	0.90	1.90	2.0

Lubricant Composition

H. Shaub and W.E. Waddey; U.S. Patent 4,105,571; August 8, 1978; assigned to Exxon Research & Engineering Co. describe a lubricating composition containing a combination of additives comprising (1) a zinc dihydrocarbyl dithiophosphate; (2) an ester of a polycarboxylic acid and a glycol; and (3) an ashless dispersant containing a high molecular weight aliphatic hydrocarbon oil solubilizing group and wherein either one of the zinc or ester components or both separately are predispersed in the ashless dispersant prior to adding the other of the zinc or ester components to the lubricating composition.

By keeping the zinc and ester components separate until one of them is already predispersed, it has been found that the resulting composition overcomes the problem of incompatability and is storage stable. Additionally and significantly, such lubricant composition has excellent antifriction and antiwear properties particularly under extreme pressure or heavy load conditions.

Example 1: Several formulations were prepared using a 10W-40 SE quality automotive engine oil containing 1.5% by weight based on the total lubricating oil weight of zinc dialkyldithiophosphate (80% active ingredient in diluent mineral oil) in which the alkyl groups were a mixture of such groups having between 4 and 5 carbon atoms and made by reacting P_2S_5 with a mixture of about 65% isobutyl alcohol and 35% of amyl alcohol; 0.1% by weight, based on the total lubricating oil weight of an ester formed by the esterification of a dimer acid of linoleic acid and diethylene glycol and having the formula:

$$HOCH_2CH_2OCH_2CH_2OC(CH_2)_7HC=CH \overset{O}{\overset{\|}{}}$$

Various dispersants were used in the different lubricating formulations as described below: (A) An ashless dispersant was prepared by reacting polyisobutenyl succinic anhydride (PIBSA), the polyisobutenyl radical (PIB) having an average molecular weight ($\overline{Mn}$) of about 900, with an equimolar amount of pentaerythritol and a minor amount of a polyamine mixture comprising polyoxypropyleneamine and polyethyleneamine to form a product having a nitrogen content of about 0.35% by weight. Materials of this type are described in U.S. Patent 3,804,763 and sold by Lubrizol Corporation as Lubrizol 6401.

(B) A borated ashless dispersant was prepared by condensing 2.1 mols of polyisobutenyl succinic anhydride, the polyisobutenyl radical having an average molecular weight of about 1,300, dissolved in solvent neutral 150 mineral oil to provide a 50 wt % solution with 1 mol of tetraethylenepentamine. The polyisobutenyl succinic anhydride solution was heated to about 150°C with stirring and the polyamine was charged into the reaction vessel over a four hour period which was thereafter followed by a three hour nitrogen strip. The temperature was maintained from about 140° to 165°C during both the reaction and the subsequent stripping. While the resulting product was maintained at a temperature of from about 135° to 165°C, a slurry of 1.4 mols of boric acid in mineral oil was added over a three-hour period which was thereafter followed by a final four-hour nitrogen strip. After filtration and rotoevaporation, the concentrate (50 wt % of the reaction product) contained about 1.46 wt % nitrogen and 0.32 wt % of boron.

(C) An ashless dispersant was prepared by charging 1.0 mol of PIBSA having a PIB group with an $\overline{Mn}$ of about 1,300 dissolved in 500 ml of solvent 150 neutral, 0.36 mol of zinc acetate dihydrate as a promoter and 1.9 mols of tris(hydroxymethyl) aminomethane (THAM) into a glass reactor. Heating at about 168° to 174°C for four hours gave the expected quantity of water. After filtration and rotoevaporation, the concentrate (50 wt % active ingredient) analyzed 1.0 wt % nitrogen.

(D) An ashless dispersant was prepared in a similar manner as described in (B) above using 1.3 mols of PIBSA (PIB had $\overline{Mn}$ of about 900) and boration was not undertaken. The product had a nitrogen content of 2.1% by weight.

In preparing the final lubricating oil compositions, the ester component of each composition was first dispersed in the following amounts of the above-defined ashless dispersants: (A) 5.25% by wt of dispersant (mixture of 46.5% by wt active ingredient in mineral lubricating oil); (B) and (C) 5.25% by wt of dispersant (mixture of 50% by wt active ingredient in mineral lubricating oil); and (D) 6.3% by wt of dispersant (mixture of 50% by wt active ingredient in mineral lubricating oil).

The ester portion of each composition as described above (0.1% by wt) was dispersed in the above-defined dispersants at about 65°C and stirred for 2 hours and then added to a solution of a standard lubricating composition of 10W-40 SE crankcase oil which contained a rust inhibitor, i.e., overbased magnesium sulfonate, a detergent, a V.I. improver, i.e., an ethylene-propylene copolymer, and the aforementioned zinc dialkyldithiophosphate (1.5% by wt, 80% active ingredient in mineral oil).

As contrasted to compositions wherein the zinc dialkyldithiophosphate was added to the dicarboxylic acid/glycol ester prior to predispersing either one, all of the above exhibited storage stability over an extended period of several months at ambient temperature. The formulation containing dispersant D did show signs of somewhat poor storage stability as evidenced by additive dropout after two weeks at ambient temperature indicating that an increased amount of this type dispersant was necessary to maintain the compatability of the system.

Example 2: In this example, two compositions prepared as described in Example 1 and containing a zinc dialkyldithiophosphate and a dicarboxylic acid/glycol ester were tested for relative friction and wear using a ball on cylinder test. For comparison, a standard 10W-40 SE quality automotive engine oil containing only the zinc component was also tested for relative friction and wear.

The apparatus used in the ball on cylinder test is described in the journal of the American Society of Lubrication Engineers, *ASLE Transactions,* Vol. 4, pages 1-11, 1961.

(I) In the first composition, a standard 10W-40 SE lubricating oil composition, the same as defined in Example 1, containing dispersant D and 1.5% by weight of zinc dialkyldithiophosphate (80% active ingredient in mineral oil) and the other standard additives including a rust inhibitor, a detergent, a V.I. improver, but without the dicarboxylic acid/glycol ester, was blended together.

(II) In this composition, the ester component as defined in Example 1 was predispersed in ashless dispersent D (described in Example 1) and then combined with the standard lubricating composition containing additives including the dialkyl dithiophosphate as also described in Example 1.

(III) In this composition, the ester component was predispersed in ashless dispersant A and then combined with the standard lubricating composition containing additives including the zinc dialkyldithiophosphate as fully described in Example 1.

The following table shows the resulting relative friction and wear data for the three compositions, with composition I (without ester) being assigned relative values of 1.00:

| | Relative Ball/Cylinder Data | |
	Friction	Wear
Composition I	1.00	1.00
Composition II	0.59	0.62
Composition III	0.62	0.48

Besides the improved friction and wear properties exhibited in a lubricating oil composition containing both a zinc dialkyldithiophosphate and a dicarboxylic acid/glycol ester (Compositions II and III), Composition I (without ester) and Composition III were given a standard engine test, i.e., Sequence III C Test to determine valve train wear as shown in the following table.

| | Sequence III C Test | |
| | ....Cam and Lifter Wear.......... | |
	Max in. x 10^{-4}	Avg in. x 10^{-4}
Composition I	11	8
Composition III	7	4

The composition containing both the zinc dialkyl dithiophosphate and a dicarboxylic acid/glycol ester (i.e., Composition III) showed highly satisfactory results and this was particularly surprising in view of the expected displacement of some of the zinc component, an exceptional extreme pressure agent, by the ester.

Substituted 1,2,3-Triazoles

A process described by *K.D. Schmitt; U.S. Patent 4,115,288; September 19, 1978; assigned to Mobil Oil Corporation* is directed to a lubricant composition that is comprised of a major amount of a lubricating oil or grease and as an antiwear agent a minor amount of a substituted 1,2,3-triazole characterized by the formula:

$$
\begin{array}{cc}
N\!\!-\!\!-\!\!-\!\!-\!\!C\!-\!R_1 \\
\| \qquad \| \\
N \qquad C\!-\!R_2 \\
\diagdown_N\diagup \\
| \\
H
\end{array}
$$

where R_1 and R_2 are individually selected from the group consisting of an alkyl group containing from 1–18 carbons; an aryl group, such as phenyl; hydrogen; and

$$\begin{array}{c} N \!\!-\!\!-\!\!-\!\! C \!-\!(CH_2)_{\overline{n}} \\ \| \qquad \| \\ N \qquad C\!-\!H \\ \diagdown N \diagup \\ | \\ H \end{array}$$

and where at least one of R_1 and R_2 is other than hydrogen and where n is an integer from 5 to 15. Representative substituted triazoles which impart good antiwear properties to lubricants are 4-phenyl-1,2,3-triazole and 4-heptyl-1,2,3-triazole.

Benzotriazole and Substituted Succinic Anhydride

E.A. Swakon; U.S. Patent 4,096,077; June 20, 1978; assigned to Standard Oil Company (Indiana) describes a composition which comprises a major proportion of a natural or synthetic lubricating oil and a minor proportion of an oil-soluble wear-inhibiting additive composition. The wear-inhibiting composition is benzotriazole or C_{1-20} alkyl substituted benzotriazole and a material selected from the group consisting of at least one half-acid, half-ester, half-acid half-amide, and half-acid half-thioester of succinic or maleic acid or acid anhydride and an alcohol, amine, and mercaptan, respectively, a metal salt of at least one half-ester, half-amide, or half-thioester, and mixtures thereof.

The alcohol and mercaptan comprise acrylic aliphatic compounds containing from 7 to 50 carbon atoms, wherein the amine comprises a straight-chain primary or secondary amine having from 7 to 50 carbon atoms and is selected from the group consisting of monoamines and polyamines having at least 2 carbon atoms separating each pair of nitrogen atoms, and where the metal salt comprises a Group IIA, Group IIB, tin or lead metal salt.

Unless otherwise specified, alkyl groups in compounds used in these examples are straight-chain.

Example 1: An additive composition containing calcium dodecyl maleate was produced by first charging 210 g (1.13 mols) of dodecyl alcohol, 100 g (1.02 mols) of maleic anhydride and 200 g of toluene to a 2 liter 3-neck flask, equipped with a stirrer, thermometer, and Dean Stark trap. This mixture was heated to 120°C and refluxed at 120°C for approximately 1 hour. Then the resulting reaction product, the half-acid of dodecyl maleate, was allowed to cool to 55°C, at which point 40 g (0.54 mol) of calcium hydroxide was added to the reaction products. The temperature rose to approximately 85°C.

The mixture was then heated to reflux, and an azeotropic mixture of toluene and water containing 19 ml (1.06 mols) of water distilled over. The resulting product, calcium dodecyl maleate, was a clear yellow liquid. Next, the product was cooled, and 150 g of a solvent-extracted, dewaxed, paraffinic mineral oil, having a viscosity of 840 SUS at 100°F was added to dilute the product. Finally, the remaining toluene was stripped from the solution by heating under house vacuum.

Example 2: A second additive composition containing calcium dodecyl maleate was prepared by first charging 105 g (0.56 mol) of dodecyl alcohol, 50 g (0.51 mol) of maleic anhydride, 100 g of toluene, and 1 g of 2,6-di-tert-butyl-p-cresol, which served as an antioxidant, to a 1 liter 3-neck flask similarly equipped as the flask used in Example 1. This mixture was heated to 120°C and refluxed at about 120°C for about 2 hours. The resulting half-acid of dodecyl maleate was then allowed to cool to 60°C, and 20 g (0.27 mol) of calcium hydroxide was added to the reaction products. The temperature rose to 85°C, and the product became viscous. The mixture was then heated to reflux, and an azeotropic mixture of water and toluene containing about 8.5 ml (0.47 mol) of water distilled over. Next, the product, calcium dodecyl maleate, was cooled to 100°C, and 150 g of the same paraffinic mineral oil used in Example 1 was added to dilute the product. Finally, the remaining toluene was stripped from the solution at 140°C and under house vacuum. The resulting solution was filtered while hot and was transparent and colorless.

Example 3: An additive composition containing calcium dodecyl succinate was prepared by first charging 205 g (1.1 mols) of dodecyl alcohol, 100 g (1 mol) of succinic anhydride and 100 g of toluene to a similar flask with the same equipment as in Example 2. After refluxing this mixture at about 145°C for 2.5 hours, the resulting product, the half-acid of dodecyl succinate, was cooled to 80°C, and 37 g (0.5 mol) of calcium hydroxide was added to the mixture; The temperature rose to 90°C, and the solution became clear. The mixture was heated to reflux, and an azeotropic mixture of toluene and water containing 8.5 ml (0.47 mol) of water started to distill over at 100°C. The solution was cooled to 80°C, and 320 g of the mineral oil used in Example 2 was added to dilute the product, calcium dodecyl succinate. The remaining toluene was stripped by heating the solution under house vacuum. Finally, the solution was filtered while still hot, and, upon cooling, it solidified to a white wax.

Example 4: A second additive composition containing calcium dodecyl succinate was produced using the same equipment and similar conditions as in Example 3. 105 g (0.56 mol) of dodecyl alcohol, 50 g (0.5 mol) of succinic anhydride, and 100 g of toluene were originally charged to the flask, and 20 g (0.27 mol) of calcium hydroxide was added later to the half-acid of dodecyl succinate. The azeotropic mixture which distilled over, contained 13.5 ml (0.76 mol) of water. After cooling, 150 g of the oil added in Example 1 was added to dilute the calcium dodecyl succinate, and then the remaining toluene was stripped from the solution at 130°C and under house vacuum. The additive composition, which was viscous and light yellow, was filtered.

Example 5: A third additive composition containing calcium dodecyl succinate was produced using the same equipment, components, and amounts used in Example 4, except that the oil was charged initially to the flask instead of later in the procedure. After refluxing the starting mixture at about 140°C for between 2 and 3 hours, the resulting product, half-acid dodecyl succinate, was cooled to 80°C, and the calcium hydroxide was added. The temperature rose to 96°C. Upon refluxing, an azeotropic mixture of toluene and water containing 9 ml (0.5 mol) of water, distilled over. When the remaining toluene was stripped, a viscous, colorless transparent liquid solution was left.

Example 6: An additive composition containing a mixed calcium salt of hexadecyl maleate and dodecyl maleate was prepared by first charging 50 g (0.27 mol) of dodecyl alcohol, 70 g (0.29 mol) of hexadecyl alcohol and 50 g (0.51 mol) of maleic anhydride, 100 g of toluene and 1 g of 2,6-di-tert-butyl-4-methylphenol, which served as an antioxidant to a similar flask with the same equipment used in Example 1. After heating this mixture for 2 hours at 100°C, the resulting product mixture, the half-acid of dodecyl maleate and the half-acid of hexadecyl maleate, was cooled to 60°C, and then 20 g (0.27 mol) of calcium hydroxide was added to the mixture. The temperature rose to 85°C. The mixture was then refluxed at about 120°C, and an azeotropic mixture of water and toluene containing 8.5 ml (0.47 mol) of water distilled over. Finally, 170 g of the mineral oil used in Example 2 was added to dilute the product, and the remaining toluene was stripped from the solution at 140°C under house vacuum. The resulting product was a colorless liquid.

Examples 7 through 19: Examples 7 through 19 involve blends containing the additive compositions produced in the examples. Additive compositions were tested as antiwear agents using the Shell 4-Ball Wear Test for wear-inhibiting characteristics of oils. Blends of these additive compositions in various master blends of oils were made. The compositions of the master blends employed are presented in Table 1. The apparatus and procedure used in the Shell 4-Ball Wear Test are described in ASTM D-2266-67. The apparatus was obtained from Precision Scientific.

The tests were performed under a load of 40 kg at 200°F for 2 hours and at 600 rpm. Test results were reported as the diameter in millimeters of the circular scar produced in the Shell 4-Ball Wear Test. The components in the master blends are as follows: J—a solvent-extracted and hydrofinished, dewaxed paraffinic mineral oil having a viscosity of 380 SUS at 100°F; K—a solvent-extracted and hydrofinished, dewaxed paraffinic mineral oil having a viscosity of 840 SUS at 100°F; L—2,6-di-tert-butyl-4-methylphenol, an antioxidant; M—4,4' tetramethyl-diamino-diphenylmethane, an antioxidant; N—a polyacrylate, a viscosity index improver; O—1% silicone fluid in a light furnace oil, an antifoam agent.

Table 1: Compositions of Master Blends

Components	A (g)	B (g)
J	630	732
K	1,620	1,902
L	6.75	6.9
M	1.65	1.65
N	—	13.5
O	—	1.4*

*Milliliters

Table 2

Example No.	Master Blend*	Weight of Master Blend Added (g)	Additive from Example	Weight of Additive Added ....(g)	Weight of Benzotriazole Added	Shell 4-Ball Wear Test (mm)
7	A	—	—	—	—	0.68
8	A	375	1	0.375	—	**
9	A	375	1	0.375	0.07	0.38
10	A	375	6	1.2	—	0.52
11	A	375	6	1.2	0.07	0.43
12	A	375	2	2.25	—	0.48
13	A	375	2	2.25	0.075	0.47
14	A	375	1	0.5	0.08	0.45
15	A	375	6	1.5	0.08	0.41
16	B	880	2	4.0	0.18	0.46
17	A	375	3	1.9	—	0.48
18	A	375	3	1.9	0.08	0.47
19	A	375	5	1.9	0.08	0.46

*The master blends corresponding to the letters are identified in Table 1.
**This additive composition was insoluble, and measurements could not be made.

1,2,4-Thiadiazole Polymers

A process described by *J.P. King, E.A. Mailey, and I.C. Popoff; U.S. Patent 4,107,059; August 15, 1978; assigned to Pennwalt Corporation* relates to a polymer of 1,2,4-thiadiazole that is especially effective as an additive in lubricants that enable the lubricants to withstand extremely high pressure and yet maintain antiwear properties.

This process is thus directed to: (A) A composition having the formula:

wherein: R is selected from the group consisting of

phenylene, biphenylene, an alkylene or substituted alkylene of 2 to 50 carbons, preferably 2 to 10 carbons, cyclic alkylene or substituted cyclic alkylene of 5 to 50 carbons, preferably 6 to 10 carbons, wherein the alkylene or cyclic alkylene can contain in the chain or ring oxygen and/or sulfur atoms, or $(S)_x$-groups; m is an integer of 0 to 10, preferably 1 to 5; n is an integer of 5 to 100, preferably 10 to 40; and x is an integer of 1 to 5, preferably 1 to 2.

(B) A lubricating composition comprising a major amount of lubricating grease (or fluid) and a minor amount of the compound described in (A), above. The major amount is 80 to 99.9 parts (percent) of the lubricating grease and the minor amount is 20 to 0.1 parts, preferably 3 to 5 parts, of the polymer of 1,2,4-thiadiazole.

Triazole Derivatives

H.J. Andress, Jr.; U.S. Patent 4,144,180; March 13, 1979; assigned to Mobil Oil Corporation describes a lubricant composition comprising a major amount of an oil of lubricating viscosity or greases and a load carrying amount of the reaction product of (A) benzotriazole or a lower alkylbenzotriazole with (B) a mono- or dialkylphosphonate having 4 to 14 carbon atoms per alkyl group, a mono- or di-alkylphosphate having 4 to 14 carbon atoms per alkyl group, a primary fatty amine having 6 to 26 carbon atoms, a naphthenyloxazoline, or an alkenylsuccinyl mono- or bisoxazoline. The reaction is conducted at a temperature between about 100°C and about 200°C and in a mol ratio of (A) to (B) of 1:2, in the case of phosphonates and phosphates, and of about 1:1, in the case of amines and oxazolines.

Example 1: A mixture of 36.4 g (1.31 mols) of oleylamine and 125 g (1.05 mols) benzotriazole was stirred at 140°C for about 3 hr to form the final product.

Example 2: A mixture of 388 g (2.0 mols) dibutyl phosphonate and 133 g (1.0 mol) tolyltriazole was stirred at 140°C for about 2 hr to form the final product.

Example 3: A mixture of 210 g (0.4 mol) ditetradecyl catechol phosphonate and 26.6 g (0.2 mol) tolyltriazole was stirred at 140 to 145°C for about 3 hr to form the final product.

Example 4: A mixture of 295 g (0.8 mol) of a 50-50 mix of mono- and di-tridecyl phosphates and 54 g (0.4 mol) tolyltriazole was stirred at 140°C for 2 hr to give the final product.

Example 5: A mixture of 362 g (1 mol) of diisodecyl phosphonate and 67 g (0.5 mol) tolyltriazole was stirred at 140°C for 3 hr to give the final product.

Example 6: A mixture of 300 g (1.0 mol) naphthenic acid and 121 g (1.0 mol) tris(hydroxymethylamino)methane was refluxed in the presence of toluene diluent at about 215°C until evolution of water ceased. To this compound (a naphthenyloxazoline) was added 133 g (1.0 mol) tolyltriazole and the mixture stirred for 4 hr at about 140°C to give the final product.

Example 7: A mixture of 185 g (0.5 mol) isooctadecenylsuccinic acid and 121 g (1.0 mol) tris(hydroxymethylamino)methane was refluxed in toluene at a temperature of about 250°C until evolution of water ceased. The product is an isooctadecenylsuccinyl bisoxazoline.

Example 8: A mixture of 269 g (0.5 mol) of isooctadecenylsuccinyl bisoxazoline (made according to Example 7) and 66.5 g (0.5 mol) tolyltriazole was stirred at about 140°C for 6 hr to form the final product.

The composition was tested in the Timken Load Test. The test is a known test used to determine the load carrying properties of additives in lubricating oil compositions.

Timken OK Load Test — Additives were blended in a conventionally refined mineral oil containing a sulfurized hydrocarbon, metal passivator, pour point depressant, antioxidant, demulsifier, antirust agent, and defoamant.

Additive	Percent	Load (lb)
None	0	50
Example 1	0.1	60
Example 2	0.1	65
Example 3	0.1	60

(continued)

Additive	(%)... Percent	Load (lb)
Example 4	0.1	60
Example 5	0.1	60
Example 6	0.1	65
Example 8	0.1	60

N-Butoxydodecenylsuccinimides

K.J. Chou and W.W. Hellmuth; U.S. Patent 4,104,182; August 1, 1978; assigned to Texaco Inc., describe a lubricating oil composition which comprises a hydrocarbon oil of lubricating viscosity and a metal-containing additive characterized by promoting the formation of hard deposits in an internal combustion engine and an effective deposit softening amount of a hydrocarbon-substituted succinimide represented by the formula:

$$R-CH-C \overset{\displaystyle O}{\diagup} \diagdown N-O-R'$$
$$CH_2-C \diagdown O$$

in which R is a hydrocarbon radical having from 1 to 50 carbon atoms and R' is a hydrocarbon radical having from 3 to 20 carbon atoms. The hydrocarbon radical represented by R' can be an aliphatic, aromatic or a mixed aliphatic-aromatic hydrocarbon radical.

The hydrocarbon-substituted succinimide described above which is effective for softening the deposits formed in the combustion zone of an internal combustion engine can be readily prepared from an N-hydroxy-hydrocarbon-substituted succinimide. The preparation of the N-hydroxy-hydrocarbon-substituted succinimide precursor has been described in U.S. Patent 3,796,663.

In the preparation of the additive of the process, an N-hydroxy-hydrocarbon-substituted succinimide is reacted with an aliphatic halide, conveniently an aliphatic chloride or bromide, in the presence of an organic base or a hydrogen halide acceptor in a suitable solvent for the reactants. This reaction proceeds at room temperature with the formation of the hydrogen halide salts and the prescribed N-alkoxy-hydrocarbon-substituted succinimide. The solvent is removed from the reaction product preferably under vacuum and the reaction product is washed with water until the prescribed succinimide is washed free of the salts.

The following examples illustrate the preparation of the deposits modifier or softener for the lubricant of this process. Yields are in weight percent based on the N-hydroxy hydroxysuccinimide reactant.

Example 1: 0.14 mol of butyl bromide and 0.14 mol of N-hydroxydodecenylsuccinimide are dissolved in 100 ml of dimethylformamide. 50 ml of triethylamine are added and the mixture is stirred overnight at room temperature. The solvent is then removed under vacuum and the reaction product is water-washed until salt free and then dried. A yield of 85% of N-butoxydodecenylsuccinimide is recovered.

Example 2: 0.14 mol of propyl bromide and 0.14 mol of N-hydroxydodecenylsuccinimide are dissolved in 100 ml of dimethylformamide. 50 ml of triethyl-amine are added and the mixture is stirred overnight at room temperature. The solvent is then removed under vacuum and the reaction product is water-washed until salt free and then dried. A substantial yield of N-propoxydodecenylsuccin-imide is recovered.

Example 3: 0.14 mol of pentyl bromide and 0.14 mol of N-hydroxydodecenyl-succinimide are dissolved in 100 ml of dimethylformamide. 50 ml of triethyl-amine are added and the mixture is stirred overnight at room temperature. The solvent is then removed under vacuum and the reaction product is water-washed until salt free and then dried. A yield of 85% of N-pentoxydocenylsuccinimide is recovered.

Example 4: 0.14 mol of hexyl bromide and 0.14 mol of N-hydroxyoctadecenyl-succinimide are dissolved in 100 ml of dimethylformamide. 50 ml of triethyl-amine are added and the mixture is stirred overnight at room temperature. The solvent is then removed under vacuum and the reaction product is water-washed until salt free and then dried. A substantial yield of N-hexoxyoctadecen-ylsuccinimide is recovered.

Example 5: 0.14 mol of triphenylchloromethane and 0.14 mol of N-hydroxy-tetrapropenylsuccinimide are dissolved in 100 ml of dimethylformamide. 50 ml of triethylamine are added and the mixture is stirred overnight at room tempera-ture. The solvent is then removed under vacuum and the reaction product is water-washed until salt free and then dried. A yield of 90% of N-trityloxytetra-propenylsuccinimide is recovered.

An essential feature of the lubricant of the process is the presence of an organic metal-containing lubricant additive which under the conditions of use promotes the formation of hard deposits in an internal combustion engine. The most harm-ful lubricating oil additives in this respect are the organic zinc compounds, such as the zinc hydrocarbyl dithiophosphates.

Other conventional organic metal-containing lubricating oil additives which may promote hard deposits in an internal combustion engine include calcium- or mag-nesium-containing phenolates or sulfonates.

The improvement of reducing deposits hardness caused by a lubricating oil com-position was demonstrated in a bench test and in an engine test in which the hardness of the deposits produced was measured against comparison lubricants. Hardness is reported as the Knoop Hardness Number, the higher numbers indicat-ing increasingly harder deposits.

A base oil employed in the examples was a blend of paraffinic mineral oils having the following inspection tests.

Gravity, °API	27.2
Viscosity, SUS at	
100°F	183.3
210°F	15.18

This base oil was blended with a conventional organic metal containing additive to form the following Base Blend A, parts being in weight percent: base oil,

99.0; and zinc dialkyldithiophosphate, 1.0. The zinc dialkyldithiophosphate was prepared by reacting 2.7 mols of an alcohol mixture (consisting of 70% heptanols, 10% hexanols, 10% octanols and 10% butanols and minor components), 2.3 mols of isopropanol and 1.0 mol P_2S_5 in a conventional manner and then reacting the dialkyldithiophosphonic acid with an excess of zinc oxide to form the zinc dialkyldithiophosphate.

Base Blend B was a conventional diesel engine oil. It contained a calcium carbonate overbased [300 TBN (Total Base Number)] calcium sulfonate, an overbased (5 TBN) calcium sulfonate, a 2/1 overbased sulfurized calcium alkylphenolate, a zinc dialkyldithiophosphate and a zinc diaryldithiophosphate in addition to a conventional dispersant, rust inhibitor and foam inhibitor. Its inspection values were as follows:

Gravity, °API	25.2
Viscosity, SUS at	
100°F	137.0
210°F	12.76
Percent phosphorus	0.17
Percent calcium	0.24
Percent zinc	0.10

The oil compositions of the process as well as comparison oils were employed under deposit forming conditions in a bench test and in an engine test in order to evaluate the effectiveness of the lubricating oil compositions of the process. In the bench test, a burner nozzle was positioned midway inside of a 6 inch heat shielded well which rested on a hot plate as a source of heat. An aluminum panel was placed at the bottom of the well 3 inches below the vertically positioned, downwardly directed nozzle. The temperature of the aluminum panel was controlled by a thermocouple.

In operation, the temperature of the aluminum panel was maintained at 675°F. The test oil was passed through the nozzle at a rate of 8 liters per minute in conjunction with an air jet. This test was run for a period of 2 hr. On completion of this high temperature oil oxidation run, the aluminum panel with the adhering oil deposits was cooled. The deposits formed were removed from the panel and their hardness determined.

In the engine test, a Caterpillar-1-G engine was used with the test oil employed as the crankcase lubricant for this engine. This engine was operated under standard operating conditions for a period of 180 hr. On completion of the running time, the engine was disassembled and the deposits formed were removed from the piston rings, grooves and skirts and from the cylinder walls and their hardness determined.

The effectiveness of the deposit softening additive of the process was determined by adding same to the foregoing base blends and comparing the results in the abovedescribed Bench and Caterpillar Engine Tests. The results are given in the table below.

Both the Bench Test and the Engine Test demonstrate that there was a substantial reduction in the hardness of the deposits formed when the lubricating oil composition of the process was employed in these tests.

Lubricant Composition

Run No.	Additive Composition	Knoop Hardness No.* Bench Test	Caterpillar 1-G Engine Test
1	Blend A + 2 wt % N-hydroxydodecenyl succinimide	227-268	—
2	Blend A + 2 wt % N-butoxydodecenyl succinimide	194-227	—
3	Blend B	268-755	268-355
4	Blend B + 2 wt % N-butoxydodecenyl succinimide	155-194	155-194
5	Blend B + 2 wt % N-trityloxytetrapropenyl succinimide	155-194	—

*Knoop Hardness Numbers were determined by direct comparison to standard specimens of known hardness.

Azole Aminopolysulfides

A process described by *T. Colclough; U.S. Patent 4,104,179; August 1, 1978; assigned to Exxon Research & Engineering Co.,* provides a lubricating oil or petroleum fuel oil composition comprising: (1) a lubricating oil or a petroleum fuel oil; (2) an azole amino polysulfide or an azine amino polysulfide of the structure:

$$-N\diagdown \atop -X \diagup C-S_n-N\diagup{R^1} \atop \diagdown{R^2}$$

where the group

$$-N\diagdown \atop -X \diagup C-$$

forms part of a 5-membered azole heterocyclic ring or a 6-membered ring; n is at least 2, X is O, S or NR^3 where R^3 is H or a lower alkyl group and R^1 and R^2 which are the same or different are either hydrogen atoms or hydrogen- and carbon-containing groups or they are part of a hydrogen-and carbon-containing ring; and (3) an ashless dispersant.

Example 1: Preparation of Morpholinobenzothiazole Disulfide — tert-Butyl aminobenzothiazole (480 g, 2 mols) was heated with sulfur (649 g, 2 g atoms) and morpholine (180 g) in ethanol (1,700 ml) for 2 hr at 75° to 80°C. The reaction mixture, which contained much solid was cooled, solvent stripped, filtered and washed with petroleum ether to remove morpholine. The product, morpholinobenzothiazole disulfide (540 g) contained sulfur 34.5%, nitrogen 9.7% (theory, sulfur 33.7, nitrogen 9.9).

This morpholinobenzothiazole disulfide was added to a 10W-30 lubricating oil containing a conventional ashless dispersant and a V.I. improver to give a Petter W-1 bearing weight loss (BWL) of 65 mg at 1 weight percent concentration. A similar lubricating oil composition without any disulfide gave a BWL of about 4,000 mg on the same test. It is clear by these results that the disulfide has good antiwear properties.

The results of bench tests using the same 10W-30 lubricating oil containing a conventional ashless dispersant and a V.I. improver was as follows:

Additive (wt % in 10W-30 oil)	4 Ball Wear Test (120 kg, 1 min) Scar Diameter (mm)	Lead Corrosion (mg wt loss)
Nil	3.0	6,000
Morpholinobenzothiazole disulfide (0.5)	1.9	22

The morpholinobenzothiazole disulfide was a solid which was found to be soluble in a highly paraffinic base oil of 150 SSU viscosity at 210°F at between 70° and 80°C although slight sedimentation occurred at 20°C. It was found, however, that 15 parts of the product were soluble in 75 parts of zinc dialkyldithiophosphate and that a clear mobile liquid was retained at temperatures as low as 0°C when 10 parts of nonylphenol were added to this solution.

Example 2: The antiwear and corrosion properties of the morpholinobenzothiazole disulfide prepared in Example 1 were compared to those of morpholinobenzothiazole monosulfide. Equimolar quantities of the two compounds were incorporated into separate samples of a highly refined paraffinic lubricating oil containing 6% by weight of a conventional polyamine type ashless dispersant. These oils were then subjected to the standard 4 ball wear scar diameter test and SOD test in which standard samples of copper and lead are attached to a stirrer in a vessel containing the oil sample which is then subjected to air oxidation by blowing for 20 hr at 325°F and measuring the weight loss of the metal.

An oil containing no sulfide but 1.0% by weight of a zinc dialkyldithiophosphate (ZDDP) was also subjected to the same 4 ball test. The results were as follows:

Additive, wt %	4 Ball Wear Test (120 kg, 1 min) Scar Diameter (mm)	Lead Weight Loss in SOD Test (mg)
Morpholinobenzothiazole disulfide, 0.83	1.65	2
Morpholinobenzothiazole monosulfide, 0.75	2.05	310
ZDDP, 1.0	2.50	—

Sulfurization of Alcohol-Epichlorohydrin Reaction Products

K.P. Michaelis and H.O. Wirth; U.S. Patent 4,147,666; April 3, 1979; assigned to Ciba-Geigy AG, Switzerland have found that compounds of the general formula (1), and mixtures thereof,

$$(1) \qquad R-X-(CH_2-CH-O-)_nH$$
$$\underset{CH_2SR^1}{|}$$

in which R is linear alkyl having 12 to 24 C atoms, branched alkyl having 8 to

30 C atoms or alkylated phenyl or phenylalkyl having 8 to 24 C atoms, X is an oxygen or sulfur atoms, n is a value from 0.5 to 8 or an integer from 1 to 8 and R^1 is a hydrogen atom or a radical of the formula

$$R-X-(CH_2-CH-O)_n H$$
$$\quad\quad\quad\quad\; |$$
$$\quad\quad\quad CH_2-S_m-$$

in which m is 0 or integers from 1 to 3, are excellent extreme-pressure additives in lubricants.

The compounds of formula (1) are obtained by processes similar or analogous to those described in U.S. Patent 3,361,723, by reacting compounds of the formula (2), or mixtures thereof,

$$R-X-(-CH_2-CH-O)_n-H$$
$$\text{(2)}\quad\quad\quad\quad |$$
$$\quad\quad\quad\quad CH_2Cl$$

in which R, X and n are as defined, with hydrogen sulfide or dialkali metal sulfide or dialkali metal sulfides Na_2S_x (x = 1-4) in at least stoichiometric amounts and removing from the reaction mixture the hydrogen chloride or the sodium chloride which is formed. The dialkali metal sulfides are obtained in a known manner by dissolving sulfur in Na_2S.

Sulfurized Tallow Oil

A process developed by *J.M. Wakim; U.S. Patent 4,134,845; January 16, 1979; assigned to Shell Oil Company* relates to a sulfurized material obtained by sulfurizing (a) at least one of an inedible tallow oil and an alcoholysis product thereof, or (b) a mixture of (a) and an ester of an unsaturated fatty acid having 12 to 30 carbon atoms and an alkanol which ester is different from the alcoholysis product and/or a triglyceride different from inedible tallow oil.

The inedible tallow oil is preferably obtained by dissolving inedible tallow in a selective solvent, cooling the solution to a suitable temperature, filtering and evaporating the solvent from the filtrate.

The inedible tallow is preferably purified to remove extraneous materials and to reduce the free fatty acid content. Suitable purification methods include extraction with a lower alcohol having, e.g., 1 to 5 carbon atoms. Methanol is particularly suitable. Another suitable method is that of neutralizing the free fatty acids with alkali and removal of the soaps produced, by conventional techniques such as filtration, centrifugation, etc. The neutralization step can be performed on the tallow, as received, or as a solution in a selective solvent.

The alcoholysis product of the inedible tallow oil is preferably obtained by transesterification of inedible tallow oil with an alkanol having 1 to 30 carbon atoms. Alkanols having 1 to 5 carbon atoms are preferred and methanol is particularly preferred.

The sulfurized material may contain an ester of an unsaturated fatty acid having 12 to 30 carbon atoms and an alkanol, suitably an alkanol having from 1 to 30 preferably 1 to 5 carbon atoms; methanol is preferred. A suitable unsaturated fatty acid is oleic acid. Esters of mixtures of unsaturated acids and/or of mixtures of alkanols are also suitable.

Examples 1 through 9: 100 pbw inedible tallow oil having an acid number of 7.2 mg KOH/g and an iodine value of 62 was dewaxed with acetone at 0°C which resulted in 58.0 pbw tallow oil having an acid number of 10.8 mg KOH/g and an iodine value of 75 and 42.0 pbw solid phase having an acid number of 2.6 and an iodine value of 42.

107.5 pbw inedible tallow was extracted with methanol which resulted in 100 pbw inedible tallow having an acid number of 1.1 mg KOH/g and an iodine value of 56. Dewaxing with acetone at 0°C resulted in 44.1 pbw tallow oil having an acid number of 1.9 mg KOH/g and an iodine value of 72 and 55.9 pbw solid phase having an acid number of 0.6 mg KOH/g and an iodine value of 30.

Mixtures containing this methanol extracted tallow oil or the methyl ester prepared therefrom were sulfurized to a sulfur content of 9.5 wt % and 5 wt % of the sulfurized products were added to a SAE 90 lubricating oil. The results are indicated in the following table.

A 5 wt % solution of sulfurized sperm oil, sulfur content 9.5 wt %, in the same base oil, showed a Timken OK Load of 40 lb and a 4 ball wear scar diameter of 0.63 mm.

	. . . . .Examples								
	1	2	3	4	5	6	7	8	9
Ester, wt %	*	**	**	**	–	**	**	**	***
(all as methyl esters)	55	55	55	55		100	30	70	55
Oil, wt %	**	*	**	***	**	–	**	**	**
	45	45	45	45	100	–	70	30	45
Sulfur, wt %	9.5	9.5	9.5	9.5	9.5	9.5	9.5	9.5	9.5
Viscosity at 100°F, cs	94.2	147	71.0	110	2040	18.1	179	55.1	96.2
Properties of 5% solution in base oil									
Timken OK load, lbs	50	45	50	45	35	40	35	50	40
4-ball wear, scar diameter, mm	0.54	0.54	0.57	0.54	0.53	0.55	0.55	0.55	0.54
CCS viscosity at 0°F, poise	185	204	204	199	215	177	204	199	177
RBOT Life, min ASTM D-2272	315	365	355	280	305	235	260	255	245
Pour point, °F	5	15	15	15	0	15	5	10	5

*Canbra.
**Tallow.
***Soya.

Sulfurization of Dicyclopentadiene and Alloocimene

G.J.J. Jayne, D.R. Woods, and J.P. O'Brien; U.S. Patent 4,147,640; April 3, 1979; assigned to Edwin Cooper and Company Limited, England describe a lubricating oil additive which is prepared by reacting about one mol of a first reactive olefin hydrocarbon containing about 6 to 18 carbon atoms and 1 to 3 olefin double bonds concurrently with 0.1 to 5 mols of elemental sulfur and about 0.1 to 1 mol of hydrogen sulfide to obtain an intermediate. Then the intermediate is reacted at about 100° to 210°C with about 0.2 to 1 mol of a second reactive olefinic hydrocarbon containing about 6 to 18 carbon atoms and 1 to 3 olefinic double bonds to obtain an oil soluble lubricating oil additive.

Examples of the reactive olefinically unsaturated hydrocarbons include isoprene, cyclopentene, methylcyclopentene, cyclohexene, 1-octene, limonene, norbornene, norbornadiene, octadecene, styrene, and α-methylstyrene.

The more preferred olefinically unsaturated hydrocarbons are alloocimene, i.e., 2,6-dimethyl-2,4,6-octatriene, and cyclopentadiene dimers including dicyclopentadiene and lower C_{1-4} alkyl substituted cyclopentadiene dimers such as methylcyclopentadiene dimer.

The initial reaction is preferably carried out using a sulfurization catalyst. These are well-known and include quaternary ammonium salts, guanidines, thiuram sulfides and disulfides, sodium dialkyldithiocarbamates, alkyl and cycloalkyl amines, such as n-butylamine, di-n-butylamine, n-octylamine, triethylamine, diisopropylamine, dicyclohexamine, and cyclohexylamine.

A catalyst is preferably used in the second stage. For instance, it has been found that 2,5-dimercapto-1,3,4-thiadiazole, 2,5 bis(tert-octyldithio)-1,3,4-thiadiazole and 2-(tert-dodecyldithio)-5-mercapto-1,3,4-thiadiazole may be employed in the second stage with particular advantage.

Accordingly, a preferred catalyst for use in the second stage is a thiadiazole having the formula

$$R_1S-C \overset{N\overset{\|}{}----N\overset{\|}{}}{\underset{S}{\diagdown \diagup}} C-SR_2$$

where R_1 and R_2 are independently selected from hydrogen or $-SR_3$, wherein R_3 is alkyl (e.g., methyl, tert-octyl, tert-dodecyl and the like).

The following examples illustrate the manner of preparing additives of the process.

Example 1: In a reaction vessel was placed 264 g (2 mols) of dicyclopentadiene, 128 g (4 mols) of sulfur and 3.9 g of di-n-butylamine. The mixture was stirred at 80° to 90°C while hydrogen sulfide was injected into the liquid phase. An exothermic reaction occurred. After 4 hr hydrogen sulfide up-take had about stopped.

In a second stage 132 g (1 mol) of dicyclopentadiene was added and while stirring the solution was heated to 160°C. Stirring was continued one hour at 160° to 170°C. The product was heated in hot petroleum ether (BP 62° to 68°C) and filtered. It was vacuum stripped of volatiles to yield 487 g of a viscous product analyzing 29.2% sulfur.

Example 2: In a reaction vessel was placed 264 g of dicyclopentadiene, 128 g of sulfur and 3.9 g of di-n-butylamine. The mixture was stirred at 80° to 90°C while hydrogen sulfide was injected. The reaction exotherm was sufficient to maintain 80° to 90°C temperature requiring some cooling. Hydrogen sulfide injection was continued until temperature dropped. A total of 92.7 g of hydrogen sulfide was passed into the liquid phase of which 30.9 g reacted, the remainder being trapped in the off-gas.

The mixture was blown with nitrogen to remove hydrogen sulfides and 132 g of dicyclopentadiene was added. The solution was heated to 160°C and stirred at that temperature for 30 min. It was then cooled, diluted with petroleum ether and filtered. Nothing appeared to be removed on the filter. The product was stripped of volatiles under vacuum yielding 507 g of a dark viscous oil soluble liquid analyzing 29.7% sulfur.

Example 3: A stock batch of intermediate was prepared by reacting 528 g of dicyclopentadiene, 256 g of sulfur, 7.8 g of di-n-butylamine and 64 g of hydrogen sulfide at 90° to 110°C for 2 hr.

In a separate reaction vessel was placed 106.5 g of the above intermediate and 34 g alloocimene. The mixture was heated at 150°C for 2 hr giving a viscous liquid oil-soluble product analyzing 29% sulfur. It gave a 2b to 3b rating in the Copper Strip Corrosion Test.

Example 4: In a reaction vessel was placed 264 g of dicyclopentadiene, 128 g of sulfur, 3.9 g of di-n-butylamine and 3.9 g of 2,5-bis(octyldithio)-1,3,4-dithiadiazole. The mixture was reacted at 95° to 100°C during which 29 g of hydrogen sulfide was taken up.

In a second stage 145 g of dicyclopentadiene was added and the solution reacted at 170°C for 15 min. It was then filtered giving 550.1 g of a viscous oil-soluble liquid analyzing 27.6% sulfur.

Example 5: In a reaction vessel was placed 264 g of dicyclopentadiene, 128 g of sulfur, 3.9 g of di-n-butylamine and 3.9 g of 2,5-bis(octyldithio)-1,3,4-thiadiazole. This mixture was stirred at about 95°C while hydrogen sulfide was injected. A total of 26 g reacted.

Following this, 136 g of alloocimene was added and the mixture stirred 2 hr at 150°C. It was filtered hot to give 540.6 g of liquid product analyzing 26.1% sulfur. It gave a 1b rating in the Copper Strip Corrosion Test.

Tests have been carried out to demonstrate the utility of the additives. One test was the 4-Ball in which an EN 31 steel ball is rotated in loaded contact with 3 fixed similar balls. The contact is lubricated with a mineral oil solution of the test additive (IP 239/73T). Test criteria are the initial seizure load at which collapse of the oil film between the balls occurs, weld load and scar diameter at different loads. The test oil contained sufficient additive to provide 0.85% S in the oil blend. Table 1 gives the test results.

Table 1

Additive	Initial Seizure Load (kg)	Weld Point (kg)
Base oil	65	135
Example 3	90	340
Example 4	80	350
Example 5	85	390

A second test was the Timken OK Load Test (IP 240/74). Additive concentration was sufficient to provide 0.3% sulfur in the base oil which was a 150 solvent neutral mineral oil. In this test, a test block bears against a rotating cup. The OK load is the maximum load at which no scoring or seizure occurs. Table 2 gives the results of this test.

Table 2

Test Additive	Timken OK Load (lb)
None	12
Example 3	40
Example 4	35
Example 5	50

Sulfurized Fatty Acid Esters

D.A. Lee, and J.A. Boslett; U.S. Patent 4,149,982; April 17, 1979; assigned to The Elco Corporation describe extreme pressure additives which are prepared by sulfurizing a mixture comprised of (a) 50 to 85% by weight of an ester of a higher fatty acid and glycerol, or a mono lower aliphatic ester of fatty acid, or mixtures thereof and (b) 50 to 15% by weight of a mono alpha unsaturated olefin having about 15 to 20 carbon atoms. The extreme pressure additives are especially useful in improving the high pressure characteristics of greases, gear oils, and way lubricants.

Example 1: Mixtures containing various percentages of cottonseed oil and alpha olefin were sulfurized. The cottonseed oil employed had an iodine number of 194. The alpha olefin employed was a mixture of C_{15-18} alpha olefins, comprised of 91% by weight of alpha olefins with 8% by weight of diolefins and 1% by weight of paraffins. Both the cottonseed oil and the alpha olefin were clear and free of sediment.

The mixtures were sulfurized with 11% by weight of the total weight of the mixture using flowers of sulfur. The process condition included initial blending of the starting material together. The reaction mixtures were then heated to 320° to 360°F for 6 hr with continuous stirring. The resulting mixtures were cooled to 100° to 200°F and air was blown through the mixtures for an additional 6 hr.

The amount of cottonseed oil, alpha olefin and sulfur employed in each mixture is shown in the table below. Test results are also given.

Sample No.	Percent Alpha Olefin	Percent Cottonseed Oil	Viscosity at 210°F (SUS)	Appearance* A	B	Mean Hertz (kg)
1	0	100	847	**	**	35.2
2	22.8	77.2	272	clear	clear	37.5
3	59.0	59.0	89.0	clear	clear	36.1
4	63.8	36.2	183	clear	clear	36.0
5	87.5	12.5	403	clear	hazy	34.5
6	94.9	5.1	522	clear	**	37.9
7	100	0	50.2	clear	clear	29.8

Sample No.	Seizure (kg)	Weld (kg)	Wear (mm)***
1	71	316	0.560
2	71	316	0.539
3	71	282	0.587
4	71	282	0.574
5	71	282	0.553
6	79	316	0.560
7	56	224	0.595

(continued)

| | | | | . . . . . . Reverse Spring . . . . . . | | |
Sample No.	1,600 cpm/20 lb	2,100 cpm/11 lb	2,600 cpm/2 lb	1,600/ 2 lb	2,100/ 11 lb	2,600/ 20 lb
1	1.63	1.39	0.88	0.51	1.41	1.61
2	1.82	1.59	0.84	0.59	1.13	1.47
3	1.74	1.59	1.15	0.86	1.78	1.95
4	1.80	1.72	1.09	0.81	1.87	2.02
5	1.97	1.75	0.76	0.80	1.53	1.93
6	1.84	1.88	0.93	0.92	1.57	1.80
7	1.85	2.00	1.69	0.88	1.88	2.12

(. Barry Test . . . .)

*A, appearance of 10% in solvent extracted neutral 100/100 oil. B, appearance of 10% midcontinental solvent extracted base stock SAE 90.

**Hazy with precipitate.

***Shell Four Ball Test, 20 kg, 130°F, 1,000 rpm, 1 hr (scar diameter, mm)

Each mixture was evaluated for viscosity and solubility. The extreme pressure properties of the sulfurized mixtures were evaluated using Mean Hertz and Weld tests in accordance with ASTM D 2783-69T. A wear test was conducted using the Shell 4 Ball Test. These tests did not disclose the significant difference between the additives which was observed in actual field tests.

A further test was run in accordance with the test procedures reported by Barry and Binkleman in the paper entitled "A Wear Tester For The Evaluation Of Gear Oil", published in the January 1971 issue of *Lubrication Engineering* which was presented at the 25th ASLE Annual Meeting, May 4 to 8, 1970. This test, referred to in the chart as the Barry test, clearly demonstrated the criticality of the combinations of oils for the extreme pressure additive. The results obtained (wear scar diameter, mm) are shown in the table.

Sulfur-Containing Molybdenum Dihydrocarbyldithiocarbamate Compounds

T. Sakurai, A. Nishihara, T. Handa, H. Katoh, Y. Tomoda, K. Aoki, and M. Yoto; U.S. Patent 4,098,705; July 4, 1978; assigned to Asahi Denka Kogyo KK, Japan describe a compound having the following general formula (1);

(1)
$$\left[\begin{array}{c} R_1 \\ \\ R_2 \end{array} \!\!\! \diagdown\!\!\!\diagup N{-}\overset{\displaystyle}{\underset{\underset{\displaystyle S}{\parallel}}{C}}{-}S \right]_2 \!\!\!{-}Mo_2O_xS_{4-x}$$

where R_1 and R_2 stand for a hydrocarbyl group having from 1 to 24 carbon atoms, x is a number of 0.5 to 2.3, useful as an additive for lubricants.

The compound having the general formula (1) is obtained, with reliability, by reaction between carbon disulfide and a secondary amine having the following general formula (2)

(2)
$$\begin{array}{c} R_1 \\ \\ R_2 \end{array}\!\!\!\diagdown\!\!\!\diagup NH$$

where R_1 and R_2 have the same meanings as defined in the general formula (1), under a temperature of higher than 80°C, in water medium containing a molybdenum compound selected from the group consisting of molybdenum trioxide, alkaline metal molybdates, ammonium molybdate and their mixtures, and containing a sulfide compound selected from the group consisting of an alkali metal hydrogen sulfide, ammonium hydrogen sulfide, an alkali metal sulfide, ammonium sulfide and their mixtures, in the molar ratio of molybdenum compound to sulfide compound in the range between 1:0.05 and 1:4, having a pH of 0.5 to 10.0.

For the purpose of more reliable production, simpler operation of the reaction and quality control, the molybdenum compound is at least partially, preferably almost completely, reacted with the sulfide compound before addition of carbon disulfide and the secondary amine, to the reaction system. The reaction easily takes place with agitation under a temperature of 10° to 60°C for 30 to 60 min. In the case of using molybdenum trioxide, it is preferable to use the powdered one in order to react faster with the sulfide compound, which will increase its solubility in water. Preferable alkali metal molybdates are sodium molybdate and potassium molybdate. Sodium molybdate dehydrate was used in the examples.

Example 1: Into a reactor were added 100 ml of water, 0.05 mol of sodium sulfide nonahydrate and 0.05 mol of sodium molybdate. The pH of the mixture was adjusted to 2.8 by adding 20% sulfuric acid while agitating the mixture. Into the mixture were added 0.1 mol of di-n-butylamine and 0.1 mol of carbon disulfide, and the mixture was agitated for 30 min at room temperature and then was reacted for 4 hr at 95° to 102°C. Then, the precipitate was filtered and washed by methanol and dried. 16.8 g of fulvescent solid having a melting point of 271° to 272°C, was obtained. The yield is 95.5%. The results of elementary analysis are as follows: C = 31.5%, H = 4.7%, N = 4.2%, S = 29.4%, Mo = 28.4%. The number of x in general formula (1) was determined as 1.55.

Example 2: Into a reactor were added 100 ml of water, 0.05 mol of sodium sulfide nonahydrate and 0.05 mol of sodium molybdate. The pH of the mixture was adjusted to 2.8 by adding 20% sulfuric acid while agitating the mixture. Into the mixture were added 0.055 mol of di-n-butylamine and 0.075 mol of carbon disulfide, and the mixture was agitated for 30 min at room temperature and then was reacted for 4 hr at 95° to 102°C. Then, the precipitate was filtered and washed by methanol and dried. Yellowish orange solid having a melting point of 269° to 270°C was obtained. The yield is 96.3%. The results of elementary analysis are as follows: S = 30.6%. The number of x in general formula (1) was determined as being 1.35.

Example 3: Into a reactor were added 100 ml of water, 0.15 mol of sodium sulfide nonahydrate and 0.05 mol of sodium molybdate. The pH of the mixture was adjusted to 7.1 by adding 20% sulfuric acid while agitating the mixture. Into the mixture, were added 0.1 mol of di-n-butylamine and 0.1 mol of carbon disulfide, and the mixture was agitated for 30 min at room temperature and then was reacted for 4 hr at 95° to 102°C. Then, the precipitate was filtered and washed by methanol and dried. Yellowish orange solid having a melting point of 215° to 217°C was obtained. The yield is 83.0%. The results of elementary analysis are as follows: S = 32.5%, Mo = 28.4%. The number of x in general formula (1) was determined as being 0.75.

Example 4: Into a reactor were added 100 ml of water, 0.05 mol of potassium sulfide nonahydrate and 0.05 mol of sodium molybdate. The pH of the mixture was adjusted to 2.8 by adding 20% sulfuric acid while agitating the mixture. Into the mixture, were added 0.1 mol of dilaurylamine and 0.1 mol of carbon disulfide, and the mixture was agitated for 30 min at room temperature and then was reacted for 4 hr at 95° to 102°C. Then, the precipitate was filtered and washed by methanol and dried. Pale yellow solid having a melting point of 172° to 173°C was obtained. The yield is 93.2%. The results of elementary analysis are as follows: S = 18.0%. The number of x in general formula (1) was determined as being 1.70.

Example 5: Into a reactor were added 100 ml of water, 0.05 mol of potassium sulfide nonahydrate and 0.05 mol of sodium molybdate. The pH of the mixture was adjusted to 2.8 by adding 20% sulfuric acid while agitating the mixture. Into the mixture were added 0.1 mol of dicyclohexylamine and 0.1 mol of carbon disulfide, and the mixture was agitated for 30 min at room temperature and then was reacted for 4 hr at 95° to 102°C. Then, the precipitate was filtered and washed by methanol and dried. Pale yellow solid having a melting point of 290° to 291°C, was obtained. The yield is 90.0%. The number of x in general formula (1) calculated from the results of elementary analysis was determined as being 1.63.

Example 6: Into a reactor were added 100 ml of water, 0.05 mol of sodium sulfide nonahydrate and 0.05 mol of sodium molybdate. The pH of the mixture was adjusted to 2.8 by adding 20% sulfuric acid while agitating the mixture. Into the mixture were added 0.055 mol of stearylbenzylamine and 0.055 mol of carbon disulfide, and the mixture was agitated for 30 min at room temperature and then was reacted for 4 hr at 95° to 102°C. Then, the precipitate was filtered and washed by methanol and dried. Yellow solid having a melting point of 90° to 91.5°C was obtained. The yield is 95.0%. The number of x in general formula (1) calculated from the results of elementary analysis was determined as being 1.67.

Comparative Production Example 1: 0.1 mol of sodium molybdate dihydrate is dissolved in 100 ml of water, and neutralized with sulfuric acid. 0.2 mol of dibutylamine and 0.2 mol of carbon disulfide are added to the neutralized mixture and agitated for 30 min. Then the mixture is refluxed for 5.5 hr at 97° to 100°C. Then the reaction mixture is filtered and washed with toluene. 23.8 g of yellowish solid having a melting point of 259°C is obtained. The results of elementary analysis are as follows: C = 31.5%, H = 5.28%, N = 4.1%, S = 25.1%, and Mo = 28.0%. The number of x in general formula (1) was determined as 2.5.

The reaction product refluxed for 8 hr is the same product as the above solid product.

Examples 7 through 12 and Comparative Examples 1 through 3: The compounds obtained in the above examples are used in an amount of 1 weight percent in a solution or suspension of turbine oil. The diameters of the wear tracks created by the tests of the Shell 4 ball machine at 80 kg of load and 1,800 rpm are measured. The results are shown in Table 1.

Table 1

	Compound	Value of x*	Wear Tracks Diameter (mm)
Ex.			
7	product of Ex. 1	1.55	0.64
8	product of Ex. 2	1.35	0.64
9	product of Ex. 3	0.75	0.63
10	product of Ex. 4	1.70	0.40
11	product of Ex. 5	1.63	0.43
12	product of Ex. 6	1.67	0.38
Comparative ex.			
1	molybdenum disulfide	–	1.13
2	**	0	0.63
3	–	–	2.41

*In the general formula (1).   **$[(n\text{-}C_4H_9)_2NCS_2]_{\frac{1}{2}}Mo_2S_4$.

Examples 13 through 18 and Comparative Examples 4 and 5: The compounds obtained in the above examples are mixed with lithium 12-hydroxystearate grease at the ratio of 0.005 mol per 100 g of the grease. The welding loads measured by the tests of the Shell 4 ball machine are shown in Table 2.

Table 2

	Compounds	Welding Load (kg)
Ex.		
13	product of Ex. 1	320
14	product of Ex. 2	350
15	product of Ex. 3	380
16	product of Ex. 4	330
17	product of Ex. 5	320
18	product of ex. 6	360
Comparative ex.		
4	molybdenum disulfide	250
5	–	160

Mercapto-Substituted Boron-Containing Compounds

According to a process described by *B.A. Baldwin; R.P. Williams, and R. Rohde; U.S. Patent 4,115,286; September 19, 1978; assigned to Phillips Petroleum Company* the antiwear properties of a lubricating composition are improved, as compared to lubricating compositions without the named additives, by incorporating therein a minor amount of at least one mercapto-substituted boron-containing compound selected from the group consisting of 2-hydroxy-4-(mercaptohydrocarbyl)-1,3,2-dioxaborolanes and alkylammonium bis[(mercaptohydrocarbyl)ethylenedioxy] borates. The 2-hydroxy-4-(mercaptohydrocarbyl)-1,3,2-dioxaborolanes are provided by contacting boric acid with at least one mercapto-substituted vicinal diol in a molar ratio of substantially 1:1, with removal of by-product water.

The hydrogen bis[(mercaptohydrocarbyl)ethylenedioxy] borates are provided by contacting boric acid with at least one mercapto-substituted vicinal diol in a molar ratio of substantially 0.5:1, respectively, with removal of by-product water. The alkylammonium bis[(mercaptohydrocarbyl)ethylenedioxy] borates are provided by contacting at least one hydrogen bis[(mercaptohydrocarbyl)-ethylenedioxy] borate with at least one alkylamine.

Examples of some 2-hydroxy-4-(mercaptohydrocarbyl)-1,3,2-dioxaborolanes which can be employed in lubricating compositions include 2-hydroxy-4-(mercapto-methyl)-1,3,2-dioxaborolane, 2-hydroxy-4-(2-mercaptoethyl)-1,3,2-dioxaborolane, 2-hydroxy-4-(3-mercaptopropyl)-4,5-dimethyl-5-ethyl-1,3,2-dioxaborolane, and 2-hydroxy-4-(3-mercaptobutyl)-4-isopropyl-5-isobutyl-1,3,2-dioxaborolane.

Example 1: 2-Hydroxy-4-(mercaptomethyl)-1,3,2-dioxaborolane was prepared in the following manner.

A mixture of 6.2 g (0.1 mol) boric acid and 10.8 g (0.1 mol) 3-mercapto-1,2-pro-panediol was stirred at 80°C for 2 hr. Then 150 ml toluene was added, and the mixture was heated to remove water in an azeotropic distillation step. A total of 4.3 ml water was thus removed. A small amount of white solid was allowed to settle from the resulting reaction mixture, the toluene solution was decanted from the solid, and toluene was distilled under reduced pressure from the toluene solution, leaving 9.1 g of 2-hydroxy-4-(mercaptomethyl)-1,3,2-dioxaborolane as a viscous, hazy liquid.

Example 2: Octadecylammonium bis[(mercaptomethyl)ethylenedioxy] borate was prepared as follows.

A mixture of 6.2 g (0.1 mol) boric acid and 21.6 g (0.2 mol) 3-mercapto-1,2-propanediol was stirred at 80°C for 30 min under a nitrogen atmosphere. A portion of the boric acid remained undissolved. Then 200 ml toluene was added, and the mixture was heated to remove water in an azeotropic distillation step. A total of 5.8 ml water was thus removed. To the resulting reaction mixture comprising hydrogen bis[(mercaptomethyl)ethylenedioxy] borate was added 26.9 g (0.1 mol) octadecylamine in 100 ml toluene, and the mixture thus formed was stirred at 80° to 90°C until it became clear. The toluene was then distilled from the toluene solution, leaving 50.5 g of octadecylammonium bis[(mercapto-methyl)ethylenedioxy] borate as a pale yellow liquid which solidified on cooling.

Example 3: 2-Hydroxy-4-(mercaptomethyl)-1,3,2-dioxaborolane, prepared in Example 1, and octadecylammonium bis[(mercaptomethyl)ethylenedioxy] borate, prepared in Example 2, were evaluated as antiwear additives in an ashless lubricat-ing oil formulation. Also evaluated, for the purpose of comparison, was the ash-less lubricating oil formulation without antiwear additive.

The composition of the ashless lubricating oil formulation, without antiwear additive of this process was as shown in Table 1.

Table 1

. . . . Component, wt %.		Purpose
Lubricating oil*	66.62	—
Phil-Ad VII solution**	24.08	viscosity index improver
Lubrizol 925 additive***	7.50	dispersant

(continued)

Table 1: (continued)

.... Component, wt %.		Purpose
Acryloid 152 additive†	0.20	pour point depressant
Vanlube PN additive††	0.10	antioxidant
Vanlube SS additive†††	1.00	antioxidant
Ethyl 702 additive§	0.50	antioxidant

   *A refined, generally paraffinic midcontinent lubricating oil blend.
  **A 10 wt % solution of a hydrogenated butadiene-styrene copolymer
    in a refined, generally paraffinic midcontinent lubricating oil.
 ***A mixture of polyisobutenyl succinimide and polyisobutenyl
    succinamide.
   †A polymethacrylate-based resin.
  ††Phenyl-β-naphthylamine.
 †††Mixture of octylated diphenylamines.
   §4,4'-Methylenebis(2,6-di-tert-butylphenol).

The wear tests were carried out using the well-known Falex test machine in accordance with a slight modification of the ASTM D 2670-67 procedure. The results are summarized in Table 2.

Table 2

Antiwear Additive	Additive Level, (wt %)*	Wear, Number of Teeth
None	0	>>100**
A***	2.0	29
B†	2.0	13

   *Based on total weight of ashless lubricating oil formulation
    plus antiwear additive.
  **Excessive wear led to catastrophic failure.
 ***2-Hydroxy-4-(mercaptomethyl)-1,3,2-dioxaborolane.
   †Octadecylammonium bis[(mercaptomethyl)ethylene-dioxy]
    borate.

Thus, 2-hydroxy-4-(mercaptomethyl)-1,3,2-dioxaborolane and octadecylammonium bis[(mercaptomethyl)ethylenedioxy] borate were each effective as antiwear additives, each of them improving the antiwear properties of the lubricating oil into which the antiwear additive was incorporated.

Alkyl Beta-Thiopropionic Acid Esters

R.F. Bridger and R.J. Cier; U.S. Patent 4,076,639; February 28, 1978; assigned to Mobil Oil Corporation have found that a lubricant composition containing a minor amount sufficient to improve the antiwear properties thereof of an alkyl β-thiopropionic acid ester selected from the group consisting of:

$$R_1-S-CH_2-CH_2-\overset{O}{\overset{\|}{C}}O-R_2$$

$$R_1-\overset{O}{\overset{\|}{O}}C-CH_2-CH_2-S-\overset{S}{\overset{\|}{C}}-N\underset{R_3}{CH_2}-CH_2\underset{R_4}{N}-\overset{S}{\overset{\|}{C}}-S-CH_2-CH_2-\overset{O}{\overset{\|}{C}}O-R_2$$

wherein R_1 and R_2 are alkyl groups, straight or branched, containing from 1 to 30 carbon atoms and R_3 and R_4 are alkyl groups, straight or branched, containing from 1 to 12 carbon atoms provides compositions of improved antiwear properties. Especially suitable are lubricant compositions wherein R_1 and R_2, as above identified, contain from 2 to 12 carbon atoms and R_4 and R_3 contain from 1 to 6 carbon atoms.

The subject sulfur-containing esters provide effective antiwear properties when minor amounts are incorporated into various oils, greases, and functional fluids of lubricant viscosity; they are suitable for use in steel-on-steel, and steel-on-bronze applications. They are especially useful in steel-on-bronze applications, e.g., in devices having steel and bronze worm gears.

Alkene-Sulfur Monochloride Reaction Products

A process described by *B.W. Hotten; U.S. Patent 4,132,659; January 2, 1979; assigned to Chevron Research Company* comprises heating a mixture of 1-alkene, sulfur, and sulfur monochloride in the molar ratio of 1 mol 1-alkene per 0.6 to 0.9 mol sulfur and 0.05 to 0.2 mol sulfur monochloride to incorporate sulfur and chlorine into the 1-alkene. Preferably, a ratio of n mol S to 1 – n/2 mol S_2Cl_2 (n <1 for 1 mol alkene) is used to provide a stoichiometric balance of reagents. Under normal circumstances the reaction will be complete after 5 to 15 hr at 150° to 170°C.

The products of this process can be used as lubricant additives without any additional treatment. If desired, the process may be carried out under a slow stream of inert gas, such as nitrogen gas, to remove any hydrogen sulfide or hydrogen chloride that might be generated.

The product generally contains from 5 to 20% sulfur and from 0.5 to 5% chlorine. For a most effective lubricating oil additive, 8 to 16% sulfur content and 1 to 3% chlorine content is preferred.

Particularly preferred because of their availability and price are the 1-alkenes prepared by cracking wax. These 1-alkenes are often referred to in the art as cracked wax olefins. Various cracked wax olefin fractions may be used as starting materials for the reaction of this process. Particularly preferred are the 1-decene (including C_{9-10}) fraction and the C_{15-18} fraction.

Example A: To a 4-liter flask was added 2,240 g (16 mols) of C_{9-10} cracked wax olefin and 512 g (16 mols) sulfur. The reaction mixture was heated to 160°C for 10 hr under nitrogen.

The mixture was filtered through diatomaceous earth at room temperature to yield 2,719 g of product containing 18.2% sulfur.

Example 1: To a 4-liter flask was added 2,095 g (15 mols) of C_{9-10} cracked wax olefin, 384 g (12 mols) sulfur and 202.5 g (1.5 mols) sulfur monochloride. The reaction mixture was stirred for 10 hr at 160°C under nitrogen. The mixture was then stripped at 100°C under vacuum to yield 2,596 g product containing 17.9% sulfur and 2.7% chlorine.

Examples 2 through 5 and B: Following the general procedure of Example 1, the products shown in Table 1 below were prepared. The term CWO means cracked wax olefin.

Table 1

Ex. No.	Time (hr)	Temperature (°C)	1-Alkene (mol)	Sulfur (mol)	Sulfur Monochloride (mol)	. Product . . (% S)	(% Cl)
2	10	160	C_{9-10} CWO (1)	0.8	0.1	18	2.7
B	10	160	C_{15-18} CWO (1)	1	0	11	0
3	10	160	C_{15-18} CWO (1)	0.8	0.1	12	1.8
4	10	160	C_{15-18} CWO (1)	0.9	0.05	12	3.1
5	10	160	C_{15-18} CWO (1)	0.7	0.15	12	3.1

The following examples illustrate the effectiveness of the products of this process (Examples 1 to 5) as compared to those falling outside the scope of the process (Examples A and B), as well as mixtures designed to mimic in chlorine and sulfur content of the products of this process.

Example 6: Table 2 below illustrates the advantages with respect to oxidation control of the additives of this process over closely related types of additives.

The oxidation test measures the resistance of the test sample to oxidation using pure oxygen with a Dornte-type oxygen absorption apparatus (R.W. Dornte, "Oxidation of White Oils," *Industrial and Engineering Chemistry,* Vol 28, p 26, 1936).

The conditions are: an atmosphere of pure oxygen exposed to the test oil, the oil maintained at a temperature of 171°C, and oxidation catalysts, 0.69% Cu, 0.41% Fe, 8.0% Pb, 0.35% Mn, and 0.36% Sn (as naphthenates) in the oil. The time required for 100 g of the test sample to absorb 1,000 ml of oxygen is measured.

Table 2: Oxidation Inhibition by Thioethers

. . . . Thioether . . . . Additive, %. . . .		. . . . Source of Chlorine . . . . Added Compound, % . . .		Cl in Oil (%)	Life (hr)
. . . . .In 480 SSU/100°F Neutral Oil					
None	—	none	—	0.00	0.4
Example A*	1	none	—	0.00	3.4
Example A*	2	none	—	0.00	6.7
Example A*	1	chlorinated wax**	0.068	0.027	3.4
Example A*	2	chlorinated wax**	0.014	0.054	6.9
Example 2***	1	none	—	0.027	8.5
Example 2	2	none	—	0.054	12
. . . . In 480 Neutral Oil†.					
None	—	none	—	0.00	4.6
Example B††	1	none	—	0.00	6.1
Example B††	1	chlorinated wax**	0.1	0.04	7.3
Example 3†††	1	none	—	0.016	13.13

*Sulfidized 1-decene.
**40% Cl.
***Sulfidized-sulfochlorinated 1-decene.
†Containing 6% succinimide dispersant and 9 mmol/kg zinc dialkyl dithiophosphate.
††Sulfidized C_{15-18} 1-alkenes (11% S).
†††Sulfidized-sulfochlorinated C_{15-18} 1-alkenes.

Table 3 shows the results of the same oxidation test as used above and illustrates the effective ranges of molar proportions of sulfur and sulfur monochloride in the preparation of additives by the process.

Table 3: Effect of S_2Cl_2:S Ratio Within Preferred Range on Oxidation Inhibition by Sulfidized-Sulfochloridized 1-Alkene (SSA)

Product of Example	Molar ... Ratio.... S	S_2Cl_2	Analysis ...%.... S	Cl	SSA Concentration (%)	Oxidation Time* (hr)
A	1	–	13	–	1.5	2.2
3	0.8	0.1	12	1.8	0.5	0.8
					1.0	6.4
					2.0	11
4	0.9	0.05	12	1.1	0.5	0.6
					1.0	2.6
					2.0	9.1
5	0.7	0.15	12	3.1	0.5	0.7
					1.0	4.4
					2.0	10

*To absorption of 1 l/100 g at 171°C.

Wax Esters of Vegetable Oil Fatty Acids

E.W. Bell; U.S. Patent 4,152,278; May 1, 1979; assigned to The United States Secretary of Agriculture, has found an effective lubricant composition comprising a mixture of wax ester compounds which can be easily derived entirely from renewable vegetable oil sources. This composition is prepared from a free fatty acid mixture where the fatty acids are characterized by the structural formula RCO_2H where R is a radical selected from the group of:

$$(1) \quad CH_3(CH_2)_x CH{=}CH(CH_2)_y CH{=}CH(CH_2)_z-$$

where x is 1 to 4, y is 1 to 4, z is 7 to 8, and x + y + z = 12;

$$(2) \quad CH_3(CH_2)_x CH{=}CH(CH_2)_y-$$

where x is 0 to 9, y is 5 to 14, and x + y = 14; (3) $CH_2{=}CH(CH_2)_{15}-$; (4) $CH_3(CH_2)_{16}-$ and (5) $CH_3(CH_2)_{14}-$. The distribution of the radicals in the fatty acid mixture includes from a trace to about 50 mol percent of radical (1), from 35 to 80 mol percent of radicals (2) and (3) combined, from 3 to 20 mol percent of radical (4), and from 5 to 15 mol percent of radical (5), and the distribution of radicals in the mixture also contains from a trace to 60 mol percent of isolated trans double bonds, from 0 to about 16 mol percent of conjugatable double bonds, and from a trace to about 45 mol percent of nonconjugatable double bonds.

The process of preparation comprises the following steps: (a) selectively reducing a first portion of the fatty acid mixture to substantially the corresponding alcohols; (b) esterifying a second portion of the fatty acid mixture with the alcohols obtained in step (a) in order to yield wax esters having the structural formula:

$$RC\underset{\textstyle OCH_2R'}{\overset{\textstyle O}{\Big\langle}}$$

where both R and R' are radicals independently selected from the group of radicals set forth above; and (c) recovering the wax esters for use as the lubricant composition.

It has also been found that the sulfurized derivatives of these compounds are comparable to sulfurized sperm oil as EP/AW additives.

Thus, the process provides a replacement for sperm whale oil and substitutes for sulfurized sperm whale oil.

Silicone Composition for Lubricating Copper and Bronze

A process described by *E.D. Brown, Jr.; U.S. Patent 4,138,349; February 6, 1979; assigned to General Electric Company* relates to a silicone lubricant composed of an organopolysiloxane polymer and an effective amount of chlorinated phosphite or phosphonate. Such a lubricant composition is especially effective for lubricating soft metals such as copper or bronze and for improving dimethylpolysiloxane lubricants.

The most preferred phosphites and phosphonates for use in the composition are trischloroethylphosphite and bischloroethylchloroethyl phosphonate.

Example: In all of the test results tabulated below, there was utilized as the basic lubricating fluid, a fluid identified as Fluid X in the table below, which Fluid X is a linear polysiloxane having a 50 cp viscosity at $25°C$ and was a methyldecyl-substituted polysiloxane polymer having trimethylsiloxy terminal groups which has 1 to 3% coupled into it of tert-hydroxyphenyl units. All the percentages in the table are by weight of the additives added.

The test was carried out in a Shell 4-ball tester which was run with a 40 kg load on the rotating ball which was made of 52/100 steel and rotated at a rotation of 1,200 rotations per minute for 1 hr while impinging on 3 balls constructed of bronze all of which balls were immersed in the Fluid X. The amount of the additives added is indicated in the table below in which the lubricant and additive mixture was maintained at a temperature of $167°F$ for the 1 hr length of the test. Results of the tests are as follows. In addition, other conventional additives, such as dialkyl zinc dithiophosphite, were tried. These failed completely.

	Wear Scar (mm)
Fluid X	1.7
Fluid X + 4% dibutylhexyltetrachlorophthalate	0.84
Fluid X + 0.5% trischloroethylphosphite	0.82
Fluid X + 0.25% trischloroethylphosphite	0.93
Fluid X + 1.0% trischloroethylphosphite	0.84

As noted from the results set forth in the table above, the process phosphite and phosphonate additives were much more effective as antiwear additives for soft metals than were other conventional additives which failed completely and were at one-fourth the concentration just as effective as dialkyltetrachlorophthalates as antiwear additives for soft metals. It should be noted that the phosphites and phosphonates as the above results in the table indicate, even in trace amounts, that is at a concentration of 0.25 weight percent in the methyl higher-alkyl fluid markedly improved the antiwear properties of the methyl higher-alkyl fluid toward soft metals as indicated above. Accordingly, at the same concentrations the chlorinated phosphite and chlorinated phosphonate additives of the process are the most effective additives for organopolysiloxane polymers as far as optimizing the antiwear properties of silicone lubricant polymers; they are more effective than prior art phthalate and benzoate additives and other prior art additives for this purpose.

Hybrid Lubricant Containing Colloidal Dispersion of Polytetrafluoroethylene

A process developed by *F.G. Reick; U.S. Patent 4,127,491; November 28, 1978;* provides a hybrid lubricant in which a stabilized colloidal dispersion of solid lubricant particles (PTFE) is uniformly dispersed in a fluid lubricant carrier to form a hybrid lubricant which when diluted with a major amount of a conventional fluid lubricant (oil or grease) functions in the environment of rubbing surfaces to develop a layer of solid lubricant on these surfaces.

A salient feature of the process is that rubbing surfaces to which the hybrid lubricant is applied have the continuing benefit of both solid and fluid lubrication, thereby minimizing friction under all operating conditions, regardless of their severity.

The hybrid lubricant includes a small but effective amount of a halocarbon oil which acts to impregnate the microscopic voids and rough spots on a typical rubbing surface (even one that is highly polished), with polytetrafluoroethylene particles of submicronic size to create an integrally-bonded solid lubricant layer thereon that is supersmooth and extraordinarily slippery.

The use of a hybrid lubricant as an additive for standard crankcase oil in a diesel or internal combustion engine brings about distinctly better performance, increased mileage for a given amount of fuel, faster cold starts and an absence of hesitation. The additive reduces friction and wear, yet it never coagulates and does not clog oil filters. And because the hybrid lubricant makes it possible to operate at lower idling speeds and with very lean air/fuel mixtures, the emission of unburned hydrocarbons and carbon monoxide from the exhaust is sharply reduced, thereby minimizing the discharge into the atmosphere of pollutants.

A preferred procedure for producing a hybrid lubricant which includes a small but effective amount of halocarbon oil is as follows.

Step A: The following substances are thoroughly intermixed — 1,200 g Halocarbon Oil (oil 10-25 has limited solubility in mineral oils) and 1,500 g Monoflor 52 (nonionic fluorochemical surface active agent which is oil-soluble).

Step B: The mixture produced by Step A is thoroughly intermingled with 1 gal Quaker State lubricating oil (10W-40 SAE) to produce a nonaqueous emulsion, hereinafter referred to as Component 1.

Step C: To produce a dilute, stabilized PTFE aqueous dispersion, use is made of 2,400 cc of a PTFE dispersion (ADO/38 of ICI, and T-42 of DuPont) and 2.5% Monoflor 32. The Monoflor 32 acts as a charge-neutralizing agent, and the resultant stabilized dispersion is then diluted with distilled water to reduce its solid content to 17%.

Step D: The stabilized PTFE dispersion produced in step C is then thoroughly intermingled with 2 gal Quaker State lubricating oil (10W-40 SAE). The resultant emulsion of the stabilized aqueous PTFE dispersion in oil produces Component 2.

When mixing the PTFE dispersion in oil, it is important that the mixing action be thorough and yet not excessively violent, for this would disturb stability of the dispersion. For this purpose, use is preferably made of a rotating wire brush operating at high speed (i.e., 3,600 rpm) within a mixing vessel. The brush is provided with an annular array of upstanding bristles, oil being fed into the core of the brush and being centrifugally hurled toward the periphery through the thicket of bristles which serves to work the dispersion into the oil without undue impact or shear forces. Collectively, the wire bristles forming the brush bring about a very thorough intermingling of the constituents.

Step E: Components 1 and 2 are then blended together and thoroughly intermingled (low shear) with 4 gal of Acryloid 956 (warm). This polymeric dispersant serves to uniformly homogenize the emulsion and to prevent the formation of large globules.

Step F: Added to the homogenized emulsion produced by step E is 1,000 cc Surfy-nol (mixture of 104/440 in 2 to 1 ratio). Surfy-nol 104 is solid at room temperature, whereas Surfy-nol 440 is then liquid.

These surfactants have an affinity for metal and serve as a wetting agent, facilitating adhesion of the PTFE particles to the rubbing metal parts.

Step G: When the Surfy-nol has been uniformly mixed into the homogenized emulsion, 3 lb of Neutral Barium Petronate (50-S) are added. This constituent is a synthetic barium sulfonate with a low viscosity, providing ease of handling coupled with a high barium sulfonate concentration. It is oil soluble and possesses the ability to increase the spreading coefficient. It also improves the long term stability of PTFE dispersion and inhibits settling thereof.

Step H: Finally, the above is dispersed in 3 gal Quaker State Oil (10W-40 SAE). This produces a hybrid lubricant in accordance with the process which may be added to a standard lubricant to improve its lubricity and to cause the formation of a PTFE coating on the rubbing surfaces being lubricated.

Metal Flakes and Molybdenum Disulfide Particles

R.J. Soucy; U.S. Patent 4,155,860; May 22, 1979; describes a lubricant additive composition for a petroleum base lubricant medium. The composition is formed of a mixture of copper metal flake or powder particles, chromium metal flake or powder particles, zinc metal flake or powder particles, and molybdenum disulfide particles. All of the particles are finely divided, being micron size, generally from 1 to 2 microns. These materials are readily available.

The additive composition is made up of from 1 to 3 parts by weight copper metal particles, from 2 to 4 parts by weight chromium metal particles, from 1 to 3 parts by weight zinc metal particles and from 2 to 4 parts by weight molybdenum disulfide particles. A composition mixed in the foregoing ratios is then added to the desired lubricant in the ratio of 0.001 to 0.015 part by weight additive composition to each part by weight oil or other petroleum base medium. More specifically, it has been found that the preferred range is from 0.0019 to 0.0038 part by weight additive to each part by weight petroleum base lubricant medium.

The following examples illustrate the process. The particulate materials utilized in each example are as described above.

Example 1: An additive composition composed of 2 g copper metal flake, 4 g chromium metal flake, 4 g zinc metal flake, and 6 g molybdenum disulfide, were added to 5 qt of oil in the engine crankcase of a Honda 750 4 cylinder internal combustion engine. Prior to addition of the additive, the engine cylinders showed compression figures of 75, 95, 87 and 91. Following the addition of the composition, the cylinder compression figures were 110, 114, 118 and 120 respectively. Engine noise was reduced, engine temperature was lowered and decreased oil use was noted.

Example 2: An additive composition composed of 2 g copper metal flake, 3 g chromium metal flake, 1 g zinc metal flake, and 2 g molybdenum disulfide were added to 5 qt of oil in the internal combustion engine of a 1968 Duster. An increase in power and gas mileage was noted. The additive stopped a leak in the rear main seal, quieted the valve lifters and increased cylinder compression.

ESTER LUBRICANTS

Organic Sulfonic Acid Ammonium Salts

D.S. Bosniack; U.S. Patent 4,079,012; March 14, 1978; has found that organo sulfonic acid ammonium salts containing a C_1 to C_{26} sulfonic acid group and an ammonium group derived from an organic nitrogen containing base selected from the group consisting of mono-, di- or tertiary amines, or acid addition salts or quaternary ammonium salts of such amines, can be incorporated in synthetic ester lubricating oils to provide useful lubricating oil compositions. The ammonium group has the general formula $NR^1R^2R^3R^4{}_x$ wherein x is 0 or 1, and R^1, R^2, R^3 and R^4 are each hydrogen or a C_1 to C_{36} hydrocarbyl group, provided that at least one of R^1, R^2, R^3 and R^4 is a C_1 to C_{36} hydrocarbyl group.

The lubricating compositions are particularly useful in lubricating gas turbine engines containing silicone or fluorosilicone seals because they do not cause deterioration or shrinking of such seals and have a high lead carrying capability. The organosulfonic acid ammonium salts prepared from quaternary ammonium salts are especially useful and exhibit improved oxidation stability over organosulfonic acid ammonium salts prepared from other nitrogen bases. The amount of organosulfonic acid ammonium salt incorporated in the lubricating oil is between about 0.05 to 1.0 weight percent, based on the total amount of synthetic ester lubricating oil.

Representative of the organosulfonic acids that can be used in preparing the additives are p-toluenesulfonic acid, 2-naphthalenesulfonic acid, dodecylbenzenesulfonic acid, DL-camphorsulfonic acid, and methanesulfonic acid.

Many of the nitrogen bases are commercially available as mixtures, such as amines derived from C_{12} oxo alcohol bottoms, for example, a mixed branched chain isomeric, 1,1-dimethyl C_{12} to C_{14} primary aliphatic amine composition (Primene 81-R) and a mixed branched chain isomeric 1,1-dimethyl C_{18} to C_{22} primary amine composition (Primene JMT). Moreover, many of the amine mixtures (Armeen) are likewise suitable compositions to be employed in producing the sulfonic acid ammonium salts. These amines contain alkyl radicals ranging from C_{12} through C_{18} or mixtures thereof which are derived from fatty acids. The corresponding C_2 to C_{20} polyalkylene polyamines are also useful.

The following examples illustrates the process. All percentages and parts are by weight unless otherwise specifically indicated.

Example 1: This example illustrates the preparation of an additive prepared in accordance with the process. 12.5 g of p-toluenesulfonic acid are dissolved in 100 ml of 95% ethanol. The solution is stirred at room temperature and 23.4 g of Primene JMT are added dropwise to it until a pH of 7.6 is reached. The alcohol is then removed by distillation under 100 to 10 mm of mercury. The product is finally stripped at 120°C under a pressure of 0.5 mm of mercury. The yield of product was 90%. Infrared analysis of the product shows that it contains primarily an ammonium sulfonate salt.

Example 2: The general procedure of Example 1 is repeated except that the p-toluenesulfonic acid is replaced with 10.0 g of 2-naphthalenesulfonic acid and the Primene JMT is replaced with 9.0 g of Primene-81R. The Primene-81R is added dropwise until a pH of 7.6 is reached. The yield of product was 90%.

Example 3: The general procedure of Example 1 is repeated except that the p-toluenesulfonic acid is replaced with 10 g of 2-naphthalenesulfonic acid and the Primene JMT is replaced with 9.5 g of tri-n-butylamine. The tri-n-butylamine is added dropwise until a pH of 8.2 is reached. The yield of product was 16.3 g (84%).

Example 4: This general procedure of Example 1 is repeated except that the p-toluenesulfonic acid is replaced with 20 g of dodecylbenzenesulfonic acid and the Primene JMT is replaced with 20.7 g of tri-n-butylamine. The tri-n-butylamine is added dropwise until a pH of 7.6 is reached. The yield of product was 36.2 g (89%).

Example 5: This general procedure of Example 1 is repeated except that the p-toluenesulfonic acid is replaced with 10 g of camphorsulfonic acid and the Primene JMT is replaced with 4.2 g of triethylamine. The triethylamine is added dropwise until a pH of 7.6 is reached. The yield of product is 13.8 g (97%).

Example 6: This general procedure of Example 1 is repeated except that the p-toluenesulfonic acid is replaced with 10.0 g of camphorsulfonic acid and the Primene JMT is replaced with 8.0 g. The tri-n-butylamine is added dropwise until a pH of 7.5 is reached. The yield of product is 12.1 g (67%).

Example 7: The general procedure of Example 1 is repeated except that the p-toluenesulfonic acid is replaced with 20.0 g of methanesulfonic acid and 54.4 g of Primene JMT are added dropwise until a pH of 7.6 is reached. The yield of product is 63.1 g (85%).

Example 8: The general procedure of Example 1 is repeated except that the p-toluenesulfonic acid is replaced with 20.0 g of methanesulfonic acid and the Primene JMT is replaced with 16.3 g of tri-n-butylamine. The tri-n-butylamine is added dropwise until a pH of 7.6 is reached. The yield of product is 28.0 g (77%).

Example 9: The additives of Examples 1, 2, 4, 6 and 8 are formulated in a series of synthetic ester oil base stocks prior to being evaluated in certain standard lubricant tests. The following synthetic base stocks are used in preparing the lubricating compositions.

Base stock Oil A is a tetraester of pentaerythritol with a mixture of C_4 to C_7 normal alkanoic acids (Hercolube A).

Base stock Oil B is a mixture of about 90 wt % of pentaerythritol and 10% of dipentaerythritol esterified with a mixture of C_5 to C_9 normal alkanoic acids.

The extreme pressure properties of a number of lubricant compositions formulated from the organosulfonic acid ammonium salts of Examples 1, 2, 4, 6 and 8 and the above base stocks are evaluated in the Ryder gear test using the Ryder gear machine described in ASTM Method D-1947.

Briefly, this test subjects a set of gears lubricated by the test oil to a series of load increments under controlled conditions. The amount of tooth-face scuffing occurring at each load is measured. The percentage of tooth-face scuffing is plotted against the load to determine the load-carrying ability of the test oil.

The load-carrying ability of the oil is defined as the tooth load, in pounds per inch of gear tooth, at which an average tooth face scuffing at 22.5% of the tooth area has been reached.

The following table lists the components of each lubricating composition and the results of the tests.

| . . . . . . Lubricating Composition. . . . . . | | | . . . . . .Ryder Gear Testing . . . . . . | | | |
Additive	Oil Base Stock	Additive Concentration*	Side A	Side B	Average (ppi)	Δ Load (ppi)
Example 1	A	0.1	3,015	2,756	2,886	486
Example 2	A	0.2	3,220	3,102	3,161	761
Example 2	A**	0.2	3,302	2,500	2,901	501
Example 4	B	0.3	3,592	3,717	3,655	855
Example 4	A	0.25	3,039	2,851	2,945	545
Example 4	B	0.2	3,591	3,228	3,410	610
Example 4	B	0.1	3,038	3,118	3,078	278
Example 6	B	0.3	3,469	3,089	3,279	479
Example 8	B	0.3	4,183	4,088	4,136	1,336

*Parts of additive per 100 parts base stock.
**Contains 1.0 part by weight of tricresyl phosphate per 100 parts by weight oil.

From an examination of these results, it can be seen that the lubricant compositions of this process increase the average load-carrying capacity of the oil from 300 to 1,300 lb per inch of gear tooth. For example, from the data in the table it can be seen that the methanesulfonic acid-tri-n-butyl product of Example 8 in a concentration of 0.3 part by weight per 100 parts by weight of oil gave an increase in the average load-carrying capacity of the oil of over 1,300 pounds.

Organoamine Salts of Phosphate Esters

H. Shaub and S.J. Metro; U.S. Patent 4,130,494; December 19, 1978; assigned to Exxon Research & Engineering Co., describe a synthetic lubricating oil composition comprising a base oil consisting of one or more carboxylic acid esters and a unique combination of additives. The additive combination comprises, essentially, an organoamine salt of an alkyl phosphate ester, and an organosulfonic acid ammonium salt.

The additive combination will, generally, also comprise an oxidation inhibitor such as a di(alkylphenyl)amine, a phenyl-α or β-naphthylamine, an alkylphenyl-α or β-naphthylamine or a mixture of any of these amines. The synthetic lubricating oil composition may, and generally will, also contain other additives which are compatible with the aforementioned additives. Such other additives may include a dispersant, an alkylphenyl ester of phosphoric acid, such as the tri(alkylphenyl) ester, a hydrolytic stabilizer and/or a storage stabilizer. It is, of course, essential that all additives used be oil soluble and compatible at the concentrations actually employed.

The preparation of the lubricating compositions of the process involves no special techniques. Generally, the lubricants are formed by adding an appropriate amount of the organoamine salt of a phosphate ester and an appropriate amount of the organosulfonic acid ammonium salt additive to the synthetic ester oil base stock and heating and stirring the composition until the additive is dissolved.

All percentages and parts referred to herein are by weight unless otherwise specifically indicated.

Examples 1 through 6: The following chart compares the results of Ryder Gear Load tests and the silicone elastomer compatibility of an oil with a variety of additives.

	. . . . Example.					
	1	2	3	4	5	6
Base A, pbw	100	100	100	100	100	100
Neutral aromatic amine salt of phosphate ester, pbw	0.04	0.04	–	0.04	0.04	0.05
Primary alkylamine, pbw	–	0.014	–	0.014	0.014	–
Methane sulfonic acid ammonium salt, pbw	–	–	0.04	0.04	0.05	0.04
Ryder Gear Load, lb/in	3,023	3,263	3,005	3,814	4,080	3,990*
Silicone compatibility, % swell after 96 hr at 250°F**	8.4	8.1	9.9	8.2	8.9	7.7

*Estimated value obtained with simulated test.
**No deterioration or cracking of the rubber sample was detected after this test.

Base stock Oil A is a tetraester of pentaerythritol with a mixture of C_5 to C_9 normal alkanoic acids. The neutral aromatic amine salt of a phosphate ester was a mixture which can be represented by the following formulas:

$$(CH_3O)_2\overset{\displaystyle O}{\overset{\|}{P}}O^- \ H^+NH_2C_6H_4R^1$$

$$CH_3O\overset{\displaystyle O}{\overset{\|}{P}}(O^- \ H^+NH_2C_6H_4R^1)_2$$

wherein R^1 is a mixture of C_9 to C_{13} alkyl groups. The methane sulfonic salt which was used can be represented by the formula:

$$CH_3\overset{\displaystyle O}{\underset{\displaystyle O}{\overset{\|}{\underset{\|}{S}}}}O^- \ \overset{+}{H}NH_2R^2$$

wherein R^2 is a mixture of C_{12} to C_{14} alkyl groups and was the salt of the primary amine used in each formulation.

The U.S. Navy XAS-2354 specification requirement for Ryder Gear Load is 3,300 lb/in minimum. Thus, only Examples 4, 5, and 6 meet the minimum requirement.

OXIDATION AND CORROSION INHIBITORS

ANTIOXIDANTS AND CORROSION INHIBITORS

Hydrocarbon oils are partially oxidized when contacted with oxygen at elevated temperatures for long periods. The internal combustion engine is a model oxidizer, since it contacts a hydrocarbon motor oil with air under agitation at high temperatures. Also, many of the metals (iron, copper, lead, nickel, etc.) used in the manufacture of the engine and in contact with both the oil and air, are effective oxidation catalysts which increase the rate of oxidation. The oxidation in motor oils is particularly acute in the modern internal combustion engine which is designed to operate under heavy work loads and at elevated temperatures.

The oxidation process produces acidic bodies within the motor oil which are corrosive to typical copper, lead, and cadmium engine bearings. It has also been discovered that the oxidation products contribute to piston ring sticking, the formation of sludges within the motor oil and an overall breakdown of viscosity characteristics of the lubricant.

The most recent developments relating to specialty additives and additive combinations having improved antioxidant properties are described in this chapter.

Substituted Triazine-Mercapto Compound Reaction Products

R.L. Sowerby; U.S. Patent 4,081,386; March 28, 1978; assigned to The Lubrizol Corporation describes reaction mixtures made from reacting one or more substituted di- and triazines of the formula:

(A)

$$\underset{\underset{\displaystyle R_2'C\diagdown_{X'}\diagup CR_2'}{\overset{\displaystyle |\qquad\quad|}{QCR_2'N\diagdown\underset{\displaystyle C}{\overset{\displaystyle \overset{X}{\|}}{}}\diagup NCR_2'Q'}}}{}$$

where Q and Q' are each independently halo, hydrocarbyloxy of the formula —OR and hydrocarbyl of the formula —R, where each R independently contains up to 20 carbon atoms, with the proviso that at least one of Q and Q' is either hydrocarbyloxy or halogen; X and X' are each independently oxygen, divalent sulfur, =NH, and =NR$_1$ where R$_1$ is hydrocarbyl up to 20 carbon atoms; and each R' is independently hydrogen or hydrocarbyl of up to 10 carbon atoms; with (B) one or more mercapto compounds such as mercaptans, thiolphosphorus acids, thiocarbamates, and thiocarbonates.

In the following examples all percentages are expressed as percent by weight and all parts are expressed as parts by weight unless otherwise indicated. Likewise, all temperatures are expressed in degrees centigrade (°C) unless otherwise indicated and all pressures are expressed in pounds per square inch unless otherwise indicated.

Example 1: A mixture of urea (60 g, 1.0 mol), formaldehyde (480 g of a 37.7% formalin solution) and 10 ml of a 40% sodium hydroxide solution is held at 60°C for 2 hr. The pH of the reaction mixture is adjusted to 5 with approximately 5 ml formic acid and then the solution is stripped to 60°C under a vacuum of 25 torrs. To the syrupy residue is added methanol (576 g, 18.0 mols) and 16.7 ml of a 37.5% hydrochloric acid solution and the mixture is kept below 30°C for 1 hr using a water bath.

The material is neutralized with a 40% sodium hydroxide solution, stripped to 40°C under a vacuum of 5 torrs and then filtered through diatomaceous earth. The filtrate contains the desired, 3,5-dimethoxymethyl-4-oxotetrahydro-1,3,5-oxadiazine, (i.e., formula A where X and X' are oxygen; Q is methoxy; and R' is hydrogen) having a % N of 12.9.

Example 2: The procedure of Example 1 is followed except thiourea (76 g, 1.0 mol) is substituted for the urea and propionaldehyde (232 g, 4.0 mols) is substituted for the formalin solution. The filtrate contains the desired product (i.e., X is sulfur; R' is ethyl; and X' and Q are as in Example 1).

Example 3: The procedure of Example 1 is followed except guanidine hydrochloride (95.5 g, 1.0 mol) is substituted for the urea, methyl ethyl ketone (288 g, 4.0 mols) is substituted for the formalin solution and 110 ml of the sodium hydroxide is used. The filtrate is the desired product (i.e., X is =NH; R' is ethyl; and X' and Q are as in Example 1).

Example 4: Dodecylthiol (199 g), the filtrate of Example 1 (95 g) and 1.5 ml of a 37.5% hydrochloric acid solution is held at 60°C for 3 hr. After standing about 72 hr at 25°C, the material is held at 80° to 90°C for 10 hr. The material is added to 500 ml textile spirits (an aliphatic petroleum naphtha having a distillation range of 63° to 79°C at 760 torrs), washed with water, neutralized with a potassium carbonate solution and washed again with water, then dried with magnesium sulfate.

The material is filtered, the filtrate is stripped to 100°C under a vacuum of 25 torrs and the residue is filtered through diatomaceous earth. The filtrate contains the desired product having a % N of 4.77 and a % S of 12.6.

The compositions [that is, the reaction mixtures of (A) and (B), optionally sulfurized] of this process are soluble and/or stably dispersible in the normally liquid media (e.g., oil, fuel, etc.) in which they are intended to function. Thus, for example, reaction mixtures formed by reacting (A) with (B), optionally sulfurized, intended for use in oils are oil-soluble and/or stably dispersible in an oil in which they are to be used.

The reaction products of (A) and (B), optionally sulfurized, of this process are useful as additives for lubricants, in which they function primarily as oxidation inhibitors, antiwear agents, and extreme pressure (EP) agents.

Benzotriazole-Allyl Sulfide Reaction Products

M. Braid; U.S. Patent 4,153,563; May 8, 1979; assigned to Mobil Oil Corporation has found that the reaction product of a benzotriazole compound and an organic sulfur-containing compound selected from the group consisting of α and β substituted allyl sulfides and disulfides, and sulfurized olefins having reactive unsaturated bonds, imparts improved antioxidant and anticorrosion properties to the oleaginous compositions to which it is added.

The benzotriazole compounds which are utilized to form the reaction products of the process are described by the formula:

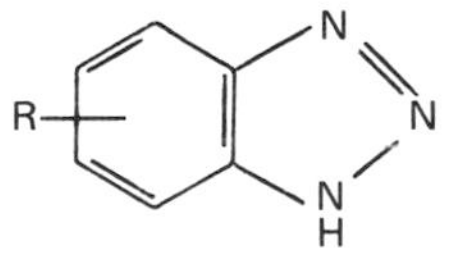

where R is hydrogen or a hydrocarbyl group containing from 1 to 12 carbon atoms. Preferred are benzotriazoles in which R is hydrogen or an alkyl group containing from 1 to 8 carbon atoms. Particularly preferred are benzotriazole and toluotriazole.

The α and β substituted allyl sulfide and disulfide compounds which are utilized to form the reaction products have the formulas:

$$R'CH=CR-CH_2-S-CH_2-RC=CHR' \quad \text{and} \quad R'CH=CR-CH_2-S-S-CH_2-RC=CHR'$$

where R and R' are selected from the group consisting of hydrogen, alkyl having from 1 to 8 carbon atoms, aryl and aralkyl having from 6 to 14 carbon atoms.

Particularly preferred are allyl sulfide, allyl disulfide, methallyl sulfide, methallyl disulfide, propallyl sulfide, propallyl disulfide, phenallyl sulfide and phenallyl disulfide.

In general the reaction products are formed by reacting the benzotriazole compound with the organic sulfur-containing compound in proportions (expressed as the molar ratio of benzotriazole compound:organic sulfur-containing compound) of from 0.25:1 to 10:1. Preferred are molar ratios of from 0.5:1 to 2.5:1.

Reaction temperature of from 5° to 150°C are utilized, with from 25° to 125°C being preferred. Generally, the reactants are contacted for 0.5 to 2 hr.

The reaction is catalyzed by the presence of an acidic material. Suitable materials include hydrogen sulfide, hydrogen chloride, methanesulfonic acid, and p-toluenesulfonic acid. A preferred catalyst is p-toluenesulfonic acid.

Polydisulfide and Arylamines

R.F. Bridger, and P.S. Landis; U.S. Patent 4,123,372; October 31, 1978; assigned to Mobil Oil Corporation have found that compositions of improved antioxidant characteristics are provided when a mixture of certain polydisulfides which are soluble in organic media, such as lubricating oils and greases, and arylamines or hindered phenols are added in appropriate amounts.

In general, the polydisulfides comprise a compound having at least one structural unit therein defined as follows:

$$-\left[CH_2-\underset{\underset{R_1}{|}}{\overset{\overset{R}{|}}{C}}-S-S\right]_n-$$

where n is from 2 to 20 and preferably is in the range of 4 to 10 and R and R_1 are hydrogen of C_{1-10} alkyl.

The process of making the polydisulfides may be conveniently carried out in accordance with U.S. Patents 3,925,414 and 3,697,499.

The polydisulfides are mixed with either a hindered phenolic compound or an arylamine to obtain a synergistic mixture of improved antioxidant properties. The weight ratio of polydisulfide to amine or phenol in the admixtures is usually from 0.75–2.5:1 of polydisulfide to amine or phenol. Preferred is a weight ratio of 1–2:1 or substantially equimolar ratios.

Methylene Bis(Dibutyldithiocarbamate) and 4-Methyl-2,6-Di-tert-Butylphenol

R.P. Chesluk, J.D. Askew, Jr., and C.C. Henderson; U.S. Patent 4,125,479; November 14, 1978; assigned to Texaco Inc. describe a lubricating oil comprising a base oil and an ashless combination of additives comprising methylene bis(dibutyldithiocarbamate) and 4-methyl-2,6-di-tert-butylphenol.

Example: The Rotary Bomb Oxidation Test (RBOT), ASTM D 2272 was performed to demonstrate the advantages of this process. This test provides a rapid means for estimating the oxidation stability of new turbine oils having the same composition (base stock and additives). In addition, the test is used to assess the remaining oxidation test life of such oils in service.

The test oil, water and copper catalyst coil, contained in a covered glass container, are placed in a bomb equipped with a pressure gauge. The bomb is charged with oxygen to a pressure of 90 psi, placed in an oil bath which is held at a constant temperature of 150°C, and rotated axially at 100 rpm at an angle of 30° from the horizontal. The time it takes the test oil to react with a given volume of oxygen is measured. Test completion time is indicated by a specific drop in pressure (more than 25 psi drop below the maximum pressure).

Tables 1 and 2 demonstrate the synergistic effect on oxidation inhibition resulting from an additive combination comprising 4-methyl-2,6-di-tertiary-butylphenol and methylene bis(dibutyldithiocarbamate). As Table 2 shows, no synergism is noted when 2-benzothiazyl-N,N-diethyldithiocarbamate is substituted for methylene bis(dibutyldithiocarbamate).

Table 1

	A	B	C	D	E	F	G	H
Composition, wt %								
Paraffinic base oil*	99.85	98.85	99.00	99.83	98.83	98.98	98.85	98.85
MDBP**	0.15	0.15	—	0.15	0.15	—	1.15	—
Methylenebis(dibutyl-dithiocarbamate)	—	1.0	1.0	—	1.0	1.0	—	1.15
Tetrahydrobenzotriazole	—	—	—	0.02	0.02	0.02	—	—
Test results RBOT, min	141	1,125	240	284	703	205	205	580

*Contains rust and foam inhibitors.
**4-methyl-2,6-di-tert-butylphenol.

Table 2

	A	B	C	D	E	F
Composition, wt %						
Paraffinic base oil*	99.950	99.950	99.9565	99.9565	99.93	99.93
MDBP	0.0065	0.0065	—	—	0.0065	0.0065
Ethylac**	0.0435	—	0.0435	—	0.0435	—
Methylene bis(dibutyl-dithiocarbamate)	—	0.0435	—	0.0435	—	0.0435
Tetrahydrobenzotriazole	—	—	—	—	0.020	0.02
Test results RBOT, min	37	63***	32	33	47	49

	G	H	I	J	K
Composition, wt %					
Paraffinic base oil*	99.93	99.93	99.95	99.95	99.99
MDBP	—	—	—	—	0.0065
Ethylac**	0.05	—	0.05	—	—
Methylene bis(dibutyl-dithiocarbamate)	—	0.05	—	0.05	—
Tetrahydrobenzotriazole	0.02	0.02	—	—	—
Test results RBOT, min	38	36	33	33	55

*Contains rust and foam inhibitors.
**2-Benzothiazyl-N,N-diethyldithiocarbamate (not soluble in base oil in larger concentrations).
***Average of two runs (55 and 70).

Sulfur-Based Antioxidant Compositions

According to a process described by *W. Lowe; U.S. Patent 4,161,451; July 17, 1979; assigned to Chevron Research Company* a lubricating oil additive composition which imparts improved oxidation properties to lubricants comprises an antioxidant selected from aromatic and alkyl sulfides and polysulfides, sulfurized olefins, sulfurized carboxylic acid esters and sulfurized ester-olefins, and an oil-soluble secondary amine of the formula HNRR' wherein R and R' are aliphatic radicals.

It has been found that the antioxidant defined above in combination with the secondary amine as described above complement each other in a synergistic manner resulting in a combination having antioxidant properties superior to either additive alone. The secondary amine component has virtually no antioxidant effect. However, when the combination of secondary amine and antioxidant is added to a lubricating oil, less antioxidant is needed to control oxidation than when the secondary amine compound is not present.

Preferably, an oil-soluble zinc salt is present in the lubricating oil composition. While this zinc salt is not required to achieve the syhergistic effect from the combination of the antioxidant and the secondary amine compound, a lubricating oil composition having improved performance characteristics results from the use of all three additive components.

Example 1: This example is presented to illustrate the effectiveness of the combination of the secondary amines with the antioxidant in improving the antioxidation properties of a lubricating oil over the use of either of the components individually. The oxidation test employed herein measures the resistance of the test sample to oxidation using pure oxygen with a Dornte-type oxygen absorption apparatus [R.W. Dornte, "Oxidation of White Oils," *Industrial and Engineering Chemistry*, Vol. 28, p. 26 (1936)].

The conditions are an atmosphere of pure oxygen exposed to the test oil maintained at a temperature of 340°F. The time required for 100 g of test sample to absorb 1,000 ml of oxygen is observed and reported in the following Tables 1 and 2. The midcontinent test oil in Table 1 contains 6% of a conventional succinimide dispersant, 0.05% terephthalic acid, 0.4% of a conventional rust inhibitor and 9 mmol/kg of a zinc dihydrocarbyldithiophosphate. The test oil in Table 2 comprises a midcontinent neutral oil (50% Citcon 350 neutral + 50% Citcon 650 neutral), and contains 6% of a conventional succinimide dispersant, 0.05% terephthalic acid, 0.4% of a conventional rust inhibitor and 9 mmol/kg of a zinc dihydrocarbyldithiophosphate.

Table 1

Antioxidant* (%)	. Secondary Amine (%) .		Oxidation Life (hr)
none	dicocoamine	0.1	0.48**
1	—	—	6.1***
1	dicocoamine	0.1	8.9
1	dioctylamine	0.1	8.6

*Diparaffin polysulfide.
**Base oil without zinc dithiophosphate.
***Average of three tests.

Table 2

Antioxidant* (%)	. . . Secondary Amine (%) . . .		Oxidation Life (hr)
none	—	—	5.0
1	—	—	7.6**
none	disoyamine	0.1	5.05
1	disoyamine	0.1	7.9

(continued)

Table 2: (continued)

Antioxidant* (%)	. . . Secondary Amine (%) . . .		Oxidation Life (hr)
none	di-n-hexylamine	0.2	5.25
1	di-n-hexylamine	0.2	8.6
none	di(ethylhexyl)amine	0.1	5.0
1	di(ethylhexyl)amine	0.1	8.7

*Diparaffin polysulfide.
**Average of three tests.

Example 2: The resistance to increase in viscosity of lubricating oils containing an antioxidant composition of this process as compared to antioxidant alone is illustrated by the following engine test. In this test, an Oldsmobile engine is charged with 10.65 lb of the oil to be tested. The engine is then started and run for 2 min at 850 rpm with no load. The rpm's are then increased to 1,500 with a 50 lb load and 450 psi oil pressure, and the engine is run for 8 min. The engine is shut down with the oil circulating for 10 min. The oil pump is then shut down and the oil sampled after 5 min. This procedure is repeated until the viscosity of the oil increases 500%. The number of hours elapsed during the test is recorded.

The base oil used in this test is a midcontinent neutral oil having a viscosity of 350 SUS at 100°F, 6% wt of a conventional succinimide dispersant, 0.05% wt of terephthalic acid, and 0.4% wt of a conventional rust inhibitor. The results of this test are reported in Table 3.

Table 3

Antioxidant (%)	. . Secondary Amine (%) . .		Hours to Increase Viscosity**
none	—	—	28
2	—	—	52
1	dicocoamine	0.1	***

*Diparaffin polysulfide.
**Viscosity increased 500% at 100°F.
***232% viscosity increase after 142 hr. Engine test was terminated because of electronic problems.

The test procedure of Example 2 was repeated in a formulation containing 6% of a conventional succinimide dispersant, 40 mmol/kg of a calcium phenate and 18 mmol/kg of a zinc dithiophosphate. The results are shown in Table 4.

Table 4

. . . . Antioxidant (%)		Secondary Amine (%)		Hours to Increase Viscosity*
none	—	—	—	44.56
Diparaffin polysulfide	2	dicocoamine	0.2	123
Sulfurized CWO** (13% S)	2	dicocoamine	0.2	72
Sulfurized CWO** (24% S)	2	dicocoamine	0.2	>166

*Viscosity increased 500% at 100°F.
**CWO is cracked wax olefin.

W. Lowe; U.S. Patent 4,086,172; April 25, 1978; assigned to Chevron Research Company describes a lubricating oil additive composition which imparts improved oxidation properties to crankcase lubricants and comprises an antioxidant selected from aromatic or alkyl sulfides and polysulfides, sulfurized olefins, sulfurized carboxylic acid esters and sulfurized ester-olefins, and a hydroxy amine of the formula:

$$\left[R \underset{\underset{Y}{|}}{\overset{}{+}} N-A' \underset{y}{\overline{\big)}} \right]_{3-x} N + A-OH)_x \qquad \text{or} \qquad \overset{R}{\underset{\smile}{N \diagdown N-AOH}}$$

where A and A' are C_{2-10} alkylene, R is C_{1-30} alkyl, Y is H, C_{1-6} alkyl or $-A-OH$, y is 0, 1 or 2 and x is 1 or 2.

In related studies, *W. Lowe; U.S. Patent 4,089,792; May 16, 1978; assigned to Chevron Research Company* describes a lubricating oil additive composition which imparts improved oxidation properties to crankcase lubricants and comprises an antioxidant selected from aromatic or alkyl sulfides and polysulfides, sulfurized olefins, sulfurized carboxylic acid esters and sulfurized ester-olefins, and a primary amine of the formula:

$$R-NH_2 \qquad \text{or} \qquad R \underset{\underset{A}{|}}{\overline{+}} N-(CH_2)_x \underset{y}{\overline{\big]}} NH_2 \qquad \text{or} \qquad Het \overset{\frown}{\underset{\smile}{\hspace{1em}}} N-R'-NH_2$$

where each A is independently hydrogen or alkyl, R is alkyl or alkenyl of at least 6 carbon atoms, R^1 is an alkyl group of at least 2 carbon atoms, x is an integer from 2 to 4, y is 1 to 4, and Het forms with the N atom, a 5- or 6-membered heterocyclic ring optionally containing an additional N or O hetero atom. Zn dithiophosphates may, also, be used in the additive composition.

W. Lowe; U.S. Patent 4,097,386; June 27, 1978; assigned to Chevron Research Company describes a lubricating oil additive composition which imparts improved oxidation properties to crankcase lubricants and comprises an antioxidant selected from aromatic or alkyl sulfides and polysulfides, sulfurized olefins, sulfurized carboxylic acid esters and sulfurized ester-olefins, and a tertiary amine of the formula $(R)_3N$ where each R is independently C_{3-10} alkyl, phenyl or phenyl substituted by 1 or 2 alkyl groups of up to 12 carbon atoms.

Adducts of Benzotriazole Compounds and Alkyl Vinyl Ethers

M. Braid and P.S. Landis; U.S. Patent 4,153,565; May 8, 1979; assigned to Mobil Oil Corporation have found that adducts of benzotriazole compounds and alkyl vinyl ethers or vinyl esters impart antioxidant, metal-corrosion prevention and antiwear properties to the lubricant compositions to which they are added.

The adducts of the process are formed by reacting the benzotriazole compound with the alkyl vinyl ether or vinyl ester of a carboxylic acid in proportions, expressed as molar ratios of benzotriazole compound to alkyl vinyl ether or vinyl carboxylate, of from about 1:1 to 1:1.5.

The adduct products may comprise several isomers, i.e., the vinyl ethers and esters may connect to the benzotriazole in either the 1-H or 2-H position. Also, both Markownikow and anti-Markownikow additions may occur. It has been found that each isomer is individually effective in imparting the improved anti-oxidant and anticorrosion properties to the lubricant compositions.

Alkenyl Succinimide Dispersant and Aminobenzoic Acid Compounds

W.W. Woods; U.S. Patent 4,107,054; August 15, 1978; assigned to Continental Oil Company describes lubricating oil compositions having improved corrosion inhibition. The lubricating oil compositions comprise: (a) a major amount of a base lubricating oil, (b) alkenyl succinimide dispersant, and (c) corrosion-inhibiting amount of certain specific acidic derivatives of benzene (e.g., 2-amino-5-methylbenzoic acid).

Example 1: This example shows that an alkenyl succinimide-type dispersant is required to solubilize the acidic derivative of benzene. In the test 40 g of dispersant, 30 g of midcontinent base oil having a viscosity of 100 SSU at 100°F (37.8°C) and 5 g of anthranilic acid were heated to approximately 130°C. The results are shown in Table 1.

Table 1

Dispersant	Solubility Behavior
Alkenyl succinimide (Oronite OLOA 1200)	Soluble, bright and stable liquid at room temperature for months; easily soluble in additional mineral oil
Alkenyl succinimide (Copper E644)	Soluble, bright and stable liquid at room temperature for months; easily soluble in additional mineral oil
Alkenyl succinimide (Monsanto 5070B)	Soluble, bright and stable liquid at room temperature for months; easily soluble in additional mineral oil
Partial succinimide (Oronite OLOA 373C)	Soluble hot and initially at room temperature, but recrystallization starts by 24 hr and is marked at 1 wk
Alkyl succinic ester (Lubrizol 936)	Partially soluble hot, major recrystallization on slight cooling to solid 2-phase matrix
Nonsuccinimide (Lubrizol 6401)	Soluble hot, major recrystallization on minor cooling
Nonsuccinimide (Amoco 9250)	Turbid hot, slow recrystallization on cooling

Example 2: Example 1 was repeated substituting a variety of other types of lubricant additives for the alkenyl succinimide dispersant. The anthranilic acid was not soluble in viscosity index improvers, zinc dithiophosphates, or overbased sulfonates.

Example 3: Tests were run to determine which acidic derivatives of benzene are soluble in alkenyl succinimide dispersant and which are not. The procedure was the same as for Example 1. The following derivatives had suitable solubility: o-aminobenzoic acid (anthranilic acid), m-aminobenzoic acid, N-methylanthranilic acid, N-dimethyl-3-aminobenzoic acid, 2-amino-5-methylbenzoic acid, catechol, 4-methylcatechol, resorcinol, 2-methylresorcinol, salicylic acid, thiosalicylic acid, and 5-methylsalicylic acid.

The following were insoluble or recrystallized: o-aminophenol, m-aminophenol, p-aminophenol, p-aminobenzoic acid, 2-amino-p-cresol, 6-amino-m-cresol, N-dimethyl-3-aminophenol, 2-hydroxypyridine, m-hydroxybenzoic acid, p-hydroxybenzoic acid, phthalic acid, isophthalic acid, terephthalic acid, hydroquinone, and 2-amino-3-hydroxypyridine.

Example 4: This example shows the results obtained using a bench-scale rust test on most of the acidic derivatives of benzene shown to be soluble in Example 3 plus the results on 2-naphthol; the test is described in U.S. Patent 3,897,350.

The test employed 600 ml of oil blend, 150 ml of gasoline and 100 ml of standardized acid mix (sulfuric, nitric, and hydrochloric acids). The admixture was placed in an Erlenmeyer flask with magnetic stirrer and reflux condenser to avoid gasoline loss. A polished hydraulic valve lifter was suspended in the stirred mix for 23 hr and then inspected for rust and corrosion. A series of standards were used for rating the lifter on a 0 to 10 merit scale. Prior correlations had indicated that ratings of 8.3 or better were required to pass Sequence II–C double length (64 hr) engine rust tests.

The bench tests were run on an oil blend containing: 9.0 wt % alkenyl succinimide dispersant, 1.9 wt % zinc dithiophosphate (9.1% Zn), 2.0 wt % calcium phenate/phosphonate (1.65% Ca), 2.0 wt % overbased sulfonate (11.7% Ca), 7.5 wt % methacrylate V.I. improver, 0 to 0.5% benzenoid compound with a balance of Mid-Continent base oil having a viscosity of 100 SSU at 100°F (37.8°C). The compounds tested and the results were as follows in Table 2.

Table 2

Benzenoid Compound	Concentration Percent		
	0	0.125	0.5
None	8.3		
2-Amino-5-methylbenzoic acid		8.4	9.1
4-Methylcatechol		8.6	8.8
Anthranilic acid		7.1	8.9
2-Naphthol (U.S. 3,897,350)		8.4	8.6
N-methylanthranilic acid		8.2	5.9
N-dimethyl-3-aminobenzoic acid		7.4	7.2
Salicylic acid		8.1	6.0
Thiosalicylic acid		7.1	6.8
Catechol		8.4	6.0
Resorcinol		8.4	8.3

Phenylated Naphthylamine, Sulfoxide Compound and Copper

A process developed by *F.C. Loveless and W. Nudenberg; U.S. Patents 4,122,021; October 24, 1978; 4,110,234; August 29, 1978; both assigned to Uniroyal, Inc.* is directed to a synergistic antioxidant system for use in oils comprising a phenylated naphthylamine, a sulfoxide compound, and optionally an oligodynamic amount of copper.

The following examples illustrate the process. In these examples, NN means neutralization number, as determined using ASTM test procedure D974–55T, and % ΔV 100 means percent change in Saybolt viscosity at 100°F, where the Saybolt viscosity at 100°F is determined using ASTM test procedure D445-53T. In all examples, the neutralization number of the unaged oil was essentially zero.

Example 1: This example shows the outstanding synergistic effect of using the stabilizer system of the process to protect a low-unsaturation synthetic hydrocarbon oil against oxidation degradation. The oil used was a polyoctene-based oil having 0.20 mol of unsaturation per 1,000 g of oil and an average molecular weight of about 600.

Various samples were prepared in order to evaluate the effectiveness of the stabilizer system. Sample A was prepared by adding phenyl-alpha-naphthylamine and diphenyl sulfoxide in the amounts set forth in Table 1 to 100 g (about 125 ml) of the polyocetene-based oil and heating to about 100°C in order to facilitate the dissolution of the additives. The copper metal was added in the form of a metal washer, as described below. Other samples contained the amine, diphenyl sulfide or diphenyl sulfone, and copper metal. The amounts used in each case are set forth in Table 1.

Each of the samples was tested according to the following test procedure: A 100 ml sample having the compositions set forth in Table 1 was poured into a Pyrex glass test cell and aged by inserting one end of a glass air delivery tube into the test cell while the remaining 25 ml portions of each original oil sample were set aside and analyzed for neutralization number and Saybolt viscosity at 100°F.

Around this tube immersed in the oil were placed from zero to four washers of various metals (Mg, Cu, Ag, and Fe). Multiple washers were separated from each other by glass spacers. These washers remained in the oil during the aging process and served to indicate the extent of metal corrosion caused by the aging oil. The test cell was fitted with a reflux condenser, and the entire assembly was placed in a constant temperature aluminum block. Air was flowed through the oil at the rate of 5 l/hr while the assembly was held at a temperature of 370°F.

These aging conditions were maintained for 72 hr, after which the oil was filtered hot and the sludge which had formed was collected and measured. The filtered oil was analyzed to determine changes in neutralization number and Saybolt viscosity at 100°F. The metal washers, which had been weighed initially, were carefully washed and reweighed to determine the weight change.

The data in Table 1 show that when a sulfoxide, such as diphenyl sulfoxide, and a phenylated naphthylamine, such as phenyl-alpha-naphthylamine, and copper metal are added to a synthetic hydrocarbon oil, the properties of the aged oil are excellent. There is very little change in the viscosity or neutralization number, very little sludge formation, and essentially no weight change in the metals.

In contrast, when diphenyl sulfide or diphenyl sulfone is used with the amine and copper, the oil is totally unprotected against oxidative degradations as shown by its inability to be filtered because of excessive sludge formation and viscosity increase.

Table 1

	A	B	C
Oil*, g	100	100	100
Sulfur compound, g	0.25**	0.25***	0.25†
PAN††, g	0.50	0.50	0.50
Cu, ppm	1–5†††	1–5†††	1–5†††
% ΔV 100	3.5	§	§
NN	0.31	§	§
Sludge, mg	5.3	§	§
Weight change of washers, g			
Magnesium	0	–	–
Iron	+0.0001	–	–
Copper	–0.0002	–	–
Silver	0	–	–

  *Polyoctene-based oil.
  **Diphenyl sulfoxide.
 **Diphenyl sulfide.
 †Diphenyl sulfone
††Phenyl-α-naphthylamine.
†††Estimated from washer.
  §Failed test, would not filter.

Example 2: This example demonstrates the synergism between the amine and the sulfoxide compounds. The procedure of Example 1 was repeated using the compositions listed in Table 2, below. Sample D contained the amine without the sulfoxide; Sample E contained the sulfoxide without the amine and Sample F contained both the amine and the sulfoxide.

The results show that if the amine or sulfoxide is used individually, even with copper, essentially no protection is afforded the oil, whereas the use of the amine and the sulfoxide together, according to the process, provides substantial protection to the oil.

Table 2:

	D	E	F
Oil*, g	100	100	100
Diphenyl sulfoxide, g	–	0.25	0.25
PAN**, g	0.5	–	0.5
Cu, ppm	1–5***	1–5***	1–5***
% ΔV 100	22	25.5	3.5
NN	3.46	6.1	0.31
Sludge, mg	1,320	3,011	5.3
Weight change of washers, g			
Magnesium	–0.1511	–0.0208	0
Iron	+0.0002	+0.0006	+0.0001
Copper	–0.0014	–0.0075	–0.0002
Silver	+0.0004	+0.0001	0

  *Polyoctene-based oil.
  **Phenyl-α-naphthylamine.
 ***Estimated from washer.

Example 3: This example shows the protection provided by the process to mineral and ester oils. The mineral oil used was a highly refined white mineral oil identified as Ervol from Witco Chemical with a Saybolt viscosity of 137.7 SUS at 100°F. The ester oil used was trimethylol propane triheptanoate. The samples were prepared and tested in accordance with the procedure of Example 1.

The results of the tests, in Table 3, show that both oils were protected when the synergistic antioxidant system of this process was used, whereas the unprotected mineral oil (without the antioxidant system) could not even be filtered (sample K).

Table 3

	K	L	M
Oil type	mineral	mineral	ester
Oil unsaturation level, mol/1,000 g	0.03	0.03	0
PAN*,**	0	0.50	0.50
Diphenyl sulfoxide**	0	0.25	0.25
% ΔV 100	***	+27.8	+1.8
NN	***	0.45	0.24
Sludge, mg	***	83	12.6
Weight change of washers, g			
Magnesium	dissolved	−0.0207	0
Iron	−0.0019	−0.0002	0
Copper	−0.0063	−0.0022	−0.0010
Silver	−0.0001	0	+0.0002

*Phenyl-α-naphthylamine.
**Parts per hundred parts of oil.
***Failed test, would not filter.

Antioxidants and Halogen-Containing Compounds

T.V. Liston and W. Lowe; U.S. Patent 4,148,737; April 10, 1979; assigned to Chevron Research Company describe a lubrication oil additive composition which imparts improved oxidation properties to lubricants which comprises: (a) an antioxidant selected from aromatic or alkyl sulfides and polysulfides, sulfurized olefins, sulfurized carboxylic acid esters and sulfurized ester-olefins, and (b) an oil-soluble brominated hydrocarbon containing at least three carbon atoms.

It has been found that the defined antioxidants in combination with the oil-soluble brominated hydrocarbons of the process complement each other in a synergistic manner, resulting in a combination having antioxidant properties superior to either additive alone. The brominated hydrocarbon component alone has virtually no antioxidant effect. However, when the defined combination of brominated hydrocarbons and antioxidant is added to a lubricating oil, less of the antioxidant is needed to obtain oxidation control than when the brominated hydrocarbon is not present.

Preferably, from 2 to 40 mmol of an oil-soluble zinc salt is present per kilogram of the lubricating oil composition. While this zinc salt is not required to achieve the synergistic effect from the combination of the antioxidant and brominated hydrocarbons, an improved lubricating oil composition results from the use of all three additive components.

The compositions of this process are highly stable additives for crankcase lubricating oils and impart excellent antioxidant properties to these oils.

Example 1: The combination of brominated hydrocarbons with sulfur-containing antioxidants in improving the antioxidation properties of a lubricating oil over the use of either of the components individually is illustrated by the following test. The oxidation test uses the resistance of the test sample to oxidation using pure oxygen with a Dornte-type oxygen absorption apparatus for "Oxidation of White Oils," *Industrial and Engineering Chemistry*, Vol 28, p. 26 (1936)].

The conditions are an atmosphere of pure oxygen exposed to the test oil maintained at a temperature of 340°F. The time required for 100 g of test sample to absorb 1,000 ml of oxygen is observed and reported in the following Table 1. The base oil formation for base oil A comprises 6% of a conventional succinimide dispersant, 0.05% terephthalic acid, 0.4% of a conventional rust inhibitor, and 9 mmol/kg of a zinc dithiophosphate in Cit-Con 30. Base oil B comprises 6% of a conventional succinimide dispersant, and 9 mmol/kg of a zinc dithiophosphate in Cit-Con 30.

Table 1

Test No.	Base Oil Formulation	Antioxidant (%)	1-Bromodecane (%)	Oxidation Life (hr)
1	A	none	none	5.2
2	A	none	0.1	6.0
3	A	none	0.2	6.3
4	A	1*	none	6.6, 6.4
5	A	1*	0.1	17.3
6	A	1*	0.2	21.0, 21.5
7	A	1*	0.4	21.3
8	A	0.5**	none	8.9
9	A	0.5**	0.5	10.9
10	A	1**	none	12.9, 12.1
11	A	1**	0.1	16.5
12	B	none	none	4.5, 4.9
13	B	1*	none	8.2
14	B	1*	0.1	14.5
15	B	0.25*	0.1	7.2
16	B	1**	0.1	15.4
17	B	0.5**	0.1	10.9
18	B	1**	none	11.3
19	B	1**	0.1	15.4
20	B	1***	none	4.5
21	B	1***	0.1	9.6
22	B	1†	none	4.2
23	B	1†	0.1	9.3
24	B	1††	none	7.1
25	B	1††	0.1	14.0

*Diparaffin polysulfide.
**Sulfurized cracked wax olefin. Reaction of cracked wax olefin with sulfur and sulfur monochloride.
***Sulfurized butyl acrylate, 20% S.
†Sulfurized 2-ethylhexyl acrylate, 16% S.
††Sulfurized C_{9-10} cracked wax olefins.

The above data demonstrates the synergistic effect of the combination of sulfur-containing antioxidants and brominated hydrocarbons.

Example 2: The combination of 1,4-dibromobutane with sulfur-containing anti-oxidants in improving the antioxidation properties of a lubricating oil over the use of either of the components individually is illustrated by the data in Table 2. The oxidation test is described in Example 1. In tests 1 through 9, the base oil comprised a mixture of 50% Cit-Con 350N and 50% Cit-Con 650N. In test 10 through 20, the base oil comprised Cit-Con 30.

Table 2

Test No.	Succinimide Dispersant (%)	Zinc Dithiophosphate (mmol/kg)	Antioxidant (%)	1,4-Di-bromobutane (%)	Oxidation Life (hr)
1	3.5	—	—	—	0.5
2	3.5	—	—	0.05	0.4
3	—	9	—	—	4.5
4	3.5	9	—	0.05	5.5
5	3.5	—	1*	—	3.0
6	3.5	—	1*	0.05	8.7
7	3.5	9	—	—	4.2
8	3.5	9	1*	—	7.9
9	3.5	9	1*	0.05	14.7
10	6.0	9	—	—	4.9
11	6.0	9	0.5*	0.25	27.1
12	6.0	9	1.0*	0.10	19.6
13	6.0	9	0.5*	0.10	17.1
14	6.0	9	0.5*	0.05	10.5
15	6.0	9	0.25*	0.05	9.0
16	6.0	9	0.5*	0.025	10.9
17	6.0	9	0.25*	0.025	9.0
18	6.0	9	1**	—	12.1
19	6.0	9	0.5**	—	8.9
20	6.0	9	1.0**	0.10	20.3
21	6.0	9	0.5**	0.10	13.8
22	6.0	9	0.5**	0.05	12.0

*Diparaffin polysulfide.
**Cracked wax olefin. Reaction of cracked wax olefin with sulfur and sulfur mono-chloride, 11 to 12% S and 1 to 2% Cl.

The above data demonstrates the synergistic effect of the combination of sulfur-containing antioxidants and 1,4-dibromobutane.

T.V. Liston and W. Lowe; U.S. Patent 4,148,738; April 10, 1979; assigned to Chevron Research Company have developed another lubricating oil additive composition which imparts improved oxidation properties to crankcase lubricants which comprises: (a) an antioxidant selected from aromatic or alkyl sulfides and polysulfides, sulfurized olefins, sulfurized carboxylic acid esters and sulfurized ester-olefins, and (b) an oil-soluble quaternary ammonium halide.

T.V. Liston and W. Lowe; U.S. Patent 4,148,739; April 10, 1979; assigned to Chevron Research Company describe a related lubricating oil additive composition which imparts improved oxidation properties to lubricants which comprises:

(a) an antioxidant selected from aromatic or alkyl sulfides and polysulfides, sulfurized olefins, sulfurized carboxylic acid esters and sulfurized ester-olefins, and (b) an oil-soluble iodo-containing hydrocarbon.

It has been found that the defined antioxidants in combination with an iodo-containing hydrocarbon complement each other in a synergistic manner, resulting in a combination having antioxidant properties superior to either additive alone. The iodo-containing component alone has virtually no antioxidant effect. However, when the defined combination of the iodo-containing hydrocarbon and antioxidant is added to a lubricating oil, less of the antioxidant is needed to obtain oxidation control than when the iodo-containing hydrocarbon is not present.

Mercaptan-Carbonyl Compound Reaction Products

A process described by *D.M. Smith; U.S. Patent 4,129,510; December 12, 1978; assigned to The Lubrizol Corporation* concerns sulfur-containing reaction mixtures made by reacting certain mercaptans with certain carbonyl compounds to yield an intermediate which is subsequently reacted with certain carboxylic acids. These reaction mixtures are useful as additives in lubricants and fuels. The process provides compounds of the formula:

$$R \!-\!\!\left(\!S\!-\!\underset{\underset{A}{|}}{\overset{\overset{R_1}{|}}{C}}\!-\!\underset{}{\overset{\overset{R_2}{|}}{C}}\!=\!\underset{\underset{R_4}{|}}{\overset{\overset{R_3}{|}}{C}}\!\right)_{m'}$$

where R is a hydrocarbyl radical; each of R_1, R_2, R_3 and R_4 is independently hydrogen or a hydrocarbyl radical; m' is from 1 to 5; and each A is an alkyl radical containing at least one carboxyl group.

The process comprises first reacting, in the presence of a catalytic amount of an acid of the following: (a) at least one mercaptan of the formula $R-(SH)_{m'}$ where R is a hydrocarbyl radical; and m' is from 1 to 5, with (b) at least one carbonyl compound of the formula:

$$R_1 - \overset{\overset{\textstyle O}{\|}}{C} - R_8$$

wherein R_8 is

$$-\underset{\underset{H}{|}}{\overset{\overset{R_2}{|}}{C}}-\underset{\underset{R_4}{|}}{\overset{\overset{R_3}{|}}{C}}-H$$

and each of R_1, R_2, R_3 and R_4 is independently hydrogen or a hydrocarbyl radical, to form an intermediate which is subsequently reacted with (c) at least one olefinic carboxylic acid or functional derivative thereof.

In the following examples all percentages and parts are by weight (unless otherwise stated expressly to the contrary) and the molecular weights are number average molecular weights as determined by vapor phase osmometry (VPO).

Example 1: A mixture of 1,830 parts (9 mols) n-dodecyl mercaptan, 720 parts (10 mols) of isobutyraldehyde, 500 parts of toluene and 2 parts of p-toluenesulfonic acid is heated at reflux for 20 hr. During the reflux period, 155 parts of water is removed by azeotropic distillation. The reaction mixture is stripped to 160°C under vacuum and filtered to yield 2,304 parts of the desired intermediate as the filtrate.

A mixture of 279 parts (1.09 mols) of the abovedescribed intermediate, 109 parts (1.09 mols) of maleic anhydride and 100 parts of xylene is refluxed at 200°C for 7½ hr, then stripped at 190°C under vacuum and filtered. The filtrate is the desired product; it contains 10.14% S and has a saponification number of 202 as determined by ASTM D–94 specifications available from the American Society for Testing Materials (ASTM), 1916 Race Street, Philadelphia, PA 19103.

Example 2: The general procedure of Example 1 is repeated except the n-dodecyl mercaptan, isobutyraldehyde and maleic anhydride are replaced on an equimolar basis by the corresponding mercaptans, aldehydes and/or ketones and olefinic carboxyl acids or functional derivatives thereof as shown in the following table.

Mercaptan	Aldehyde and/or Ketone	Olefinic Carboxyl Acid or Derivative
(A) butyl	stearaldehyde	acrylic acid/methacrylic acid (50/50) w*
(B) hexyl	tetradecanal	methyl methacrylate
(C) tert-nonyl	3-phenyl-2-methylpropanal	pentaerythritol acrylate
(D) tert-dodecyl	2-ethylbutyraldehyde	maleic anhydride
(E) n-dodecyl	2-ethylbutyraldehyde	maleic anhydride
(F) n-dodecyl	2-butanone	citraconic acid
(G) tert-dodecyl	3-hexaone	maleic anhydride
(H) n-dodecyl	4-heptanone/2-heptanone (1/1) m**	acrylic acid
(I) tert-nonyl	2-methyl-3-butanone/2-ethyl-butyraldehyde (75/25) w*	maleic acid
(J) cyclohexyl	hexahydrobenzaldehyde	cinnamic acid/fumaric acid (1/1) m**
(K) polybutenyl ($\overline{M}n$ = 300 VPO)	propionaldehyde/n-butyraldehyde (50/50) w*	maleic acid
(L) tetrapropenyl thiophenol	stearaldehyde/tetradecanol (50/50) w*	***
(M) tert-octyl-/tert-decyl (50/50) w*	3,3-dimethylpentanal	acryloyl chloride
(N) pentyl-/tetra-decyl (1/1) m**	di-n-propylacetaldehyde	acrylamide

*Proportion by weight.
**Molar proportion.
***The amide prepared from the reaction of one mol of maleic anhydride and one mol of diethylene triamine.

Example 3: An acylated amine derivative (predominantly a mixture of amides and imides) of the product of Example 2(K) is prepared by heating a mixture of 478 parts of the product of Example 2(K), 300 parts of mineral oil and 189 parts of a commercial ethylene polyamine mixture substantially corresponding in empirical formula to tetraethylene pentamine at 155° to 165°C for 8 hr while water is removed. The mixture is stripped at 165°C under vacuum and filtered to yield as the filtrate an oil solution of the desired product.

The condensation products of this process are useful as additives in preparing lubricant compositions where they function primarily as oxidation and rust inhibitors, particularly where the oil is subjected to high-temperature environments or to cyclic stresses such as those encountered in stop-and-go automobile driving.

Sulfur-Bridged Hydrocarbon Ring Compounds

A process described by *G.J.J. Jayne and H.F. Askew; U.S. Patents 4,141,846; February 27, 1979; 4,144,247; March 13, 1979; both assigned to Edwin Cooper and Company Limited, England* relates to compounds which are derived from intramolecular and intermolecular, in the case of a polymer, sulfur-bridged hydrocarbon rings containing from 6 to 12 carbon atoms. The hydrocarbon ring is preferably derived from 1,5-cyclooctadiene. The compounds also contain the residue of an alkoxyalkyl or aryloxyalkyl xanthate and may also contain the residue of a further nucleophilic group. The compounds are useful in lubricating oil compositions.

The sulfur-bridged compounds may be prepared by reacting sulfur dichloride with the unsaturated ring compound, preferably in an inert solvent at a temperature between $-20°$ and $100°C$ to give a dichloro intramolecular sulfur-bridged derivative. Details of the preparations of these compounds are given in *J. Org. Chem.* 33, p. 2627 (1966); *J. Org. Chem.* 31, pp. 1679 and 1669 (1966).

The compounds of the process may then be prepared by reacting the metal, preferably the alkali metal, e.g., sodium or especially potassium, salt of an alkyl, alkenyl, aralkyl, aryloxyalkyl or alkoxyalkyl xanthate with the dichloro sulfur-bridged compound.

Example 1: In a reaction flask was placed 600 ml of the monoethyl ether of ethylene glycol and 144.9 g of 85% KOH. While stirring, this mixture was heated to $80°C$ to dissolve the KOH. The solution was cooled to about $40°C$ and 167.2 g of CS_2 added dropwise with cooling. A precipitate formed and toluene was added as required to maintain fluidity.

Following this a solution of 211 g of 2,6-dichloro-9-thiabicyclo[3.3.1]nonane, (prepared by reacting equal mol amounts of SCl_2 and 1,5-cyclooctadiene in methylene chloride solvent at $-5°$ to $0°C$) in 800 ml of toluene was added and the mixture stirred at $75°C$ for 3 hr. The reaction product was washed with water to remove precipitated salts and then dried over anhydrous $MgSO_4$. The solvent was distilled out in a rotary evaporator to give the product 2,6-bis-(ethoxyethoxythiocarbonylthio)-9-thiabicyclo[3.3.1]nonane in 86% yield; analysis: 31.5% sulfur, 0.46% chlorine.

Examples 2 through 12: Following the above general procedure, a series of compounds were prepared using the following glycol ethers.

Table 1

Example	Glycol Ether
2	Ethyleneglycol monomethyl ether
3	Ethyleneglycol monoisopropyl ether
4	Ethyleneglycol monobutyl ether
5	Diethyleneglycol monoethyl ether

(continued)

Table 1: (continued)

Example	Glycol Ether
6	Diethyleneglycol monobutyl ether
7	Propyleneglycol monomethyl ether
8	Dipropyleneglycol monomethyl ether
9	Tripropyleneglycol monomethyl ether

These examples gave compounds having the following structure:

(A)

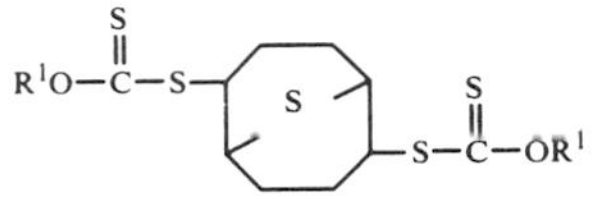

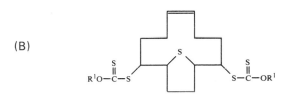

R^1, yield and analysis were as follows:

Table 2

Example	R^1	Yield	S (%)	Cl (%)
2	$-CH_2CH_2OCH_3$	69	33.6	0.42
3	$-CH_2CH_2OCH(CH_3)_2$	80	31.1	0.28
4	$-CH_2CH_2OC_4H_9$	88	29.3	0.38
5	$-(CH_2CH_2O)_2-C_2H_5$	78	29.5	0.36
6	$-(CH_2CH_2O)_2C_4H_9$	56	24.9	0.29
7	$-CH_2CH(CH_3)OCH_3$	87	32.1	0.60
8	$-[CH_2CH(CH_3)O]_2CH_3$	64	26.1	1.55
9	$-[CH_2CH(CH_3)O]_3CH_3$	88	20.6	0.26
10	1-methylheptyl:ethoxyethyl (1:1)	74	28.3	0.34
11	1-methylheptyl:methoxypropyl (1:1)	87	28.9	0.44

Further compounds were prepared having the structure:

(B)

wherein R^1 is as previously described. In Example 12, R^1 was the group ethoxyethyl. The yield was 72% and the product analyzed 26.8% S and 0.5% Cl.

Example 13: In a reaction flask were placed 52.8 g (0.25 mol) of 2,6-dichloro-9-thiabicyclo[3.3.1]nonane, prepared as in Example 1, 7.8 g (0.125 mol) of ethylene glycol and 100 ml of toluene. The mixture was heated in a current of nitrogen for two periods of 7 hr at 80°C, followed by two periods of 7 hr at 100°C, by which time evolution of HCl had virtually ceased.

Potassium isopropyl xanthate was prepared by reacting 22.8 g (0.3 mol) of CS_2 with a solution of 19.8 g, 85% KOH (0.3 mol) in 100 ml of isopropanol and 100 ml of toluene. This was then added to the first reaction product and the mixture was heated to 78°C for 3 hr. More toluene was then added, and the product washed with water, dried and stripped, yielding a brown viscous liquid in 85.5% yield. This product, which was believed to have the formula:

(C)

contained 31.3% S (theory 31.5%) and 0.72% Cl.

Several of the above compounds were subjected to a Copper Corrosion Test (Copper Strip Test ASTM–D 130-168; IP 154/69). Examples 1 to 10 and 12 and 13 were subjected to the test and all gave a pound rating.

Several of the compounds were subjected to a 36 hr Petter W-1 engine test (IP 176/69) but the procedure was slightly amended in that the oil sample removed at 16 hr was not replaced by new oil. In this test the lubricating oil was blended to contain 1% by weight of the test additive in a formulated mineral lubricating oil which contained other conventional oil additives (e.g., succinimide dispersant, overbased magnesium sulfonate, zinc dialkyldithiophosphate, etc.).

All the test additives except Examples 12 and 13 were dixanthates of Formula (A) and were derived from 2,6-dichloro-9-thiabicyclo[3.3.1]nonane but differed in the group R^1 attached to the xanthate oxygen. Several alkyl xanthates are included for comparison. Test criteria are the bearing weight loss and the piston rating on a scale from 0 to 10 (10 = clean).

Table 3

Xanthate Oxygen Substituent	Bearing Weight Loss (mg)	. . . . Piston Rating.	
		Skirt	Undercrown
Bisisopropyl	6	8.5	5.1
Ethyl/isopropyl	8	10.0	7.9
Ethyl/sec-butyl	10	9.4	0.1
Bis(4-methylpentyl-2)	21	8.3	2.0
Bis(sec-octyl)	14	9.6	6.6
Example 1	8	9.9	9.0
Example 2	18	9.9	9.0
Example 3	14	9.6	7.1
Example 4	16	9.8	9.1
Example 7	25	10.0	10.0
Example 8	8	9.9	8.7
Example 9	4	10.0	8.0
Example 10	6	10.0	9.9
Example 11	13	9.9	7.0
Example 12	11	9.8	7.5
Example 13	12	9.9	5.4

As the above test results show, the alkoxyalkyl xanthates of this process are exceptionally effective in maintaining piston cleanliness. They are substantially more effective than similar alkyl xanthates, especially in regard to piston undercrown rating. Appearance of the bearings in all cases was very good, being light straw to brown with no sign of pitting or corrosion. In addition, they do not have a disagreeable odor as usually is encountered with lower alkyl xanthates.

The additives were also tested by the Rotary Bomb Oxidation Test (IP 229/73) to further determine their antioxidant effectiveness. Results obtained with the blank oil and the same oil containing the additives are shown in the following table.

Table 4

Additive Of	Minutes to Failure
Blank	40
Ex. 1	100
Ex. 2	105
Ex. 3	85
Ex. 5	120
Ex. 6	90
Ex. 7	42
Ex. 8	20
Ex. 9	54

The EP properties of the additives were demonstrated by subjecting several of them to the Timken O.K. Load Test. In this test, one number is rotated against another number under increasing load. Their contact area is lubricated with oil containing the test additive. The Timken O.K. load is the maximum load at which no scoring or seizure occurs.

Table 5

Additive Of	Timken O.K. Load (lb)
Blank	12
Ex. 1	45
Ex. 5	24
Ex. 9	18

The additives are useful in a broad range of lubricating oils including both mineral and synthetic oils.

Sulfurized Dihydroxybenzene Compounds

T. Colclough and B. Swinney; U.S. Patent 4,115,287; September 19, 1978; assigned to Exxon Research and Engineering Company describe a lubricating oil containing a sulfurized dihydroxy benzene of formula:

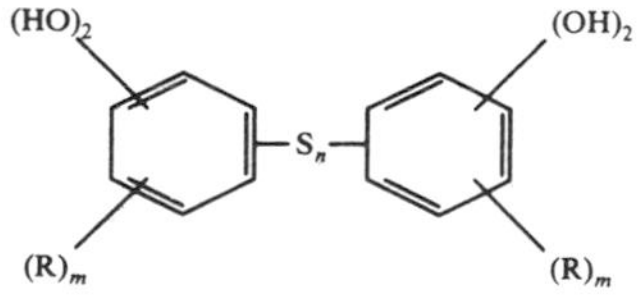

where n is 1 to 4, m is 1 to 3 and R is an alkyl group containing at least 7 carbon atoms. These compounds impart antirust and antioxidant properties to the oil with reduced copper corrosion.

Example 1: Nonyl catechol was prepared by charging 15 kg of catechol, 17 kg of nonene and 0.818 kg of Amberlyst 15 as catalyst into a 50-liter glass reactor and stirred at 100°C for 21 hr. The product was filtered and stripped at 150°C under a pressure of 43 mm of mercury.

10 kg of the product were then charged together with 10 liters of carbon tetra-chloride into a 50-liter glass reactor. The temperature of the solution was cooled to below 10°C and 2.86 kg of sulfur monochloride added over a period of 4 hr ensuring that the temperature did not rise above 15°C. Finally the carbon tetra-chloride was distilled off to yield 11.35 kg of sulfurized nonyl catechol. The product was soluble in mineral oil yielding a black solution.

Example 2: Nonyl catechol was prepared by the same process as Example 1 and distilled at 136° to 156°C at 0.1 mm mercury to purify the product. 9 kg of this distilled nonyl catechol were then charged to the glass reactor together with 9 liters of carbon tetrachloride. 2.58 kg of sulfur monochloride were then added over a period of 4 hr while the reactor was held at a temperature of from 9° to 11°C. Finally the carbon tetrachloride was distilled off at 102°C under a pressure of 52 mm of mercury.

The product was soluble in mineral oil to give a clear stable solution. The solution was also stable when an ashless dispersant was present.

Example 3: 15 kg of catechol and 25.77 kg of nonenes were heated together with 0.414 kg of Amberlyst 15 as catalyst. Initially 70% of the nonenes were added together with the catechol and the mixture heated to 96°C, the exotherm took the temperature to 144°C, was then cooled to 100°C and the remainder of the nonenes added and the vessel then heated to 130°C for 4¾ hr. The product was separated and stripped at 154°C at atmospheric pressure and then at 150°C at 62 mm mercury pressure for 6 hr.

Analysis of the product showed less than 4% by weight of catechol, 17.5% by weight of dinonyl catechol, 63% of 4-nonyl catechol and 0.9% of 3-nonyl cat-echol.

9 kg of this product were charged to a glass vessel together with 9 liters of car-bon tetrachloride. 2.58 kg of sulfur monochloride was added slowly over a period of 4 hr while the temperature was held between 9° and 11°C. Hydro-chloric acid was removed by a caustic scrubber and when all the sulfur mono-chloride had been added the product was allowed to warm up to room tempera-ture and was then stirred for 1 hr. Finally the carbon tetrachloride was distilled off at 102°C at 52 mm mercury pressure.

Analysis of the product showed: 19% by wt nonyl catechol (not sulfurized), 16.0% by wt dinonyl catechol (not sulfurized), 11.0% by wt sulfur, and 1.6% by wt chlorine.

It was found that while most of the original mononony! material was sulfurized the dinonyl material remained unsulfurized. The product was soluble at 1% in mineral oil although the solution was somewhat hazy; a clear stable solution was obtained when 3% by weight of a conventional polyamine dispersant was added.

Example 4: Various sulfurized alkyl catechols were prepared using in one pro-cess a crude alkyl catechol as used in Example 1 and reacting the crude alkyl catechol with a sulfur chloride in a toluene solution in the presence of an iron catalyst (Process A). In a second series of preparations purified alkyl catechol as used in Example 2 was reacted with a sulfur chloride in a carbon tetrachloride solution (Process B). The reaction may be summarized as follows:

Samples of each of the sulfurized alkyl catechols were incorporated into a 10W/30 0.5% sulfonated ash lubrication oil formulated to MIL-L-46152 specification containing 6.0% by wt of a polyamine ashless dispersant, zinc dithiophosphate antiwear agent and an overbased metal dispersant.

Each oil was then subject to the MS IIC rust test (published in the ASTM STP 315F) in which an Oldsmobile V8 engine is operated for 32 hr with the lubricating oil and the engine then dismantled and the rusting of the parts evaluated and given a merit rating from 0 to 10, 10 representing no rusting. The results obtained were as follows, the values of n and x referring to the formula given above.

n	x	Process Used	Alkyl Catechol/ $S_x Cl_2$	Amount Sulfurized Alkyl Catechol in Oil (%)	MS IIC Rating
—	—	—	—	0	5.5
9	1	A	1.2	1.0	7.0
9	1	B	2	1.0	8.5
9	2	B	1.2	1.0	8.5
9	2	A	2	1.0	8.1
12	2	B	2	1.0	8.5
12	1	B	2	1.0	7.8
9	2	B	2	0.8	7.8
8	2	B	1.2	0.8	8.4
9	2	B	2	1.0	8.6

Example 5: 1.0% by weight of the product of Example 2 was included in a 10W/50 lubricating oil 0.5% ash containing 6.0% by weight of a polyisobutylene succinic anhydride/polyamine ashless dispersant and the bearing corrosion of the oil determined by the L38 corrosion test (SAE publication 680538).

The performance of these oils was compared with the performance of oils containing 0.46% by wt of a 70% active material dodecenyl succinic acid and 0.8% by wt of the reaction product of styrene and the condensation product of P_2S_5 and C_9 alkyl phenol 8.0% by wt active ingredient containing about 7% by wt sulfur and about 3.5% by wt phosphorus.

The bearing weight loss of the oil containing the sulfurized alkyl catechol was 76 mg while that of the other oil was 494 g.

The tests described above were repeated using as dispersant the same product as was used previously but reacted with boric acid. In this instance the bearing weight loss of the oil containing the sulfurized alkyl catechol was 54 mg while that of the other oil was 170 mg.

Example 6: The antiwear effect of sulfurized alkyl catechol was evaluated in a 10W/30 oil formulation containing 6.0% by wt of an ashless dispersant, and 1.0% by wt of the sulfurized nonyl catechols by the Hertz 4 ball wear test under a load of 120 kg for 1 min. The performance was compared with a similar oil containing a zinc dialkyldithiophosphate and one that is free of antiwear additives. The results were as follows.

Additive	Wear Scar Diameter (mm)
Zinc dialkyldithiophosphate	2.55
Nonyl catechol/S_2Cl_2	
2:1	1.90
Nonyl catechol/SCl_2	
2:1	2.40
None	3.0

Alkaline Earth Metal Sulfonate and Phosphonate-Phenate Sulfide

E.L. Miller and R.A. Krenowicz; U.S. Patent 4,123,369; October 31, 1978; assigned to Continental Oil Company have found that incorporation of a combination of highly overbased alkaline earth metal sulfonate and alkaline earth metal phosphonate-phenate sulfide provides greatly improved antirust properties over an identical composition utilizing only overbased alkaline earth metal sulfonate as the rust inhibitor.

Thus the process provides an extended service lubricating oil composition comprising a base oil and an additive package including a specific combination of additives to give superior rust inhibition properties.

The improved rust inhibition properties provided by this process are illustrated in the following example, in which a standard Sequence II-C engine rust test was performed on the oil composition, followed by a duplicate test on the same oil sample to provide a service test. The standard Sequence II-C procedure involves the operation of a 425 in³ 1967 Oldsmobile V-8 engine under low-speed, low-temperature conditions.

Upon completion of the test (32 hr), the engine is inspected for evidence of rust and valve lifter sticking. The test is carried out under carefully controlled conditions of engine speed, load, oil and coolant temperatures, air fuel ratios, carburetor air temperature and humidity levels, exhaust back pressure levels, etc. In order to provide a severe test of the composition of this process, the composition was subjected to a double Sequence II-C test in which the standard Sequence II-C was performed twice on the same oil sample.

Two compositions, essentially identical except that in one case magnesium sulfonate was utilized as the rust inhibitor and in the other case a combination of magnesium sulfonate and calcium phosphonate-phenate sulfide was utilized, were evaluated, and the average rust rating (scale 1 to 10) obtained was 8.7 for the composition of the process, whereas the rust rating for the composition utilizing only magnesium sulfonate as a rust inhibitor was 5.8. The specific compositions of the two oils tested are shown below.

Ingredient	Process Composition Weight Percent	Comparison Composition Weight Percent
Ashless dispersant	8.0	8.0
Magnesium sulfonate (30% active ingredient)	2.4	2.4
Calcium phosphonate-phenate sulfide (45% active ingredient)	2.0	—
Zinc dithiophosphate	1.9	1.9
Viscosity index improver	7.5	7.5
100 SSU base oil	65.1	65.1
170 SSU base oil	13.0	15.0
Rust rating (scale 1-10) Double sequence II-C	8.7	5.8

The difference in rust rating obtained in this experiment is very significant, and the rating of 8.7 obtained from the composition of the process is particularly impressive in view of the severity of the experiment.

The composition which gave a rust rating of 8.7 in the above test is well suited as an extended surface crankcase lubricating oil composition. Such a composition, however, in certain conditions exhibits increased viscosity after a substantial service life, and an improved composition which is not subject to this deficiency is exemplified by the following composition.

Ingredient	Weight Percent
Ashless dispersant	5.0
Magnesium sulfonate (30% active ingredient)	1.5
Calcium phosphonate-phenate sulfide (45% active ingredient)	2.0
Zinc dithiophosphate	1.9
Viscosity index improver	7.5
Base oil	Balance

Condensation Products of Dithiophosphoric Acid Esters and Styrene

J.J. Jaruzelski; U.S. Patent 4,153,562; May 8, 1979; assigned to Exxon Research & Engineering Co. has found that effective antioxidants for hydrocarbon lubricating oils can be prepared by condensation of certain unsaturated materials with dithiophosphoric acids that have been obtained from alkyl phenol sulfides having from about 3 to 20 carbon atoms in the alkyl groups. The unsaturated materials that are condensed with the dithiophosphoric acids include styrene, alkylated styrene having from 1 to 6 carbon atoms in the alkyl groups, highly reactive olefins of 7 to about 12 carbon atoms, vinyl esters having a total of from 4 to about 20 carbon atoms, and acrylate or methacrylate esters having a total of from 4 to about 20 carbon atoms.

To prepare the additives of this process it is desirable to use an excess of the alkyl phenol sulfide over the amount theoretically required to form the desired dithiophosphoric acid ester when reacting the phenol sulfide with P_2S_5, so as to leave up to 50% of those initially available hydroxyl groups as free OH groups in the phenol sulfide moiety of the dithiophosphoric acid ester. Thus in the reaction of the dithiophosphoric acid ester with styrene the reaction and the general structure of the product could be represented as follows.

In the above formula the symbol $-(S)_x-$ indicates that the alkyl pehnol sulfide can be either the mono- or the disulfide, x being 1 or 2. The monosulfide usually includes small quantities of the disulfide. It will be seen from the above that condensation occurs at the double bond of the unsaturated material.

The antioxidants of the process are particularly useful in compounded lubricants that contain detergents or dispersants of a conventional nature that can be either of the metal-containing type or of the ashless or nonmetallic type, or both.

Alkyl Phenol, Aromatic Amine and Triesters of Dithiophosphoric Acid

K. Sugiura, T. Miyagawa and H. Seki; U.S. Patent 4,116,874; September 26, 1978; assigned to Nippon Oil Co., Ltd., Japan describe a compressor oil composition, which comprises a major proportion of lubricating base oil, and a minor proportion of an aromatic amine, an alkylphenol and a triester of dithiophosphoric acid as essential ingredients.

The composition is characterized in that its acid number does not increase appreciably after use, it has a long life under oxidation conditions, and the color of the oil after use is satisfactory.

Examples 1 through 3: Various additives in the amounts shown in Table 1 were added to a mineral lubrication oil base having a viscosity of 35.0 cs at 37.8°C to form compressor oil compositions.

In this table, 2,5-dimercapto-1,3,4-thiadiazole and benzotriazole are metal deactivators; a hemiester of an alkenylsuccinic acid is a rust inhibitor; and a silicone oil is an antifoamer.

Table 1

Compositions (percent by weight) and Test Results

	Examples			Comparative Examples									
	1	2	3	1	2	3	4	5	6	7	8	9	10
Component (a)													
Diphenylamine	—	0.1	0.5	1.0	—	—	—	—	—	—	0.5	0.3	—
Phenyl-α-naphthylamine	0.2	—	—	—	—	—	—	0.5	—	—	—	—	0.2
Component (b)													
2,6-Di-tert-butyl-p-cresol	—	0.1	—	—	1.0	—	—	—	—	0.3	—	—	0.3
4,4'-Methylenebis(2,6-di-tert-butylphenol)	0.1	—	0.2	—	—	—	—	0.5	0.3	—	—	0.3	—
Component (c)													
Zinc di-4-methylpentyl dithiophosphate	—	—	—	—	—	1.0	—	—	—	0.3	—	0.1	0.1
O,O'-bis(propylphenyl)-S-benzyl dithiophosphate	—	—	0.2	—	—	—	—	—	0.5	—	0.3	—	—
O,O'-bis(propylphenyl)-S-styryl dithiophosphate	0.05	0.1	—	—	—	—	0.5	—	—	—	—	—	—
Other additives													
2,5-Dimercapto-1,3,4-thiadiazole	0.05	0.05	—	0.05	—	0.01	—	0.05	0.05	—	—	0.05	—
Benzotriazole	0.005	0.3	—	—	0.05	0.05	0.05	—	—	0.01	0.01	—	—
Hemiester of alkenylsuccinic acid	0.3	0.3	—	0.05	0.05	0.05	0.05	0.03	0.03	0.03	0.03	0.05	—
Silicone oil	0.002	0.002	—	0.002	0.002	0.002	0.002	0.002	0.003	0.002	0.002	0.002	—
Test Results													
Rotary vane-type air compressor (2,000 hr) (mg KOH/g)*	0.05	0.11	0.06	2.0	3.0	1.5	2.4	0.5	0.10	0.22	0.20	0.11	0.15
Indiana stirring oxidation test (48 hr at 165.5°C) (mg KOH/g)*	0.12	0.22	0.10	7.5	10.0	0.25	0.15	0.45	0.15	0.35	0.15	0.11	0.18
Rotating bomb-type oxidation test (life in min)	570	480	650	550	250	50	50	580	200	180	600	600	550
Turbine oil oxidation stability test Time until the acid value reached 1.0 mg KOH/g (hr)	5,300	4,900	5,600	2,900	2,500	1,000**	1,100**	2,900	2,000	1,800**	2,700	3,200	2,800
Color at 2,000 hr (union value)	4	4½	4	8	4	6***	6***	7	6	6½***	6½	6½	7

* Acid value after service test.
** Breakdown time.
*** The life did not reach 2,000 hr. Therefore, the color of the composition immediately before breakdown is given.

The performances of the lubricating oil compositions were evaluated by an oxidation test in a rotary vane-type air compressor, an Indiana stirring oxidation test, a rotating bomb-type oxidation test (ASTM D2272), and a turbine oil oxidation stability test stipulated in JIS (Japanese Industrial Standards) K2515. The results are shown in Table 1.

When a service test is performed for 2,000 hr using a rotary vane-type air compressor, the acid number of the lubricating composition after testing is preferably not more than 0.15 mg KOH/g, particularly not more than 0.10 mg KOH/g.

When the Indiana stirring oxidation test is performed at 165.5°C for 48 hr, the acid number of the lubricating oil composition after testing is preferably not more than 0.25 mg KOH/g, particularly not more than 0.10 mg KOH/g.

In the rotating bomb-type oxidation test, the life of the lubricating oil composition under oxidation conditions is preferably at least 450 min, particularly longer than 600 min.

In the turbine oil oxidation stability test (JIS K2515), the lubrication oil composition preferably attains an acid number of 1.0 mg KOH/g in at least 4,000 hr, particularly in at least 5,500 hr, after initiation of the test. On the other hand, the color (union value) of the lubricating oil composition at the end of 2,000 hr in this test is preferably 5 or below.

It can be seen from Table 1 that the compositions in accordance with Examples 1 through 3 exhibited superior results in all of these tests, and the composition of Example 3 gave especially superior results.

Comparative Examples 1 through 10: Lubricating oil compositions were prepared in accordance with the recipes shown in Table 1, and tested in the same manner as in Examples 1 through 3. The results are also shown in this table.

The results shown in Table 1 demonstrate the following facts. The lubricating oil compositions containing only one antioxidant component (Comparative Examples 1 through 4) showed unsatisfactory results in all the tests. The compositions containing two of the antioxidant components (Comparative Examples 5 through 8) were improved over the compositions in Comparative Examples 1 through 4, but were still unsatisfactory. The lubricating oil compositions containing three components but using zinc di-4-methylpentyl dithiophosphate as a component corresponding to component (c) in the process exhibited considerably improved results over the use of two antioxidant components, but proved unsatisfactory in the turbine oil oxidation stability test (JIS K2515).

In contrast, the lubricating oil compositions in accordance with this process (Examples 1 through 3, particularly Example 3) exhibited superior performances in all of these tests.

Bispiperazidophosphorus Compounds

A process described by *J.C. Hermans; U.S. Patents 4,142,979; March 6, 1979; 4,081,445; March 28, 1978; assigned to S.A. Texaco Belgium NV, Belgium* concerns bispiperazidophosphorus and trispiperazidophosphorus compounds, and methods for the preparation of these compounds.

The process provides compounds of the formula:

(1)

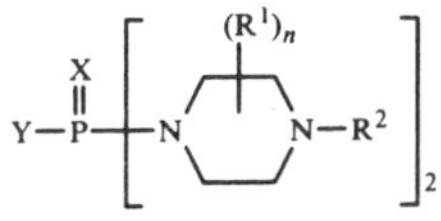

and salts of such compounds, where X represents an oxygen or sulfur atom or is absent, Y represents (a) an aliphatic, cycloaliphatic, or aromatic hydrocarbon group or a heterocyclic group; (b) a group of the formula $-NR_2$ in which each group R represents a hydrogen atom, an aliphatic, cycloaliphatic, or aromatic hydrocarbon group or a heterocyclic group, or the two groups R, together with the nitrogen atom to which they are attached, represent a N-containing heterocyclic ring; (c) a group of the formula $-OR$ in which R has the meaning given above; or (d) a group of the formula:

(2)

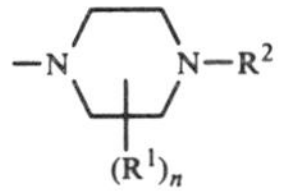

where R^1 represents a substituent on the piperazine ring; R^2 represents a hydrogen atom or a substituted or unsubstituted aliphatic, cycloaliphatic or heterocyclic radical or heterocyclic group, an acyl group, a sulfonyl group, a substituted phosphonyl group or a substituted carbanoyl group; and n represents 0 or an integer, and salts of such compounds in which at least one of the groups R^2 represents a hydrogen atom; but with the proviso that Y does not represent dimethylamino when X represents oxygen and R^1 and R^2 both represent hydrogen.

Preferred lubricating compositions and concentrates comprise p-(p'-dodecylanilino)-phenyl bispiperazidothiophosphate, and p-(p'-ethylanilino)phenyl bis-(N'-β-carbethoxyethylpiperazido)thiophosphate.

This process also provides methods for the synthesis of the compounds described above. According to a preferred method, a compound of the formula:

(3)
$$X=P(Hal)_{3-y}$$
$$(Y)_y$$

where y is an integer 0 to 1 (when y is 0, the compound has the Formula 4; when y is 1, the compound has the Formula 5:

(4) $X=P(Hal)_3$ or   (5) $Y-\overset{\overset{X}{\|}}{P}-(Hal)_2$

in which X and Y have the meanings given above and Hal represents chlorine or bromine) is reacted with an excess of a piperazine derivative of the formula:

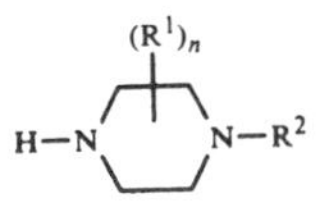

in which R^1, R^2 and n have the meanings given above.

Compounds of the formula XP(Hal)$_3$ and YPX(Hal)$_2$ are known in the art. Some of them will be commercially available, and others may be prepared by the methods described in, for example Houben-Weyl, *Methoden der Organischen Chemie*, Vol. XII/2, or by Olah and Oswald in *Ann.* 625 pp 92-94 (1959).

The compounds of the process have a variety of uses. They constitute multifunctional lubricating oil additives, exhibiting an anticorrosion oxidation and an antiwear action, and are mild extreme pressure agents. Since the compounds in which R^2 represents hydrogen are difunctional and trifunctional secondary amines, they are able to participate in polymer-forming and crosslinking by polyaddition and polycondensation reactions. The compounds in which R^2, for example, represents a β-carboalkoxyethyl group constitute flame-retardant plasticizers.

Example 1: 0.20 mol of phenyldichlorophosphate dissolved in 50 ml of benzene was added dropwise to a mixture of 4.0 mols of piperazine and 0.50 mol of triethylamine dissolved in 800 ml of benzene at 50° to 80°C. The mixture was mechanically stirred. When all the acid chloride was added, the mixture was refluxed for one to two hours and the precipitated triethylamine hydrochloride was filtered off. The filtrate was cooled, and more precipitated hydrochloride was filtered off. The solvent and the excess of piperazine were distilled off at normal pressure. 250 g of piperazine were recovered.

The remaining piperazine was removed under vacuum at 60° to 90°C and a crude product was obtained, which could be crystallized from benzene, from acetone or from a mixture of both, yielding colorless white crystals of phenylbispiperazidophosphate, MP 114.5° to 115°C. The crude product can also be purified by vacuum distillation, BP 235°C/0.4 mm. The total yield was more than 60% of the theoretical amount.

A standard procedure used for the synthesis of bispiperazido compounds is as follows. To 1 mol of anhydrous piperazine dissolved in 600 ml of dry benzene at 50° to 80°C is slowly added 0.10 mol of the phosphorus acid dichloride (YPXCl$_2$; X = O, S or absent) dissolved in 50 to 100 ml of benzene, while the mixture is gently stirred. The mixture is then refluxed for 15 to 30 min and the precipitated piperazine hydrochloride is filtered off. Benzene and the excess of piperazine are distilled off at reduced pressure (up to 80° to 100°C at 1 mm). This process easily occurs on a thin-film rotating evaporator. The residue is redissolved in benzene (400 ml) and the remaining piperazine hydrochloride is filtered off. Benzene and the remaining piperazine are then distilled off under vacuum (up to 100°C at 0.5 to 1 mm) which yields the crude piperazidophosphorus compound.

Example 2: Phenyl bispiperazidophosphate, having the structure shown below, was prepared by the following standard procedure.

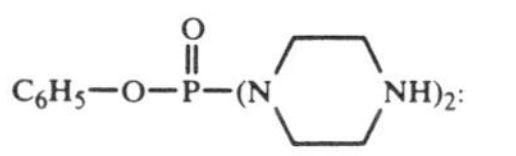

$$C_6H_5-O-\overset{\displaystyle O}{\overset{\displaystyle \|}{P}}-(N\overbrace{\qquad}NH)_2:$$

0.5 mol (105.5 g) of C$_6$H$_5$-O-POCl$_2$ was reacted with 5 mols (430 g) of piperazine dissolved in 3 liters of benzene. The standard work-up procedure, including filtration of the hydrochloride and distillation of the solvent and excess piperazine, yielded 143 g (92%) of a crude product which was first recrystallized from benzene (yield: 115.5 g; 74.3%) and then from benzene:acetone (1:1) which yielded 92.2 g (59.9%) with MP of 117° to 118°C. From the remaining viscous oil another fraction of pure compound could be isolated by vacuum distillation (BP 235°C/0.4 mm). Other compounds, indicated in the following table, were prepared by similar methods.

Bispiperazidophosphoryl Compounds

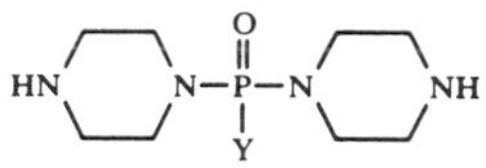

Y	Yield (%)	MP (°C)
$-O-C_2H_5$	100	oil
$-O-(CH_2)_{17}-CH_3$	86.4	66.5-68.5
$-O-C_6H_5$	74.3	117-118
$-NH(CH_2)_{17}-CH_3$	95	46-47
$-NH-(C_{10-14})$	82.7	oil
$-N(C_3H_7)_2$	100	oil
$-NH-C_6H_5$	56	155-157
$-NH-C_6H_4C_{12}H_{25}$	88.8	glassy solid

Substituted Trialkanolamines and Zinc Phosphates

A process described by *H. Haugen and D.G. Weetman; U.S. Patent 4,089,791; May 16, 1978; assigned to Texaco Inc.* pertains to a low-ash, antiwear, rust-inhibiting lubricating oil composition comprising a mineral oil base and minor amounts of an overbased alkaline earth metal compound, a substituted trialkanolamine represented by the formula:

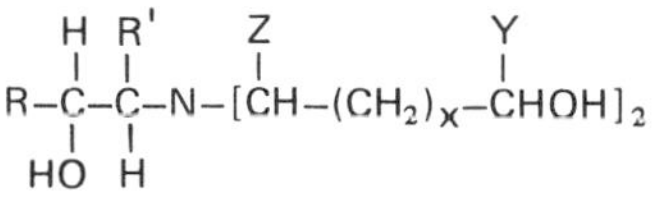

in which R is a hydrocarbyl radical having from 1 to 24 carbon atoms, R', Y and Z represent hydrogen or a hydrocarbyl radical having from 1 to 10 carbon atoms, and x is 0 to 1, and a zinc dihydrocarbyl dithiophosphate represented by the formula:

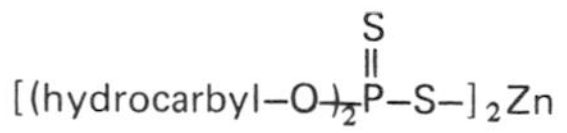

in which at least 50% of the hydrocarbyl radicals are alkaryl radicals having the formula:

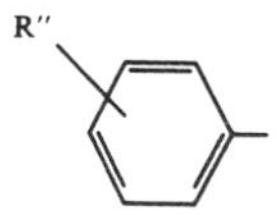

in which R'' is an alkyl radical having from about 4 to 18 carbon atoms.

The substituted trialkanolamine employed is most conveniently obtained by reacting an alpha-olefin epoxide compound with a dialkanolamine. This reaction is conducted by reacting approximately one mol of the alpha-olefin epoxide with a mol of dialkanolamine.

The following examples illustrate the preparation of the specific ashless substituted trialkanolamine additives employed in the process.

Example 1: 2-[2-(C_{13-15}Alkyl)]-2',2''-Nitrilotriethanol — 484 g (2.0 mols) of C_{15-18} straight-chain alpha-olefin epoxide mixture and 210 g (2.0 mols) of diethanolamine were charged to a reaction vessel. The stirred mixture was gradually heated to 100° to 120°C at which point an exothermic reaction occurred. External heating was discontinued while the heat of reaction carried the temperature of the mixture to 170° to 190°C. The reaction mixture was kept within this temperature range for about 0.25 to 0.50 hr. The mixture was then cooled to room temperature and 685 g of product were recovered.

Example 2: 2-[2-(C_{9-12} Alkyl)]-2',2''-Nitrilotriethanol — 827 g (4.25 mols) of C_{11-14} straight-chain alpha-olefin epoxide mixture and 445 g (4.25 mols) of diethanolamine were charged to a reaction vessel and reacted and recovered as in Example 1. 1,204 g of product were realized.

Example 3: 2-(2-Hexyl)-2',2''-Nitrilotriethanol — 390 g (3.0 mols) of 1,2-epoxyoctane and 215 g of diethanolamine (3.0 mols) were charged to a reaction vessel and reacted as described in Example 1. 705 g of product were realized.

The bearing weight loss was determined in the CLR L-38 Oxidation-Bearing Corrosion Test, Federal Test Method STD No. 791a, Method 3405. According to this test, the lubricant being tested is employed in a single-cylinder Labeco CLR Oil Test Engine equipped with copper-lead connecting rod bearings of known weight. The engine is run for 40 hr at 3,150±25 rpm. The copper-lead bearings are weighed a second time at the end of the test and the bearing weight loss determined.

The base oils employed for preparing the lubricant were essentially paraffinic base mineral oils. Base Oil A had a SUS viscosity at 210°F of about 41.5 and Base Oil B had a SUS viscosity at 210°F of about 54. The base oils are used singly or in blends to produce the mineral oil substrate for the lubricant of the process. The mineral oil substrate is the major component of the lubricant composition. In general, the oil base comprises from 85 to 95% by wt of the lubricating oil composition.

A Base Blend was prepared from the above base oils containing minor amounts of conventional lubricating oil additives to provide viscosity index improvement, antioxidant, dispersant and antifoaming properties.

This Base Blend has no material effect on bearing weight loss as measured in the L-38 Oxidation-Bearing Corrosion Test. The SAE limits for bearing weight loss is a maximum of 40 mg. This Base Blend was combined with the essential additive components of the process as set forth in the following table.

The low-ash, rust-inhibiting, antiwear lubricating oil composition of the process and comparison oil were tested for bearing weight loss in the L-38 Test and the results are set forth in the table below.

Lubricating oil composition, wt %	A	B	C
Calcium carbonate overbased calcium sulfonate, 300 TBN (% Ca)	0.16	0.16	0.16
Substituted trialkanolamine, example 1	–	–	0.25
Substituted trialkanolamine, example 2	0.25	0.25	–
Zinc di-C_{7-9}-alkyldithiophosphate (% Zn)	0.07	0.05	–
Zinc di-dodecylphenyldithiophosphate (% Zn)	0.03	0.05	0.10
Base blend	99.49	99.49	99.49
L-38 bearing wt loss, mg	97.7	17.9	24.9

Lubricating Oil Composition A described above, which contained a mixture of zinc dithiophosphate compounds but less than 50% of a zinc dialkaryldithiophosphate, failed the L-38 Bearing Weight Loss Test with a weight loss of 97.7 mg.

In contrast, Lubricating Oil Compositions B and C both passed the L-38 Bearing Weight Loss Test by wide margins as shown by the bearing weight losses of 17.9 and 24.9 mg respectively. These results demonstrate criticality in the composition of an effective lubricant according to the process, namely that the zinc dihydrocarbyldithiophosphate must consist of at least 50 wt % of a zinc dialkaryldithiophosphate.

A lubricating oil composition comparable to those described above but containing 0.10 wt % of zinc derived from zinc di-C_{7-9} alkyldithiophosphate and no zinc dialkaryldithiophosphate also failed the L-38 Bearing Weight Loss Test with a weight loss of 275 mg.

Titanate Dithiophosphate Compositions

G. Caspari; U.S. Patent 4,137,183; January 30, 1979; assigned to Standard Oil Company (Indiana) has found that hydrocarbyl titanate dithiophosphates constitute a special class of additives suitable for use as lubricant oil additives. These compounds consist of covalently bonded quadrivalent titanium derivatives in which the substituents of the titanium are either alkoxy, aroxy, alkaroxy substituents or dithiophosphate substituents. Lubrication oil compositions containing hydrocarbyl titanate dithiophosphate additives derive antiwear properties from the hydrocarbyl titanate moiety and antioxidant properties from the dithiophosphate moiety.

The compounds of this process can be represented by the following structural formula.

(1)
$$(R_1O)_x Ti - [S - \overset{\overset{\displaystyle S}{\|}}{P} - (OR_2)_2]_y$$

Briefly, the hydrocarbyl titanate dithiophosphate compounds can be prepared by reacting a tetrahydrocarbyl titanate compound or mixed hydrocarbyl titanium halide, with a dihydrocarbylphosphoric acid compound. The equations (2) and (3) are examples of the reactions producing the above compound.

(2)
$$Ti(OR_1)_4 + (n)HS-\overset{\overset{\displaystyle S}{\|}}{P}-(OR_2)_2 \longrightarrow (n)HOR_1 + (R_1O)_{(4-n)}Ti-[S-\overset{\overset{\displaystyle S}{\|}}{P}-(OR_2)_2]_n$$

(3)
$$Ti(OR_1)_{m}X_{(4-m)} + (4-m)Na^+-S^--\overset{\overset{\displaystyle S}{\|}}{P}-(OR_2)_2 \longrightarrow (R_1O)_m-Ti[S-\overset{\overset{\displaystyle S}{\|}}{P}-(OR_2)_2]_{(4-m)} + (4-m)NaX$$

In equations (1), (2), and (3) m, n, x, and y are numbers from 1 to 3; x + y = 4; R_1 is an alkyl, aryl, or alkaryl group of about 1 to 18 carbon atoms; R_2 is a hydrocarbyl group of about 1 to 24 carbon atoms and X is a halide (chloro, bromo, etc.) atom. The number of hydrocarbyl and dihydrocarbyldithiophosphate substituents on the titanium can be altered by changing the reactant molar ratios.

Antioxidant and Phosphorus-Containing Compound

W. Lowe; U.S. Patent 4,088,587; May 9, 1978; assigned to Chevron Research Company has developed a lubrication oil additive composition which imparts improved oxidation properties to lubricants and industrial oils which comprises an antioxidant selected from oil-soluble, sterically hindered phenols or thiophenols, oil-soluble aromatic amines and organic sulfur compounds containing from 3 to 40 wt % sulfur which is present within the compound as an organic sulfide or polysulfide or mixtures thereof, and a phosphorus-containing composition.

The latter is prepared by reacting phosphorus oxychloride or phosphorus thiochloride with a 1,2-substituted imidazoline in which the 1-substituent is hydroxyalkyl of 1 to 10 carbon atoms and the 2-substituent is an aliphatic hydrocarbon of 6 to 24 carbon atoms or an oil-soluble N-aliphatic-substituted amino alkylene amide of a higher fatty acid containing 6 to 24 carbon atoms and containing a terminal hydroxy group in the aliphatic substituent, the alkylene group containing from 2 to 5 carbon atoms and the aliphatic radical containing 1 to 10 carbon atoms or mixtures thereof.

While the precise effect of the combination of the antioxidant and phosphorus-containing composition in imparting antioxidation properties to the lubrication oil is unknown, it has been found that these two components complement each other in a synergistic manner resulting in a combination having properties superior to either additive alone. With this combination, the amount of phenolic, aromatic amine, or sulfurized antioxidant necessary in order to impart the desired properties to a lubricating oil, functional fluid or industrial oil blend is significantly less than that amount needed when the synergistic phosphorus-containing compound is not present.

Example 1: This example is presented to illustrate the preparation of a representative phosphate triester phosphorus-containing composition of this process.

A reaction vessel is charged with 525 g of 1-hydroxyethyl-2-heptadecenyl imidazoline in 500 ml of benzene and 231 g of triethylamine. The mixture is stirred until homogeneous and 76.8 g of phosphorus oxychloride ($POCl_3$) are added over a period of 30 min. The reaction flask is maintained at a temperature of 65°C during the addition of the phosphorus oxychloride. Thereafter, the reaction vessel was maintained at room temperature for 1.5 hr and then heated to reflux (82° to 83°C) for 16 hr.

The reaction mixture is diluted to 3.8 liters by the addition of benzene and allowed to stand for four days. The product is filtered to remove the triethylamine hydrochloride. The filtrate is stripped of the benzene by heating to 150°C under reduced pressure. The product weighed 573 g and had the following analysis: P, calc. 2.8%, found 3%; N, calc, 7.6%, found 7.3%. The primary product has the following formula.

$$[\overset{\overset{\displaystyle C_{17}H_{35}}{|}}{\underset{\underset{\displaystyle CH_2\text{—}CH_2}{|\quad\quad|}}{\overset{C}{N}}}\diagdown N-CH_2-CH_2-O\hspace{-2pt}\big]_{\hspace{-1pt}3}P{=}O$$

Example 2: The example is presented to illustrate the effectiveness of the combination of antioxidant and a phosphorus-containing composition in improving the antioxidation properties of a lubricating oil over the use of either component individually. The oxidation test as employed herein uses the resistance of the test sample to oxidation of pure oxygen with a Dornte-type oxygen absorption apparatus (R.W. Dornte, "Oxidation of White Oils," *Industrial & Engineering Chemistry*, Vol. 28, p 26 (1936). The conditions are an atmosphere of pure oxygen exposed to the test oil, maintained at a temperature of 340°F. The time required for 100 g of test sample to remove 1,000 ml of oxygen is observed and reported in the following table.

In order to simulate the oxidation occurring in an internal combustion engine, a mixture of various soluble metal naphthenates, typifying the metal analysis frequently encountered in crankcase oils, is mixed with the test oil.

Experimental samples subjected to the above oxidation test consist of the following. Sample A, referred to herein as the base oil, is composed of a midcontinent neutral oil having a viscosity of 350 SUS at 100°F, 6 wt % of a conventional succinimide dispersant, 0.05 wt % terephthalic acid, and 0.4 wt % of a conventional rust inhibitor. The remaining test samples contained varying amounts of a conventional diparaffin polysulfide antioxidant, a diisobornyl diphenylamine antioxidant and 0.1% of a phosphate triester produced by the method of Example 1 and, in some tests, a conventional zinc dialkyl dithiophosphate. The results of the oxidation test for the above samples are reported in the following table.

. . . . .Base Blend plus Additives . . . .

. .Antioxidant (% wt) . .		Calcium Polypropylene Phenate (mmol/kg)*	Zinc Dialkyl Dithiophosphate (mmol/kg)	Product of Example 1 (wt %)	Oxidation Life (hr)
Diparaffin Polysulfide	Diisobornyl Amine				
—	—	—	9	—	5.1
—	—	—	18	—	6.8
—	—	—	9	0.1	5.3
—	—	—	—	—	2.6
1	—	—	9	—	6.4
1	—	—	—	0.1	7.9
1	—	—	9	0.1	10.3, 11.0**
—	1	—	9	—	8.1
—	1	—	9	0.1	11.0
0.5	0.5	—	9	—	8.4
0.5	0.5	—	9	0.1	13.0
—	—	20	9	—	4.8
—	—	40	9	—	4.8
1	—	20	9	0.1	9.5
1	—	40	9	0.1	9.2

*Carbonated and sulfurized.
**Duplicate runs.

Triketone

B. Swinney and R. Scattergood; U.S. Patent 4,081,389; March 28, 1978; assigned to Exxon Research & Engineering Co. describe a lubricating oil containing a minor amount of a compound of the general formula:

where R^1 and R^2 are hydrogen or hydrocarbyl and may both form part of a cyclic structure.

Triketones themselves are known compounds and various methods may be used for their production. For example the triketone in which R^1 and R^2 are part of the same cyclohexyl group may be prepared by condensing two mols of methyl benzoate with one mole of cyclohexanone in the presence of a base such as sodium hydride, as in the following equation.

Alternatively the triketones may be prepared by the condensation of acetophenones with ethyl carbonate, again in the presence of a base. In each instance the reaction is preferably carried out under anhydrous conditions in an inert solvent. Triketones with substituents on the aromatic nucleus may be prepared from substituted acetophenones which may themselves be prepared from detergent alkylates.

Example 1: 1,5-diphenyl pentanetrione was prepared by dissolving acetone (11.6 g) and methyl benzoate (81.6 g) in glyme (150 ml) and adding this to a slurry of sodium hydride (30 g) in glyme (200 ml) at reflux temperature. After refluxing for 6 hr the solvent was removed, ether added and the excess sodium hydride destroyed by water. The product was extracted with water and precipitated by addition of concentrated hydrochloric acid.

The product after crystallization from ethanol was a yellow solid melting at 110°C which could be dissolved in mineral oil at 100°C and when blended with a conventional lubricating-oil ashless dispersant was soluble in the mineral oil at 60°C.

An oil containing 0.5 wt % of this diphenyltriketone with 5.5 wt % of a conventional dispersant was subjected to the ASTM D-665 rust test with N/5 hydrochloric acid in which a steel pin is submerged in a sample of oil which is held at 140°F for 24 hr. The pin remained clean and bright while a similar pin submerged in a similar oil without the triketone rusted under the same conditions.

Example 2: 2:6-dibenzoyl cyclohexanone was prepared by dissolving 80 g of methyl benzoate and 30.4 ml of cyclohexanone in 50 ml of glyme and adding this to a slurry of 35.2 g of sodium hydride in 500 ml of glyme and the mixture refluxed for 7 hr. The product was then dissolved in ether, excess sodium hydride reacted with water and the product extracted from the ethereal solution with water and precipitated with hydrochloric acid.

The product was soluble in mineral oil at 100°C and at 60°C when blended with a conventional mineral oil dispersant.

Oils containing 0.5 and 1 wt % of this 2:6-dibenzoyl cyclohexanone with 5.5 wt % of a conventional dispersant were subjected to the same ASTM D-665 test as used in Example 1 and the pin remained clean and bright.

SYNTHETIC ESTER COMPOSITIONS

Dibenzothiophene and Other Sulfur-Containing Additives

R. Yaffe; U.S. Patent 4,124,514; November 7, 1978; assigned to Texaco Inc. describes a lubricating oil composition which comprises a major portion of an aliphatic ester base oil formed from the reaction of pentaerythritol and an organic monocarboxylic acid having from 2 to 18 carbon atoms per molecule containing:

(a) from 0.3 to 5 wt % of the lubricating oil composition of alkyl or alkaryl derivatives of phenyl naphthylamines in which the alkyl radical contains from 4 to 12 carbon atoms, and the alkaryl radical has from 7 to 12 carbon atoms;

(b) from 0.3 to 5 wt % of a dialkyldiphenylamine in which the alkyl radicals contain from 4 to 12 carbon atoms;

(c) from 0.001 to 1 wt % of a polyhydroxyanthraquinone;

(d) from 0.25 to 10 wt % of a hydrocarbyl phosphate ester in which the hydrocarbyl radical contains an aryl ring and has from 6 to 18 carbon atoms, and;

(e) from 0.1 to 2.5 wt % of an aromatic substituted thiophene.

The lubricating oil composition provides substantial improvements in oxidative stability, particularly excellent control of acidity and viscosity increase under severe oxidizing conditions.

The lubrication oil compositions of the process were evaluated in the Pratt and Whitney Aircraft Specification PWA-521B Oxidation-Corrosion Test 425°F/48 hr and the Navy MIL-L-23699B Specification 400°F/72 hr Oxidation Corrosion Test and were found to satisfy these specification requirements.

These oxidation inhibitor compositions have been further studied in considerable detail by *R. Yaffe; U.S. Patents 4,096,078; June 20, 1978; 4,119,551; October 10, 1978; 4,124,513; November 7, 1978; 4,141,844; February 27, 1979; 4,141,845; February 27, 1979; all assigned to Texaco Inc.; and R. Yaffe and R.R. Reinhard; U.S. Patent 4,157,971; June 12, 1979; assigned to Texaco Inc.*

Thus, N-(alkyl)-benzothiazole-Z-thione (U.S. Patent 4,124,513), bis(dialkylthio-carbamyl)sulfide (U.S. Patent 4,119,551), S-alkyl-2-mercaptobenzimidazole (U.S. Patent 4,141,844), S-alkyl-2-mercaptobenzothiazole (U.S. Patent 4,141,845), alkyl thioacid esters (U.S. Patent 4,157,971), 4,4'-dithiodimorpholine (U.S. Patent 4,096,078) have all been found to be effective additives to provide improved oxidation stability in synthetic ester lubricants.

Phosphorus Pentasulfide Adduct of Polycyanoethylated Keto Fatty Esters

H.E. Kenney and E.T. Donahue; U.S. Patent 4,069,163; January 17, 1978; assigned to U.S. Secretary of Agriculture describe lubricants which have above-average antiwear and oxidative thermal properties when formulated with synthetic diester oils and lubricants that have above-average antiwear, corrosion, and oxidative thermal properties when formulated with petroleum oils.

The lubricant compositions contain a major proportion of a base lubricating oil and from about 0.06 to 2.0 wt % of an additive that is a partial phosphorus pentasulfide adduct of a polycyanoethylated keto fatty ester.

The phosphorus pentasulfide adduct is prepared by reacting a keto fatty ester with acrylonitrile and reacting the resultant polycyanoethylated keto fatty ester with phosphorus pentasulfide until about one-half of the cyanoethyl groups are reacted with the phosphorus pentasulfide. The keto fatty ester may be a pure compound or it may be a mixture of ketostearates or keto esters.

The commonly used base lubrication oils such as di(2-ethylhexyl) sebacate (DOS), dipropylene glycol dipelargonate (DPDP), diisooctyl azelate (DIOA), di(2-ethylhexyl) azelate (DOA), and paraffin mineral oil (standard test petroleum oil), are compatible for use in preparing the lubricant compositions of this process.

Esters of Arylaminophenoxyalkyl Carboxylic Acids

A process described by *M. Braid; U.S. Patent 4,136,044; January 23, 1979; assigned to Mobil Oil Corporation* is directed to a lubricant composition comprising a major amount of a lubricating oil or grease and a minor and sufficient amount of an arylaminophenoxyalkyl carboxylic acid ester to inhibit the oxidation of the lubricant composition.

Thus, the compounds of this process are the arylaminoaryloxyalkyl carboxylic acids represented by the structure (1) below, the esters thereof represented by the structure (2) below, and the polyesters thereof represented by the structure (3) below.

(1) $\underset{H}{ArN}Ar'O(CH_2)_n\underset{O}{C}OH$

(2) $\underset{H}{ArN}Ar'O(CH_2)_n\underset{O}{C}OR$

(3) $[\underset{H}{ArN}Ar'O(CH_2)_n\underset{O}{C}O]_mR$

wherein Ar and Ar' are individually selected from the phenyl and naphthyl groups that may contain substituent groups such as alkyl and alkoxy groups; R is selected from the group consisting of alkyl, aryl, alkaryl, and aralkyl hydrocarbyl groups containing 1 to 20 carbon atoms in any isomeric constitution and may contain substituent groups such as alkoxy, alkoxyalkyl, acyloxy, acyloxalkyl, and carbalkoxy; m is a whole number within the range of 1 to 6; and n is a whole number within the range of 1 to 12.

Compounds that are esters of arylaminophenoxyalkylcarboxylic acids as illustrated by Structure (2) are prepared by the reaction of an ω-haloalkylcarboxylic acid ester with a hydroxy-substituted diarylamine in a suitable solvent such as N,N-dimethylformamide, N,N-dimethylacetamide or dimethylsulfoxide in the presence of anhydrous potassium or sodium carbonate. The corresponding acids, as illustrated by the Structure (1), are prepared by hydrolysis of the esters of Structure (2). The polyesters of arylaminoaryloxyalkylcarboxylic acids represented by the Structure (3) are formed by esterification or transesterification of the appropriate arylaminoaryloxyalkylcarboxylic acid, Structure (1), or esters thereof, Structure (2), by using polyhydric alcohols such as ethylene glycol, trimethylolpropane, pentaerythritol and partially or completely esterified derivatives of these, di- and tri-methylolphenol, and benzenedimethanol.

The compounds of this process are particularly applicable as oxidation inhibitors in ester base lubricants such as the C_5 and C_9 carboxylic acid esters of hydrocarbons selected from the group consisting of pentaerythritol and trimethylolpropane. The compounds may be incorporated by transesterification of the arylaminoaryloxyalkylcarboxylic acids, Structure (1), with an ester base stock and the acids, for example, the C_5 and C_9 acids which are liberated, are removed by distillation. Lubricants are thus provided wherein the oxidative inhibitor esters of this process are incorporated as an integral part of the ester base stock.

Catalytic oxidation tests were carried out to evaluate the compounds of this process as oxidation inhibitors. These tests involved comparing the stability of a base stock of C_5 and C_9 esters of pentaerythritol with and without the compounds of this process when exposed to oxidizing conditions at test temperatures of 450°F for a test period of 24 hr in the presence of metal catalysts.

In carrying out the catalytic oxidation test, a 25 ml test sample in a glass apparatus is placed in a heating bath at the desired temperature. Present in the sample are the following materials which are either known to catalyze oxidation of organic substances, or are commonly used materials of construction, in an amount sufficient to provide the specified exposed surface area as indicated by the following: (a) 15.6 in^2 of sand-blasted iron wire; (b) 0.78 in^2 of polished copper wire; (c) 0.87 in^2 of polished aluminum wire; and (d) 0.167 in^2 of polished lead surface.

Dry air is passed through the heated sample at the rate of about 5 l/hr for the specified duration of the test. At the conclusion of the test, the increase in the acidity (NN) and kinematic viscosity (KV) resulting from the oxidation is measured. In addition, the loss in weight of the lead specimen is determined as an indication of corrosion and relative amounts of visual sludge are observed. The results of these tests are given in the following table.

Additive of Example No.	Concentration (wt %)	ΔNN	ΔKV (%)	Pb Loss (mg)	Sludge
None	—	8.25	586	13.7	trace
Methyl 4-(p-anilinophenoxy)butyrate	2	3.5	139	4	light
	1	5.1	225	3.7	trace

Note: 24 hr, 450°F, C_5 and C_9 esters of pentaerythritol base stock.

OTHER FLUIDS AND PROCESSES

Fluorinated Polyether Fluids

C.E. Snyder, Jr. and C. Tamborski; U.S. Patent 4,097,388; June 27, 1978; as-signed to U.S. Secretary of the Air Force describe a lubricant composition comprising (a) a base fluid consisting essentially of a mixture of linear fluorinated polyethers having the following formula:

$$R_fO(CF_2CF_2O)_m(CF_2O)_nR_f$$

where R_f is CF_3 or C_2F_5, m and n are integers whose sum is between 2 and 200 and the ratio of n to m is between 0.1 and 10; and (b) a corrosion-inhibiting amount of a perfluoroalkylether-substituted aryl phosphine (fluorinated phosphine) having the following formula:

where one of the R's is a perfluoroalkylether group $(CF_2R_fOR_f)$, two of the R's are fluorine, and n is 1, 2 or 3.

A more detailed description of the synthesis of the fluorinated phosphines is contained in U.S. Patent 4,011,267.

Example 1: A series of runs was conducted for the purpose of determining the effectiveness of lubricant compositions of this process. Lubricant compositions were formulated by mixing (a) a base fluid having the following formula:

$$R_fO(CF_2CF_2O)_m(CF_2O)_nR_f$$

where R_f is CF_3 or C_2F_5, m and n are integers having values such that the fluid has a kinematic viscosity of about 17.8 cs at 100°F with (b) various weight percentages, based upon the weight of the base fluid, of a fluorinated phosphine having the following formula:

The base fluid used was Fomblin Z fluid, a product of Montedison, SpA, Milan, Italy.

In the runs a specimen of steel, titanium alloy or titanium was immersed in the formulations that were prepared. The compositions of the steel and titanium alloys are described in the literature. For comparison purposes, runs were also carried out in which specimens were immersed in polyether fluid which did not contain the anticorrosion additive. The materials were contained in an oxidation test tube having a take-off adapter coupled to an air entry tube. An aluminum block bath provided the means for heating the test tube and an "overboard" test procedure (no reflux condenser) was followed.

Air was bubbled through the formulations, or in the case of the control test through the polyether fluid, at the rate of one liter of air per hour for a period of 24 hr. The runs were conducted at a constant temperature of 550°F. The specimens as well as the apparatus used were weighed prior to and after completion of each run. The data obtained in the runs are set forth below in Tables 1 and 2.

Table 1

Additive (wt %)	Kinematic Viscosity Change at 100°F (%)	Fluid Loss (wt %)	.........Weight Change (mg/cm^2).........				
			4140 Steel	52100 Bearing Steel	410 Stainless Steel	M-50 Tool Steel	440C Stainless Steel
...................................550°F.....................................							
None	*	83.75	+0.024	+0.48	-5.54	-2.37	-3.10
0.5	+3.99	0.57	-0.87	+0.51	+0.01	+0.68	+0.12
1.0	+0.22	0.31	+0.042	+0.031	+0.05	+0.01	0.00
2.0	+0.85	0.69	+1.22	+0.84	+0.13	+1.02	+0.16
...................................600°F.....................................							
None	*	100	-3.54	+1.60	-8.58	+0.60	-9.89
0.5	0.0	0.53	-3.61	+1.38	-0.01	+2.25	-0.01
1.0	+0.1	0.25	+1.43	+0.41	-0.35	+0.44	-0.02
2.0	-0.22	0.45	14.65	+0.46	0.00	+2.74	+0.01

*Insufficient fluid to measure.

Table 2

Temperature (°F)	Additive (wt %)	Kinematic Viscosity Change at 100°F (%)	Fluid Loss (wt %)	.....Weight Change (mg/cm^2).....		
				Ti (6Al-4V)	Ti (pure)	Ti (4Al-4Mn)
550	none	-97.22	59.87	+0.06	-0.28	-0.28
550	0.5	+3.87	0.57	+0.06	0.00	+0.03
550	1.0	+0.16	0.10	+0.01	+0.01	+0.01
550	2.0	+0.39	0.17	+0.07	+0.05	+0.10

Phosphate Ester Turbine Lubricant Formulation

H. Dounchis; U.S. Patent 4,146,490; March 27, 1979; assigned to FMC Corporation describes a turbine lubricant having outstanding oxidation stability, corrosion resistance and good viscosity stability which is based on a liquid mixed isopropylphenyl/phenyl phosphate ester and contains small amounts of bis[4-(dimethylamino)phenyl] methane, benzotriazole and a mixed mono- and di-alkylphosphate of the formula RH_2PO_4 and R_2HPO_4, where R is an alkyl group of 8 to 12 carbon atoms.

The remarkable benefits that can be achieved by the use of the additive combination of the process with such isopropylphenyl/phenyl phosphate esters will be shown by the following example.

Example: A prototype gas turbine lubricant is prepared using as the base oil a liquid mixed isopropylphenyl/phenyl phosphate ester obtained by the phosphorylation of an alkylate having the following analysis.

Ingredient	Weight Percent
Phenol	44
2-isopropylphenol	33
3-isopropylphenol plus	
4-isopropylphenol	12.5
2,6-diisopropylphenol	3
2,4-diisopropylphenol	5
2,5-diisopropylphenol plus	
3,5-diisopropylphenol	<1
2,4,6-triisopropylphenol	<1
2,3,5-triisopropylphenol	<1

To this base oil is added 0.001 wt % of a dimethyl silicone polymer antifoam composition (Antifoam A, manufactured by the Dow Chemical Co.); 1 wt % bis[4-(dimethylamino)phenyl]methane (Ortholeum 304, manufactured by E.I. De Pont de Nemours & Co.); 0.01 wt % benzotriazole and 0.025 wt % of a mixed mono- and dialkyl phosphate of the formula RH_2PO_4 and R_2HPO_4, where R is an alkyl group of 8 to 12 carbon atoms (Ortholeum 162).

The lubricant so obtained passes the ASTM Rusting Test D 665-IP 135, 24 hr with distilled water (part A) and 24 hr with synthetic salt water (part B). When tested in accordance with ASTM Test D 445, the viscosity change at 100°F is 11.57%, and the acid number is 1.08 mg KOH/g.

The lubricant of this example is evaluated by means of the 72 hr 175°C five metal corrosion-oxidation stability test. This test, which is finding increasing use in the evaluation of high-temperature lubricants and hydraulic fluids, is described in Federal Test Method Standard No. 791B, Method 5308 and is carried out as follows.

Weighed, polished 1 in² specimens of copper, steel, aluminum, magnesium, and silver are tied together into a box, with the silver specimen as diagonals separating the copper and steel on one side and the aluminum and magnesium on the other. The box is immersed in 100 ml of the test oil in an oxidation tube fitted with a reflux condenser, and air is bubbled through at a rate of 5 l/hr while the oil is maintained at 175°C for the 72 hr. When the test period is completed, the oil and metals are examined for evidence of oxidative degradation, for example, a large increase or decrease in oil viscosity, a large increase in the acid number, a large deposition of sludge, and corrosive attack on one or more of the metal specimens.

There is no substantial change in the color of the lubricant. There is an increase of 0.117 mg/cm² in the weight of the copper sample, 0.044 mg/cm² in the weight of the steel sample, and 0.022 mg/cm² in the weight of the silver sample. The weights of both the magnesium alloy and aluminum alloy samples increase by 0.051 mg/cm².

Organosulfur-Containing Nickel Complexes as UV Stabilizers

M. Braid and S.J. Leonardi; U.S. Patent 4,101,430; July 18, 1978; assigned to Mobil Oil Corporation have found that oxidative degradation of lubricant compositions, for example, by ultraviolet light, present in sunlight or other sources of actinic radiation, can be effectively inhibited by the incorporation of organosulfur-containing nickel complexes in the lubricant compositions.

The organic sulfur-containing nickel complexes include nickel alkyl phenolate sulfides having the following structure:

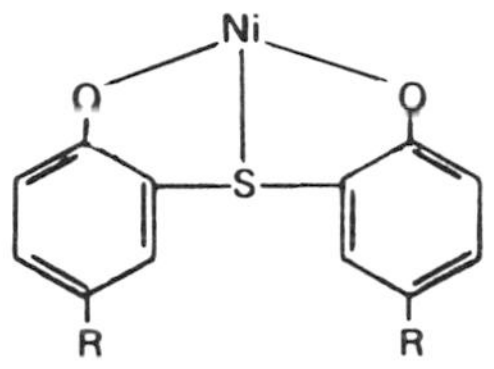

in which R is either hydrogen or an alkyl group having from 1 to 30 carbon atoms and preferably from 4 to 12 carbon atoms.

Representative of the nickel phenolate sulfides is nickel 2,2'-thiobis-(4-tert-octyl)-phenolate having the structure:

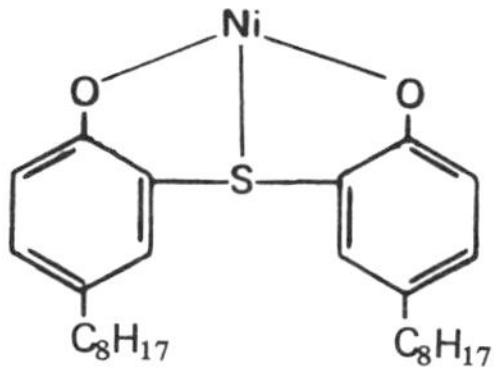

and nickel phenol sulfide having the structure:

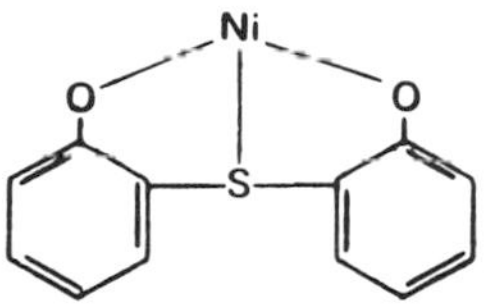

The phenolates of the process can be conveniently prepared by known methods; see for example, U.S. Patent 2,971,941.

METALWORKING LUBRICANTS

Metalworking operations, for example, rolling, forging, hot-pressing, blanking, bending, stamping, drawing, cutting, punching, spinning and the like, generally employ a lubricant to facilitate the same. Lubricants greatly improve these operations in that they can reduce the power required for the operation, prevent sticking and decrease wear of dies, cutting bits and the like. In addition, they frequently provide rust-inhibiting properties to the metal being treated.

Since it is conventional to subject the metal to various chemical treatments (such as the application of conversion coating solutions) after working, a cleaning operation is necessary between the working step and the chemical treatment step. In addition to the above properties, therefore, it is preferred that the working lubricant be easily removable from the metal surface by ordinary cleaning compositions.

This chapter provides a number of formulations and recent developments for lubricants which provide to the metal being worked a unique combination of properties including lubricity, corrosion resistance, extreme pressure properties and protection against wear of working parts, and which, as required in specific applications, are relatively easy to remove from the surface of the metal by cleaning after the working operation is completed.

WATER-BASED LUBRICANTS

Mixture of Salts of Alkylaryl Sulfonic Acids

J.-P. Kistler and P.-D. Marin; U.S. Patent 4,140,642; February 20, 1979; assigned to Exxon Research & Engineering Co. describe emulsifier compositions which are suitable for mixing with mineral oil to form metalworking lubricants. These compositions comprise a mixture of salts of alkylaryl sulfonic acids, the acids having a molecular weight distribution with two distinct peaks, one peak being preferably in the range of 270 to 400, while the other peak is in the range of 350 to 600, which peaks differ by at least 80. Mixtures of 5 to 95 wt % sodium salt of branched chain C_{12-16} alkyl o-xylenesulfonic acids with 95 to 5 wt % sodium salts of branched chain C_{20-28} alkyl benzenesulfonic acids are especially preferred compositions, particularly when blended with naphthenic mineral oil, to thereby form stable emulsifiable metalworking lubricants.

Example 1: A concentrate X1 of alkylaryl sulfonates was prepared according to the process by mixing 73% of a concentrate A and 27% concentrate B.

Concentrate A contained 60% sodium alkylbenzene sulfonates whose mean molecular mass was 370 and which had been prepared from the alkylation product of o-xylene and a technical tetrapropylene.

Concentrate B contained 60% sodium alkylbenzene sulfonates whose mean molecular mass was 520, and which had been prepared from the alkylation product of benzene and a propylene oligomer. The mean carbon condensation of this oligomer was 24 atoms per molecule.

The concentrate X1, therefore, contained 60% alkylbenzene sodium sulfonates whose mean molecular mass was 405 and the distribution of whose molecular weight displayed two maxima, corresponding to the molecular weights 370 and 520.

Example 2: A concentrate X2 was prepared of alkylaryl sulfonates according to the process by mixing 75% of concentrate A defined in Example 1 and 25% concentrate C.

Concentrate C contained 60% sodium alkylbenzene sulfonates whose mean molecular weight was 580, and which had been prepared from the alkylation product of o-xylene and a propylene oligomer. The mean carbon condensation of the latter was 27 atoms per molecule.

Concentrate X2, therefore, contained 60% sodium alkylbenzene sulfonates whose mean molecular weight was 405, and the distribution of whose molecular masses displayed two maxima corresponding to the molecular weight 370 and 580.

Example 3: A concentrate X3 was prepared of alkylaryl sulfonates according to the process, by mixing 65% of a concentrate D and 35% of concentrate B defined in Example 1.

The concentrate D contained 60% sodium alkylbenzene sulfonates whose mean molecular weight was 300 and which had been prepared from the alkylation product of benzene and a technical tripropylene.

The concentrate X3, therefore, contained 60% sodium alkylbenzene sulfonates whose mean molecular weight was 353 and the distribution of whose molecular weight displayed two maxima corresponding to the molecular masses 300 and 520.

Example 4: The emulsifying power of the alkylbenzene sulfonates X1, X2 and X3, defined in the foregoing examples, in relation to a mineral oil dispersed in water, was compared with that of the various commercial alkylaryl sulfonates.

To conduct these tests an oil with naphthetic tendency was used, derived from crude petroleum by means of the usual processes of distilling and refining. This oil had a viscosity of 27 cs at 37.8°C. The hardness of the water used was 22° (French hardness).

Each test was performed according to the following method of operation: One part by weight of the concentrate being tested and 4 parts by weight of oil are mixed. In 95 parts by weight of water there are dispersed 5 parts by weight of the emulsifiable composition thus prepared. An emulsion is obtained which is allowed to stand for 7 days at 20°C, in a stoppered graduated test tube. The emulsion is considered to be stable if the volume of the salted-out phase (oil or cream) after 7 days is less than 1% of the total volume of the emulsion. The emulsions obtained with products X1, X2 and X3 are stable under these conditions.

Under the same conditions were tested five commercial sodium alkylaryl sulfonates, derived from the sulfonation of mineral oils, whose mean molecular weights were individual peaks in the range 350 to 550. None of these alkylaryl sulfonates furnished a stable emulsion.

Also tested under the same conditions were the alkylbenzene sulfonates A, B, C and D defined in Examples 1 through 3. All these sulfonates, used separately, resulted in salting-out exceeding 4% by volume.

Of the foregoing concentrates it was found that concentrate X1 of Example 1 gave the most desirable emulsion properties. When, however, the C_{24} benzene sulfonate was replaced by a C_{18} benzene sulfonate, the stability of the emulsion produced was much less. This illustrates the highly preferred feature that the molecular weight difference referred to above should be at least 80.

The concentrate X2, giving less desirable properties than X1, indicates the unexpected advantages of employing components of the type represented in X1 (i.e., a high molecular weight benzene type and a low molecular weight o-xylene type).

Complex Phosphate Surfactants

A process described by *R.J. Verdicchio and L.J. Nehmsmann III; U.S. Patent 4,132,657; January 2, 1979; assigned to GAF Corporation* relates to complex phosphate surfactants that are derived from quaternary dihydroxy compounds of the formula:

$$
\begin{array}{c}
R_1 \quad \overset{\displaystyle R_2}{(CH_2CHO)_n\!-\!CH_2CHOH} \\
\overset{\displaystyle \diagdown}{N^+} \diagup \qquad\qquad\qquad X^- \\
R \diagup \quad \diagdown \underset{\displaystyle R_3}{(CH_2CHO)_n\!-\!CH_2CHOH}
\end{array}
$$

where R and R_1 represent the same or different alkyl moieties containing 1 to 22 carbon atoms, the sum of R and R_1 not being less than 7 carbon atoms; R_2 and R_3 each represent H or CH_3; n is an integer having an average value of from 1 to 50; X represents an anion of halogen, sulfate, or alkyl sulfate; and where the resultant complex phosphate ester contains at least one-third gram-atom of P per mol of quaternary dihydroxy compound; which can be made through reaction (phosphating) of such dihydroxy quaternary compound with a conventional phosphating agent such as P_2O_5, polyphosphoric acid, or $POCl_3$ to form amphoteric surface active agents having utility as metal lubricants, heavy-duty cleaners, detergents, etc.

Example 1: Charge into a 1-liter flask equipped with an agitator, thermometer and gas inlet, 320 parts by weight (0.5 mol) of cocoamine + 10 ethylene oxide and 2 parts by weight of 50% hypophosphorous acid. Dry the mixture under vacuum (10 to 15 mm) at 80° to 100°C. Cool under dry nitrogen at 40° to 50°C and add 80 parts by weight (0.52 mol) of diethyl sulfate at 40° to 50°C over 2 hours.

Strip any unreacted diethyl sulfate at 90° to 100°C under good vacuum. Cool to 30° to 40°C and, under a nitrogen blanket, add 35.5 parts by weight of phosphorus pentoxide and stir at 100°C for 5 hours. Cool to 80° to 85°C, add 5 parts by weight water, and stir for 2 hours. There is obtained 433 parts by weight of active surfactant. This represents a 99% yield.

Example 2: Operating as in Example 1, 460 parts by weight (0.4 mol) of tallowamine (Armeen-TD, a distilled tallowamine made by and available from Armak) + 20 ethylene oxide and 2 parts by weight of 50% hypophosphorous acid are charged to a 1-liter flask. The mixture is dried at 90° to 100°C under a good vacuum, cooled to 30° to 40°C, and 60 parts by weight of diethyl sulfate are added over 2 hours. The mixture of quaternized amine ethoxylate is phosphated in the following manner. Hypophosphorous acid (50%), 1 part by weight, is added at 40° to 50°C, followed by 29 parts by weight of phosphorus pentoxide at 50° to 60°C. The mixture is heated to 100°C for 5 hours under a nitrogen blanket, cooled to 80° to 85°C, and 5 parts by weight water are added. The system is stirred for 2 hours at 80° to 85°C. There is obtained 553 parts by weight of 100% active surfactant.

Example 3: In this example there were prepared, according to the procedure of Example 1 (except for the identity and amount of amine reactant used), (a) the sodium phosphate of tallowamine + 5 ethylene oxide diethyl sulfate, and (b) the sodium phosphate of oleylamine + 5 ethylene oxide diethyl sulfate. With respect to (a), 240 parts by weight (0.5 mol) of tallowamine + 5 ethylene oxide (Ethomeen T/15 of Armak) was used (instead of the 0.5 mol of cocoamine + 10 ethylene oxide of Example 1), and, with respect to (b), 250 parts by weight (0.5 mol) oleylamine + 5 ethylene oxide (Ethomeen O/15) was used (instead of the 0.5 mol of cocoamine + 10 ethylene oxide of Example 1).

Then, 0.4% by weight aqueous solutions of (a) and (b) were prepared, their pH values were adjusted to pH 8.5 with triethanolamine, and the pH-adjusted solutions evaluated for use as water-based lubricants. The Falex Load Test was used to determine the lubricating properties of the respective solutions, and the results are summarized below. The results indicate excellent lubricant properties.

Sample	Load at Failure (lb)
0.4% aqueous sodium phosphate of tallowamine + 5 ethylene oxide diethyl sulfate	4,000
0.4% aqueous sodium phosphate of oleylamine + 5 ethylene oxide diethyl sulfate	4,250
Water (control)	550

Methacrylic Acid Grafted Polyoxyalkylene

W.H. Martin; U.S. Patent 4,146,488; March 27, 1979; assigned to Union Carbide Corporation has found that metals may be lubricated with a liquid medium consisting essentially of water having dissolved therein a salt obtained by neutralizing an acrylic or methacrylic acid graft copolymer of a polyoxyalkylene compound having the formula:

$$R'' + (OC_n H_{2n} + z OR')_a$$

where R'' is a hydrocarbon radical free of aliphatic unsaturation and having a valence of a, a is an integer having a value of 1 to 4, R' is a member selected from a group consisting of a monovalent hydrocarbon radical free of aliphatic unsaturation, a hydrogen atom or an acyl radical free of aliphatic saturation, n has a value of 2 to 4 inclusive, z is an integer having a value of from 8 to 800 inclusive, and preferably 12 to 500 with an alkanolamine having the formula:

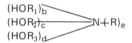

where R is hydrogen or alkyl having 1 to 4 carbons each of R_1, R_2 and R_3 is an alkylene radical having 2 to 4 carbon atoms, e is an integer having values of 0, 1 or 2, b, c, and d are integers each having a value of 0 or 1 with the proviso that when b, c and d are each 1 then e is 0, and wherein the graft copolymer contains about 3 to 15% by weight of acrylic or methacrylic acid graft copolymerized therein.

Examples 1 through 3:   A graft copolymer was prepared by feeding 640 g of a butanol started ethylene oxide/propylene oxide (50/50 by weight) polyalkylene oxide having a viscosity of 5,100 SUS at 100°F (37.5°C) and a molecular weight of about 4,500 containing 2.5 g of azobisisobutyronitrile as a stream into one neck of a 3-necked round-bottom reaction flask fitted with a stirrer and thermometer together with a stream of 30 g of acrylic acid monomer into a second neck of the flask over a period of 2.5 hours while maintaining the flask at a temperature of about 150°C.

The reaction mass was then postheated for 1 hour at 150°C and then transferred to a larger flask where it was stripped with a rotary evaporator at 100°C at 11 mm for 1 hour to remove unreacted starting materials. An acrylic acid/polyalkylene oxide graft copolymer containing 3.1% by weight of acrylic acid graft polymerized therein was thus obtained (Graft Copolymer A).

The above procedure was repeated except that the amounts of acrylic acid monomer charged used were 50 g and 80 g, respectively. There was thus obtained acrylic acid/polyalkylene oxide graft copolymer containing 5.3% by weight (Graft Copolymer B) and 8.7% by weight (Graft Copolymer C) copolymerized therein.

Then 15 g samples of each of Graft Copolymers A, B and C were dissolved in 15 g of water. The resultant solutions were neutralized with 14.3 g, 16.5 g and 24.2 g respectively of triethanolamine (99% pure). The homogeneous solutions which resulted had a pH of 9.

The examples demonstrate that all three triethanolamine salts are readily soluble in water, one criterion for a metalworking lubricant.

Examples 4 through 6: As an extension of the properties demonstrated in Examples 1 through 3, samples of the Graft Copolymers A, B and C were neutralized with amounts of triethanolamine in excess of the stoichiometric amount in bulk without the benefit of the mutual cosolvent water. The amounts used were 33, 35 and 51 parts by weight of triethanolamine per 100 parts by weight of Graft Copolymers A, B and C respectively. In each case a clear, homogeneous mixture having a pH of 8.5 was obtained. By way of contrast, when 40 parts of triethanolamine was blended with 8 parts by weight of a commercial dimer acid (prepared by heating linoleic acid and having an acid value of 186 to 194 g KOH/g of acid and a saponification value of 191 to 199 g KOH/g of acid) and 100 parts by weight of the polyalkylene oxide used in Examples 1 through 3, the product (Control B) was hazy and separated into two layers.

When 35 g of triethanolamine was blended with 100 g of the polyalkylene oxide used to make the graft copolymer, the product was hazy and separated into two layers (Control A).

The homogeneity of the alkanolamine salt provides another advantage over some of the prior art additives used in formulating metalworking compositions.

Examples 7 through 9: The evaluation of the triethanolamine salts prepared in Examples 4 through 6 was effected in a Falex Extreme Pressure Lubrication Tester.

	. . . . Control. . . .		. . . . Examples		
	A	B	7	8	9
Maximum jaw load before failure, lb*	<500	>4,500	3,900	4,200	4,100
Scar width, in	0.081	0.049	0.076	0.063	0.059
Film strength of lubricant at failure, psi	<8,700	130,000	72,500	94,300	98,300

*Maximum jaw load in lb is $\sqrt{2}$ x 0.5 scar width.

The lubricants were tested as 1% by weight aqueous solutions. A 5-minute break-in at 500 lb was used in the Falex test followed by continuous loading until failure.

The data delineated in the table demonstrate that the extreme pressure characteristics of polyalkylene oxide (Control A) could not even be tested. Failure occurred at the lowest possible jaw load and during the break-in period. The film strength, 8,700 psi, is extremely low. Control B, the combination of equal parts of triethanolamine neutralized dimer acid and polyalkylene oxide, was greatly superior to Control A in extreme pressure characteristics as shown by the full jaw load of 4,500 lb without galling or seizure of the pin blocks. Scar wear was reduced and the film strength exceeded 100,000 psi. However, triethanolamine neutralized dimer acid alone did not show these extreme pressure capabilities from a 1% aqueous solution.

It should also be recalled that the combination in Control B is not a very satisfactory commercial formulation because it separates into two layers in the absence of water or other mutual cosolvents.

Examples 7 through 9 representing the triethanolamine salts of the graft copolymers prepared in Examples 4 through 6 showed excellent extreme pressure capabilities in all of the three measurements.

In addition, these salts also exhibit the requisite compatability, reduced foaming characteristics and stability in hard waters needed for a commercial water-soluble metalworking lubricant.

Amino-Amides Derived from Polymeric Fatty Acids

R.J. Sturwold and M.A. Williams; U.S. Patent 4,107,061; August 15, 1978; assigned to Emery Industries, Inc. have discovered that amino-amides derived from polymeric fatty acids (dimer acids) and polyoxyethylene diamines are excellent synthetic lubricants. The amino-amides are hydrophilic and readily compatible with water in all proportions to form clear, aqueous lubricating solutions. Aqueous lubricating solutions containing the amino-amides have superior extreme pressure properties and exhibit rust-preventive properties not commonly possible with water-soluble dimer-based lubricants.

To obtain the amino-amide products of this process, a polymeric fatty acid, preferably a dimer acid obtained by the dimerization of an unsaturated C_{18} monocarboxylic acid, is reacted with a polyoxyethylene diamine. Useful diamines will have an average molecular weight between 500 and 1,000. About 1.7 to 2.3 equivalents of the polyoxyethylene diamine is reacted per equivalent dicarboxylic acid. Optionally, a short-chain dicarboxylic acid having from 2 to 16 carbon atoms can be included in the reaction. Short-chain saturated aliphatic dicarboxylic acids having from 6 to 12 carbon atoms are preferably employed for this purpose. The resulting amino-amides are compatible with water in all proportions but will typically constitute from 0.1 to 40 wt % of the aqueous lubricant solution.

The aqueous lubricants are useful with a wide variety of metals, including both ferrous and nonferrous metals and their alloys. They can be employed for working nonferrous metals such as aluminum, copper, manganese, titanium, brass and bronze or with other metals of this type which are susceptible to staining. They are particularly useful, however, with ferrous metals where the development of rust is a particular problem.

Example 1: To a glass reactor equipped with a stirrer, thermometer, nitrogen inlet and recycle trap equipped with a water-cooled condenser were charged 194.0 g (1 equivalent) dimer acid (Empol 1018 Dimer Acid containing 83% C_{36} dibasic acid) and 408 g polyoxyethylene diamine having an average molecular weight of about 600. The reaction mixture was then slowly heated over a 2-hour period with agitation under an atmosphere of nitrogen to 210°C while removing water of reaction. The reaction temperature was then maintained at 210°C for an additional 4 hours until the theoretical amount of water was collected. Heating was terminated at this point and the clear viscous product discharged from the reactor.

The amino-amide product had the following properties:

Acid value	0.85
Amine value	56.7
210°F viscosity, cs	114.4
100°F viscosity, cs	1718.1
Pour point	15°F
Flash point	555°F
Fire point	595°F

Lubication properties of the above-prepared amino-amide product were determined using the Falex machine (modified ASTM D 2670-76). The aqueous lubricant solution prepared with the amino-amide of this example withstood the maximum load of 4,500 lb without failure.

The amino-amide product was also evaluated in a rust test. For this test, a small amount of cast iron fillings is placed on a filter paper covering the bottom of a 4-inch Petri dish. Aqueous lubricant solution (5%) is added so that the cast iron fillings are partially immersed. The amount of rust developed under ambient conditions is visually rated (no visible rust, slight, moderate, or heavy) after 1 hour and again after 2 hours. A control containing no lubricant develops heavy rust within 15 to 30 minutes whereas the lubricant solution prepared with the amino-amide of this example showed only a slight trace of rust even after 2 hours.

Example 2: To demonstrate the ability to use a short-chain dibasic acid with the dimer acid, the following experiment was conducted. For this example, the amino-amide was prepared in accordance with the procedure of Example 1, except that 2 equivalents of the polyoxyethylene diamine, 0.5 equivalent dimer acid and 0.5 equivalent adipic acid were charged. The resulting amino-amide product had an acid value of 1.7 and an amine value of 64 and formed clear lubricant solutions when combined with water. A 5% aqueous solution of the product was evaluated in the rust test and showed only slight rust development after 2 hours. In the Falex test, the aqueous lubricant solution gave only 11 units wear and withstood 4,500 lb without failure in the extreme pressure cycle.

Phosphate Esters and Sulfur Compounds

A. Nassry and J.F. Maxwell; U.S. Patent 4,138,346; February 6, 1979; assigned to BASF Wyandotte Corporation found that a composition useful as a hydraulic fluid or metalworking composition can be prepared having a desirable lubricity and anitwear properties. A synergistic combination of a sulfur-containing compound and a phosphate ester in a water-base consisting of 70 to 95% and higher water in the base results in the improved properties. The composition contains a phosphate ester selected from

$$RO-(EO)_n-\overset{\overset{\displaystyle O}{\|}}{\underset{\underset{\displaystyle OX}{|}}{P}}-OX \quad \text{and} \quad RO-(EO)_n-\overset{\overset{\displaystyle O}{\|}}{\underset{\underset{\displaystyle OX}{|}}{P}}-(EO)_n-OR$$

and a mixture thereof where EO is ethylene oxide; R is selected from the group consisting of linear or branched chain alkyl groups having about 6 to 30 carbon atoms, preferably about 8 to 20 carbon atoms, aryl or arylalkyl groups wherein the arylalkyl groups have about 6 to 30 carbon atoms, preferably about 8 to 18 carbon atoms, and X is selected from the group consisting of the residue of hydrogen, ammonia or an amine and an alkali or alkaline earth metal or mixtures thereof and n is a number from 1 to 50. Metals such as lithium, sodium, potassium, rubidium, cesium, calcium, strontium and barium are examples of X.

The phosphate ester compositions utilized in the compositions of this process are more fully described in U.S. Patents 3,004,056 and 3,004,057.

Stable concentrates of the hydraulic fluids and metalworking fluids can be prepared. These can be made up completely free of water or contain up to 20 wt % water to increase fluidity and provide ease of blending at the point of use.

Representative concentrates are as follows:

Table 1: Hydraulic Fluid Concentrates

Ingredients			Weight Percent.			
Process phosphate ester	50	75	25	40	60	20	25
Process sulfur compound	50	25	75	40	20	60	25
Corrosion inhibitor (i.e. morpholine)	–	–	–	–	–	–	50
Water	–	–	–	20	20	20	–

The sulfur-containing compound useful in the compositions can be at least one of the ammonia, amine or metal salts of 2-mercaptobenzothiazole or 5-, 6- and 7-substituted 2-mercaptobenzothiazole, the salts being formed upon neutralization of the free acid form of 2-mercaptobenzothiazole of the formula:

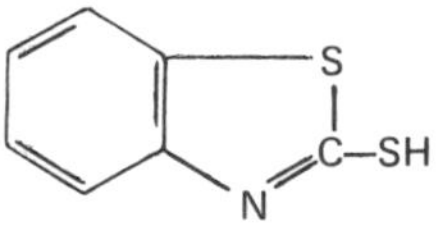

with a base. Ammonia or an amine or an alkali or alkaline earth metal hydroxide or carbonate are suitable bases for the formation of the salts wherein the metal is selected from groups I-A and II-A of the Periodic Table. Representative substituted compounds are selected from the group consisting of the chloro, bromo, sulfonic acid, amido, methyl, carboxylic acid and ethoxy substituted compounds. Examples of such substituted compounds are the following: 5-chloro-2-mercaptobenzothiazole, 5-bromo-2-mercaptobenzothiazole, 5-amido-2-mercaptobenzothiazole, 5-methyl-2-mercaptobenzothiazole, 5-methyl-2-mercaptobenzothiazole, 5-carboxylic acid-2-mercaptobenzothiazole, 6-ethoxy-2-mercaptobenzothiazole, and 6-chloro-2-mercaptobenzothiazole, etc.

Example 1:   A hydraulic fluid was prepared by blending 83.6 parts by weight water, 1.0 part by weight morpholine, 0.5 part by weight phosphate ester, 0.25 part by weight potassium 2-mercaptobenzothiazole (K-2-MBT), and 14.2 parts by weight of a polyglycol thickener having a viscosity of 95,000 SUS at 100°F. Such thickening materials are available commercially and are made by copolymerizing about 70 mol % of ethylene oxide with about 30 mol % of propylene oxide to obtain a product sufficiently high in molecular weight so as to act as a thickener. The phosphate ester utilized is produced by the reaction of 1 mol of P_2O_5 with a condensation product of 1 mol of oleyl alcohol and 4 mols of ethylene oxide in accordance with the methods disclosed in U.S. Patents 3,004,056 and 3,004,057. A clear to slightly hazy, free-flowing water-based hydraulic fluid is obtained which is stable to storage at room temperature.

Example 2: Using the same procedure and proportions of ingredients as in Example 1 except as follows, a hydraulic fluid was prepared using 0.38 part by weight potassium 2-mercaptobenzothiazole.

Example 3: Using the same procedure and proportions as in Example 1 except as follows, a hydraulic fluid was prepared omitting potassium 2-mercaptobenzothiazole.

Using the hydraulic fluid compositions, as described in Examples 1, 2 and 3, Vickers vane pump wear tests were performed. The results are described in Table 2.

Table 2: Vickers Vane Pump Wear Tests at 100°F and 1,000 psi Load

Ex. No.	Ratio Phosphate Ester/ Sulfur Compound	Run Time (hr)	Cam Ring	Wear (mg) Vanes	Total
1	2/1	20	470	2	472
2	2/1.5	20	589	1	590
3	2/0	20	670	4	674

Examples 4 through 13: Hydraulic fluids were prepared by blending indicated proportions of various phosphate esters prepared by esterifying 1 mol of P_2O_5 with various amounts as indicated by molar ratio of the surface active agent condensation products indicated in Table 3 below with the potassium salt of 2-mercaptobenzothiazole (Examples 4 through 8) and the same proportions of water, thickener and morpholine used in Example 1. These fluids were evaluated using the well-accepted Shell four-ball test and results are shown in Table 3. Comparison with similar fluids prepared omitting the potassium salt of 2-mercaptobenzothiazole, (Examples 9 through 13), shows no indication of synergism and in fact, a general trend toward reduced wear in the examples in which the potassium salt of 2-mercaptobenzothiazole is omitted from the hydraulic fluid.

It is, therefore, unexpected that the Vickers van pump test results shown in Table 2 above, indicated a synergistic improvement in wear reduction where the phosphate ester of the process is combined with the sulfur-containing compound of the process.

Table 3: Shell Four-Ball Test*

Ex. No.	Phosphate Esters (PF) Surface Active Agent	Molar Ratio Agent:P_2O_5	0.50% PE With 0.25% K-2-MBT** Scar Diameters (mm, 30 min) 7.5 kg	40 kg	Without K-2-MBT** Scar Diameters (mm, 30 min) 7.5 kg	40 kg
4	Dinonylphenol + 7.0 mols EO	2.7:1	0.611	0.775	—	—
5	Nonylphenol + 4.0 mols EO	4:1	0.627	0.735	—	—
6	Phenol + 6.0 mols EO	4:1	0.914	0.927	—	—
7	Tridecyl alcohol + 10.0 mols EO	2.7:1	0.666	0.748	—	—
8	Oleyl alcohol + 4.0 mols EO	2:1	0.720	0.752	—	—
9	Dinonylphenol + 7.0 mols EO	2.7:1	—	—	0.588	0.610
10	Nonylphenol + 4.0 mols EO	4:1	—	—	0.590	0.611
11	Phenol + 6.0 mols EO	4:1	—	—	0.758	0.899
12	Tridecyl alcohol + 10.0 mols EO	2.7:1	—	—	0.638	0.923
13	Oleyl alcohol + 4.0 mols EO	2:1	—	—	0.604	0.622

*52100 steel balls, 1,800 rpm, room temperature, 30 min running times, in a base-stock high-water-content hydraulic fluid.

**Potassium salt of 2-mercaptobenzothiazole.

1,3,5-Tris(Furfuryl)Hexahydro-s-Triazine

N. Grier and B.E. Witzel; U.S. Patent 4,149,983; April 17, 1979; assigned to Merck & Co., Inc. have found that compositions of 1,3,5-tris(furfuryl)hexahydro-s-triazine are useful as antimicrobial agents particularly when added to metalworking compositions subject to fungal and bacterial attack.

The biocidal metalworking compositions of this process comprise 0.1 to 5% by weight 1,3,5-tris(furfuryl)hexahydro-s-triazine; from 40 to 99.9% by weight of a carrier fluid; and from 0 to 30% by weight of a surfactant.

Some representative cutting fluid compositions are:

	Parts by Weight
Triethanolamine	20
Caprylic acid	3
Polyoxyethylene glycol	7
Sodium 2-mercaptobenzothiazole	3
1,3,5-Tris(furfuryl)hexahydro-s-triazine	0.5
Water	66.5

Note: For use the fluid is diluted with water.

	Parts by Weight
Mineral oil (d$_{20}$, 0.915)	80
Petroleum sulfonate (MW 400) as 70% aqueous solution	14
Ethoxylated (5 mols ethylene oxide) oleyl alcohol	3.5
1,3,5-Tris(furfuryl)hexahydro-s-triazine	2
2,6-Di-tert-butyl-p-cresol	0.5

Note: For use one part of the fluid is diluted with 20 to 40 parts of water.

Disodium Monocopper(II) Citrate

A process described by *S.K. Shringarpurey and G.L. Maurer; U.S. Patent 4,129,509; December 12, 1978; assigned to National Research Laboratories* is directed to metalworking fluid compositions. According to this process, metalworking emulsions, i.e., oil and water dispersions, are stabilized by the addition of an effective stabilizing amount of a metal complex of a metal ion and a polyfunctional organic ligand.

The stabilizing metal complex is characterized by an unexpected aqueous proton-induced dissociation property which causes the controlled release of metal ion into the oil and water dispersions to impart metalworking stability to the dispersions. This dissociation property is represented by a sigmoidally shaped curve on a Cartesian coordinate plot of the negative log of the metal ion concentration versus the negative log of hydrogen ion concentration, i.e., a pM-pH diagram.

It has been discovered that metalworking fluids can be stabilized against attack and deterioration by different causes. Thus, in one aspect of the process, metalworking fluids have been stabilized against deterioration by a general purpose metal complex stabilizer additive. For instance, certain metal complexes are effective as antimicrobial agents in metalworking emulsions. It has further been developed that these metal complexes impart other stabilizing characteristics to the metalworking fluids and achieve other improvements. This stability is not only achieved against bacteria, but the fluids are stabilized against degradation by physical, chemical, and physicochemical causes associated with metalworking conditions including heat, pressure, metalworking compositional environment of metalworking particles, polyvalent ions, etc. For example, in industrial coolant areas, service life of fluids has been extended from a period of, for example, a few weeks to nearly a year or more, and results indicate that service life may be extended even indefinitely.

The metal complex stabilizers themselves are very stable at relatively high alkaline pHs on the order of about 9 or 10 to about 12. However, in an alkaline pH range above 7 to 9, where metalworking fluids are required to function, the stabilizing complexes very advantageously impart stabilizing effects to the metalworking fluids.

A specific example of the metal complex is dialkali monocopper(II) citrate represented by disodium-, dipotassium- or dilithium monocopper(II) citrate. These dialkali monocopper(II) citrates have a dissociation property represented by a sigmoidal plot wherein the curve of two directions meets at a point within the pH range of about 7 to 9.

It has been established that these monocopper(II) complexes in basic media, on the order of about pH 9-12, are very stable, i.e., having an effective stability constant, K_{eff}, on the order of about 10^{12} to 10^{13}. However, K_{eff} of these monocopper(II) citrate complexes at a pH of about 7 to 8 is on the order of about 10^5 to 10^8. Therefore, at or about a pH around 7 to 8, the effective stability constant of the monocopper(II) citrate complex is considerably lower (a thousand to several hundred thousand times lower) and a significant free Cu^{++} concentration is available for stabilizing activity. For example, about 10% of the copper in the complex is in the ionized state at or about pH 7 while approximately 0.1% of the copper is ionized at or about pH 9.

This would not be true for an EDTA or polyamine complex of a multivalent metal such as copper, since its stability constant (10^{14} to 10^{16}) will vary only slightly in the normal pH of 7 to 9. Such EDTA complexes do exhibit a pH effect on the stability constant, but it is represented by a smooth, monotonic curve reaching a limiting effect by proton-induced dissociation at pH values from about 7 to 9, yielding only from about 0.001% ionized species at or about pH 7 to as little as 0.00001% ionized species at or about pH 9.

Example: A cutting fluid composition is prepared by mixing the following ingredients on a volume basis. 1% sodium xylene sulfonate, 9% naphthene sulfonate, and 90% mineral oil, viscosity approximately 300.

This mixture is then used to prepare a 3% (volume) emulsion by blending with water and the pH is adjusted to about 8.5 to 8.9 with the addition of HCl. The disodium monocupric citrate is added to the emulsion to provide 100 mg/l of Cu^{++} in the aqueous phase. When such a metalworking fluid is employed in metal cutting operations, it has been found that all the advantages discussed can be achieved.

Zinc Phosphate, Dimer Acid and Alkenyl Succinic Anhydride

A.H. Birke; U.S. Patent 4,101,429; July 18, 1978; assigned to Shell Oil Company
has developed lubricating oil compositions, which comprise a major amount of
an oil of lubricating viscosity and containing dissolved therein from 0.25 to 1.75
weight percent of a zinc primary dihydrocarbyl dithiophosphate; from 0.06 to
0.5 part per part of the zinc dithiophosphate of a mixture of

(a)   a dimeric acid produced by the condensation of unsaturated aliphatic monocarboxylic acids having between about 16 and 18 carbon atoms per molecule;
and

(b)   the reaction product obtained by reacting a monocarboxylic acid, a polyalkylene polyamine having more than one nitrogen atom per molecule, and
an alkenyl succinic acid or anhydride.

The weight ratio of (a) to (b) is from about 0.001:50 to 15:0.001 and from 0.035
to 0.25 part of a substantially neutral zinc salt of a dihydrocarbyl sulfonic acid.
These fluids are particularly valuable since they provide excellent lubrication,
are fully satisfactory as hydraulic fluids and show excellent water compatibility
and filtration properties.

The mixture of dimer acid and reaction product serves both to act as an antirust and an emulsion depressant in the oil compositions of this process.

Examples 1 and 2: Blends were prepared employing as base stocks hydrotreated
Gulf Coast petroleum lubricating oil fractions blended to a viscosity in the range
of 61 to 67 cs at 40°C and solvent-extracted neutral lubricating oil fractions orig-
inating from a South American crude oil blended to the same viscosity range.
The blends each contain 1.0 wt % of zinc di(ethyl-hexyl primary)dithiophosphate
(ZDDP), 0.10% by weight of an additive containing 50% by weight active ingredient, a neutral zinc hydrocarbyl sulfonate (NaSulZS available from R.T. Vanderbilt), and 0.09% by weight of a mixture of dimer acids and the reaction product
of a monocarboxylic acid, a polyalkylene amine having more than one nitrogen
atom per molecule and a succinic acid or succinic anhydride (Hitec E-536,
Edwin Cooper).

The blends further contain a conventional polymethacrylate pour point depressant at 0.2% by weight and a conventional silicon antifoam agent at 0.0003% by
weight.

The two finished oils were tested for various properties. The results of these
tests are shown in Table 1.

Table 1: Performance Properties

	1	2
Base oil source	Gulf Coast	South American
Type	hydrotreated	solvent-extracted
Pump test ASTM D-2882		
Ring and vane wear loss, mg	23.0	18.3
Four-ball wear ASTM D-2266		
600 rpm, 175°F, 1.5 kg, 2 hr, mm	0.282*	0.271
1,800 rpm, 200°F, 40 kg, 2 hr, mm	0.578*	0.619

(continued)

Table 1: (continued)

	1	2
Denison filterability		
Dry oil, 1.2 μ filter, sec/75 ml	177	218
Wet oil, 1.2 μ filter, sec/75 ml	224	250
Emulsion characteristics		
ASTM D-1401, separation time, min	30	30
Hydrolytic stability ASTM D-2619, mg/cm^2	0.22	0.15*
Rust test ASTM D-665B		
Synthetic seawater, 24 hr	none	none
Oxidation characteristics ASTM D-943		
TOST, hr	3,024+*	2,000+
TAN-C, when stopped	0.23*	0.62*
Appearance of oil	clear*	laden with insolubles

*Average of duplicate results.

The above results demonstrate that an excellent lubricating oil composition, particularly suited as a hydraulic fluid is provided by this process.

Examples 3 through 5: In order to demonstrate the synergistic effect of the additive components, a series of lubricant fluids having compositions substantially identical to that of Example 1, but omitting one or more of the additives was prepared. Test results are shown in Table 2.

Table 2: Effect of Additive Components on Oil Performance

	. . . . Examples			
	3	4	5	1
Composition, % by wt				
Base oil	98.8	98.71	98.7	98.61
ZDDP	1.0	1.0	1.0	1.0
Reaction product	—	0.09	—	0.09
Zinc hydrocarbyl sulfonate*	—	—	0.10	0.10
Pour point depressant	0.20	0.20	0.20	0.20
Antifoam agent	0.0003	0.0003	0.0003	0.0003
Performance				
Denison filterability, sec to filter 75 ml				
Dry oil, 1.2 μ filter	—	322	298	234
2% water, 1.2 μ filter	600+**	470	297	284
Emulsion characteristics, ASTM D-1401, time for separation, min	***	60†	***	30
Hydrolytic stability, ASTM D-2619, Cu loss, mg/cm^2	***	0.07	2.73††	0.16

*Amount shown includes 50 wt % diluent in commercial additive.
**Results in excess of 600 sec are considered to have inadequate filterability.
***Test not run due to failure in another test.
†Times in excess of 30 min are considered to have failed the test.
††Results in excess of 0.5 indicate unsatisfactory hydrolytic stability.

Sulfurized Molybdenum Compound and Water-Soluble Fatty Ester

According to a process described by *A. Nassry and J.F. Maxwell; U.S. Patent 4,151,099; April 14, 1979; assigned to BASF Wyandotte Corporation* a water-based hydraulic fluid or metalworking composition can be obtained by blending water, a sulfurized molybdenum or antimony compound and a water-soluble C_{8-36} ester of an oxyethylated aliphatic alcohol and an oxyethylated aliphatic acid wherein either or both the acid or alcohol can be oxyethylated.

Alternatively, a water-based hydraulic fluid or metalworking composition can be obtained by blending water and

(1) a phosphate ester obtained by esterifying 1 mol of phosphorus pentoxide with 2 to 4.5 mols of a nonionic surface active agent obtained by condensing at least 1 mol of ethylene oxide with 1 mol of a compound having at least 6 carbon atoms and a reactive hydrogen atom with

(2) a water-soluble ester obtained by reacting an oxyethylated C_{8-36} aliphatic alcohol or aliphatic acid and

(3) a sulfurized molybdenum or antimony compound.

Stable concentrates of these ingredients can be prepared both with and without water, and where desirable the compositions can be thickened with a polyglycol type thickener, a polyacrylate thickener, or other thickeners known to those skilled in the art such as sorbitol, polyvinyl pyrrolidone, and polyvinyl alcohol. Corrosion inhibiting agents can also be added to the compositions to obtain increased corrosion resistance.

The molybdenum compound utilized can be oxymolybdenum phosphorodithioate. Antimony compounds of similar structure are useful. Representative concentrates are as follows:

Ingredient	. . . . Weight Percent.					
Polyoxyethylene 20 sorbitan monostearate	20	16	38	30	19	9
Alkyl phosphate ester of surface active agent*	5	4	8	6	4	2
Sulfurized oxymolybdenum or antimony organophosphorodithioate at 40% solids	15	12	31	26	15	8
Sodium 2-mercaptobenzothiazole	20	16	8	6	4	2
Isopropylaminoethanol or morpholine	40	32	15	12	8	4
Water	—	20	—	20	50	75

*From the reaction of 2 mols of phosphorus pentoxide with the surface active agent condensation product obtained by reacting 1 mol of oleyl alcohol and 4 mols of ethylene oxide.

Example 1: A comparative hydraulic fluid representing the best available water-based hydraulic fluid of the prior art was prepared by mixing 10% of a water-soluble polyoxyethylene ester of sorbitan monostearate having 20 mols of ethylene oxide per mol of ester and sold as Emsorb 6905 (Emery Industries) with 12% of a polyglycol thickener, sold as Pluracol V-10 (BASF Wyandotte Corporation). Morpholine in the amount of 1% was added as a vapor-phase corrosion inhibitor together with 1.8% of the triethanolamine salt of 2-mercaptobenzothiazole. The salt was prepared by reacting 5 parts by weight of triethanolamine with 1 part by weight of 2-mercaptobenzothiazole. The balance of the composition was deionized water. The fluid was clear in appearance and had a viscosity of 140 to to 150 SUS at 100°F. When tested according to the Vickers V-104C Vane Pump Test, for a period of 20 hours under a 750 psi load at 100°F and 1,200 rpm speed, the total weight loss was found to be 848 mg.

Example 2: A water-based hydraulic fluid of the process was prepared by blending 10% of a water-soluble polyoxyethylene ester of sorbitan monostearate, sold as Emsorb 6905, with 12% of a polyglycol thickener, sold as Pluracol V-10. To this mixture there was added 1% morpholine and 1.8% of the triethanolamine salt of 2-mercaptobenzothiazole prepared as in Example 1. There was then added 2% of a 40% solids emulsion of a sulfurized oxymolybdenum-organophosphorodithioate, sold as Vanlube 723 (R.T. Vanderbilt Company). The balance of the composition was deionized water.

The fluid had a viscosity of about 145 SUS at 100°F and was clear, amber colored and was tested in the Vickers V-104C Vane Pump Test for a period of 20 hours at 750 psi load at 100°F, and 1,200 rpm speed. Test results were obtained indicating a total wear weight loss of 566 mg.

Example 3: A water-based hydraulic fluid was prepared by mixing 10% of a water-soluble polyoxyethylene ester of sorbitan monostearate, sold as Emsorb 6905, 2% of a 40% solids emulsion of a sulfurized oxymolybdenum-organophosphorodithioate, sold as Vanlube 723, and 0.5% of a straight chain alkyl phosphate ester, sold as Antara LB-400, with 12% of a polyglycol thickener sold as Pluracol V-10. To this mixture there was added 1% of morpholine and 1.8% of the triethanolamine salt of 2-mercaptobenzothiazole as liquid-vapor corrosion inhibitors, the preparation of the triethanolamine salt of 2-mercaptobenzothiazole being described in Example 1.

The hydraulic fluid obtained had a viscosity of about 145 SUS at 100°F and was clear and amber colored. When tested in the Vickers V-104C Vane Pump Test, this fluid afforded excellent wear performance. Under performance testing at conditions of 750 psi load at 100°F and 1,200 rpm speed over a period of 20 hours, the unexpected excellent wear loss result of 117 mg was obtained. In a second test of the same hydraulic fluid, at 1,000 psi load, the wear weight loss was 120 mg.

Example 4: The fluid of Example 3 was diluted with deionized water using 1 part by weight water to 4 parts by weight of the hydraulic fluid of Example 3. A clear fluid was obtained which had a viscosity of about 85 SUS at 100°F and when evaluated in the Vickers Vane Pump Test using a 750 psi load at 100°F, 1,200 rpm speed and 20 hours test time, a wear loss of 134 mg was obtained.

Comparative performance of the hydraulic fluids prepared in Examples 1 through 4 is presented in the table below. As indicated in the table, the hydraulic fluid of Example 2 provides a marked improvement over the results obtained for the fluid of Example 1, but upon the addition of the phosphate ester to the fluid of Example 2 a completely disproportionate reduction in wear weight loss is obtained which improvement is substantially retained where the composition of Example 3 is diluted with 20% additional water (Example 4).

Vickers Vane Pump Wear Results

Hydraulic Fluid Example Number	Total Wear Weight Loss, mg
1	848
2	566
3	117
4	134

Modified Triglycerides

R.J. Sturwold; U.S. Patents 4,075,393; February 21, 1978; 4,108,785; Aug. 22, 1978, and 4,067,817; January 10, 1978; all assigned to Emery Industries, Inc. has found that mixed ester products obtained by treatment of a triglyceride under transesterification conditions with a polyoxyalkylene glycol and a high molecular weight dicarboxylic acid, such as polymer acids, are useful metalworking fluids. The modified triglycerides exhibit enhanced thermal stability and can be used either in neat form or in solution and are readily compatible with water to provide stable aqueous emulsions and dispersions having superior lubrication properties.

The mixed ester lubricants are especially useful with nonferrous metals and particularly those metals and metal alloys which are susceptible to lubricant and oxidative staining such as aluminum, copper, titanium and magnesium and their alloys. Aluminum and aluminum alloys containing copper, silicon, magnesium, zinc, lithium, beryllium, and the like derive particular benefit from the modified triglycerides of this process. It has been found that by the use of the modified triglycerides it is possible to minimize and in many cases completely eliminate the formation of undesirable lubricant stains on the surface of the aforementioned metals.

It is also possible to provide a protective hydrophobic coating on the surface of these metals, particularly aluminum metal, which is resistant to the formation of water staining and other similar forms of oxidative attack upon exposure to atmospheric conditions during shipment, storage, etc.

The nonstaining ability of these lubricants and aqueous formulations thereof make them particularly useful as lubricant/coolants (rolling oils) for both the hot and cold rolling of aluminum and its alloys. In this regard, the modified triglycerides in addition to providing the desired lubrication and cooling, also minimize "pickup" on the working rolls, do not foam excessively or have an offensive and irritating odor and are capable of producing a bright stain-free sheet. These ester products have additional advantage if the aluminum is annealed.

Example 1: A glass reactor equipped with a stirrer, thermometer, nitrogen inlet and water-trap connected to a condenser was charged with 288 g (1.0 equivalent) soybean oil, 60 g (0.3 equivalent) polyethylene glycol (PEG) having an average molecular weight of 400 and 85.5 g (0.3 equivalent) Empol 1014 dimer acid (95% C_{36} dibasic acid). The weight percentages of the respective reactants, based on the total charge, was 66.4, 13.8 and 19.8. To dry the system the mixture was heated with agitation while pulling a vacuum before addition of the stannous oxalate catalyst (0.03 wt % based on the total reactant charge). The reaction mixture was then heated to 200°C for about 9 hours while removing water of reaction. After cooling the reaction product was filtered using 0.5% diatomaceous earth filtering aid. The modified triglyceride (acid value 16.9) exhibited good lubricity and was readily emulsifiable in cold tap water with moderate agitation. The resulting aqueous emulsions had good stability. The modified triglycerides also exhibited markedly improved thermal stability as compared to unmodified soybean oil. Thermal stability was determined by thermal gravimetric analysis (TGA) by heating the samples in a vacuum while increasing the temperature at a rate of 10°C/min. Unmodified soybean oil was 90% decomposed at 275°C whereas the modified triglyceride showed only 35% weight loss at 275°C and only after heating to 425°C was 90% weight loss obtained.

Example 2: A series of modified triglycerides was prepared from top white tallow and PEG 400 employing varying amounts of dimer acid. The procedure employed was similar to that described in Example 1 with the exception that tetrabutyl titanate was used as the catalyst. Composition of the various products (equivalents/weight percent) and other pertinent properties are set forth below.

	. . . . Sample No.		
	2A	2B	2C
Tallow	1.0/76.5	1.0/71.1	1.0/66.4
PEG 400	0.3/15.9	0.3/14.8	0.3/13.8
Empol 1018 dimer acid*	0.1/7.6	0.2/14.1	0.3/19.8
Acid value	5.2	8.5	13.6
Hydroxyl value	36.2	24.3	19.1
Smoke point, °F**	370	380	380
Flash point, °F***	550	570	530
Fire point, °F***	600	620	590

*83% C_{36} dibasic acid and 17% C_{54} tribasic acid.
**First visible signs of smoke.
***ASTM D 92-66.

All of the above compositions were readily emulsifiable with water and exhibited enhanced thermal stability as compared to unmodified tallow.

Example 3: Following the above-described procedures, soybean oil was modified with PEG 400 and Empol 1018 dimer acid. Product 3A was obtained by reacting one equivalent refined soybean oil (bleached prior to use), 0.3 equivalent PEG and 0.3 equivalent dimer acid. The resulting modified triglyceride had the following properties:

	. . . . . . Sample No.	
	3A	3B
Acid value	8.1	16.3
Hydroxyl value	42.3	19.7
Viscosities, cs*		
100°F	62.6	113
210°F	12.1	19.5
Flash point, °F	540	570
Fire point, °F	650	610
Thermal stability	excellent	excellent
Emulsifiability in water	excellent	excellent

*ASTM D 445-65.

Example 4: To demonstrate the necessity of reacting the polyoxyalkylene glycol and high molecular weight dicarboxylic acid with triglyceride and the ability to obtain a variety of useful products by varying the reaction conditions, the following experiment was conducted. A reaction mixture consisting of 66.4 wt % soybean oil, 13.8 wt % PEG 400 and 19.8 wt % Empol 1018 dimer acid was heated at 220°C in the presence of 0.03 wt % tetrabutyl titanate catalyst. Samples were taken from the reaction mixture initially and after 15, 30 and 360 minutes of reaction. The acid value of each of the products was determined and the product was then evaluated for emulsifiability and thermal stability. The results were as follows:

	. . . . Sample.			
	Initially	15 Min	30 Min	360 Min
Acid value	38.8	31.3	26.9	13.6
Ability to form emulsions with water	no	very slight	excellent	excellent
Temp, °C, at which wt loss occurred				
50%	270	280	275	300
90%	280	320	415	425

The above results clearly show the improvement in emulsifiability and thermal stability as the PEG and dicarboxylic acid are reacted with the triglyceride.

Example 5: The effectiveness of the modified triglycerides to function as metal-working lubricants was demonstrated using a Falex machine. The test was conducted on both the neat oils and aqueous emulsions in accordance with ASTM test procedure D 2670-67. Results obtained with the products of Examples 2 and 3 were:

	. . . . Neat Oil		5% Aqueous Emulsion	
	Units		Units	
Product	Wear	Failure	Wear	Failure
2A	0	1,500	0	4,250
2B	2	1,250	1	3,500
2C	1	1,250	2	3,750
3A	7	1,250	0	3,000
3B	1	1,250	0	3,000

Example 6: To demonstrate the nonstaining character of the modified triglycerides the product was volatilized to determine the amount and type of residue remaining. Prior to use the aluminum weighing dishes (1¼" diameter) were heated 6 to 8 hours at 800°F to remove any residual oils. The dishes containing 0.1 ml sample (uniformly spread over the bottom) were then heated in a muffle furnace at 650°F for 30 minutes and visually inspected and rated for staining from 1 (no stain or very light tan stain) to 5 (heavy brownish-black stain). An average of at least four tests is reported as the stain rating. When a 5% aqueous emulsion of Product 2A was evaluated using this test procedure a stain rating of 1 was obtained.

R.J. Sturwold; U.S. Patent 4,132,662; January 2, 1979; assigned to Emery Industries, Inc. also describes additional, improved, lubricant compositions suitable for use as rolling oils for aluminous metals which produce bright, unstained metal sheet free of surface defects. In addition to their nonstaining character, the rolling oil compositions also exhibit superior lubricating properties and provide a protective coating on the surface of the aluminum or aluminum alloy so that the surface of the metal is resistant to water staining during subsequent storage and/or shipment.

The rolling oil lubrication compositions are comprised of a petroleum oil base stock containing from 1 to 20, and more preferably 3 to 10, weight percent of an additive consisting of:

(a)   from about 25 to 65% by weight of a polymeric fatty acid and prefer-ably a dibasic acid containing greater than 75% by weight C_{36} dibasic acid and having a maximum iodine value of 35;

(b)   from about 15 to 45% by weight of a fatty alcohol having 8 to 20 carbon atoms and more preferably a saturated alcohol having from 10 to 18 carbon atoms, or mixtures thereof; and

(c)   from about 15 to 45% by weight of a lower alkyl ester of a fatty acid having from 12 to 18 carbon atoms and more preferably a methyl ester of C_{14-18} fatty acids or mixtures of the fatty acids.

Pasty Lubricant for Nonchip Metal Forming

H.D. Grasshoff; U.S. Patent 4,138,348; February 6, 1979; assigned to Deutsche Texaco AG, Germany describes a lubricant for nonchip metal forming, the lubri-cant containing no mineral oil and being water-soluble and pasty. The following components are employed in producing the lubricant: a neutral fat and/or vege-table oil as a base material, a fatty acid or a mixture of fatty acids, an alkali metal hydroxide, water, and an alkali metal salt of a boric acid.

The finished lubricant comprises a neutral fat and/or vegetable oil, an alkali metal soap, water and a borate of an alkali metal.

The lubricants are produced such that the fatty acid and the fat and/or oil are charged to a container and heated to a temperature of about 80°C. At this tem-perature, the hydroxide dissolved in water is added to the batch to form soap. After completion of the soap formation, the remaining water and borate are added under intensive stirring.

For the purpose of testing the process lubricant, the penetration measurement according to DIN Standard No. 51 804 is carried out as well as measuring the pressure loading capacity according to DIN Standard No. 51 350 on a 4-ball apparatus.

The tests and the results are summarized in the table below. Test A is carried out without the addition of borate and serves as an example for comparison. The results of Tests B, C and D illustrate the high pressure loading capacity of the lubricant of the process as compared to Test A.

As a result of using the lubricant of the process in deep drawing, the number of perfect worked pieces, i.e., pieces without scoring, cracking and the like, is very high, and in wire drawing the wear of die and wire is found to be strongly re-duced.

	. . . . Examples			
	A	B	C	D
Starting materials, kg				
Fatty acid	180*	170**	200***	100*
Fat				
Tallow	400	380	—	250
Oil				
Rapeseed oil	—	—	400	—
Hydroxide				
KOH	37	35	—	21
NaOH	—	—	29	—
Borate				
$Na_2B_4O_7$	—	20	—	50
$K_2B_4O_7$	—	—	20	—
H_2O	383	395	351	579
Test results				
Penetration	180 mm/10	200 mm/10	220 mm/10	240 mm/10
Pressure loading capacity	1,200 N	1,600 N	1,800 N	

*Fish oil fatty acid, hardened.
**Tallow fatty acid.
***Stearic acid.

Forging Lubricant

A process developed by *S.C. Jain and C.A. Morris; U.S. Patent 4,104,178; Aug. 1, 1978; assigned to Wyman-Gordon Company* involves a water-based lubricant for hot forging metal. The composition is virtually nonflammable and nonpolluting. It comprises water, graphite, an organic thickener, sodium molybdate, and sodium pentaborate. Other additives are sodium bicarbonate and ethylene glycol, or mica. The water-based lubricant is exemplified by the following preferred composition.

	Weight Percent	Weight for 50 Gallons (lb)
Sodium carboxymethylcellulose (CMC)	0.77	4
Aqueous 30% graphics suspension (Quaker LQ-405 or Acheson 147)	38.60	200
Sodium molybdate	5.0	26
Sodium pentaborate	3.18	16.5
Sodium bicarbonate	4.83	25
Ethylene glycol	9.02	46.6
Water	38.60	200
Total	100.00	518.1

To prepare 50 gallons of the mixture, dissolve 4 lb of CMC (carboxymethylcellulose) in 16 gallons (133.34 lb) of water and mix thoroughly. Then, dissolve 26 lb of sodium molybdate and 16.5 lb of sodium pentaborate in the same container. Add 200 lb of 30% by weight graphite and stir. Dissolve 25 lb of sodium bicarbonate in the mixture. Add 46.55 lb (5 gallons) ethylene glycol and stir until the mixture is uniform. Finally, add sufficient water (8 to 10 gallons) to adjust viscosity for the method of application to dies. The resulting composition is a nonpolluting or minimal-air-polluting, water-based forging die lubricant for use on steel, stainless steel, nickel-base, and titanium-base alloys. It does not flame, has minimal smoke, and contains low sulfur (a requirement for many nickel-base alloys). It is particularly adapted for hammer (impact) forging, while substitution of 2 to 5 wt % mica for the sodium bicarbonate and ethylene glycol results in a formula particularly adapted for press (pressure) forging.

Lubricating Composition Applied over Primer Coat

G.A. McIntosh and R. Smith; U.S. Patent 4,148,970; April 10, 1979; assigned to Diamond Shamrock Corporation describe a lubricant that will maintain consistency of excellent coating adhesion. Further, in addition to lubrication, the composition also provides excellent die maintenance. The lubricating composition shows desirable stability including freedom from settling, layering or gumming. Moreover, it is readily applied by conventional techniques and has been found to provide ease of clean-up after application and metal deformation. When applied to substrates such as one-side primer-coated coil steel, the composition can provide desirable corrosion resistance to the bare steel.

The composition comprises metallic stearate, emulsifier, titanium dioxide and *Xanthomonas* hydrophilic colloid, all in aqueous medium.

Example: To a mixer there is charged 79 parts by weight of water, 0.23 part by weight sodium nitrite, 0.05 part by weight ethoxylated propoxylated alkylphenol having a viscosity in centipoises at 25°C of 180 and a density at 25°C of 8.7 lb/gal, and 3.45 parts by weight of polyethylene glycol 400 monooleate. As these ingredients are being mixed under vigorous agitation, there is next added to the mixture 1.09 parts by weight of xanthan gum hydrophilic colloid, which is a heteropolysaccharide prepared from the bacteria *Xanthomonas campestris* and having a molecular weight in excess of 200,000. To this resulting mixture, as mixing continues, there is then added 1.28 parts by weight formaldehyde, 13.07 parts by weight zinc stearate and 1.83 parts by weight pigmentary titanium dioxide. Mixing is continued until a thick mixture results.

A portion of this thick mixture is then diluted with deionized water in a weight ratio of 1 part by weight of mixture to 7 parts by weight of water. The resulting die lubricant is readily flowable. Steel test panels, which are cold-rolled low-carbon steel panels, are then coated with lubricant applied with an open-cell urethane foam paint brush. Panels are thereafter removed from the lubricant and air dried.

An additional set of steel panels is similarly coated with a comparative lubricant composition that has met with commercial acceptance as a drawing compound in the automotive industry, as for example, in stamping operations for automotive body parts. This comparative drawing compound contains a ground calcite pigment and emulsifier plus water in a mineral oil base.

After all panels are dried, they are placed in a condensing humidity test. In the condensing humidity test, the water is heated and mechanically circulated with the water temperature being maintained at 140°F. The panels are placed in the cabinet at about a 15° angle to the water surface, the bottom edge of the panels being 8" above the water surface and the top edge about 10" above the water surface. The back of the panels away from the water surface is exposed to the ambient air to permit condensation of the cabinet humidity on the test panel surface. In terminating the test, panels are simply removed from the cabinet, dried, visually inspected and rated during such inspection, comparing panels one with the other, for the amount of red rust on the panel face.

After 17 hours of such testing, the panels coated with the drawing compound of the process showed 2.7% by weight red rust, for an average of three panels, while those bearing the comparative drawing compound displayed 28% by weight red rust, again for an average of three panels.

In further testing, primer-coated test panels that are first lubricated, are subjected to shear testing. The panels are first precoated in the manner of U.S. Patent 3,990,920. The compositions used are the pH-adjusted compositions. The panels further have the primer top coating that has been discussed in U.S. Patent 3,990,920. Some test panels are then further prepared for testing by dipping in the lubricating composition of the process.

Panels are then tested in a draw test (shear adhesion) such as has been described in U.S. Patent 3,990,920. However, in the present test the clamp load or side pressure is 2,800 lb. Further, rather than the test panel being pulled through the die and bent around the radius of the ram as described in U.S. Patent 3,990,920, the test panel is pulled at a 90° angle to the die. In such testing, the unlubricated panels show failure, sometimes by panel breakage, and other times by severe coating removal. On the other hand, the lubricated panels all show no breakage and display desirable coating retention whereby they are judged to pass the draw test.

OIL-BASED LUBRICANTS

Organic Hydroperoxides and Sulfur

R.H. Davis and J.W. Schick; U.S. Patent 4,080,302; March 21, 1978; assigned to Mobil Oil Corporation have found that improved load carrying, antiwear and extreme pressure properties are imparted to lubricant compositions by the presence of an organic hydroperoxide and sulfur.

The organic hydroperoxide suitable for use in the compositions of this process are, in general, of the formula: R—OOH where R is a hydrocarbyl group which contains from about 2 to 46 carbon atoms. In particular R may be alkyl, including cycloalkyl, aryl, aralkyl and alkenyl. The R group may be straight or branched and the hydroperoxide attachment may be to a carbon atom which is either primary, secondary or tertiary.

Most particularly preferred as the organic hydroperoxides: cumene hydroperoxide, and tert-butyl hydroperoxide.

The compositions were tested by a Tapping Efficiency Test. This test is described by C.D. Fleming and L.H. Sudholz in *Lubrication Engineering,* 12, 3, pp 199-203, (May-June 1956), and also in U.S. Patent 3,278,432.

It should be noted, in accordance with the Tapping Efficiency Test, that if the test fluid torque values exceed the reference value, tapping efficiency is below 100%. Criteria for product acceptance are evaluated as shown in the table on the following page.

Tapping Efficiency, %	Comments
>100	Fluid considered outstanding and should outperform reference product in severe cutting operations.
80–100	Acceptable range for moderate cutting fluids.
<80	All products with tapping efficiencies below 80% are considered unacceptable. Torque values are erratic, frequently due to tap sticking and/or breakage.

Examples 1 through 7: Small amounts of organic hydroperoxides were added to a base oil and the resulting compositions were tested for tapping efficiency in the tapping test described above. The table below gives the composition and results.

				Composition, wt %		
Ex. No.	Cumene Hydro-peroxide	tert-Butyl Hydro-peroxide	Paraffinic Oil* 100 SUS at 100°F	Sulfurized Mineral Oil** 150 SUS at 100°F	Tapping Efficiency
1	–	–	100	–	***
2	–	–	–	100	87
3	1.0	–	99	–	***
4	0.1	–	–	99.9	93
5	1.0	–	–	99.0	102
6	2.0	–	–	98.0	100
7	–	1.0	–	99.0	102

*Contains 0.2 wt % sulfur which was not removed by the refining process.
**68 wt % elemental S.
***Cannot tap.

Referring to the data presented in the table, Example 1 indicates that it is virtually impossible to tap SAE 1020 steel with a paraffinic oil which contains 0.2 wt % "natural" sulfur, i.e., not removed by the refining process. Example 3 shows that the addition of 1.0 wt % cumene hydroperoxide to that oil provides no improvement. The composition tested in Example 3 is similar to those described in U.S. Patent 2,470,276 in that the sulfur present is mainly natural sulfur. The composition of Example 3 does not exhibit the improved extreme pressure properties as do the compositions of this process.

A sulfurized oil, i.e., an oil to which elemental sulfur has been added, is acceptable only for moderate duty machining, as illustrated in Example 2. However, a vast improvement in tapping efficiency is observed when organic hydroperoxides are added to the sulfurized oils, as shown by Examples 4 through 7. Thus, the lubricant compositions of the process, represented by Examples 4 through 7 in the above table, are suitable for the most severe metal-cutting operations.

Sulfurized Oil and Polyalkenylsuccinic Anhydride-Polyglycol Reaction Product

H.J. Andress, Jr.; U.S. Patent 4,072,618; February 7, 1978; assigned to Mobil Oil Corporation describes a cutting oil composition comprising a mineral oil and minor amounts, sufficient to provide improved extreme pressure properties thereto of (A) a sulfurized oil and (B) the reaction product of a polyalkenyl succinic anhydride with a polyethylene glycol or a polypropylene glycol or this reaction product further reacted with sulfur or phosphorus pentasulfide.

Example 1: A mixture of 500 g (0.33 mol) of polybutenyl succinic anhydride (MW 1,500) and 200 g (0.66 mol) of polyethylene glycol (MW 300) was heated with stirring and held at that temperature for about 4 hours. Filtration gave the final product.

Example 2: A mixture of 1,050 g (0.70 mol) of polybutenyl succinic anhydride (MW 1,500) and 420 g (0.70 mol) of polyethylene glycol (MW 600) was heated with stirring to about 275°C and held at that temperature for about 2 hours. Filtration gave the final product.

Example 3: A mixture of 368 g (0.25 mol) of polybutenyl succinic anhydride (MW 1,472) and 150 g (0.375 mol) of polypropylene glycol (MW 400) was heated with stirring to about 250°C and held at that temperature for about 8 hours. Filtration gave the final product.

Example 4: A mixture of 380 g (0.5 mol) of an alkenyl (dimerized C_{18-26} 1-olefin mixture) succinic anhydride (MW 760) and 150 g (0.5 mol) of polyethylene glycol (MW 300) was heated with stirring to about 250°C and held at that temperature for about 3 hours. Filtration gave the final product.

Example 5: A mixture of 1,375 g of a product prepared as described in Example 2 of 30 g of powdered sulfur was stirred at about 200°C for about 7 hours. Filtration gave the final product.

The ability of a cutting oil to operate efficiently is measured in the tapping test. In the tapping test, a series of holes are drilled in a test metal such as SAE 1020 hot-rolled steel. The holes are tapped in a drill press equipped with a table which is free to rotate about the center on ball bearings. A torque arm is attached to this "floating table" and the arm in turn activates a spring scale, so that the actual torque during the tapping, with the oil being evaluated, is measured directly. The same conditions used in evaluating the test oil are used in tapping with a strong oil which has arbitrarily been assigned an efficiency of 100%. The average torque in the test oil is compared to that of the standard and a relative efficiency is calculated on a percentage basis.

For example, the torque with standard reference oil is 19.3, the torque with test oil is 19.8 and the relative efficiency of the test oil is 97.4 [(19.3/19.8) x 100]. The test is described by C.D. Fleming and L.H. Sudholz in *Lubrication Engineering,* 12, 3, pp 199-203 (May-June 1956).

A base oil comprising a paraffinic mineral oil having a viscosity of 150 SUS at 100°F and 3 wt % sulfurized lard oil (15% sulfur) was subjected to the tapping test. Also, blends of the base oil with each of the products of Examples 1 through 6 (Component B) were prepared and subjected to the tapping. Pertinent data are set forth in the following table.

Component B, Product of	Concentration (wt %)	Tapping Efficiency (%)
None	—	87
Example 1	3	114

(continued)

Component B, Product of	Concentration (wt %)	Tapping Efficiency (%)
Example 2	3	109
Example 3	3	101
Example 4	3	118
Example 5	3	117
Example 6	3	100

Partially Neutralized Aluminum Acid Alkylorthophosphates

J.W. van Hesden; U.S. Patent 4,115,285; September 19, 1978; assigned to Borg-Warner Corporation has found that liquid compositions comprising a hydrocarbon and an aluminum acid alkylorthophosphate are useful as cutting fluids in machining operations. The addition of minor amounts of aluminum acid alkylorthophosphates to hydrocarbon cutting fluids greatly decreases tool wear and consequently improves tool life.

Alkylorthophosphoric acids are readily prepared by the reaction of phosphorus pentoxide with alcohols, according to the classical formula:

$$P_2O_5 + 3ROH \longrightarrow ROPO(OH)_2 + (RO)_2POOH$$

wherein R is a C_{1-22} alkyl or alkenyl radical or a mixture thereof. The product mixture thus contains both mono and dialkyl acid phosphates, with three reactive acidic groups for every two atoms of phosphorus. The acidic mixture will thus require three equivalents of reactive base to completely neutralize the remaining acidity.

The aluminum acid alkylorthophosphates are prepared by reacting less than a stoichiometric quantity of a basic aluminum compound such as hydrated alumina or aluminum isopropoxide with the alkylorthophosphoric acid. The amount of basic aluminum compound may be varied between 20 and 70% of the stoichiometric amount, i.e., the amount required to fully neutralize the acidity of the alkylorthophosphoric acid.

The preparation of full aluminum salts of alkylorthophosphoric acids is described in U.S. Patent 2,329,707.

Example 1: An aluminum acid alkylorthophosphate was prepared in the following manner: 112 g of P_2O_5 were mixed with 74.7 g of butyl alcohol and 384.2 g (1.7 mols) of a commercial mixture of C_{12-22} n-alkanols. The reaction mixture was heated to a gentle reflux and stirred until all the P_2O_5 had reacted.

The resulting mixture of alkylorthophosphoric acids was reacted with 26.5 g (42% of stoichiometry) of hydrated alumina by mixing the components and heating at 110°C for 1 hr. The resulting aluminum acid alkylorthophosphate was a viscous oil.

Example 2: A cutting oil base was prepared by mixing 10 gallons of a commercial naphthenic process oil having a viscosity of 67 cs at 100°F with 5 gallons of a second commercial naphthenic process oil having a viscosity of 23.4 cs at 100°F.

To 18 liters of the base oil were then added with stirring 65.35 g (0.4 wt %) of the aluminum acid alkylorthophosphate prepared in Example 1. The final composition was clear and showed no visible thickening.

Example 3: An 18-liter batch of base oil containing 65.35 g (0.4 wt %) of the aluminum acid alkylorthophosphate of Example 1 was prepared as in Example 2. The acidity was neutralized by stirring into the batch 16 ml of 15 wt % aqueous sodium hydroxide. The final composition was clear and showed only a slight increase in viscosity, from 100 to 132 cp, measured with a Brookfield viscometer using a No. 3 spindle at 50 rpm.

Example 4: An 18-liter batch of base oil containing 65.35 g (0.4 wt %) of aluminum acid alkylorthophosphate was prepared as in Example 2. The batch was then made alkaline by stirring in 34 ml of 15 wt % aqueous sodium hydroxide. There was no visible thickening, and the viscosity was 95 cp, measured as before.

Examples 5 through 8: The cutting oils prepared were evaluated in a comparison test by milling cold-finished 1018 steel bar stock, using a 14-tooth cutter at 160 rpm, fed 0.100-inch deep at 2 inches per minute, for a total of 600 lineal inches per test. The wear for each of the 14 teeth was then measured by inspection at 100 times magnification and averaged. The wear data are presented in tabular form in the table below.

Ex.		. . . . .Wear at			
No.	Oil*	Top	Corner	Front	Combined
5	Control	56	157	128	113
6	Ex. 2	34	46	106	62
7	Ex. 3	43	75	111	80
8	Ex. 4	42	65	104	70

*Control equals base oil prepared as in Example 2, without additive.

It will be apparent from these data that 0.4 wt % of aluminum acid alkylorthophosphate in a base cutting oil (Example 6) markedly decreases tool wear in three key areas, as compared with uncompounded base cutting oil (Example 5). Neutralizing the acidity as in Example 7 or making the composition distinctly alkaline as in Example 8 does not destroy the wear reducing effect of the additive, even though wear is slightly increased over the acidic composition of Example 6.

Hot Melt Drawing Lubricant

According to a process described by *R.W. Jahnke; U.S. Patent 4,116,872; Sept. 26, 1978; assigned to The Lubrizol Corporation* metalworking operations, especially drawing, are facilitated by applying to the metal a composition which provides lubricity thereto and which melts within the range of 30° to 100°C. The composition comprises at least one neutral ester, and preferably a mixture of esters, prepared from polyalkylene glycols and saturated aliphatic alcohols having at least 10 carbon atoms, and C_{12-25} aliphatic monocarboxylic acids and C_{4-20} aliphatic polycarboxylic acids.

The preferred ester mixtures are prepared from polyethylene glycols, C_{14-20} predominantly straight chain alkanols, stearic acid and adipic, azelaic or sebacic acid. Optional ingredients include phosphorus acid salts and antioxidants. The composition may be applied in liquid form and solidifies on cooling to normal ambient and storage temperatures.

In the following table are listed typical hot melt compositions suitable for use in the process.

Ingredients	A	B	C	D
	.... pbw			
Neutral adipic acid ester of C_{14-18} 1-alkanols*	10	–	–	–
Neutral azelaic acid ester of C_{16-18} 1-alkanols*	–	8.5	–	–
Stearic acid ester of Carbowax 1540	90	76.5	–	–
Ester mixture prepared from 0.75 eq Carbowax 1540, 0.25 eq of C_{14-18} 1-alkanols*, 0.75 eq stearic acid and 0.25 eq adipic acid	–	–	100	90
Zinc salt of a mixture of isobutyl and primary amylphosphorodithioic acids	–	15	–	10

*Commercial mixture of predominantly straight chain alkanols.

Any metal to be worked may be treated according to the process; examples are ferrous metals, aluminum, copper, magnesium, titanium, zinc and manganese as well as alloys thereof and alloys containing other elements such as silicon.

Drawing Lubricant

According to a process described by *R.W. Jahnke; U.S. Patent 4,118,331; Oct. 3, 1978; assigned to The Lubrizol Corporation* lubricants useful in metalworking processes, especially drawing, comprise an oil of lubricating viscosity; a carboxylic acid or derivative thereof, especially an anhydride of a hydrocarbon-substituted succinic acid in which the substituent contains about 6 to 30 carbon atoms; and a phosphorus acid salt, usually a zinc salt of a phosphorodithioic acid. The lubricant may also contain other ingredients, especially chlorinated wax or mixtures of aliphatic alcohols and hydrocarbons. The lubricant provides improved lubricity, corrosion resistance, cleanability, antiwear properties and extreme pressure properties when used in the working operation. The constitution of typical lubricants suitable for use in the process is given in Table 1 below.

Table 1

Ingredients	A	B	C	D	E	F	G	H	J	K	L
	.... pbw										
Mineral oil	75.65	13.65	26.65	88.5	14.9	39.3	43.55	12.9	93.1	92.5	18.8
Bright stock mineral oil	–	62	45	–	64	28.8	32.1	58.9	–	–	46.0
Tetrapropenylsuccinic anhydride	20	20	19	8	8	–	20	–	–	–	–
Polyisobutene*-substituted succinic anhydride	–	–	–	–	–	18	–	19	–	–	5.8
Lithium salt of polybutenyl*-substituted succinic acid in which the polybutenyl units are principally polyisobutenyl	–	–	–	–	–	–	–	–	4	4	–
Zinc salt of mixed isobutyl**- and primary amyl***- phosphorodithioic acid	4.35	4.35	4.35	3.5	13.1	3.9	4.35	–	–	3.5	9.4
Zinc salt of tetrapropenyl-phenylphosphorodithioic acid	–	–	–	–	–	–	–	–	2.9	–	–

(continued)

Table 1: (continued)

Ingredients	A	B	C	D	E	F	G	H	J	K	L
	. . . . pbw										
Lead isooctylphosphoro-dithioate	–	–	–	–	–	–	–	4.2	–	–	–
Chlorinated wax (~42% chlorine)	–	–	5	–	–	10	–	5	–	–	–
Epal 20+ alcohol-hydrocarbon mixture	–	–	–	–	–	–	–	–	–	–	20.0

*Mol wt about 1,000.
**65 mol %.
***35 mol %.

Any metal to be worked may be treated according to the process; examples are ferrous metals, aluminum, copper, magnesium, titanium, zinc and manganese as well as alloys thereof and alloys containing other elements such as silicon.

The lubricity properties of the compositions used in the process are demonstrated in a test in which a cold-rolled steel strip, 2 x 13½ inches, is drawn between two dies in an Instron Universal Tester, Model TT-C. Prior to drawing, the edges of the strip are deburred and the strip is vapor degreased and wiped with a clean cloth. It is then coated with the drawing lubricant and mounted in the testing machine. The dies are tightened by means of a torque wrench set at 40 ft-lb torque and the strip is pulled through the die for 2 inches at the rate of 5 inches per minute.

The force or "load" required to pull the strip through the die is recorded on a chart; if there is "chattering" and irregular movement due to friction, the deviation from a uniform load is also recorded on the chart. When a number of lubricants are being compared, the tests are all run on the same day on strips from the same sheet of steel.

The results of several lubricity tests are given in Table 2. The load readings are those at the end of the 2-inch drawing period. All tests were run at room temperature. The entry marked Control 1 is a commercial chlorinated wax drawing lubricant sold by D.A. Stuart Oil Co., Ltd, as T13-B; that marked Control 2 is a rust-inhibiting drawing oil sold by Quaker Chemical Corp. as Ferro Cote; and a fatty acid ester drawing lubricant sold by Quaker Chemical Corp. as Quaker 693M.

Table 2

Test	Lubricant	Load, lb	Deviation
1	1*	1,125	40
	C	1,125	25
2	2*	1,750	75
	A	1,925	0
3	3*	1,750	0
	A	1,825	0
4	A	1,750	0
	B	1,100	0
	G	1,500	0
5	B	1,350	0
	C	1,500	0
6	L	1,450	0

*Controls.

The extreme pressure properties of the lubricants are measured by means of the SAE test, in which two Timken cups on shafts are operated in contact with each other at a relative speed of 500 rpm in a bath of test lubricant as a load is gradually and automatically applied. The load at which scoring occurs is the test result. When tested by this method, the results listed in Table 3 were obtained.

Table 3

Lubricant	Load, lb
Control 1	147
B	102
C	111

Oilless Fluid for Scoring Glass

H.R. Gorman, J.R. Dahlberg and J.L. Oravitz, Jr.; U.S. Patent 4,084,737; April 18, 1978; assigned to PPG Industries, Inc. describe an improved glass cutting fluid which contains no oil and leaves no residue on the glass surface, thereby eliminating the need for washing. More particularly, this process relates to a method of cutting glass employing an oilless cutting fluid which produces a spall-free cut glass edge having increased edge strength over an edge cut by dry scoring and which also extends the useful life of the cutting tool.

The primary component of the improved glass cutting fluid is a halogenated hydrocarbon. The cutting fluid may consist essentially of a liquid halogenated hydrocarbon or may comprise a halogenated hydrocarbon and a paraffin, naphtha or aromatic solvent. The cutting fluid need not contain an emulsifying agent to render the fluid water-washable since the cutting fluid is essentially completely removed by evaporation. Solutions or blends are adjusted for evaporation rates to meet the specific conditions imposed by production procedures and both automatic and manual operation equipment.

The method of cutting glass involves effecting a score on the surface of the glass in the presence of the cutting fluid and propagating the score to produce a cut through the glass. The cut edge of glass scored by the method of the process has increased edge strength compared with a cut edge of glass scored in the absence of the cutting fluid of the process.

SOLID LUBRICANT COMPOSITIONS

SOLID LUBRICANTS

Zinc-Based Polysulfides

J.M. Longo and J.J. Steger; U.S. Patent 4,130,492; December 19, 1978; assigned to Exxon Research & Engineering Company have discovered that materials of the formula MXY_3 wherein M is selected from the group consisting of Mg, V, Mn, Fe, Co, Ni, Zn, Cd, Sn, Pb and mixtures thereof, preferably Fe, Zn and mixtures thereof, most preferably Zn; X is a pnictide selected from the group consisting of phosphorus, arsenic, antimony, and mixtures thereof, preferably phosphorus, arsenic and mixtures thereof, most preferably phosphorus; and Y is a chalcogenide selected from the group consisting of sulfur, selenium, and mixtures thereof, most preferably sulfur, are superior lubricants. They exhibit resistance to oxidation and thermal degradation, low friction, excellent antiwear activity and long effective life.

Surfaces coated with these compositions resist gauling and damage due to adhesive or corrosive wear. As lubricants they can be used either dry or in conjunction with conventional lubricants selected from the group consisting of lubricating oils and greases.

Example: 4 g of $ZnPS_3$, $FePS_3$ and MoS_2 were added to 96 g of an aluminum complex soap-salt grease made from the following components: 90 g animal fatty acid; 30 g benzoic acid; 50 g Kolate (aluminum alcoholate, isopropyl); 1,742 g Coray 80/50 (unextracted naphthenic mineral lubricating oil); and 6 g UOP 225 (commercial antioxidant).

The grease was prepared by heating the mixture of acids and alcoholate in the oil, with the removal of methyl alcohol, to form the complex of the aluminum with the fatty acid and benzoic acid, and adding the antioxidant. The test materials, i.e., the $ZnPS_3$, $FePS_3$, and MoS_2 were mixed into the grease components at room temperature and the resulting formulations were milled so as to uniformly mix the solid components into the grease.

302

In addition, two standard extreme pressure formulations, base grease + 2% Elco 114 (zinc dialkyl dithiophosphate) and base grease + 3% tricalcium phosphate + 1% sulfurized polybutene were formulated and milled. Each grease formulation was subsequently tested for extreme pressure properties following the re-approved 1974 ASTM procedure D 2596-69 (Four Ball Method). For the base grease and for each formulation the wear scar diameter (mm) was measured as a function of the applied load (kg). The last nonseizure load and the weld point were recorded and the standard load wear index was calculated. The results are shown below for these tests.

Results of Four Ball Extreme Pressure Tests*

Sample Description	Load Wear Index
Base grease	20
Base grease + 2% Elco 114	25
Base grease + 3% tricalcium phosphate + 1% sulfurized polybutene	33
Base grease + 4% MoS_2	37
Base grease + 4% $FePS_3$	39
Base grease + 4% $ZnPS_3$	44

*ASTM D 2596-69

From this data it is seen that addition of $ZnPS_3$ or $FePS_3$ to an aluminum complex soap grease improves the extreme pressure characteristics relative to the base grease and also with respect to standard extreme pressure formulations. In addition, the performance of the grease with 4% $ZnPS_3$ or $FePS_3$ is superior to that found for a grease formulated with an equivalent amount of the well known solid lubricant MoS_2.

Finely Divided Metal Carbonate and Halogenated Compounds

P. Wainwright, M.J. Green and B.S. Charlston; U.S. Patent 4,159,252; June 26, 1979; assigned to Rocol Limited, England have discovered that compositions containing in combination a halogenated organic lubricant and a Group II-A metal carbonate, optionally with molybdenum disulfide also, give excellent results, comparable to or in some circumstances better than those given by conventional molybdenum disulfide compositions. The best results are obtained when an alkaline earth metal sulfate or other inorganic sulfate is also present. Preferred halogenated lubricants are halogenated hydrocarbons, particularly chlorinated paraffins.

The most preferred materials are calcium carbonate (whiting) and calcium sulfate hemihydrate, preferably in combination with the chlorinated paraffins, but other materials are successful, for example other carbonates; other sulfates such as magnesium sulfate·$7H_2O$, calcium sulfate mono- and dihydrates, anhydrous sodium sulfate, potassium sulfate, potassium aluminum sulfate, zinc sulfate, sodium hydrogen sulfate, and sodium thiosulfate·$5H_2O$.

Among preferred halogenated lubricants are materials such as Cereclor chlorinated long chain paraffin hydrocarbons grades 70 (powder), 70 L, 63 L and 50 LV (ICI); similar bromoparaffins; fluorinated graphites of formula $(CF_x)_n$ (Air Products); Monoflor 53 and 91 fluorocarbons, which are liquids of formula

Four-Ball Test Machine Results for Various Lubricating Compositions (Scar Diameters in mm)

Blend	Composition	Applied Load (kg)										
		56	100	158	200	251	316	355	398	447	501	562
1*	White petroleum jelly	1.58	2.59	Welds at 141 kg								
2*	White petroleum jelly (a) + 2% $CaSO_4 \cdot \tfrac{1}{2}H_2O$ (b) + 20% $CaSO_4 \cdot \tfrac{1}{2}H_2O$	0.33 –	1.31 0.46	2.20 –	2.48 1.24	Weld –	1.48	Weld				
3*	White petroleum jelly + 2% Cereclor 63L	0.43	0.66	2.07	2.55	Welds at 224 kg						
4*	White petroleum jelly + 20% whiting (Snocal 8/SW)	0.43	0.66	0.86	0.96	1.45	1.66	1.76	1.78	1.96	2.15	2.27
5*	White petroleum jelly + 20% whiting + 2% $CaSO_4 \cdot \tfrac{1}{2}H_2O$	0.41	0.61	0.93	1.10	1.12	1.51	1.52	1.57	1.68	2.27	2.24
6	White petroleum jelly + 20% whiting + 2% Cereclor 63L	0.33 0.34	0.39 0.42	0.72 0.66	0.90 0.78	0.97 0.93	1.06 1.04	1.36 1.22	1.53 1.44	1.66 1.49	1.79 1.54	1.84 1.73
7	White petroleum jelly + 20% whiting + 2% Cereclor 63L + 2% $CaSO_4 \cdot \tfrac{1}{2}H_2O$											(Welds at 708 kg)
10	Blend 7 + 20% MoS_2	–	0.42	–	0.60	–	1.04	–	–	–	–	(Weld at 631 kg)
9	Blend 7 + 8% 'Tiona G' anatase TiO_2	–	0.39	–	0.90	–	1.28	–	–	–	–	(No weld at 794 kg)
8*	Rocol ASP amber petroleum jelly + 50% MoS_2	0.35	0.42	0.96	1.14	1.43	1.47	1.44	Weld			
11*	White petroleum jelly + 50% graphite	0.36	0.46	0.71	1.23	2.00	Weld					

*(Comparative)

$(C_2F_4)_n$ made by ionic polymerization of tetrafluoroethylene (ICI); Fluon L 169 polytetrafluoroethylene (ICI); oligomer-based fluorochemical waxes such as RDPE and RDPE-S Wax (ICI); and low molecular weight chlorotrifluoroethylene polymers of formula $(CF_2 \cdot CFCl)_n$, such as Halocarbon Products' Oil 14-25.

The inorganic materials are in finely divided form, for example, the carbonate is suitably 99% less than 25 μ, 93% less than 10 μ.

The successful results of the process are specific to the combination of components, as is shown by the results in the table of tests of various blends in white petroleum jelly as a lubricating base. The tests were done in the known Seta-Shell four-ball test machine, used for assessing lubricant performance under extreme pressure. The smaller the scar diameter found, the better the lubricant. The compositions are by weight, the amounts of additives being relative to the composition as a whole.

In addition to the results which are shown in the table, the mean Hertz loads (a figure corrected for indentation of the balls and indicating wear properties over a range of loads) of Blends 7, 10, 9, 8 and 11 were determined at 104.7, 118.1, 99.9, 85.0 and 68.5 kg respectively.

Polyphenylquinoxaline Antifriction Fillers and Modifiers

V.V. Korshak, I.A. Gribova, A.P. Krasnov, E.S. Krongauz, A.M. Berlin, O.V. Vinogradova, G.V. Mamatsashvili, S.-S.A. Pavlova, P.N. Gribkova, N.I. Bekasova, L.G. Komarova, V.D. Vorobiev, I.V. Vlasova and A.V. Vinogradov; U.S. Patent 4,076,634; February 28, 1978 describe a composition which contains from 5 to 98 pbw of polyphenylquinoxaline, 1 to 94 pbw of an antifriction filler, 0 to 40 pbw of a reinforcing filler, and 0.1 to 10 pbw of a modifying dopant such as terephthalaldehyde, tetranitrile of pyromellitic acid, metal polyphosphinates or carborane-containing compounds.

The antifriction material on the basis of the composition features a low and stable coefficient of friction and provides for the operation of dry-friction assemblies constructed therefrom at temperatures up to 350°C.

Polyphenylquinoxalines feature excellent thermal and chemical stability and high moldability. Thus, for instance, polyphenylquinoxaline produced from 3,3',4,4'-tetraaminodiphenyl ether and 1,4-bis(phenylglyoxalyl)benzene and having a molecular weight of approximately 80,000 and viscosity in m-cresol η = 0.83 dl/g, has a softening point of 280°C (by thermomechanical curves). Thermogravimetric testing of the polymer in air at a temperature gradient of 4.5°C/min indicates that its thermal destruction point lies at 500°C.

The antifriction fillers for the proposed composition may be both mineral and man-made substances with antifriction properties, the two major solid lubricants being graphite and molybdenum disulfide.

The composition for antifriction material is prepared by mixing, e.g., in a vibratory mill, powdered polyphenylquinoxaline with modifying dopants and fillers. The components are mixed until a homogeneous mixture has been obtained. The composition thus produced is subjected to pressing at a temperature of from 350° to 450°C and a pressure between 600 and 2,800 kg/cm^2.

Under dry-friction conditions, the antifriction material experiences both thermal and mechanical loading. The polyphenylquinoxaline component together with the modifying dopants provide for a low and stable coefficient of friction of the antifriction material in a broad temperature range. Thus, the antifriction materials on the basis of the proposed composition employed in dry-friction assemblies easily withstand operating temperatures of up to 350°C, whereas the prior art antifriction materials have a thermal ceiling of only 220° to 240°C.

The modifying dopants cause crosslinking of the polyphenylquinoxaline, thereby raising the thermal stability of the antifriction material and improving its mechanical properties. Thus, for instance, a material on the basis of polyphenylquinoxaline, graphite and carborane-containing polyamide exposed for 7 hours to a temperature of 250°C lost only 0.0170 g through wear, whereas a similar material having no carborane-containing polyamide lost 0.0223 g.

The antifriction materials exhibit very satisfactory structural properties: Brinell hardness, from 20 to 40 kg/mm^2; bending strength, from 500 to 1,500 kg/cm^2; and impact strength, from 5 to 35 kg-cm/cm^2.

Example 1: The composition of this example was prepared by mixing the components in a vibratory mill. The composition was converted by pressing to specimens in the form of bushings of external diameter 22 mm and internal diameter 12 mm. The pressed specimens were tested for friction behavior under end face friction conditions against steel at a linear velocity of 2 m/sec and a load of 2 kg/cm^2.

The composition for antifriction material contained 40 pbw of polyphenylquinoxaline of the formula

where n = 120, produced from 3,3',4,4'-tetraaminodiphenyl ether and 1,4-bis-(phenylglyoxalyl)benzene, as well as 2 pbw of terephthalaldehyde and 58 pbw of graphite.

The composition was pressed at 400°C and 1,500 kg/cm^2. At a temperature of 350°C, the coefficient of friction was equal to 0.056; in the temperature range from 150° to 350°C it was between 0.05 and 0.06. The antifriction material had an impact strength of 5 kg-cm/cm^2 and a Brinell hardness of 20 kg/mm^2; the material was employed for manufacturing antifriction bearing retainers and linings.

Example 2: The composition contained 20 pbw of the polyphenylquinoxaline of Example 1, 0.2 pbw of tetranitrile of pyromellitic acid and 79.8 pbw of molybdenum disulfide. The specimens were prepared and tested in procedures duplicating those of Example 1. At 350°C the coefficient of friction was 0.06, while in the temperature range between 150° and 350°C from 0.048 to 0.06.

Example 3: The composition contained 20 pbw of the polyphenylquinoxaline of Example 1, 10 pbw of manganese polydiphenylphosphinate and 70 pbw of molybdenum disulfide. The specimens were pressed at a temperature of 400°C and a pressure of 1,000 atm. The specimens were tested in a procedure duplicating that of Example 1. The coefficient of friction was 0.06 at 350°C and between 0.038 and 0.062 in the temperature range from 150° to 350°C.

Molybdenum Disulfide, Hydroxyethyl Cellulose and Silicate

According to a process described by *W.L. Karpen; U.S. Patent 4,088,585; May 9, 1978; assigned to Carpenter Technology Corporation* finely divided MoS_2 powder is mixed in water and with water-soluble hydroxyethylcellulose and sodium or potassium silicate. A small amount of pine oil may also be added as a biocide and defoaming agent because of the susceptibility of hydroxyethylcellulose to bacterial attack and its tendency to foam excessively, particularly since this mixture is preferably continuously stirred to maintain homogeneity. The mixture is applied to workpieces which are then dried to leave a tough dry lubricant film strongly adhered to the surfaces thereof.

As an example of this process, the following ingredients were combined in a tank.

	Pounds	Kilograms	Weight Percent
MoS_2	301.1	136.6	19
Sodium silicate solution*	159.3	72.3	10**
Hydroxyethylcellulose	17.5	7.9	1
Pine oil	1.6	0.7	0.1
Water	1,113.5	505.1	69.9
Total	1,593.0	722.6	100.0

*RU solution (Philadelphia Quartz Co.) is about 47.05 wt % solutes.
**This is equivalent to about 4.7 wt % dissolved silicate in the bath.

To ensure homogeneity, the bath was continuously recirculated. Coils of stainless steel wire feedstock of various diameters from 0.077" to 0.480" (0.196 to 1.22 cm), and with variations in surface smoothness and cleanliness were cleaned with acid to remove oxides and foreign matter. The acid was then washed off the coils before dipping to avoid acidifying the bath which was maintained at a pH of from 11 to 13, since lowering the pH may cause some of the components to precipitate or agglomerate.

The coils were immersed into the bath, with no required residence time as long as each coil was thoroughly wetted, and with care not to touch the coils to the sides or bottom of the tank. The coils were spread as much as possible to allow maximum surface coverage. After dipping, the coils were allowed to drain over the tank for a short time, about 15 seconds being adequate, and then dried, using forced ambient air to accelerate the drying time. Heated air can be used, but care must be taken to dry the coils only until the silicate reaches hydration equilibrium, at about 9 to 11 wt % water for the RU sodium silicate. Further drying with heated air will embrittle the silicate and possibly cause cracking or flaking of the coating in cold working. It was found that by discontinuing the drying when the coils seemed visibly dry, over-drying was avoided.

Once the coils were visibly dry, they were rotated 180° for the purpose of changing the drainage patterns and altering loop contact points, and then dipped and dried a second time by the same method. Following this procedure, additional dips could also be made.

The resultant coating densities ranged from 8.36 to 18.29 mg/in² (1.30 to 2.83 mg/cm²), averaging at about 14 mg/in² (2.2 mg/cm²). This feedstock was then used in a variety of cold working operations in which the lubricant adhered to the substrate even in multistage operations while providing consistently good lubrication.

Self-Lubricating Bearings

A process described by *J.R. Rumierz; U.S. Patent 4,146,487; March 27, 1979; assigned to SKF Industries, Inc.* provides lubricating compositions capable of providing lubrication for prolonged periods of time at operating temperatures above 105°C (221°F).

The compositions consist essentially of about 50 to 90% by wt of an oil of lubricating viscosity and from 50 to 10% by wt of polymethylpentene (PMP) having an average molecular weight in the range from 3 to 5 million. These compositions are provided in the form of firm tough, solid gels having an oily surface provided by the exudation of oil from the gels. The preferred compositions are made from synthetic hydrocarbon oils having a viscosity in the range of from 20 to 260 mm²/s measured at a temperature of 38°C (100°F).

Other known lubricating oils of comparable viscosity may also be employed including the diester oils described in Military Specifications MIL-L-23699B and MIL-L-7808G. These refer to aircraft turbine engine lubricants. Products qualified under these specifications are Exxon ETO 2380 and ETO 2389.

Example 1: To illustrate the preferred practice of the process a shaped mass of lubricating gel was prepared containing 70% oil and 30% PMP by wt of the total composition. More specifically, 40 g of PMP was mixed with 93 g of lubricating oil in a conventional blender for about one minute until a homogeneous mixture was obtained. The PMP was in the form of a 60 to 120 mesh powder which is TPX Polymer from Mitsui Petrochemical Industries. This PMP has an average molecular weight of 4 million. The oil was a synthetic hydrocarbon oil SHC624 from Mobil Oil Corporation and had a viscosity of 33 mm²/s at 38°C (100°F).

After blending, the oil-PMP mixture was charged to a suitable mold with provision for heating, heated to 218°C (425°F) and maintained at that temperature for 60 minutes. The end was reached when the mixture became transparent and self-cohesive.

Heating was then discontinued and the mold and its contents allowed to cool to ambient temperature. When the mold was opened, a self-supporting, shaped mass of lubricating composition was obtained in the form of a firm, tough, solid gel having an oily surface caused by exudation of oil from the gel.

Example 2: The general procedure of Example 1 was repeated using the same PMP but substituting Mobil SHC 629 for the oil used previously. This oil differs primarily in having a viscosity of 160 mm^2/s at 38°C (100°F). No end point of the cure cycle was reached after 240 minutes using a 218°C (425°F) cure temperature. No shaped mass was formed. This is attributable to the greatly increased viscosity of the lubricant which hinders intimate mixing during the cure cycle at 218°C (425°F).

Example 3: The general procedure of Example 2 was repeated except a cure temperature of 254°C (490°F) was employed. An end point was reached in 45 to 50 minutes. The resulting shaped lubricating gel was similar to that obtained in Example 1.

Example 4: A functional test was designed in order to assess the relative merits of the lubricating compositions and similar oil-polyethylene lubricating gels of the prior art. A standard 6205 ball bearing having a fixed outer ring and a rotatable inner ring was provided with an intermediate lubricant mass to be tested having a surface in contact with the rotating inner ring. This apparatus was then used in a series of tests as follows.

The bearing was loaded with a lubricating mass consisting of an oil-polyethylene gel of the prior art such as those described in U.S. Patent 3,541,011. The exact composition of the gel was 30% polyethylene (Hercules UHMW 1900) and 70% Mobil DTE XH oil.

The speed of rotation of the inner ring of the bearing was increased stepwise in 3,600 rpm increments, allowing the apparatus to run until the operating temperature had stabilized at each step. No extraneous heat was supplied, i.e., the test was run under ambient temperature conditions.

The bearing speed reached 7,200 rpm at which speed the oil-polyethylene gel lubricant failed by the lubricant mass being expelled from the bearing. The bearing temperature was 49°C (120°F) at 7,200 rpm.

The foregoing procedure was repeated after substituting an oil-PMP lubricant gel of Example 3 in the bearing. The bearing was run at 3,600 rpm, 7,200 rpm, and 8,500 rpm with the bearing temperature allowed to stabilize at each step. Lubricant failure occurred after 1.5 hours running at 8,500 rpm and a bearing temperature of 110°C (230°F), indicating that the oil-PMP lubricating compositions are as good as or better than, the oil-polyethylene gels of the prior art under ambient conditions.

The same standard 6205 ball bearing was charged with an oil-polyethylene lubricating mass of 30% polyethylene and 70% Mobil DTE XH oil and a test was conducted at a constant bearing speed of 3,600 rpm under a thrust load of 665 newtons (150 lb). In this test, however, the operating temperature was increased in discrete steps after ambient running for 72 hours. The extraneous heat was supplied by mounting the bearing in a housing containing electrical cartridge heaters which raised the temperature of the bearing outer ring.

Under these conditions the oil-polyethylene lubricant composition of the prior art failed upon reaching a temperature of 100°C (212°F).

Self-Lubricating Antifriction Material

A process described by *J.N. Vasiliev, A.V. Petrenko, G.N. Bagrov, G.N. Gordeeva, T.I. Kazintsev, V.D. Telegin, and V.N. Goldfain; U.S. Patent 4,093,578; June 6, 1978* provides self-lubricating antifriction material. This material comprises carbon filler in the form of a powder of fired oil coke, a powder of artificial graphite or carbon cloth; a binder in the form of epoxy resin having a molecular weight of 300 to 1,560 obtained through condensation of epichlorohydrin with diphenylolpropane, and polyaluminophenylsiloxane resin having a molecular weight of 5,000 to 15,000 with a ratio of the silicon atoms to the aluminum atoms equal to 3:8; chlorinated paraffin having a molecular weight of 500 to 1,100 and chlorine content of 40 to 70 mol %.

The weight percent proportions of the components in the self-lubricating antifriction material are as follows: epoxy resin, 15 to 57; polyaluminophenylsiloxane resin, 3 to 10; chlorinated paraffin, 0.2 to 10; and the balance, carbon filler.

The antifriction material can also contain a dry lubricant in the form of powdery boron nitride, natural graphite or molybdenum disulfide in the amount of 5 to 10% by wt. The abovedescribed self-lubricating antifriction material has a number of valuable properties, namely, good self-lubricating capacity, low friction coefficient, high wearability, high mechanical strength and high thermal resistance. This material can be used for making components of friction units operating without lubricant, as well as for making components of friction units operating in water, kerosene, lubricating oil, and liquid oxygen.

Molten Glass Applications

A process described by *A.F. Marcantonio; U.S. Patent 4,131,552; December 26, 1978* involves a release and lubricating composition for avoiding seizure between molten glass and metal molds for forming glass articles and for minimizing wear between mold sections and for improving the pack of glassware. The composition is a dispersion of graphite in a polyphenylene sulfide.

Instead of periodic applications of the release and lubricating compositions, the compositions of the process are effective for relatively extended periods of time, but may be renewed by swabbing the compositions during production of glassware. This minimizes a loss of pack and efficiency resulting from periodic inclusion of graphite in the glassware surface and the altering of operation temperatures. Also, contamination of the surrounding machinery and the atmosphere with oily fumes and vapors are greatly reduced or completely avoided.

Example 1: A lubricating composition was prepared by taking 100 pbw of poly(p-phenylene sulfide) into 595 parts of ethylene glycol to which was added about 8 parts of octylphenoxy polyethoxyethanol. This mixture was blended for about 10 minutes. To this mixture were added 50 parts of graphite and thereafter 255 parts water. The mixture was ball milled for about 6 hours until it had an average viscosity of about 20 cp as determined by Brookfield, Spindle 4, rpm 60, 72°F. The ball milling temperature was maintained at 80°F. The average particle size was 30 μ. The resulting dispersion contained 10%

poly(p-phenylene sulfide), 5% graphite, 59% ethylene glycol, 25% water, and 1% wetting agent. This was designated Composition X.

Using the same mixing procedures as recited above, with the exception that n-butoxy poly(ethoxypropoxy)propanol was substituted for ethylene glycol, the following compositions were prepared as shown below.

	Compositions, (% by wt)			
	A	B	C	D
Polyphenylene sulfide	11.0	11.0	6.0	9.0
Graphite	7.0	7.0	4.0	6.0
n-butoxy poly(ethoxy-propoxy)propanol	64.0	56.0	63.0	63.0
Wetting agent*	1.6	2.0	0.8	1.6
Thickening agent**	0.4	0.1	0.2	0.4
Water	16.0	24.0	26.0	20.0

*octylphenoxy polyethoxyethanol
**Carbopol 934

Example 2: A blank mold and neck ring molds were lightly sandblasted to produce a satin finish, solvent cleaned, and coated with Composition X, while preheated to a temperature of 200°F. The Composition X was applied by spraying through a Binks air sprayer to produce a coating on the blank mold of between 0.002" and 0.025" thickness. The blank mold was then maintained at a temperature of 700°F for about 30 minutes to remove the carrier and to set the poly(p-phenylene sulfide) upon the surface of the molds.

After preheating at 700°F for 1 hour, the mold was placed upon an independent section glass-forming machine and used to produce commercial glassware of the 32 oz jar type. Without swabbing, the blank mold operated satisfactorily for about 23 hours. The neck rings operated satisfactorily for about 19 hours.

In related work *A. F. Marcantonio, P.J. Kress and G.A. Lee; U.S. Patent 4,140,834; February 20, 1979; assigned to Ball Corporation* describe release and lubricating compositions for avoiding seizure between molten glass and metal molds for forming glass articles comprising an aqueous dispersion containing polyphenylene sulfide, an inorganic binder, and a solid lubricant.

The inorganic binder may be represented by the general formula MPO_4 wherein M is a metal ion selected from the group consisting of aluminum, chromium, magnesium, iron and zinc. The particular solid lubricants include members selected from the group consisting of graphite, molybdenum disulfide, and polytetrafluoroethylene.

A preferred inorganic binder is aluminum phosphate. Other useful inorganic binders include magnesium phosphate and zinc phosphate. Generally, about 30 to 80 wt % of the inorganic binder is required.

GREASE COMPOSITIONS

In modern practice, it has become increasingly important that grease compositions be able to provide adequate lubrication at high temperature, e.g., temperatures of 350° to 450°F or higher. Further, because many of the newer high temperature grease applications, e.g., high speed sealed bearings, require that the grease maintain a high level of lubricant activity for extended time periods, it is also essential that the thickener impart a high degree of mechanical stability on the grease formulation at such high temperatures.

A variety of thickening agents have been proposed for use in such high temperature applications including soap base thickeners, inorganic clay thickeners and organic thickening agents. Of these classes of thickening agents, the organic thickeners, specifically those containing urea or ureido functional groups, have been considered quite attractive because of their ashless nature and high temperature thickening properties.

Recent developments in this area of lubrication technology have focused on total grease composition, with efficient additive combinations being developed to satisfy many newer, and increasingly more stringent, service requirements.

THICKENERS

Benzimidazobenzisoquinolinone Ureas

H.A. Harris; U.S. Patent 4,089,854; May 16, 1978; assigned to Shell Oil Company has found that benzimidazo[2,1-a]benz[d,e]isoquinolin-7-one monourea compounds having the structural formula:

where R is a hydrocarbyl radical of 16 to 22 carbons, are excellent thickening agents for grease compositions employed in high temperature applications.

The benzimidazo[2,1-a]benz[d,e]isoquinolin-7-one monourea grease thickening agents can be conveniently prepared by the sequential processes of: (1) refluxing naphthalic anhydride and a nitrophenylene diamine in acetic acid for about 8 to 10 hours, (2) reducing the nitrobenzimidazo[2,1-a]benz[d,e]isoquinolin-7-one produced with sodium sulfide, and (3) further reacting the amine product with an equivalent of the desired isocyanate in tetrahydrofuran for about 48 to 72 hours. The procedure or reaction sequence is illustrated by the following equations:

where R is a hydrocarbyl radical as previously defined.

Example 1: The preparation of 10-octadecylureido-benzimidazo[2,1-a]benz[d,e]-isoquinolin-7-one is as follows.

A 15.8 gram portion of naphthalic anhydride was mixed with 12.2 grams of 4-nitrophenylenediamine and placed in 200 ml of glacial acetic acid in a 500 ml flask. The mixture was stirred with a magnetic stirring bar and brought to reflux. The mixture was refluxed for about 8 hours and then cooled. The mixture was filtered and the insoluble product washed with ether to remove most of the acetic acid. The product was placed in the vacuum oven overnight to remove last traces of solvent. Yield was 20.6 grams of 10-nitro-benzimidazo-[2,1-a]benz[d,e]isoquinolin-7-one.

A 22.1 gram portion of the 10-nitro-benzimidazo[2,1-a]benz[d,e]isoquinolin-7-one was placed in a 1 liter flask containing 700 ml of water and 25 grams of sodium sulfide nonahydrate. The mixture was brought to reflux and refluxed for about one hour. A second 25 gram portion of sodium sulfide was added. The mixture was refluxed for additional 2 hours, cooled, and filtered. The insoluble product was thoroughly washed with water and then dried in the vacuum oven. The yield was 18.2 grams of 10-amino-benzimidazo[2,1-a]benz[d,e]-isoquinolin-7-one.

A 5.7 gram portion of the 10-amino-benzimidazo[2,1-a]benz[d,e]isoquinolin-7-one was added to 200 ml of distilled tetrahydrofuran in a 500 ml flask. To this mixture was added 5.9 grams of Mondur O (octadecyl isocyanate) and 0.3 gram of 1,4-diazobicyclo-octane. The mixture was stirred with a magnetic stirring bar and brought to reflux.

The mixture was refluxed until no isocyanate peak was visible at 2,240 cm^{-1} on infrared spectrophotometric analysis (~48 to 72 hours). It was then cooled and the solvent removed on the rotary evaporator. The solid thus obtained was ground to a fine powder, stirred in ether, filtered and dried in the vacuum oven. The yield was 10.7 grams of 10-octadecylureido-benzimidazo[2,1-a]benz[d,e]-isoquinolin-7-one.

Example 2: A thickened grease composition according to this process was prepared from the product of Example 1 and an HVI 70/210 Neutral oil (mineral lubricating oil having a viscosity of 70 SUS at 210°F). To prepare this grease composition, 55.0 grams of the 10-octadecylureido-benzimidazo[2,1-a]benz[d,e]-isoquinolin-7-one was ground to a fine powder with a mortar and pestle and stirred into 445.0 grams of the base oil while slowly heating to 150°C. This warmed slurry was milled on a three-roll paint mill (three passes being sufficient to produce a homogenous grease) and baked in an oven for about 1 hour at 150°C. The baked grease was cooled to about 90°C and milled again at 150°C through the three-roll paint mill to afford a smooth grease having an ASTM dropping point of 570°F and an ASTM worked penetration (D 217) (60 strokes) of 302.

Triazine-Urea Compounds

T.F. Wulfers; U.S. Patent 4,113,640; September 12, 1978; assigned to Shell Oil Company has found that triazine-urea compounds having the structural formula

where R is an aliphatic hydrocarbyl radical of 16 to 22 carbons, m is 0 or 1 and n is 0 or 1, are excellent thickening agents for greases employed in high temperature applications.

Exemplary species of the compounds of the process include: 4,4'-bis[6-octadecylureido-4-amino-s-triazinyl)ureido]-3,3'-dimethylbiphenyl; bis[p-(6-octadecylureido-4-amino-s-triazinyl)ureidophenyl] methane; and 4,4'-bis[6-eicosylureido-4-amino-s-triazinyl)ureido]-3,3'-dimethylbiphenyl.

The triazine-urea grease thickening agents are conveniently prepared via sequential reaction of an alkyl isocyanate with melamine (triamino-s-triazine) to produce a ureido-s-triazine intermediate, followed by reaction of the intermediate with a dinuclear aromatic diisocyanate in a 2:1 molar ratio of intermediate to diisocyanate to produce the desired triazine-urea thickener.

Example 1: A 200 ml round bottom flask equipped with magnetic stirring bar and a reflux condenser was charged with 12.6 grams of melamine in 120 ml of distilled dimethylformamide. The resulting solution was heated to incipient boiling and 29.5 grams of octadecylisocyanate (Mondur O) was added with stirring. The resulting mixture was brought to reflux and refluxed for about 1 hour under agitation. Heating was discontinued and the hot solution was filtered through a warm funnel. The filtrate was allowed to cool slowly, and the product which crystallized from the filtrate was separated by filtration. This product was washed with ether and dried in a vacuum oven. The yield was 40 grams of product having a melting point of 181° to 185°C.

21 grams of the above product was charged to a 500 ml flask along with 300 ml of xylene. The flask was equipped with a magnetic stirring bar and a reflux condenser. 6.6 grams of 4,4'-diisocyanato-3,3'-dimethylbiphenyl was added to the flask with stirring and the mixture was refluxed for about 12 hours; completion of the reaction was indicated by the disappearance of the isocyanate absorption band at 2,240 cm^{-1} in the infrared spectrum.

Upon completion of the reflux period, the reaction mixture was cooled and the solvent removed on the rotary evaporator to yield a solid product. This solid product was ground to a powder in a mortar, stirred in 400 ml of ether, and then filtered. The filtered product was dried in a vacuum oven to yield 26 grams of 4,4'-bis[(6-octadecylureido-4-amino-s-triazinyl)ureido]-3,3'-dimethylbiphenyl.

Example 2: A thickened grease composition according to the process was prepared from the triazine-urea product of Example 1 and an HVI 70/210 Neutral oil. To prepare this grease composition, 55.0 grams of the triazine-urea was ground to a fine powder with a mortar and pestle and stirred into 445.0 grams of the base oil while slowly heating to 150°C. This warmed slurry was milled on a three-roll paint mill (three passes being sufficient to produce a homogenous grease) and baked in an oven for about 1 hour at 150°C. The baked grease was cooled to about 90°C and milled again at 150°C through the three-roll paint mill to afford a smooth grease having an ASTM dropping point of 505°F and an ASTM worked penetration (D 217) (60 strokes) of 290.

Example 3: Utilizing the procedure of Example 1, 40.5 grams of bis[p-(6-octadecylureido-4-amino-s-triazinyl)ureidophenyl] methane was synthesized from melamine (9.9 grams), octadecylisocyanate (20.0 grams) and 4,4'-diisocyanato-diphenylmethane (10.8 grams). This triazine-urea thickener (30.0 grams) was formulated with HVI 70/210 Neutral oil (170.0 grams) according to the procedure described in Example 2 to yield a smooth grease having an ASTM dropping point of 520°F and a modified (¼ scale) worked penetration (60 strokes) of 69. This penetration was determined using a modification of the grease workers' test ASTM D 217A wherein a cylindrical plug was inserted in the grease worker cup to reduce the volume of grease necessary for the test from approximately 300 to 75 ml and the penetration measured with a one-quarter scale penetrameter.

Example 4: A series of grease compositions containing various triazine-urea thickening agents according to the process were prepared using the procedure described in Examples 1 and 2. These greases, fully formulated with commercial additives, were tested in a variety of conventional grease tests and demonstrated their excellent mechanical stability at high temperatures.

Polyurea Composition

A.W. Kisselow and P.K. Wulk; U.S. Patent 4,129,512; December 12, 1978; assigned to The British Petroleum Company Limited, England describe a process for the production of polyurea lubricating greases by the reaction of a monoamino, a diamino and a diisocyanato component in a lubricating oil.

Lubricating greases of the type mentioned above with polyureas as thickeners are already known. Thus, for example, U.S. Patent 3,243,372 describes lubricating greases which contain polyureas of the general formula:

$$R'-NH\text{+}CO-NH-R'''-NH-CO-NH-R-NH\text{+}_xCO-NH-R'''-NH-CO-NHR''$$

in which R' and R'' are monovalent hydrocarbon radicals, R and R''' are divalent hydrocarbon radicals, and x is a whole number from 1 to 3.

Lubricating greases according to U.S. Patent 3,243,372 are produced by dissolving a diisocyanate in the lubricating oil which is to be thickened, possibly with the aid of a further solvent, adding amines to the mixture and then heating the total mixture so that a polyurea is formed in situ.

It has been found that lubricating greases with good mechanical/dynamic behavior, excellent stability to working, very good load-carrying capacity and excellent chemical and thermal resistance are obtained if, in the process for the production of lubricating greases according to West German Patent 2,260,496, a particular dioxa-alkane-diamine is used.

Thus, according to this process, the production of polyurea lubricating greases is accomplished by the reaction in a lubricating oil of: (a) a monoamino component, consisting of a mixture of at least one representative of each of the classes of compounds of general formulas $R-NH_2$ and $R'-NH_2$ in which R is a monovalent aliphatic and R' is a monovalent aromatic radical; (b) a diamino component of the general formula $NH_2-R-O-R'-O-R''-NH_2$ in which R, R' and R'' are divalent aliphatic radicals; and (c) a diisocyanato component, consisting of an aromatic diisocyanate or a mixture of aromatic diisocyanates, the two isocyanato groups in the molecules being connected with the same aromatic ring or with different aromatic rings, is characterized in that the diamino component is 4,17-dioxaeicosandiamine-1,20 having the following formula:

$$H_2N-(CH_2)_3-O-(CH_2)_{12}-O-(CH_2)_3-NH_2$$

The 4,17 dioxaeicosandiamine-1,20 is a known compound which can be readily prepared. The components of the grease are preferably reacted with one another in the lubricating oil at a temperature of from 20° to 150°C, most preferably at from 40° to 70°C. If desired, the mixture obtained may then be heated further to a temperature of 150° to 200°C, preferably to about 175°C. By means of this reaction under controlled conditions an additional improvement in the properties of the lubricating greases of the process may be obtained, which may be attributed to transamidation and crosslinking reactions of the polyurea.

Example: A mixture of 75.6 grams of 4,17-dioxaeicosandiamine-1,20, 25.6 grams of p-toluidine and 73.5 grams of Genamine 20/22 R 100 D (a mixture of primary aliphatic amines with chain lengths of C_{18}, C_{20}, C_{22} and C_{24} in the proportion of 4:39:54:3) and 1,500 grams of a naphthenic solvent raffinate (with a viscosity of 5°C/50°C corresponding to 27.5 cs/60°C) is heated with stirring to 60°C. The amines melt and form with the mineral oil a clear solution. 85.3 grams of Desmodur T 65 (a mixture of 65% 2,4- and 35% 2,6-toluylene diisocyanate) are added drop by drop with continuous intensive stirring within a period of 15 minutes.

The turbidity which shows at once increases rapidly and the total reactor content thickens. A further 100 grams of the same base oil are then added through the diisocyanate feed vessel. The reaction mixture is heated with further stirring to 175°C, after which the temperature is allowed to cool to 100°C. A mixture of 10 grams of Amine T (a corrosion inhibitor of Ciba-Geigy), 20 grams of Additin 30 (an oxidation inhibitor of Bayer AG) and 110 grams of base oil, is then added and the grease is allowed to cool to room temperature. The lubricating grease is then homogenized in a colloid mill.

Alkali Metal Salts of Gamma-Keto Acids

C.A. Audeh; U.S. Patent 4,079,013; March 14, 1978; assigned to Mobil Oil Corporation describes grease compositions containing, as thickeners, alkali metal salts of a gamma-keto acid.

In general, the grease thickener is produced by reacting a liquid hydrocarbon-containing material with an alkali metal at a temperature of at least about 100°C to form the corresponding alkali metal salts of hydrocarbons containing reactive hydrogen in this material, and followed by reacting the products thus obtained with a cyclic acid anhydride to form the corresponding alkali metal salts of gamma-keto acids. Particularly preferred liquid hydrocarbon-containing materials, as the vehicle, are furfural or phenol extracts derived from petroleum solvent refining processes, which are readily commercially available.

In its more specific aspects, the production of the alkali metal salts of gamma-keto acids can be carried out in accordance with the following sequence: The preparation of an alkali salt of hydrocarbons intrinsic to a lubricating oil or its extract is illustrated by the following reactions (1) or (2).

(1) $2M° + 2RH \rightarrow 2R^-M^+ + H_2\uparrow$

(2) $R_1^-M^+ + RH \rightarrow R^-M^+ + R_1H\uparrow$

More specifically, with the foregoing reactions (1) or (2) in view, an alkali metal, M°, for example, sodium, potassium or lithium, is reacted with a RH mineral lubricating oil or its extract. This reaction results in an alkali metal salt of any reactive hydrocarbon intrinsic to the oil or its extract. Reaction (1) is generally conducted with vigorous stirring and agitation at a temperature from about 100° to 200°C. When reaction (2), is employed, the alkali metal salt of a volatile hydrocarbon R_1M, such as n-butyl lithium or phenyl sodium is mixed slowly with RH, the lubricating oil or its extract in an inert atmosphere at ambient temperature.

In reaction (3) shown below, is illustrated the preparation of the alkali metal salt of a gamma-keto acid.

$$(3) \qquad R^-M^+ + R_2CH-\overset{\overset{\displaystyle O}{\|}}{\underset{\underset{\displaystyle CH_2-\overset{}{\underset{\displaystyle O}{\|}}C}{|}}{C}} \longrightarrow R_2CH-\overset{\overset{\displaystyle O}{\|}}{\underset{\underset{\displaystyle CH_2-\overset{}{\underset{\displaystyle O}{\|}}C-R}{|}}{C}}-O^-M^+$$

In this reaction, a solution of an alkyl substituted acid anhydride in a volatile solvent, for example, tetrahydrofuran, is added to the products obtained in either reaction (1) or reaction (2). After the reaction is completed, the solvent is removed and the thickener viz an alkali metal salt of a gamma-keto acid is obtained.

In the foregoing reactions (1), (2) and (3), R is a group derived from the reactive hydrocarbons intrinsic to the oil or its extract; R_1 is preferably, an alkyl group having from 3 to 5 carbons, R_2 is an alkyl or alkenyl group having from 8 to 30 carbons; and M is an alkali metal. The vehicle employed can be either a mineral oil or synthetic lubricating oil.

As shown in the table below, measured amounts of the alkali metal salts of gamma-keto acids were employed for thickening a furfural extract derived from solvent refining process to the consistency of a grease. In this respect, a measured amount of the thickener was added to the required quantity of oil and the material was heated to 150°C with stirring. The thickened oils were further homogenized by two passes through a three-roll mill. Employing different proportions of the thickener in the blend, resulted in greases of different penetrations.

In the table, a grease was employed which was produced in accordance with the sequence shown in the abovedescribed reactions (1), (2) and (3) in which R_1 was a mixture of hydroaromatic compounds, R_2 equaled $C_{18}H_{37}$, and M was lithium, as illustrated in the abovedescribed reaction (3). R, as previously indicated, was a group derived from the reactive hydrocarbons intrinsic to the oil.

Alkali-Metal Salts of Gamma-Keto Acid Thickened Furfural Extracts

Sample	Thickener (wt %)*	ASTM Penetration Unworked/Worked (60x)**	Dropping Point (°F)	Oil Separation (wt %)***
1	32	99/107	417	16.4
2	28	148/153	422	18.8
3	24	168/167	423	28.4

*Calculated from syntheses data
**One-half scale
***30 hours at 300°F

As will be apparent from the foregoing samples of the table, greases formulated with the alkali metal salts of gamma-keto acids as thickeners, exhibit satisfactory penetrations, dropping points and oil separation.

EXTREME PRESSURE ADDITIVES

Sodium and Potassium Borates

A process developed by *J.H. Adams; U.S. Patent 4,155,858; May 22, 1979; assigned to Chevron Research Company* is concerned with grease compositions containing as extreme pressure (EP) additives sodium and potassium borates of limited water content and a boron to alkali metal ratio greater than 2.5.

The additives are prepared by reacting boric acid and potassium or sodium hydroxide in an appropriate ratio of boron to alkali metal and heating the product at elevated temperature for a time sufficient to remove water to the desired extent. The reaction is carried out in aqueous medium at as high a reactant concentration as possible to minimize the amount of water that must be removed. Heating of the solid mass after liquid water is removed is preferably carried out at temperatures above 300°F and usually above 400°F.

The product is comminuted to powder form and simply dispersed in the grease by conventional means at a temperature of 100° to 250°F, usually about 140° to 180°F. The preferred limits of the water to alkali metal atom (Z/M or Z since M = 1) of 0.5 to 3 represent the most practical application in grease.

Example 1: Preparation of Borate Additive — 52 grams of KOH and 145 grams boric acid were dissolved in 200 ml water and heated to drive off water. Heating on a hot plate (surface temperature of 600°F) for 3 hours yielded 140.0 grams of $KB_3O_5 \cdot H_2O$.

Example 2: Incorporation of Borate Additive in Grease — The materials prepared as in Example I were ground with a mortar and pestle and incorporated by stirring into various base greases at a temperature of about 180°F. The greases were then passed through a three-roll mill.

The greases prepared above were subjected to Timken Test (ASTM D-2509), Penetration P-60 (ASTM D-1403) and Load Wear Index and True Weld Point (ASTM D-2596) were obtained. The greases were commercial greases.

These data are shown in the following table. For comparison, data were obtained on greases containing additives having water contents outside the preferred limits (Compositions 1 and 2) and boron to alkali metal ratios outside the limits of this process (Compositions 3 and 5), and a grease containing a conventional lead extreme pressure (EP) additive (Composition 9).

As can be seen from these data, Tests 1, 2 and 3 produced liquid materials after heating to 300°F. The extreme pressure values for the materials containing the additives of this process were excellent.

Alkali Metal Borate Extreme Pressure Greases

Composition	. . . . . .Grease Type, %.		. . . . .Borate, %.	
1	Lithium Hydroxystearate	96	$NaBO_2 \cdot 4H_2O$	4
2	Lithium Hydroxystearate	96	$Na_2B_4O_7 \cdot 10H_2O$	4
3	Lithium Hydroxystearate	96	$Na_2B_4O_7 \cdot 2.5H_2O$	4
4	Lithium Hydroxystearate	96	$NaB_3O_5 \cdot 3H_2O$	4
5	Lithium Hydroxystearate	96	$K_2B_4O_7 \cdot 3H_2O$	4
6	Lithium Hydroxystearate	96	$KB_3O_5 \cdot H_2O$	4
7	Lithium Hydroxystearate	98	$KB_3O_5 \cdot H_2O$	2
8	Lithium Hydroxystearate	98	Aqueous KB_3O_5	*
9**	Lithium Hydroxystearate	89.7	−	
10	Aluminum Complex	96	$KB_3O_5 \cdot H_2O$	4
11	Sodium n-octadecyl tere-			
	phthalamate	96	$KB_3O_5 \cdot H_2O$	4
12	Clay	96	$KB_3O_5 \cdot H_2O$	4

*Dehydrated in situ.   **Contains 10.3% lead EP additives.

Tests	. . . . .Compositions											
	1	2	3	4	5	6	7	8	9	10	11	12
Penetration, 60 strokes	349	342	342	342	334	341	345	441	313	225	347	292
Timken OK Load, lb	30	30		40	30	45	45		55			
Load Wear Index, lb			72.8	80.6	70	102	56.4		34.5	47.6	76.4	46.7
True Weld Point, kg			260	335	305	400+	250		265	175	280	150
P_{60} after heating in mixer to 300°F	*	*	*	400	346	442			368			
Thin film, dry, 275°F	25	49	29	23		35	25		8			

*Liquid

J.H. Adams; U.S. Patent 4,100,080; July 11, 1978; assigned to Chevron Research Company has found that good extreme pressure performance is provided by a grease which comprises a major portion of an oil of lubricating viscosity, a minor portion sufficient to thicken the oil to grease consistency of an organic grease thickener and a minor portion sufficient to impart extreme pressure properties of a particulate dispersion of a hydrated potassium borate having a mean particle size of less than one micron and a boron to potassium ratio of about 2.5 to 4.5. The borate dispersions are added to the greases in the form of a suspension in oil.

The potassium borate dispersions are prepared by dehydrating a water-in-oil emulsion of an aqueous solution of potassium hydroxide and boric acid to provide a boron to potassium ratio of 2.5 to 4.5. This is carried out by introducing into the inert nonpolar oil medium an aqueous solution of potassium hydroxide and boric acid (potassium borate solution) and preferably an emulsifier, vigorously agitating the mixture to provide an emulsion of the aqueous solution in the oil and then heating at a temperature and for a time which provide the desired degree of hydration of the microemulsion. The preparation of the dispersions is disclosed in U.S. Patent 3,997,454, December 14, 1976.

In related work *J.H. Adams; U.S. Patent 4,089,790; May 16, 1978; assigned to Chevron Research Company* describes a synergistic extreme pressure lubricating composition comprising an oil of lubricating viscosity having dispersed therein: (1) 1 to 60 weight percent of hydrated potassium borate microparticles having a boron to potassium ratio of about 2.5 to 4.5; (2) from 0.01 to 5.0 weight percent of an an antiwear agent selected from (a) a zinc dihydrocarbyl dithiophosphate having from 4 to 20 carbons in each hydrocarbyl group; (b) a C_{1-20} amine salt of a dihydrocarbyl dithiophosphoric acid having from 4 to 20 carbons in each hydrocarbyl group; (c) a zinc alkyl aryl sulfonate; or (d) mixtures thereof; and (3) from 0.1 to 5 weight percent of an oil-soluble antioxidant organic sulfur compound containing from 3 to 40 weight percent sulfur, which sulfur is present as organic sulfide or polysulfide.

Alkali Metal Triborate

J.L. Dreher and R.E. Crocker; U.S. Patent 4,100,081; July 11, 1978; assigned to Chevron Research Company found that excellent greases possessing outstanding extreme-pressure properties comprising a major portion of an oil of lubricating viscosity, a minor portion, sufficient to thicken the composition to grease consistency, of a polyurea grease thickener, and a minor portion of an alkali metal triborate introduced into the grease in aqueous solution. In a preferred case, the triborate is formed by reacting in the grease from about 1 to 10 mols of solid boric acid to one mol of alkali metal hydroxide in aqueous solution. Alternatively, the triborate is formed by reacting the base and the boric acid in aqueous solution and adding the product to the grease. In each case, water is substantially removed from the grease by heating. The preferred molar ratio of hydroxide to boric acid is about 1:3.

Example 1: A polyurea grease was prepared by reacting in a solvent refined West Coast oil, oleylamine, tolylene diisocyanate, and ethylenediamine. To 1,908 grams of the grease was added 132.5 grams of powdered H_3BO_3. The grease was stirred 30 minutes at 180°F. 85 grams of 50% aqueous KOH was added. The grease was stirred at 180°F for 15 minutes. 25 grams of 50% $NaNO_2$ solution was added and the temperature was raised to 300°F. The grease was cooled to room temperature, and 767.75 grams of oil was added. The grease was milled in a three-roll mill. 150 grams of oil was added, and after two additional millings, the grease had an unworked penetration of 231 and a worked penetration (P_{60}) of 314.

Example 2: The procedure of Example 1 was followed, with the exception that the boric acid and KOH were prereacted in aqueous solution before addition to the grease. The molar ratio of acid to base was 1 to 6.3.

The greases of Examples 1 and 2 were subjected to a Timken test to determine the maximum passing load. The test procedure is set forth in ASTM D-2509. The results are set forth in the table below.

Grease	Timken, Max. OK Load (lb)	ASTM D-3336 High-Speed Bearing Life, (hr at 350°F)	P_{60}	Polyurea Content, (wt %)	Borate Content, (wt %)	Irritation Score After 2 hr*
Example 1	55	325	314	6.6	6.6	3.3
Example 2	45	439	339	7.2	7.2	–
Commercial polyurea grease	<20	452	320	8.1	0	2.7
Commercial polyurea-acetate EP grease	50-60	294	320	4.0	0	2.7

(continued)

	Timken, Max. OK Load (lb)	ASTM D-3336 High-Speed Bearing Life, (hr, at 350°F)	P_{60}	Polyurea Content, (wt %)	Borate Content, (wt %)	Irritation Score After 2 hr*
Grease						
NaBO$_2$ Grease	55	—	308	6.5	6.3	6.3

*Irritation scores by the method of J.H. Draize, G. Woodward, and H.O. Calvery. J. Pharmacol. Exptl. Therap., 82, 377-390 (1944).

The greases of Examples 1 and 2, a commercial polyurea grease and a commercial extreme pressure grease (polyurea-acetate) were subjected to a Timken test (ASTM D-2509) and a High Speed Bearing Life test (ASTM D-3336). The results are shown in the above table. ASTM worked penetration (P_{60}) and the polyurea and borate contents of the greases are reported.

These data show that the greases containing the triborates have EP properties comparable to the commercial EP grease, and have higher high speed bearing lives.

The grease of Example 1, and one having the same base, but containing sodium metaborate as an EP additive were tested for skin irritation on rabbits. The grease containing the triborate produced only slight irritation; that with metaborate, severe irritation (see table). Skin staining was obtained with humans who contacted the latter grease.

Phosphorus and Sulfur Compounds

G.A. Clarke and G.L. Harting; U.S. Patent 4,107,058; August 15, 1978; assigned to Exxon Research & Engineering Company describe a grease composition having improved extreme pressure properties and comprising an additive package of an insoluble phosphorus compound and an oil soluble sulfur compound. The particularly useful insoluble phosphorus compounds include the alkali metal or alkaline earth metal salts of a phosphorus acid while the useful soluble sulfur compounds include sulfurized hydrocarbons and organometallic sulfur salts.

Example 1: 4.4 parts of 12-hydroxystearic acid triglyceride and 4.4 parts of tallow fatty acids were added to 66.3 parts of a naphthenic base oil stock having a viscosity of about 500 SUS at 100°F (18°C) and then neutralized with 1.4 parts of lithium hydroxide. The resulting formulation was then heated to a temperature of 320°F (152°C) for 5 minutes and 20.0 parts of a dewaxed paraffinic base oil having a viscosity of about 210 SUS at 210°F (99°C) was added. The resulting grease was then cooled to room temperature and milled.

3.0 parts of tricalcium phosphate and 1.0 part of a sulfurized polybutylene (Anglamol 32, Lubrizol Corp.) were added. The sulfurized polybutylene contains about 45 wt % S. When the formulation was complete, the same was subjected to a Timken test (ASTM D 2509) first at 40 lb load and then 50 lb load. The Timken test is a test wherein a ring is rotated against a fixed block at a specified loading. The test is performed at room temperature for a period of 10 minutes. Following the test, the test block is examined and if no evidence of scoring is observed, the grease is considered to "pass" at that loading stage. The grease prepared in this example containing tricalcium phosphate and a sulfurized polybutylene in a 3:1 weight ratio passed the Timken test at both the 40 and 50 lb loads.

Example 2: In this example, and for purposes of comparison, a formulation identical to that prepared in Example 1 was prepared except that the sulfurized polybutylene was not used. The resulting grease failed the Timken test at both the 40 and 50 lb loads.

Example 3: In this example, and again for purposes of comparison, a composition identical to that of Example 1 was prepared except that the tricalcium phosphate was omitted. The resulting composition passed the Timken test at the 40 lb load, but failed at the 50 lb load.

Example 4: In this example, a composition identical to that of Example 1 was prepared except that an equal amount of a zinc dialkyl dithiophosphate was substituted for the polybutylene. The dialkyl dithiophosphate had a sulfur content of 18.2%, a phosphorus content of 9.0% and a zinc content of 10.4%. The resulting composition passed the Timken test at 40 lb but failed at 50 lb.

OTHER ADDITIVES AND GREASE COMPOSITIONS

Amine Salt of Half Ester of Substituted Succinic Acid as Antifatigue Additive

W.R. Murphy and C.N. Rowe; U.S. Patent 4,100,083; July 11, 1978; assigned to Mobil Oil Corporation describe a lubricant composition comprising a lubricant and an antifatigue amount of an amine salt of a half ester of a succinic acid having the formula:

$$\begin{bmatrix} R-CHCOOR' \\ | \\ CH_2COOH\cdot \end{bmatrix}_m A$$

where m is 1 or 2, R is an alkyl (i.e., where the alkenyl group may be hydrogenated) or alkenyl group containing from 4 to 100 carbon atoms, preferably from 4 to 30 carbon atoms, R' is an alkyl group containing from 1 to 40 carbon atoms, preferably from 1 to 20 carbon atoms, and A is (1) a primary, secondary or tert-hydrocarbyl monoamine containing from 1 to 20 carbon atoms or (2) a polyamine of the formula: $NH_2CH_2CH_2(R''NH)_xNH_2$, where x is 0 to 10 and R'' is an alkylene group containing 1 to 10 carbon atoms.

Of particular significance, is the ability to counteract the accelerating effect of water on metal fatigue achieved by employing the amine salts of a half ester of an alkenyl substituted succinic acid.

Example 1: Preparation of the Half Ester — 1 mol of n-octenylsuccinic anhydride and 1 mol of methyl alcohol were placed in a suitable vessel and heated at reflux for 20 minutes while stirring.

Preparation of the Amine Salt — Into a 250 ml beaker were placed 60.9 grams (0.25 mol) of monomethyl n-octenylsuccinic acid with a magnetic stirring bar. Stirring was started and 46.5 grams (0.25 mol) of tributylamine were added slowly over a period of 10 minutes. Stirring was continued for 1 hour to give a quantitative yield (107.4 grams) of the tributylamine salt of monomethyl n-octenylsuccinic acid. The structure of the product was confirmed by the infrared spectrum.

Example 2: By the method of Example 1, a mono-sec-butyl tetrapropylsuccinic acid was prepared, and 127 grams (0.369 mol) thereof was allowed to react with 68.5 grams (0.369 mol) of tributylamine. The yield of the tributylamine salt of mono-sec-butyl tetrapropenylsuccinic acid was quantitative (195.5 grams). Infrared spectrum confirmed the structure.

The amine salts of·half esters of alkyl or alkenyl substituted succinic acids of the foregoing Examples 1 and 2 were next tested for the inhibition of water-induced metal fatigue in oil and grease lubricants employing a rotating beam fatigue tester. For this test, midly notched SAE 52100 steel specimens 18 inches long and 0.25 inch notched diameter were employed. The individual specimens were completely immersed in the test lubricants which were maintained at a temperature of 120°F.

During testing, the specimens were stressed by hanging weights and rotated at 6,000 rpm. Each lubricant was tested over a range of stress for which a range of characteristic fatigue lives were obtained. In the table below there is shown examples of the beneficial effects of the amine salts in a water-contaminated mineral oil. This mineral oil comprised a straight SAE 20 oil containing only an antioxidant and approximately 0.009%, by weight, of water.

Lubricant Formulation	Fatigue Life* (cycles)
Mineral oil	22×10^4
Mineral oil + 0.05 wt % H_2O	7.5×10^4
Mineral oil + 0.05 wt % H_2O + 0.10 wt % Ex. 1	22×10^4
Mineral oil + 0.05 wt % H_2O + 0.10 wt % Ex. 2	14×10^4
Mineral oil + 0.10 wt % Ex. 2	23×10^4

*at 110,000 psi nominal stress

As will be apparent from the comparative data of the above table, the amine salts of the process are markedly effective in counteracting the accelerating effect of water on metal fatigue in liquid lubricating compositions.

Acylated Polyamides as Rust Inhibitors

A process described by *G.P. Caruso; U.S. Patent 4,104,177; August 1, 1978; assigned to Shell Oil Company* relates to polyurea thickened grease compositions having both improved ambient temperature mechanical stability and enhanced resistance against rust formation. These compositions comprise or consist essentially of a major amount of a lubricating oil base vehicle, a polyurea gellant in a amount sufficient to thicken the base vehicle to a grease consistency and a minor amount of a specified acylated alkylene polyamine or a mixture of specified acylated alkylene polyamides.

Example 1: A polyurea thickened base grease was prepared from the following components.

Components	%
Toluene diisocyanate	4.10
Tallow amine (Armeen T)	7.06
Ethylene diamine	0.85
500 HVI oil blend	86.99
Diisooctyl diphenylamine (Vanalube 81)	0.50
Phenyl-naphthylamine	0.50

The above components were combined in the manner described in U.S. Patent 3,242,210 and, except where indicated, the gel formation reaction was allowed to go to completion before the inclusion of the acylated polyamine additive. The alkylene amines and carboxylic acids of the class described were combined in separate containers and heated at 190°C until water no longer evolved from the given reaction mixture. The particular acylated alkylene amine was then added to the gelled grease, and the mixture was stirred and heated to 88°C until the additive was uniformly incorporated.

The corrosion resistance of the acylated polyamide containing greases was then tested using the modified ASTM D1743 corrosion test, described as Test B in U.S. Patent 3,660,288. For this test, results are indicated by a rating, ranging from 1 to 3, reported for the compositions' effect with three bearings. A grade of 1 indicates no corrosion, while a rating of 2 indicates incipient corrosion with no more than three spots of size just sufficient to be visible to the naked eye. A bearing with larger or more than three spots is rated 3. In this test, in all instances, the polyamide additive comprised 4% by wt of the grease tested. Results are shown below.

Additive-Amine/Carboxylic Acid	Molar Proportion	ASTM Corrosion
A. Tallow-propylene diamine/oleic acid	1:2	1,1,1
B. Tallow-propylene diamine/naphthenic acid	1:2	1,1,1
C. Tallow-propylene diamine/ricinoleic acid	1:2	1,1,1
D. Triethylene tetramine/oleic acid	1:6	1,1,3
E. Triethylene tetramine/ricinoleic acid	1:6	1,1,1
F. Tallow-propylene diamine/oleic acid*	1:2	1,1,1
G. Tallow-propylene diamine/oleic acid**	1:2	1,1,1
H. Base grease-no polyamide	—	Fail

*amide added before tetraurea reaction complete
**amide added after tetraurea reaction complete

The results of the above table demonstrate the effectiveness of the acylated polyamides of the process as rust inhibitors. The experiments demonstrate that the rust inhibition properties are not effected by the time of addition of the additives relative to the time of gel formation.

Example 2: Polyurea thickened greases of the formulation of Example 1 were tested for penetration to demonstrate the effect the acylated polyamides have upon grease gellant efficiency.

The ASTM Unworked Penetration (P_0) and the ASTM Worked Penetration (P_{60}) after 60 strokes in the ASTM worker, are shown in the following table.

Additive	Tetraurea Gellant % by wt	P_O	P_{60}
2A Base grease	12	291	297
2B Tallow-1,3-propylene diamine dioleate, 4% by wt	12	246	268
2C Tallow-1,3-propylene diamine dioleate, 4% by wt*	10	300	345
2D Tallow-1,3-propylene diamine dioleate, 4% by wt**	9.5	232	255

*polyamide added before grease structure formed
**polyamide added after grease structure formed

Experiments 2A and 2B clearly demonstrate that less gellant is necessary to attain a given penetration with the acylated polyamide compositions of the process. Experiments 2C and 2D demonstrate that in order to attain the greatest gellant efficiency, the polyamides should be added after grease formation is complete. The grease compositions containing the additives and less gellant were visibly clearer, and lacked the cloudiness of the base grease.

Lithium Base Grease Containing Polyisobutylene for Water Resistance

According to a process described by *W.W. Bailey and B.W. Taylor; U.S. Patent 4,110,233; August 29, 1978; assigned to Gulf Research & Development Company* a lubricating grease composition specially suitable for the lubrication of heavy steel processing equipment and having improved resistance to removal by water, improved pumpability over a wide range of ambient temperatures and the ability to float on water comprises a petroleum base oil of lubricating viscosity, a lithium soap of a fatty acid, and a polyisobutylene in an amount of from 5 to 12.5 % by wt of the lubricating grease composition and a molecular weight of from 75,000 to 125,000 to confer the improved resistance to removal by water, pumpability and ability to float on the composition.

Preparation of Clay-Based Compositions

K.A. MacKenzie and A. Verhoeff; U.S. Patent 4,122,022; October 24, 1978; assigned to Shell Internationale Research Maatschap IIJ BV, Netherlands describe a process for preparing clay-based grease compositions which show an improved water resistance and mild extreme pressure properties as well as an improved response to certain additives.

It has been found that greases, based on cationically coated clay as thickener, although showing no dropping point and good pumpability, have a poor response to certain conventional grease additives, such as extreme pressure additives, anticorrosion additives and antioxidants and, furthermore, can be improved as to their water resistance and their response to low-shear stirring.

According to this process these problems can be solved if an epoxide is reacted with the clay surface bound cationic oleophilic coating agent under certain conditions at a certain point in the method of preparing the grease. This process, therefore, relates to a method for preparing a grease composition which comprises:

    (a)   forming a clay hydrogel of clay of sufficient ion exchange capacity and water;

(b) intimately mixing a conjugate acid surfactant formed from an acid and an organic amine compound;

(c) intimately mixing with the mixture formed in (b) a major proportion of lubricating oil whereby a water phase and a pregrease phase comprising curds of oil, clay, surfactant and minor amounts of water are formed;

(d) separating the water phase from the wet pregrease phase;

(e) adding to the pregrease a minor proportion of an epoxide before or after dehydration;

(f) reacting the epoxide with the unoccupied amine groups of the amine; and

(g) subjecting the resulting pregrease to a shearing action sufficient to form a grease structure.

Stated in terms of the improvement over the art, this process is an improved method for preparing a clay-bearing grease composition whereby an epoxide is added to a dewatered clay-bearing pregrease, which epoxide is cured during further dehydration of the pregrease at a temperature not greater than 250°F. The dehydration/curing step is followed by mixing the pregrease to a shearing action sufficient to form a grease structure.

Example 1: 20.1 kg of a hectorite clay hydrogel containing 2.32 wt % total solids were mixed and reacted in-line with 5.4 kg of a solution containing 5% amide-amine, being the reaction product of tall oil fatty acids (14 to 22 carbon atoms) and polyethylene polyamines, and 0.7% phosphoric acid in water. 3,078 grams of a mineral oil having a viscosity of 75 to 85 SSU at 210°F were then added to the mixture and the combined materials mixed in-line and transferred to a kettle. The material had the appearance of firm curds or pearls from which water freely drained. The separated water phase was drained from the kettle and additional water squeezed out by stirring.

At this point 190 g of a commercial diglycidyl ether of diphenylolpropane were added and the remaining water was removed by means of a vacuum distillation to dryness. After drying and cooling the grease was diluted with additional make-up oil and milled to a clay content of 5 wt %. Part of the make-up mass was 1.5 wt % lead naphthenate and 4.5 wt % sulfurized fatty oil [extreme pressure (EP) additives]. The grease was milled through a homogenizer to a final penetration of approximately 300. The composition and properties of this grease coded PP-185 are given in the tables.

Example 2: Another batch was made in the same pilot plant following the same procedure; however, at the end of the drain 0.6 wt % of sodium nitrite (basis final wt) was added along with the polyepoxide. Also, the lead naphthenate and sulfurized fatty oil were omitted from the make-up oil resulting in a nonEP epoxy resin grease marked PP-183 in the tables.

Example 3: Finally, a bath was made omitting both the polyepoxide and the EP package but incorporating sodium nitrite. This batch, marked PP-184, serves for comparison with the nonEP and EP version of the process.

The following tables show the properties of the greases described in Examples 1, 2 and 3.

Formulations in Weight Percent

	PP-184 (Ex. 3)	PP-183 (Ex. 2)	PP-185 (Ex. 1)
Clay	5.1	5.1	5.0
Amide-Amide coating agent	3.0	3.0	3.0
Mineral oil	91.0	89.2	83.8
Sodium nitrite	0.6	0.6	–
Lead naphthenate	–	–	1.5
Sulfurized fatty oil	–	–	4.5
Water	0.3	0.1	0.2
Polyepoxide	–	2.0	2.0

Test Results

	PP-184	PP-183	PP-185
ASTM D 217 penetration			
unworked	302	298	308
60X (X = strokes)	308	300	308
100,000X	335	336	336
+ 0.1% water, 60X	295	300	308
+ 10% water, 100,000X	semifluid	410	308
+ 50% water, 100,000X	semifluid	342	358
after wheel bearing test, 275°F, 60X	470	295	302
ORC Dynamic corrosion test*			
No. of cycles pass	0	0	3
ASTM D942 oxidation			
psi drop in 100 hr	12	13	6
psi drop in 500 hr	30	23	14
ASTM D2509 Timken EP test			
OK load, lb	< 20	30	65
ASTM D2265 dropping point, °C	none	none	none
ASTM D1264 water washout, 175°F, wt %	10	4	3
Bethlehem Steel Co., water spray resistance test, LT-20, wt % washed off	97	66	58
ASTM D1263 wheel bearing test, 275°F, grams bleed	5	1	1
ORC high temperature wheel bearing test**, hr to failure	20	135	212
US steel mobility test g/sec at 77°F	10	6	6
Fafnir fretting test***, mg loss	35	22	21
ASTM D2266 Four Ball Wear scar diameter, mm	0.6	0.8	0.5
ASTM D2596 Four Ball EP Test			
4 ball weld, kg	126	160	250
last nonseizure load, kg	80	100	100
load wear index	33	41	45
General Motors low temperature torque test, GM9078-P, at –40°F, in-lb			
starting	112	158	117
running	68	90	79

*Using ASTM D1263 Wheel Bearing Tester, heat to 160°F, add 55 ml of 25% Synthetic Sea Water to hub, cool and run for 6 hr. Then 18 hr cold rest. Any evidence of corrosion on greased bearing terminates test.
**This is a modification of GM test 9048P. Constant Axial load of 50 lb, temperature 300°F, rpm 1200.
***Spring load 550 psi, rpm 1600, Test Duration 22 hr.

Biodegradable Grease

A process described by *H.D. Grasshoff; U.S. Patent 4,115,282; September 19, 1978; assigned to Deutsche Texaco Aktiengesellschaft, Germany* provides a biodegradable grease for rails, rail points and crossings and rail vehicles. The grease composition comprises a biodegradable vegetable or animal oil or fat in an aqueous mixture containing a vegetable-derived thickening agent or gel-forming agent.

In a preferred composition, the lubricant according to the process consists of: 0.5 to 5% by wt of alginate or cellulose derived thickening agent; 2 to 10% by wt of natural oil and/or fat; 10 to 30% by wt of fatty alcohol (with at least 18 carbon atoms); 20 to 30% by wt of glycerol or trihydric alcohol; 0.25 to 15% by wt of graphite or molybdenum disulfide; 2 to 5% by wt of sodium tetraborate; 0.5 to 3% by wt of emulsifying agent; and 20 to 60% by wt of water.

Examples of the composition and properties of lubricants are given in the following Tables 1 and 2. Table 1 gives six basic formulations (1 through 6). They contain the component combinations essential according to the process. Comparison Example 7 gives a calcium soap lubricating grease corresponding to the German Federal Railways Specification 007.02.

The lubricant of Example 1 contains only rapeseed oil and is free of solid lubricants, which are usually used or added for increasing the load carrying capacity of greases and the like. The load capacity of 500 kp, measured by the Almen-Wieland method (AW) is correspondingly low. However, in comparison with the mineral oil containing calcium soap grease (comparison Example 7) the fat is clearly better.

By adding 10% by wt of graphite (Example 2) there is a very marked improvement in the pressure of load carrying capacity compared with Example 1. The same is true when sodium tetraborate (Examples 3 and 4) is added, and the addition of molybdenum disulfide has been found to be outstandingly effective, as can be seen from Examples 5 and 6. On adding 1.5% of MoS$_2$ (Example 6) the result (as regards the AW test) of Example 2 with an addition of 10% of graphite is greatly exceeded.

The Examples 8 through 16 of Table 2 are formulations which in addition to other thickening agents contain amounts of higher molecular weight alcohols and glycerol. Formulations can be regarded as having optimum properties if their pressure absorption or load carrying capacity, as measured by the AW test, is about 2,000 kp, or above.

The addition of molybdenum disulfide has proved to be especially advantageous, and a load-bearing capacity of 2,200 kp for the lubricating film can be achieved with an amount of only 0.5% (Example 16).

Table 1

Ingredients	1	2	3	4	5	6	7*
	. . . . Percent by Weight						
Ammonium alginate	2	2	2	2	2	2	—
Water	95	85	92	90	92	93.5	—
Rapeseed oil	2	2	2	2	2	2	—
Emulsifier**	1	1	1	1	1	1	—
Graphite***		10					—
Molybdenum disulfide†					3	1.5	—
Sodium tetraborate			3	5			—
Properties (DIN 57 804)							
Rest penetration	420	413	433	418	421	420	430

(continued)

	Examples						
	1	2	3	4	5	6	7*
Fulling penetration	445	441	450	443	435	435	450
Almen-Wieland test, kp	500	1500	1250	1650	>2500	2000	250

*Mineral oil-containing calcium soap grease (specification 077.02 of the German Federal Railways).
**Polyoxyethylene sorbitol monooleate.
***Synthetic, average grain size: <30 μm.
†Average grain size: <0.3 μm.

Table 2

	Examples								
	8	9	10	11	12	13	14	15	16
	. . . . Percent by Weight								
Ingredients									
Sodium alginate			1.5	1.5	1.5	1.5	1.5	5*	1.5*
Methylcellulose	1.0	1.0							
Water	67.5	67.5	55.5	36.5	26.5	45.0	55.0	10	65
Alcohols (>C₁₈)						10.0	30.0	50	
Glycerol	20.0	20.0	30.0	30.0	30.0	30.0			30
Rapeseed oil	10.0		2.0			2.0	2.5	25	2.0
Beef Tallow		10.0		30.0	30.0				
Emulsifier	1.5**	1.5**	1.0	2.0***	2.0***	1.5	1.0		1.0
Graphite			10.0		10.0	10.0	10.0	10	
Molybdenum disulfide									0.5
Properties (DIN 51 804)									
Rest penetration	402	360	358	295	293	407	384	271	433
Fulling penetration	423	391	403	333	338	391	386	282	458
Almen-Wieland Test, kp	1900	1800	1600	2500	2500	2200	2100	2500	2200

*Ammonium alginate.
**Potassium oleate.
***Sorbitol monooleate.

Aluminum Fatty Acid Soap Thickened Polyisobutylene Composition

G.C. Van Doorne; U.S. Patents 4,075,113; February 21, 1978; and 4,075,112; February 21, 1978; both assigned to Labofina SA, Belgium describes a lubricating grease comprised of 2 to 8% by wt of an aluminum fatty acid soap, 25 to 98% by wt of a polymer of a nonhydrogenated polyisobutylene having a mean molecular weight ranging from 300 to 2,500, 0.2% polyisobutylene having a mean molecular weight higher than 100,000, and 2 to 58% of lubricating oil.

The following examples illustrate the process and reference is made to the following tests: ASTM D 217-52T or penetration test, which is a measure of consistency and mechanical stability of the grease; and ASTM 2266 for testing the wear preventive characteristics of grease (four balls method, at 1,800 rpm, for one hour at 75°C and 40 kg/cm²).

Example 1: 5 pbw of aluminum stearate and 42 pbw of nonhydrogenated polyisobutylene having a mean molecular weight of 460 were blended at room temperature. This mixture was heated as quickly as possible with stirring. When the temperature reached 120°C, 20 pbw of naphthenic oil with a viscosity index of 70 were added. This mixture was further heated and when the temperature reached 170° to 180°C, 32.8 pbw of nonhydrogenated polyisobutylene having a mean molecular weight of 730 and 0.2 pbw of polyisobutylene having a mean molecular weight higher than 100,000 were progressively added. The temperature was kept at about 175°C during this addition with stirring. The hot grease was then withdrawn into a cooling vessel. The cooled grease was poured into a mixer and passed through a homogenizer. The grease was smooth, stringy, homogeneous, slightly colored but transparent. The hardness was 285 (in tenths of mm at 25°C, test ASTM 217-T) for the unworked grease and 297 for the 60 strokes worked grease. The wear index (ASTM 2266) was 0.72.

By way of comparison, a grease which was free from low molecular weight polyisobutylene required a higher amount of aluminum soap (8% by wt instead of 5%) to obtain the same hardness, and a higher quantity of adhesiveness improver (1% by wt instead of 0.2%). On the other hand, the wear index was 1.07 for this grease containing 8% by wt of aluminum soap, 1% of polyisobutylene with a molecular weight higher than 100,000 and 91% of mineral oil.

Thus, it is apparent that by using a polyisobutylene with a molecular weight lower than 2,500, the grease requires lower amounts of thickening agents and of additives and has better lubricating and antiwear properties.

Example 2: A grease was prepared from the following: 46.8% by wt of nonhydrogenated polyisobutylene with a mean molecular weight of 460; 48.0% by wt of nonhydrogenated polyisobutylene with a mean molecular weight of 730; 5.0% by wt of aluminum stearate; and 0.2% by wt of adhesiveness improver. This grease was smooth, homogeneous, colorless and stringy, with a penetration index of 289 and a wear index of 0.63.

By way of comparison, a grease which was also free from mineral oil but containing lithium soap instead of aluminum soap was prepared. Even with an amount of lithium soap as high as 12% by wt, this grease was less hard, the penetration being 319.

Example 3: The process of Example 1 was repeated using 39.8 pbw of polyisobutylene, nonhydrogenated with an average molecular weight of 447 and 35.0 pbw of polyisobutylene nonhydrogenated having an average molecular weight of 633. After the grease was heated to 120°C, 3% of the same naphthenic oil was added.

By way of comparison, the grease of Example 2 decomposed on heating to 550°C leaving as residue only molten aluminum stearate, the polybutene decomposing as gases. The grease of Example 3, however, also decomposed, but left a liquid residue of the mineral oil and soap still capable of functioning to a small degree as a lubricant for surfaces so highly heated.

Brake Grease Composition with Polyglycol Base

A process described by *H. Kinoshita, M. Sekiya and N. Makino; U.S. Patent 4,115,284; September 19, 1978; assigned to Nippon Oil Co., Ltd., Japan* relates to a heat resistant brake grease composition useful for the lubrication and dust sealing of sliding contact surfaces and the like in automobile braking systems.

The grease composition comprises: (A) a polyoxypropylene glycol monoether as the lubricating base oil, having a pour point of not higher than –20°C, a flash point of not lower than 200°C and a viscosity of not lower than 8 cs at 98.9°C; and (B) a gelling agent as the thickening agent, consisting of at least one diurea compound represented by the general formula

$$\underset{R_2NHCNHR_1NHCNHR_3}{\overset{\displaystyle O \qquad\quad O}{\overset{\displaystyle \|\qquad\quad \|}{}}}$$

where R_1 is a divalent aromatic hydrocarbon radical, and R_2 and R_3 are each a cyclohexyl group, a cyclohexyl derivative group having 7 to 12 carbon atoms or an alkyl group having 8 to 20 carbon atoms. It is desired that the gelling agent has a cyclohexyl or cyclohexyl derivative group content of at least 20% as calculated from the formula

$$\frac{\text{No. of cyclohexyl or cyclohexyl derivative groups}}{\substack{\text{No. of cyclohexyl or cyclohexyl derivative groups} + \\ \text{No. of alkyl groups}}} \times 100\%$$

Example 1: 410 g of 810 g of a polyoxypropylene glycol monoether having a pour point of –40°C, a flash point of 220°C, a viscosity of 10.6 cs at 98.9°C and an average molecular weight of 1,100, were incorporated with 22.7 g of diphenylmethane-4,4'-diisocyanate to form a mixture which was then heated from 60° to 70°C to be melted uniformly. The mixture so melted was incorporated with a mixture in molten state of 14.6 g of octadecylamine, 12.7 g of cyclohexylamine and the remaining 400 g of the polyoxypropylene glycol monoether and then agitated vigorously thereby producing a gel-like material immediately.

The gel-like material so produced was maintained at 120°C under continuous agitation for 30 min, after which it was incorporated with 10 g of an amine type antioxidant, 30 g of a tackiness improver (polybutene) and 100 g of ZnO and the resulting mixture was well agitated and treated on a 3-roll mill thereby to obtain a desired grease composition. The desired properties are listed in the table below.

Example 2: A polyoxypropylene glycol monoether (7,570 g) having a pour point of –40°C, a flash point of 255°C, a viscosity of 11 cs at 98.9°C and an average molecular weight of 1,150 was incorporated with 725 g of diphenylmethane-4,4'-diisocyanate to form a mixture which was heated from 60° to 70°C to be melted. The mixture so melted was then incorporated with 575 g of cyclohexylamine to produce a gel-like material immediately. The gel-like material so produced was heated to 100°C under vigorous agitation, after which it was incorporated with 130 g of an antioxidant, 500 g of a tackiness improver and 500 g of polytetrafluoroethylene powder to form a mixture which was agitated thoroughly and then milled thereby obtaining a desired grease composition. The desired properties are listed in the table below.

Example 3: A polyoxypropylene glycol monoether (1,420 g) having a pour point of –33°C, a flash point of 222°C, a viscosity of 21 cs at 98.9°C and an average molecular weight of 2,300, was incorporated with 120 g of 2,4/2,6-tolylenediisocyanate to form a mixture which was agitated to be made homogeneous. The homogeneous mixture was incorporated with 139 g of methylcyclohexylamine to produce a gel-like material immediately.

The gel-like material so produced was maintained at 120°C under vigorous agitation for 30 min, after which it was incorporated with 20 g of an antioxidant, agitated thoroughly and milled on a 3-roll mill thereby obtaining a desired grease composition. The desired properties are shown in the table below. In addition, the table also shows the properties of a commercially available brake grease (reference Example 1) prepared by blending a grease consisting of castor oil and lithium stearate, with ZnO and an antioxidant, and those of a Li type grease (reference Example 2) in which a mineral oil type base oil is used. The tests used in the table are described below.

Properties	.... Examples			.. References ..	
	1	2	3	1	2
Dropping point °C	270+	270+	270+	192	198
Consistency					
UW (Unworked penetration)	262	291	292	272	275
60W (Penetration after 60 strokes)	264	282	293	283	276
100,000W (Penetration after 100,000 strokes)	282	299	317	352	348
Oxidation stability (150°C, 100 hr) kg/cm^2	2.5	2.2	2.4	6.0	6.2
Film test (150°C, 100 hr), appearance	*	*	*	**	**
Oil bleeding (150°C, 100 hr) %	2.2	2.9	3.1	29.6	15.2
Water wash resistance					
38°C	1.0	0.8	1.2	3.4	2.4
79°C	3.2	4.0	3.9	17.0	9.6
Rubber swelling test (70°C, 70 hr)					
Chloroprene					
Change in volume %	-8.8	-7.2	-8.4	-9.2	+11.3
Change in weight %	-6.5	-5.6	-6.4	-6.7	+9.5
SBR					
Change in volume %	+4.0	+4.2	+3.9	+1.5	+5.4
Change in weight %	+3.3	+3.0	+2.8	-0.8	+4.8

Note: The properties were measured by the following test methods: dropping point, JIS (Japanese Industrial Standard) K2561; consistency (expressed by penetration), JIS K 2560.
*Substantially no change.
**Changed in color to brown, run down.

Oxidation Stability – JIS K 2569.

Film Test – Grease to be tested is coated on an iron plate so that the coated grease layer has a diameter of 45 mm and a thickness of 3 mm. The iron plate so coated is placed in a thermostatic chamber at 150°C for 100 hr and then the appearance of the grease layer is observed.

Oil Bleeding – JIS K 2570.

Water Wash Resistance – JIS K 2572.

Rubber Swelling Test – Rubber test samples are wholly immersed in grease at 70°C for 70 hr, after which they are measured for change in volume and change in weight "+" meaning increase and "-" meaning decrease.

The above table shows that the grease compositions of this process are remarkably excellent in heat resistance and water resistance as compared with the conventional ones.

The antioxidant used in Examples 1 through 3 was phenyl-α-naphthylamine and the tackiness improver was polybutene having an average molecular weight of 2,350 in Example 1 and polyisobutylene having an average molecular weight of 10,000 in Example 2.

Perfluorinated Polyether Greases

A process described by *J.B. Christian and C. Tamborski; U.S. Patent 4,132,660; January 2, 1979; assigned to the U.S. Secretary of the Air Force* resides in the discovery that the addition of a small amount of certain fluorine-containing benzoxazoles to a fluorinated polyether base fluid and a thickener therefor provides a grease having outstanding properties. Thus, the resulting grease composition inhibits rust formation when utilized as a lubricant for ferrous metals under mild temperature and high humidity conditions.

More specifically, the grease composition consists essentially of 65 to 72 wt % base fluid, 26.5 to 34.5 wt % thickener, and 0.5 to 1.5 wt % fluorine-containing benzoxazole, based upon a total of 100 wt %.

Example 1: A series of runs were conducted in which grease compositions of this process were formulated and tested. As a base fluid there was used a perfluorinated polyalkylether having the following formula:

$$C_3F_7O(\underset{\underset{CF_3}{|}}{CF}CF_2O)_nC_2F_5$$

where n is an integer having a value such that the fluid has a kinematic viscosity of 270 cs at 100°F. The base fluid was Krytox 143AC fluid. The thickener employed was a fluorinated copolymer of ethylene and propylene having a molecular weight of about 150,000.

The benzoxazole additives used in the formulations had the following structural formula:

in which R_f and Y were as indicated below in Table 1.

Table 1

R_f	Y
A	F
B	F
C	E
C	F
C	B
D	F
D	E
D	B
E	E
B	B

In the above table:

$$A = C_8F_{17}$$

$$B = C_3F_7O(\underset{\underset{CF_3}{|}}{CF}CF_2O)_4\underset{\underset{CF_3}{|}}{CF}$$

$$C = (CF_2)_8-C \overset{\displaystyle O}{\underset{\displaystyle N}{\diagdown}}$$

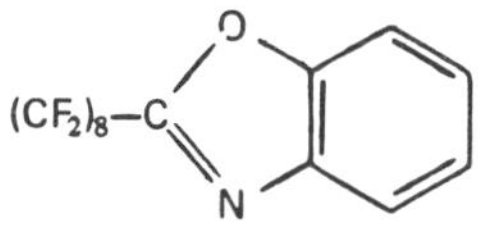

$$D = \underset{\displaystyle \underset{CF_3}{|}}{CFOCF_2}\underset{\displaystyle \underset{CF_3}{|}}{CFOCF_2}CF_2OCFCF_2OCF-C \overset{\displaystyle O}{\underset{\displaystyle N}{\diagdown}}$$

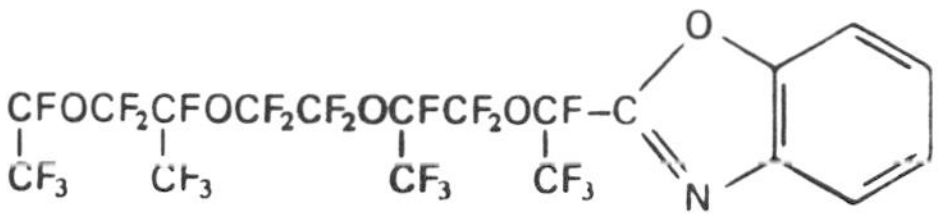

$$E = C_3F_7$$

$$F = H$$

In preparing the greases, the components were mixed and stirred until a uniform mixture was obtained. The amounts of base fluid used ranged from 65 to 72 wt % while the amounts of thickener used ranged from 27 to 34 wt %. Each grease composition contained 1.0 wt % of the abovelisted fluorine-containing benzoxazole additives. Each mixture was further blended to a grease consistency by passing it two times through a 3-roll mill with the rollers set at an opening of 0.002" at about 77°F.

The various grease compositions were tested according to several standard test procedures. The penetration test was conducted in accordance with Federal Test Method Standard 791a, method 313.2. The rust preventive properties test was carried out in accordance with method 4012 of the same standard. The high temperature corrosion was determined in accordance with the method set forth in Technical Documentary Report AFML-TR-69-290. The results of the tests are set forth in Table 2.

Example 2: A series of runs was conducted in which greases were prepared, utilizing, as described in Example 1, the same thickener and benzoxazole additives and amounts thereof as well as the same amounts of a perfluorinated polyalkylether base fluid. However, the perfluorinated polyalkylether had the following structural formula:

$$XO(C_3F_6O)_P(CF_2O)_Q(C_2F_4O)_R Y$$

where X and Y are CF_3, C_2F_5, or C_3F_7, and P, Q, and R are integers such that the fluid has a kinematic viscosity of about 90 cs at 100°F. The base fluid used was Fomblin Y fluid, (Montedison, SpA).

The greases were formulated and tested according to the procedures described in Example 1. The results of the tests are also shown below in Table 2.

Example 3: A series of runs was carried out in which greases were prepared utilizing, as described in Example 1, the same base fluid and benzoxazole additives and amounts thereof as well as the same amount of a thickener. However, the thickener used was polytetrafluoroethylene having a molecular weight of about 30,000.

The greases were formulated and tested according to the procedures described in Example 1. The results of the tests are also set forth in Table 2.

Example 4: A series of runs was carried out in which greases were prepared, utilizing as described in Example 2 the same base fluid and benzoxazole additives and amounts thereof as well as the same amount of thickener. However, the thickener used was polytetrafluoroethylene having a molecular weight of about 30,000.

The greases were formulated and tested according to the procedures described in Example 1. The results of the test are shown in Table 2.

Example 5: Control runs were conducted in which greases were prepared, utilizing the base fluids and thickener of Examples 1 and 2. The greases consisted of 71 wt % base fluid and 29 wt % thickener and did not contain any of the benzoxazole thickener.

The greases were formulated and tested according to the procedure described in Example 1. The results of the tests are included below in Table 2.

Table 2

	Greases of Example 1	Greases of Example 2	Greases of Example 3	Greases of Example 4	Grease Based on 270 cs Fluid, No Additive*	Grease Based on 90 cs Fluid, No Additive*
Penetration, dmm**	301–307	302–307	311–317	312–317	298–300	310–310
Rust preventive properties***	Pass†	Pass	Pass	Pass	Fail††	Fail
High temperature corrosion, 450°F, 72 hr†††						
52-100 93 100 steel	Pass	Pass	Pass	Pass	Fail	Fail
440C steel	Pass	Pass	Pass	Pass	Fail	Fail
M-10 steel	Pass	Pass	Pass	Pass	Fail	Fail
M-50 steel	Pass	Pass	Pass	Pass	Fail	Fail

*Control runs.
**Penetration values of the various greases formulated in the examples, using FTMS 791a, Method 313.2 Pass.
***FTMS 791a, Method 4012.
†Pass—No rusting or corrosion, a maximum of 3 spots allowed.
††Fail—More than 3 rust or corroded spots or pitting and etching.
†††AFML-TR-69-290.

The data in the foregoing table demonstrate that the grease compositions of this process do not cause ferrous metals to rust under mild temperature and high humidity conditions or to corrode under conditions of high temperature. The antirust and anticorrosion properties of the greases are directly attributable to the fluorine-containing benzoxazole additives. Thus, when the additive was omitted as in the control runs, rusting and corrosion of the ferrous metals occurred as a result of contact with greases based on perfluorinated polyalkylether fluids.

SPECIALTY LUBRICANTS
AND RELATED PROCESSES

Specialty lubricants have been developed over the years for a wide variety of end uses in both the consumer and industrial applications market. This chapter provides a large number of formulations which can be adapted for commercial use.

These formulations represent some of the latest lubrication technology as well as a number of other processes for the upgrading of hydrocarbon oils for specific applications.

FIBER AND TEXTILE LUBRICANTS

Oxidation Stable Polyoxyalkylene Compounds

D.D. Newkirk, R.B. Login and B. Thir; U.S. Patent 4,110,227; August 29, 1978; assigned to BASF Wyandotte Corporation describe lubricants for synthetic fibers such as polyester and nylon which are improved oxidation stable polyoxyalkylene lubricant compounds. Such lubricant compounds can be used alone as textile fiber lubricants or in combination with other polyoxyalkylene compounds useful as fiber lubricants but which are subject to oxidative degradation under conditions of heating at temperatures at least above 200°C. An improved textile fiber is produced by the use of such fiber lubricants in conventional processes for producing continuous filament, false twist, textured yarn as well as other type yarns.

The oxidation stable homopolymer and copolymer, i.e., block or heteric polyoxyalkylene lubricant compounds have the formulas:

(1)

$$R-\overset{\overset{\text{O}}{\|}}{C}-(O-R_1-O-R_2)_m-R_3-(O-R_1-O-R_2)_n-OH$$

(2)

$$R-\overset{\overset{\text{O}}{\|}}{C}-(O-R_1-O-R_2)_m-R_3-(O-R_1-O-R_2)_n-\overset{\overset{\text{O}}{\|}}{C}-R$$

(3)

$$R-\overset{\overset{\displaystyle O}{\|}}{C}-(O-R_1-O-R_2-O-R_5)_n-R_4$$

and can be used alone or in mixtures or in a mixture of an effective proportion thereof at least 25 wt % based on the total weight of the mixture with one or more of any prior art polyoxyalkylene fiber lubricant compounds but preferably with one or more polyoxyalkylene compounds selected from the group consisting of prior art compounds having the formulas:

(4)

$$R-\overset{\overset{\displaystyle O}{\|}}{C}-(O-R_1)_n-OH$$

(5)

$$R-\overset{\overset{\displaystyle O}{\|}}{C}-(O-R_1-O-R_2)_n-OH$$

(6)

$$R-\overset{\overset{\displaystyle O}{\|}}{C}-(O-R_1-O-R_2)_n-O-R_4$$

R is individually selected from alkyl groups of 1 to about 21 carbon atoms preferably about 7 to about 21 carbon atoms and most preferably about 12 to about 18 carbon atoms.

In lubricants 1, 3, 5 and 6, R_1 and R_2 are the residue of the same or different alkylene units and in lubricant 2, R_1 and R_2 are different alkylene units, all individually selected from the group consisting of the residue of ethylene oxide, propylene oxide, butylene oxide and an aromatic glycidyl ether, R_3 is the residue of a difunctional phenol, R_4 is hydrogen, an acyl or alkyl group and where alkyl, derived from an aliphatic monofunctional alcohol having 1 to about 21 carbon atoms, preferably about 4 to about 18 carbon atoms and most preferably about 4 to about 12 carbon atoms or where acyl derived from an aliphatic monocarboxylic acid of 2 to about 21 carbon atoms, preferably about 4 to about 18 carbon atoms and most preferably about 6 to about 12 carbon atoms, R_5 is the residue of an aromatic glycidyl ether, preferably a phenyl glycidyl ether and n or m + n have a value to produce a molecular weight of about 300 to about 2,000, preferably about 600 to about 2,000 and most preferably about 800 to about 1,800.

Example 1: This example illustrates the preparation of a fiber lubricant of the process which is the stearate ester of a hydroquinone-initiated heteric polymer consisting of 75% ethylene oxide and 25% propylene oxide by weight.

A polyoxyalkylene intermediate was prepared by adding two mols of di(β-hydroxyethyl) hydroquinone to an autoclave equipped with temperature, pressure and vacuum controls. The hydroquinone was melted under a nitrogen atmosphere at a temperature of 105° to 110°C. Thereafter, 10 g of a 90% potassium hydroxide solution were added and the autoclave heated to 125°C after evacuating to a vacuum of less than 10 mm Hg. The vacuum was broken after the removal of 6 g of volatiles and nitrogen was added to give a pressure of 3 to 7 psig. A mixture of 9.9 mols of propylene oxide and 39.2 mols of ethylene oxide was then added over a period of about 6½ hours. The mixture was held an additional 2 hours at 125°C to insure complete reaction and then the polyoxyalkylene intermediate was recovered and found to have a hydroxyl number of 96.2.

The stearate ester of this intermediate was prepared by transesterification. The polyoxyalkylene intermediate was added in the amount of 1.1 mols to a 3-liter flask equipped with a means for stirring, distillation apparatus and temperature control means. The intermediate was heated to 130°C and volatiles removed by vacuum. Methyl stearate in the amount of 0.6 mol was then added and the flask evacuated to less than 10 mm Hg and the temperature held at 130°C for about 80 minutes. Sodium methoxide in the amount of 0.5 g was added as a catalyst and the transesterification reaction was allowed to proceed at a vacuum of less than 10 mm Hg at a temperature of 130°C for an additional 105 minutes. The balance of methyl stearate, 0.5 mol, was then added and the flask was evacuated to remove volatiles.

After 60 minutes additional sodium methoxide in the amount of 0.5 g was added and the reaction continued at a vacuum of less than 10 mm Hg at a temperature of 130°C for an additional 105 minutes. The product obtained was de-ionized and the desired monostearate ester of the ethoxylated propoxylated hydroquinone-initiated polymer was obtained having a hydroxyl number of 42.3 (theoretical 38.1), an acid number of 1.32, a sodium ion concentration of 34.5 parts by weight and a potassium ion concentration of 5.5 ppm by weight.

Example 2: This example illustrates the preparation of the stearate ester of a hydroquinone-initiated heteric propylene oxide-ethylene oxide polymer having a weight ratio respectively of 70 propylene oxide and 30 ethylene oxide. The proportions and procedures of Example 1 were repeated to prepare a fiber lubricant of the process (having a theoretical hydroxyl number of 37.8) except that the deionizing process of Example 1 was eliminated and the crude product was reacted with sufficient acetic acid to neutralize the base catalyst used.

Example 3: The procedure of Example 1 was repeated except that the polyoxyalkylene intermediate has a weight ratio of 70 parts of ethylene oxide and 30 parts propylene oxide. The crude product was deionized to remove the base catalyst. The product had a hydroxyl number of 40.7 (theoretical 37.8) and an acid number of 0.5.

Example 4: This example illustrates the preparation of the laurate ester of a resorcinol-initiated heteric polyoxyalkylene compound having a weight ratio of 70 parts propylene oxide and 30 parts ethylene oxide. The same procedure and proportions are utilized as described in Example 2. The product obtained had a hydroxyl number of 52 (theoretical 38.7) and an acid number of 7.8.

Example 5: Example 4 was repeated except that a transesterification reaction utilizing methyl stearate was performed in order to obtain the stearate ester of a resorcinol-initiated heteric polyoxyalkylene having 70 parts propylene oxide and 30 parts ethylene oxide by weight. The product had a hydroxyl number of 44.5 (theoretical 36.6) and an acid number of 6.4.

Example 6: The procedure and proportions of Example 4 were repeated except that the resorcinol-initiated heteric polyoxyalkylene intermediate had a 75 parts ethylene oxide and 25 parts propylene oxide weight ratio and the base catalyst was removed by a deionization process. The product had a hydroxyl number of 51 (theoretical value of 40.2), an acid number of 1.6, a sodium ion concentration of 114 ppm and a potassium ion concentration of 17 ppm by weight.

Example 7: Following the procedure of Example 1, a fiber lubricant was prepared consisting of the laurate ester of a hydroquinone-initiated heteric polymer consisting of 70 parts of propylene oxide by weight and 30 parts of ethylene oxide by weight. The crude product was deionized to remove the base catalyst giving the desired lubricant. Hydroxyl number was found to be 57.6 (theoretical value 41.4), the acid number was 1.2, the sodium ion concentration was 92 ppm by weight and the potassium ion concentration was 1.2 ppm by weight.

Example 8: In accordance with the procedure of Example 1, a fiber lubricant was prepared consisting of the stearate ester of a resorcinol-initiated heteric polymer containing 75 parts ethylene oxide by weight and 25 parts propylene oxide by weight. The desired lubricant was found to have a hydroxyl number of 47.0 (theoretical 38.3) and an acid number of 0.6.

Example 9: In accordance with the procedure of Example 1, a fiber lubricant consisting of the laurate ester of a hydroquinone-initiated heteric polymer containing 75 parts ethylene oxide by weight and 25 parts propylene oxide by weight was prepared. The crude lubricant product, after deionization to remove the base catalyst, had a hydroxyl number of 47.2 (theoretical value 40.7), an acid number of 1.7, a sodium ion concentration of 99 ppm and a potassium ion concentration of 11 ppm by weight.

Example 10: Comparative Example — For comparative purposes, a fiber lubricant of the prior art forming no part of this process was prepared by adding by weight 75 parts ethylene oxide and 25 parts propylene oxide to stearic acid in accordance with the teaching of U.S. Patent 3,925,588 and British Patent No. 1,460,960. The final product was deionized to remove the base catalyst and obtain a product having a molecular weight of approximately 1,400, a hydroxyl number of 40.7, an acid number of 0.71, a sodium ion concentration of 19.2 ppm by weight and a potassium ion concentration of 3.6 ppm by weight.

Example 11: Comparative Example — A refined coconut oil available commercially under the name "Cobee 76" from PVO International Incorporated was utilized for comparative purposes in the tests following. A typical analysis of the product is as follows: iodine value, 9; saponification value, 255; lauric acid, 48 wt %; and unsaturated fatty acid, 8 wt %.

Example 12: Comparative Example — The fiber lubricant of Example 10, which is not a part of this process, was heat stabilized by the addition of 3% of a commercial antioxidant sold under the trade name "Topanol CA." This material is available commercially from ICI United States Incorporated and is described as a phenol condensation product.

Example 13: Comparative Example — The fiber lubricant of Example 10, forming no part of this process, was stabilized by adding 1.5% of the phenol condensation product sold under the trade name Topanol CA.

Example 14: Comparative Example — The fiber lubricant of Example 10, forming no part of this process, was stabilized by the addition of 1% of a phenol condensation product sold under the trade name Topanol CA.

In order to evaluate the physical properties of the lubricants of the process, the following test methods were utilized. The heat stability of the lubricants was evaluated by thermogravimetric analysis in which a standard quality of fiber lubricant was heated from a temperature of 25°C at the rate of 10°C per minute until 1% weight loss was obtained. The temperature at this point is recorded as the dynamic heat resistance of the lubricant. A second method of evaluating the heat resistance of the lubricants was by heating 60 mg of lubricant at a temperature of 220°C for a period of 30 minutes. The percent weight loss is recorded and is termed the "isothermal heat resistance" of the sample.

Not only weight loss at elevated temperature is important in a fiber lubricant but the remaining lubricant or residue subsequent to volatilization of a portion of the lubricant is of interest. Therefore, residue formation in the lubricants was evaluated by heating 0.2 g of the lubricant for a period of 8 hours at a temperature of 220°C in a circulating air oven. The proportion of residue remaining and the nature of the residue is recorded in this test.

A third test designed to evaluate the heat resistance of the fiber lubricants is the thin film smoke point test. In this procedure, 0.5 g of lubricant is heated at the rate of 10°C per minute until smoke is first observed rising from the surface of the sample. The temperature is recorded as the smoke point.

In order to demonstrate the resistance of the fiber lubricants to discoloration upon exposure to oxides of nitrogen, a nylon fabric was treated with approximately 1% of various lubricants of the process as well as certain prior art lubricant compositions and exposed to oxides of nitrogen in accordance with test method AATCC 75-1956 entitled "Color Fastness to Oxides of Nitrogen in the Atmosphere: Rapid Control Test."

Finally, the coefficient of friction (f) of scoured yarn to metal was determined as follows: The yarn was prepared for testing by applying the lubricant to the yarn on an apparatus made by the Precision Machine and Development Company which is entitled "Atlas Yarn Finish Applicator." In this machine, the yarn is passed at a controlled speed through a continually replenished drop of lubricant dispersed or dissolved in water. The solution is metered to the application area of the machine by an adjustable syringe pump so as to apply about 0.6 wt % lubricant (dry basis) to the yarn which is then led from the feeder globule over an adjustable canter roller which spaces the yarn for passage over the drying drum for removal of water and finally onto a winding tube. Prior to testing for the coefficient of friction, the yarn was conditioned overnight at 65% relative humidity and 70°F.

The coefficient of friction (f) was determined using a Rothschild F-Meter by passing the yarn over a 0.313-inch diameter satin-chrome pin using a contact angle of 180° and a yarn speed such as 100, 200 or 300 meters per minute. Tensiometers measured the yarn tension before and after its passage over the friction pin. The input tension was maintained at a value of 12 g.

In the evaluation of the coefficient of friction of yarn to metal by the Rothschild F-Meter approximately 1,500 meters of yarn are passed over the friction pin to obtain a reported f value. Thus, at 300 meters per minute, the evaluation would be carried out for approximately 5 minutes to obtain the average value of f from the instrument chart. At slower speeds the evaluation was carried out over a proportionally longer time.

Table 1: Heat Stability of Lubricants with 75 EO/25 PO Ratios

| | | Thermogravimetric Analysis | | | Residue | |
Ex.	Variable	Dynamic °C at 1% Weight Loss	Isothermal % Weight Loss at 220°C, ½ Hr	Smoke Point (°C)	After 8 Hours at 220°C (wt %)	Nature of Residue
1	Hydroquinone initiator	230	3.5	204	27.1	Liquid
8	Resorcinol initiator	255	5.9	192	11.6	Varnish
6	Resorcinol initiator	222	7.8	193	30.8	Liquid
9	Hydroquinone initiator	250	4.4	200	45.6	Liquid
8	Stearate ester	255	5.9	192	11.6	Varnish
6	Laurate ester	222	7.8	193	30.8	Liquid
1	Stearate ester	230	3.5	204	27.1	Liquid
9	Laurate ester	250	4.4	200	45.6	Liquid

Table 2: Heat Stability of Lubricants with Various PO/EO Ratios

| | | Thermogravimetric Analysis | | | Residue | |
Ex.	PO/EO Ratio	Dynamic °C at 1% Weight Loss	Isothermal % Weight Loss at 220°C ½ Hr	Smoke Point (°C)	After 8 Hours at 220°C (wt %)	Nature of Residue
8	75 EO/25 PO	255	5.9	192	11.6	Varnish
5	70 PO/30 EO	260	2.5	216	39.0	–
9	75 EO/25 PO	250	4.4	200	45.6	Liquid
7	70 PO/30 EO	238	–	198	20.1	Liquid

Table 3: Heat Stability of Comparative Examples

| | Thermogravimetric Analysis | | | Residue | |
Example	Dynamic °C at 1% Weight Loss	Isothermal % Weight Loss at 220°C ½ Hr	Smoke Point (°C)	After 8 Hours at 220°C (wt %)	Nature of Residue
10	228	31.5	177	0.4	Varnish
11	278	–	198	21.7	Varnish
12	297	–	–	–	–
14	276	–	–	–	–

Table 4: Frictional Properties on Polyester Yarn

Example	Tension Speed, (m/min)	Coefficient of Friction*
**	100	0.66
**	200	0.70
8	100	0.57
8	200	0.61
1	100	0.62
1	200	0.63

*Rothschild F–Meter.
**Control (no lubricant).

Upon evaluating lubricants of the process described in Examples 2, 3, and 6 through 9 for resistance to discoloration upon exposure to oxides of nitrogen, no color was found to develop. Example 10, representative of a lubricant of the prior art having poor heat resistance, also showed no color formation. A similar evaluation of Examples 12, 13 and 14, representing the prior art lubricant of Example 10 with varying amounts of antioxidant, showed brown spots on the fabric.

Silicone and Perfluoropolymer Composition

P. Huber, H. Lampelzammer, E. Pirson and F. Wimmer; U.S. Patent 4,076,672; February 28, 1978; assigned to Wacker-Chemie GmbH, Germany describe a composition containing a mixture of (1) an aqueous emulsion obtained from the emulsion polymerization of a diorganopolysiloxane having an average viscosity of at least 20,000 cs at 25°C and component (2) which is selected from the class consisting of (a) a paraffin wax, and (b) a perfluoropolymer in which some of the fluoro groups may be substituted with chlorine atoms. The composition is applied as a lubricant to organic fibers to improve their slip properties.

Example 1: A mixture containing 4 kg of a hydroxy-terminated dimethylpolysiloxane (viscosity 145 cs at 25°C) and 1 kg of refined paraffin wax (54° to 56°C) was heated to 80°C and, when the wax had melted, was homogenized in a homogenizer under a pressure of 280 kg/cm². The mixture obtained was then homogenized in the same apparatus with 200 g of water, 250 g of a nonionic emulsifier (nonylphenol and ethylene oxide in a molar ratio of 1:23), and a solution of 200 g of dodecylbenzenesulfonic acid in 500 g of water, under a pressure of 280 kg/cm². When a homogeneous mixture had been obtained, it was diluted with 3.8 kg of water, again under a pressure of 280 kg/cm². The resulting dispersion was stored at room temperature for 36 hours, after which its pH was adjusted to 7 with dilute sodium hydroxide solution.

The dispersion was very stable. The dimethylpolysiloxane in the dispersion had a viscosity of 10^5 cs at 25°C. The dispersion contained 50 wt % of dispersed materials. The weight ratio of dimethylpolysiloxane to paraffin was 4:1, and the dispersed particles were from 0.05 to 0.2 micron.

Example 2: About 500 g of a hydroxy-terminated dimethylpolysiloxane (viscosity 120 cs at 25°C), 90 g of dodecylbenzenesulfonic acid, and 100 g of water were mixed in a high-speed mixing apparatus. While stirring, 350 g of a 60 wt % aqueous suspension of polytetrafluoroethylene (molecular weight 10^6, particle size 0.1 to 0.5 micron) were added. When a homogeneous mixture had been obtained, 642 g of water were added, with agitation. The dispersion was then stored for 8 hours at room temperature, after which its pH was adjusted to 7 with 18 g ethanolamine.

The resulting dispersion was very stable and contained fine particles. It contained 50 wt % of dispersed materials, with a weight ratio of dimethylpolysiloxane to polytetrafluoroethylene of 5:2.1. The dimethylpolysiloxane had a viscosity of 10^5 cs at 25°C.

2,2,4-Trimethyl-1,2-Dihydroquinoline Polymers

A process described by *K. Katabe and T. Hirota; U.S. Patent 4,144,178; Mar. 13, 1979; assigned to Kao Soap Co., Ltd., Japan* relates to a composition for the lubricating treatment of synthetic fibers which are to be subjected to a subsequent heating process, which composition has an appropriate lubricating effect and possesses thermal stability at high temperatures.

The process involves a composition for lubricating synthetic fibers, which comprises (1) a base oil containing compounds of the following formula:

(1)

$$R + OR'\,\overline{)_x}\,O \underset{CH_3}{\overset{CH_3}{\underset{\big|}{\overset{\big|}{C}}}} O + R'O\,\overline{)_y}\,R''$$

where R and R'' each are hydrogen or acyl group having 1 to 22 carbon atoms, R' is alkylene having 2 to 4 carbon atoms, and x and y each is an integer of at least 1, with the proviso that the sum of x and y does not exceed 50, and at least one ester selected from the group consisting of esters of aliphatic monohydric alcohols with monobasic fatty acids, dibasic fatty acids or mixture thereof, and esters of aliphatic polyhydric alcohols with monobasic fatty acids, the ester having a kinematic viscosity not higher than 70 cs measured at 30°C, and (2) a polymer of 2,2,4-trimethyl-1,2-dihydroquinoline.

The polymer of 2,2,4-trimethyl-1,2-dihydroquinoline is a compound having the following formula, in which the degree of polymerization, namely, the value of n, is preferably in the range of from 2 to 5:

(2)

$$\left[\begin{array}{c} \text{structure} \end{array} \right]_n$$

Polymers of this type are marketed under the trade names "Antigene RD" (Sumitomo Kagaku), "Antage RD" (Kawaguchi Kagaku) and "Noclarck 224" (Ouchi Shinko Kagaku), and they are commercially available.

The polymer is incorporated in an amount of from 0.05 to 5.0 wt %, preferably 0.1 to 5.0 wt %, based on the total weight of a composition consisting essentially of a compound of the formula (1) an ester-type lubricating agent having a kinematic viscosity of not higher than 70 cs at 30°C, as described above, and an emulsifier and an antistatic agent.

Example: Conventional lubricating compositions Nos. 1, 2, 3, 4 and 5 are shown in Table 1, and the effects attained by the incorporation of the third component of the process, i.e., a polymer (average degree n of polymerization = 3) of 2,2,4-trimethyl-1,2-dihydroquinoline, are shown in Table 2.

In Table 1, the lubricant A is a compound of the formula (1) in which each of R and R'' is acyl having 12 carbon atoms, R' is alkylene having 2 carbon atoms, i.e., ethylene, and each of x and y is 1. The kinematic viscosity is the value measured at 30°C, and p̄ designates an average number of mols of added ethylene oxide.

Table 1

Composition	Components	Mixing Ratio (wt %)
1	bisphenol type lubricant A	60
	polyoxyethylene hydrogenated castor oil ester ($\overline{p}$ = 25)	40
2	bisphenol type lubricant A	5
	hexamethylene glycol dioleate (viscosity = 37.5 cs)	55
	polyoxyethylene hydrogenated castor oil ester ($\overline{p}$ = 25)	40
3	bisphenol type lubricant A	40
	oleyl oleate (viscosity = 23.8 cs)	20
	polyoxyethylene hydrogenated castor oil ester ($\overline{p}$ = 25)	40
4	bisphenol type lubricant A	24
	hexamethylene glycol dioleate (viscosity = 37.5 cs)	36
	polyoxyethylene hydrogenated castor oil ester ($\overline{p}$ = 25)	40
5	bisphenol type lubricant A	45
	trimethylolethane tricaprylate (viscosity = 32.5 cs)	15
	polyoxyethylene hydrogenated castor oil ester ($\overline{p}$ = 25)	40

Table 2

Composition No.	Percent Polymer Added	. Heat Resistance .		Lubricity Secondary Tension (g)
		Heating Loss (%)	Tar Forming Ratio (%)	
1-1*	0.00	33.0	0.2	>200
1-2*	1.00	31.5	0.1	>200
2-1*	0.00	58.7	36.8	110
2-2*	0.03	55.5	35.2	105
2-3*	1.00	39.8	25.0	113
3-1*	0.00	61.8	37.3	150
3-2**	0.05	61.5	7.8	152
3-3**	1.00	45.3	2.5	160
4-1**	0.00	52.3	36.0	130
4-2**	1.00	35.1	0.2	128
4-3**	2.00	34.1	0.2	135
5-1*	0.00	59.9	34.8	175
5-2**	0.1	49.0	2.4	180
5-3**	1.0	38.5	0.5	177

*Comparison
**Process

The heating loss, tar-forming ratio and secondary tension were determined according to the following methods.

Heating Loss and Tar-Forming Ratio — In a commercially available aluminum saucer, about 0.5 g of a sample was placed, and the sample was heated at 250°C for 5 hours. The weight of the sample remaining after the heating was precisely measured, and the heating loss was calculated according to the following formula:

$$\text{Heating Loss (\%)} = \frac{\text{weight before heating} - \text{weight after heating}}{\text{weight of sample before heating}} \times 100$$

After the measurement of the heating loss, the aluminum saucer was washed with acetone, and after drying, the weight of the substance left on the saucer was precisely measured. In general, the residual substance insoluble in acetone was a black resinous substance. A larger amount of this residual substance indicates a higher tar-forming ratio. The value of the tar-forming ratio was calculated according to the following formula:

$$\text{Tar-Forming Ratio (\%)} = \frac{\text{weight of acetone} - \text{insoluble residue}}{\text{weight of sample}} \times 100$$

Lubricity – A lubricating composition was applied in an amount of about 1 wt % to commercially available nylon 6 filamentary yarn, and the secondary tension of the yarn was measured under the conditions of an initial tension of 15 g, a friction pin-yarn contact angle of 180° and a yarn speed of 150 m/min by using a measurement apparatus manufactured by Eiko Sokki KK. A smaller value of the secondary tension indicates a better lubricity.

As will be apparent from the results shown in Tables 1 and 2, the lubricating composition free of the aliphatic ester-type lubricating agent of the process having a kinematic viscosity not higher than 70 cs measured at 30°C (composition No. 1) has a good heat resistance but is inferior in lubricity. The lubricating composition containing the aliphatic ester-type lubricating agent in an amount outside the range specified in the process has an improved smoothness but the heat resistance is degraded. In the case of the lubricating agent containing the aliphatic ester-type lubricating agent in an amount within the range specified in the process, the smoothness can be improved, but if the third component of the process, i.e., a polymer of 2,2,4-trimethyl-1,2-dihydroquinoline, is not incorporated in an amount in the range specified in the process, the heat resistance is drastically degraded. Thus, it will readily be understood that a composition comprising the bisphenol type lubricating agent, the aliphatic ester having a kinematic viscosity not higher than 70 cs measured at 30°C and the polymer of 2,2,4-trimethyl-1,2-dihydroquinoline has excellent heat resistance and lubricity.

Flame-Retardant Composition Containing Bromooctadecane

A process described by *J.L. Claiborne; U.S. Patent 4,135,034; January 16, 1979; assigned to Dixie Yarns, Inc.* provides a flame-retardant yarn or thread primarily for use in sewing flame-retardant apparel. The flame-retardant capability of the yarn or thread is imparted primarily by a flame-retardant lubricant which consists of one or more mono- or di-halo alkanes having from 10 to 30 carbon atoms, where the halogen is either chlorine or bromine.

Octadecane is a classic paraffin and is a lubricant in its own right because it happens to have those physical properties that make it a lubricant. It is a normal paraffin which is slick, like grease, but yet has certain other properties, such as the ability to disperse itself well along a thread structure and to be applied easily to thread to enhance the sewing capability of the thread. When bromine is attached to normal octadecane, the resulting compound becomes nonflammable because bromine is a well-known fire retardant and its presence in almost any material in sufficient amounts will render an otherwise flammable compound nonflammable. It was also found that bromine in a compound with octadecane does not detract from the lubricating value of the latter, contrary to the addition of flame-retardant materials to other lubricants.

Thus, bromooctadecane can be applied to thread as is, with no further additions, from either a solvent application, the solvent being expected to evaporate before the possible introduction of flame, or from a hot melt by a kiss roll. Since octadecane melts at a fairly low temperature, it can be applied as a 100% compound from hot melt. In a preferred case, the hot melt process was employed to topically apply the monobromooctadecane to 100% polyester thread in an amount of 10 wt % of the thread. In general, the amount of lubricant applied depends on the type of sewing operation and is not critical although ranges of 2 to 10 wt % of the thread or yarn are common.

Subsequent experiments resulted in a finding that additional brominated or chlorinated alkanes, having a carbon chain length of from about 10 to about 30 carbon atoms, also had the desired lubricity and nonflammability properties. In the course of research conducted for this process, the materials listed below were found to be particularly advantageous and useful: mixed bromo alkanes C_{10-22}; 1-bromodecane; 1-bromododecane; 1-bromotetradecane; 1-bromohexadecane; mixed bromohexadecane and -octadecane; 1-bromooctadecane; 1-bromoeicosane; 1-bromodocosane; mixed brominated alkenes (avg. C_{24-28}); 1,10-dibromodecane; 1,2-dibromodecane; mixed dibrominated alkenes (avg. C_{20-24}); mixed dibrominated alkenes (avg. C_{24-28}); 1-chlorododecane; 1-chlorooctadecane; 1-chlorodocosane; mixed partially chlorinated alkenes (avg. C_{24-28}); and 1,10-dichlorodecane.

Hot Drawing Lubricant

A.M. Fusco; U.S. Patent 4,077,992; March 7, 1978; assigned to Milliken Research Corporation describes a high temperature lubricant comprising a gem disubstituted cyclic compound in which one radical is a short chain alkyl group and the other radical is a methylene group substituted by an alkyl, alkylene or aryl amido radical or an alkyl, alkylene or aryl carboxylate radical.

Useful gem disubstituted cyclic lubricants of the process may be represented by the formula

where R_1 and R_2 are hydrogen or a lower alkyl group; R_3 is a lower alkyl group; R_4 is an alkyl, alkylene or aryl group; X is a polyalkyleneoxy chain; and Y is oxygen or NH.

Advantageously, the ring is a saturated ring structure with 6 carbon atoms, R_1, R_2 and R_3 are methyl groups, R_4 is an alkyl or alkylene group, X is a polyethyleneoxy chain and Y is NH.

Compounds of this type may be prepared from isophorone utilizing the methods described in U.S. Patents 3,270,044 and 3,352,913 to form an amino methyl cyclohexanol which is then reacted with a carboxylic acid in a fusion cook by heating stoichiometric quantities of the amine and the acid until molten and stirring the molten mixture with a nitrogen sweep to drive off water vapor. The progress of the reaction is stopped when an acid number just below theoretical is obtained. Similarly, the carboxylate derivative can be formed from a hydroxymethyl cyclohexanol.

An ethoxylated product is prepared by adding ethylene oxide to molten gem disubstituted cyclic amine in the presence of potassium hydroxide catalyst in an autoclave at 290° to 300°F. As the ethoxylation continues, the reaction mixture becomes more fluid and the product becomes water dispersible. Further ethoxylation results in a water-soluble product.

Various gem disubstituted cyclic compounds are prepared according to the above procedures and tested to determine their lubricant properties at high temperatures. The compounds are tested to determine their coefficient of friction and to determine their smoke point and flash point.

The lubricity of the compounds is determined by applying the lubricants to scoured and dried spun polyester test yarn from Test Fabrics, Inc. using an Atlab Finish Applicator. Water dispersible lubricants are applied from aqueous solutions with a 1% and 3% dry pickup. Lubricants which are not dispersible in water are applied from isopropanol solutions in the same manner. Friction testing is conducted with a Rothchild F-Meter using recommended procedures. To test the yarns at 410°F, the Rothchild Test Meter is fitted with a Fycon Type T-5 Pin Heater Assembly (Fycon Engineering Corporation). A seven-eighths inch wear sleeve without a finished surface is used with the pin heater. The yarn is conducted through the apparatus at a rate of 50 meters per minute with a contact angle of 180°.

The smoke point and flash point are determined according to American Oil Chemical Society Official Method (c-9a-48).

The following tables are listings of comparisons of lubricants of the process with a commercially available lubricant.

Table 1

	Coefficient .. of Friction ..		Smoke Point (°F)	Flash Point (°F)
	1%	3%		
Solvent treated control	0.77	0.50	–	–
TMP Ester*	0.68	0.42	185	295
TMC Oleamide**–5EO	0.73	0.43	165	277
TMC Stearamide**–5EO	0.72	0.43	140	264

*Trimethylolpropane mixed fatty ester.
**3,5,5-Trimethyl cycloamide with 5 ethyleneoxy units made from 1-hydroxy-3-aminomethyl-3,5,5-trimethylcyclohexane and a commercial grade oleic acid or a commercial grade stearic acid.

Table 2

	.Coefficient of Friction.	
	1%	3%
Solvent treated control	0.55	0.55
TMP Ester*	0.36	0.35
TMC Lauramide**-2EO	0.39	0.40
TMC Adipamide**-2EO	0.41	0.42

*Trimethylolpropane mixed fatty ester.
**3,5,5-Trimethyl cycloamide with 2 ethyleneoxy units made
    from 1-hydroxy-3-aminomethyl-3,5,5-trimethylcyclohexane
    and a laboratory grade carboxylic acid.

From the above discussion and comparisons, it is apparent that the process provides a lubricant which exhibits good friction reduction at high temperatures.
Furthermore, the lubricant of the process does not form smoke or leave residues
on filaments or yarn or on the equipment used. A particularly important advantage of the lubricant is that it is water dispersible so that it can be removed
easily from yarns or filaments. This is important since the lubricant retention
may cause problems in subsequent fabric finishing and/or dyeing due to uneven-
ness, etc. Thus, the process provides a lubricant which is useful in high temperature operations such as hot drawing, texturizing, open end spinning and the like.

TIRE LUBRICANTS

Green Tire Lubricants

*G.T. VanVleck and F.J. Traver; U.S. Patent 4,066,560; January 3, 1978; as-
signed to General Electric Company* describe a process for producing a silicone
composition useful as a green tire lubricant with improved lubricity properties.
The process comprises:

(1) mixing 20 to 55 parts by weight of a mica filler with 2 to 20 parts
    by weight of a linear diorganopolysiloxane of a viscosity varying
    from 500 to 100,000 cp at 25°C where the organo group is se-
    lected from the class consisting of monovalent hydrocarbon radi-
    cals and halogenated monovalent hydrocarbon radicals to form a
    homogeneous mixture;
(2) subsequently, adding to the homogeneous mixture in whatever order
    it is desired from 2 to 20 parts by weight of a mineral clay and
    0.5 to 10 parts by weight of a first emulsifying agent selected
    from the class consisting of anionic, cationic and nonionic emul-
    sifying agents and forming a second mixture; and
(3) finally adding the water in the second mixture and mixing the in-
    gredients thoroughly to form the final silicone green tire lubricant
    composition.

It is critical in the process that the mica filler be first mixed with the polysiloxane polymer to form a homogeneous mixture. Secondly, the water must be
the last ingredient to be added. The amount of water is added as desired to
produce a silicone lubricant with the desired end viscosity and also so that the
emulsion that is formed is stable. The viscosity of the final emulsified composition
may also be controlled by the addition of a buffering agent.

In the examples, use of ST stands for static friction and use of SL stands for sliding friction which can be determined by any published methods and particularly by the published method described in General Electric Co.'s Silicone Products Department paper CDS 2318 written and given by R.A. Moeller at Rubber Group Conference on June 15, 1973. Parts and percentages are by weight.

Example 1: To a one liter stainless steel beaker equipped with umbrella stirrer, there was added 120 g mica, 120 g talc, 20 g precipitated silica, 10 g mineral colloid, 50 g dimethylpolysiloxane of 1,000 cp at 25°C, 50 g of GP 3030, an alkylene oxide polyol, 20 g of an amine salt of an alkylarylsulfonic acid designated G-3300. The ingredients were added sequentially and blended until uniform. Once this semidry green tire lubricant was fully blended, 100 g of it was added to 100 g of water in a 400 ml stainless steel beaker and blended until uniform using an umbrella stirrer.

Once the semidry green tire lubricant was formulated with water, its lubricity properties were evaluated and are as follows: ST, 0.476 and SL, 0.356. (The above composition had outstanding properties as a green tire lubricant.)

Example 2: To a one liter stainless steel beaker equipped with umbrella stirrer there was added 20 g G-3300, an amine salt of an alkylarylsulfonic acid; 50 g GP 3030, an alkylene oxide polyol; and 50 g dimethylpolysiloxane of 1,000 cp viscosity at 25°C. These three liquid ingredients were stirred until uniform. Then 245 g of water was added followed by 180 g 160WG mica, 60 g 901A talc, and 10 g mineral colloid mineral oil. The umbrella stirrer was stirring rapidly during all additions, when practical. Once the formulation was smooth and uniformly dispersed, it was evaluated for lubricity performance and had the following properties: ST, 0.324 and SL, 0.246. It can be seen that the green tire lubricant of Example 1 has a considerably higher static friction and sliding friction.

The following example involves the use of Deriphat 170 C as the buffering agent. Darvan 7 may also be used as the buffering agent.

Darvan 7 is a 25% active aqueous solution of a polyelectrolyte claimed to be sodium methylmethacrylate or sodium polymethylmethacrylate. The pH of this material is between 9.5 and 10.5. Deriphat 170 C is an amphoteric material namely N-(laurylmyristyl)-β-aminopropionic acid.

Example 3: Deriphat only was added as needed to reduce viscosity to between 2,500 and 6,000 cp, LVT viscometer 3 spindle at 3 rpm. There was prepared an emulsion in accordance with the process having the following ingredients: mica, 40 parts; dimethylpolysiloxane of 10,000 cp viscosity at 25°C, 10 parts; H_2O, 40 parts; Veegum F (mineral clay), 0.6 part; and WS-661 (ethylene oxide, propylene oxide polyol), 8.4 parts.

The above materials were blended until homogeneous. The viscosity was greater than 30,000 cp. After adding 4 parts of Deriphat 170 C and mixing one-half hour the viscosity dropped to 2,500 cp.

Internal Tire Lubricant

J.W. Messerly and J.J. Shipman; U.S. Patent 4,096,898; June 27, 1978; assigned to The B.F. Goodrich Company have discovered that a mixture of certain kinds of synthetic hydrocarbons, having molecular weights within a particular range and certain other specific physical properties, with other synthetic or natural hydrocarbon materials of a different particular kind, will have the exact properties necessary for effective functioning for the desired purposes when applied to the inner surface of pneumatic tires. The mixture is required to be a nonflowable but flexible solid at ordinary atmospheric temperatures but to be an extremely viscous liquid at normal operating temperatures of tires. It is required also to be permanently adherent to the rubber material at the inside surface of of the tire and to be free from hardening such as would cause it to crack or flake off of the surface. In order to perform the function of sealing punctures, the composition should also be capable of swelling the material which is at the internal surface of the tire and which is usually made from a vulcanized composition consisting in large part of butyl rubber.

In accordance with this process, a composition was prepared from two commercially available materials. One of them is a particular grade of polyethylene; the other is a particular grade of polypropylene.

The polypropylene is an amorphous noncrystallizable material made by polymerizing propylene with a redox catalyst to a moderate molecular weight of about 900. This polypropylene has a softening temperature of 82° to 95°C determined by the ball and ring laboratory method, and at 190°C is a liquid with a viscosity of 49 to 90 cp.

The polyethylene is a grade which is a soft solid of average molecular weight between about 1,000 and 5,000, a density of 0.88, and which becomes liquid at 85°C. It is mostly amorphous but contains a significant proportion of crystallizable material. It is sometimes called polyolefin grease.

In this particular example, 100 parts of the polypropylene are melted with 30 to 50 parts and preferably about 35 parts of the partly crystalline polyethylene and intimately mixed.

The temperature is adjusted to about 120°C and the liquid mix is forced through a spray nozzle to produce a coarse spray of about 110° included angle directed to the inner surface of the crown portion of the tire, which requires a pressure of about 6 atmospheres. The amount applied should be sufficient to produce a coating of 1 to 2 mm thickness and preferably 1.5 mm, which requires in the neighborhood of 200 g for a medium-sized passenger automobile tire.

The composition immediately solidifies to form a smooth, soft, greasy surface on the inside of the tire, which remains flexible and tacky over the entire normal range of operating temperatures so as to retain its position unchanged. A portion of the liquid material is absorbed by the underlying rubber surface which in the preferred form of the process is the cellular rubber material. This action contributes to the firm retention of the coating in position and also somewhat swells the rubber material, which enhances the ability of the cellular rubber to seal small punctures.

A thin cellular rubber layer on the inside of a tire is very effective in sealing small punctures, particularly when coated with low molecular weight polyethylene. The puncture sealing ability is even more effective with the mixed polymers of this process.

Moreover, an internal layer of low molecular weight polyethylene is an excellent internal lubricant for preventing destructive friction when a tire becomes deflated so that the weight of the vehicle is borne by small areas of the inner surface of the tire in moving apposition. This function is also more effectively performed by the mixed polymers of this process.

Both of these benefits are obtained either with or without the presence of the cellular rubber layer.

In the absence of a cellular rubber layer, the polymer composition of this process, because of its softness and plasticity, is readily smeared into a small puncture to reduce or prevent further escape of air, and if the temperature is so high as to cause the polymer composition to be semiliquid, it tends to be absorbed rapidly by the surface of the rubber so as to swell the puncture shut. In the presence of a cellular layer the puncture closing actions are enhanced and made more certain by the expansion of the compressed gas in the closed cells toward the lower pressure of the outside atmosphere.

Again in the absence of a cellular rubber layer, the polymer composition functions as a very effective lubricant during operation of a vehicle with a flat tire, eliminating or minimizing the rubbing forces which otherwise tend to tear or shred the inner surface of the flat tire. In the presence of a cellular rubber layer, which acts as a cushion to distribute radial forces which would otherwise be concentrated in a very small area and tend to cut the tire structure, the chances of injury during operation of a vehicle on a flat tire are still further reduced.

Similar excellent results are obtained with many other specific combinations of the kind outlined above.

Thus, with the partly crystalline, low molecular weight polyethylene described above, additions of the following materials have been found to give good results. For each 100 parts of the polyethylene:

(1)　35 parts polyisobutylene of average molecular weight about 890 and density of 0.890;

(2)　25 parts of the polyisobutylene and 10 parts of paraffin base partly naphthenic petroleum oil of density 0.890, pour point $-10°C$ (+14°F) and viscosity 100 sec SUS at 38°C (100°F); and

(3)　13 parts of the petroleum oil used in (2) above and 20 parts of rosin oil of viscosity 95 to 130 sec SUS at 99°C (210°F).

Also, a mixture of 100 parts of a partly crystallizable polypropylene which becomes fluid at 95°C (200°F) with 40 parts of the same polyisobutylene used in (1) above gives similar good results both in sealing punctures and in protection against damage when run flat.

Band-Ply Lubricant Concentrate

A process described by *W.N. Fenton and J.W. Keil; U.S. Patent 4,125,470; November 14, 1978; assigned to Dow Corning Corporation* provides a band-ply lubricant concentrate which will allow tire manufacturers to formulate band-ply lubricants which meet the individual manufacturers' needs and performance standards.

The concentrates consist of (A) an alkylene oxide polymer, (B) unique siloxane copolymer dispersing agents, (C) a polydimethylsiloxane fluid, and in some instances (D) mica or talc.

The following examples illustrate the process and all parts and percents referred to are by weight and all viscosities measured at 25°C unless otherwise specified.

Example 1: A composition was prepared which consisted of (A) about 50.5% of a polyoxyethylene-polyoxypropylene random copolymer having a molecular weight of about 2,600 (Dow Chemical Polyglycol P15/200), (B) about 4.5% of a siloxane copolymer which was the reaction product derived from heating 1 part of a siloxane consisting essentially of SiO_2 units and $(CH_3)_3SiO_{1/2}$ units in which the ratio of SiO_2 units to $(CH_3)_3SiO_{1/2}$ units was in the range of 1:0.4 to 1:1.2, and 3 parts of a hydroxylated polyoxyethylene-polyoxypropylene copolymer having a molecular weight of about 6,200 and about a 1:1 mol ratio of ethylene oxide units to propylene oxide units, in 4 parts of xylene employing a stannous octoate catalyst, and (C) about 23.7% of a polydimethylsiloxane fluid having a viscosity of about 100,000 cs plus about 21.3% of a polydimethylsiloxane fluid having a viscosity of about 350 cs.

Example 2: A band-ply lubricant was prepared which consisted essentially of about 13.3% of the concentrate of Example 1, 0.4% of carboxymethylcellulose (CMC-7M), 1% of soy lecithin (thickening agent), 0.1% sodium nitrite (rust inhibitor), 42.75% water, 0.175% the sodium salt of pentachlorophenol (Dowicide G), 0.175% of a mixture of 75% 1-(3-chloroallyl)-3,5,7-triaza-1-azoniaadamantane chloride and 25% sodium bicarbonate (Dowicil 75), 32.1% mica (325 mesh), 10% talc and 0.1% of an alkyl phenoxy polyoxyethylene ethanol (Makon 10). This band-ply lubricant is designed for use where high air bleed is required.

Example 3: A band-ply lubricant was prepared which consisted essentially of about 21% of the concentrate of Example 1, 3.3% of a polyoxyethylene-polyoxypropylene random copolymer having a molecular weight of 2,600 (Dow Chemical Polyglycol P15/200), 0.25% carboxymethylcellulose (CMC-7M), 0.75% soy lecithin, 41.9% water, 0.175% of the sodium salt of pentachlorophenol, 0.175% of a mixture of 75% 1-(3-chloroalkyl)-3,5,7-triaza-1-azoniaadamantane chloride and 25% sodium bicarbonate, 0.1% sodium nitrite, and 32.35% mica (325 mesh). This band-ply lubricant is designed for use where low viscosity and high slip are required.

For purpose of comparison a band-ply lubricant similar to the one above was prepared which consisted essentially of 13.9% of a polyoxyethylene-polyoxypropylene random copolymer having a molecular weight of 2,600 (Dow Chemical Polyglycol P15/200), 0.05% of carboxymethylcellulose (CMC-7M), 0.75% soy lecithin, 25% water, 0.175% of the sodium salt of pentachlorophenol, 0.175% of a mixture of 75% 1-(3-chloroallyl)-3,5,7-triaza-1-azoniaadamantane chloride and 25% sodium bicarbonate, 0.1% sodium nitrite, 32.55% mica (325 mesh), 14.2% of a 35% aqueous emulsion of a 100,000 cs polydimethylsiloxane fluid, and 12.8% of a 35% aqueous emulsion of a 350 cs polydimethylsiloxane fluid.

To test the stability of the above prepared band-ply lubricants they were placed in a Waring blender and subjected to agitation at the "Hi-Speed" setting for various lengths of time after which samples were drawn and set aside for stability evaluation.

The band-ply lubricant prepared above for comparison showed oiling upon standing less than 1 hour after 3 minutes of agitation; showed about 5% oiling on the top immediately after 5 minutes of agitation; and showed substantial oiling and separation immediately after 8 minutes of agitation. After standing for 3 days the 3, 5 and 8 minute agitation samples all had about a 5% oil layer, about a 30% water layer, and exhibited complete mica separation or settling.

The band-ply lubricant of the process prepared above showed no oiling or settling immediately after 3, 5 or 8 minutes of agitation. After standing for 3 days the 3, 5 and 8 minute agitation samples had about a 10% water layer on top but showed no signs of oiling or mica separation or settling.

Example 4: Concentrates identical to that of Example 1 were prepared. These concentrates were subjected to various tests to evaluate stability. For example, centrifuge stability was tested by running for 60 minutes at 3,000 rpm. Shear stability was tested by placing in a Waring blender at hi-speed for 5 minutes. Oven stability was tested by placing samples in a 60°C oven for 10 days. Freeze-thaw stability was tested by alternately freezing and thawing for 12 cycles. Shelf stability was tested by letting samples stand at room temperature for 10 weeks. In all of these tests, typical results were that the concentrates remained homogeneous and exhibited no separation or oiling.

The above concentrates can be diluted with and dispersed in water or organic solvents for use as band-ply lubricants, with or without additives as disclosed above. While upon standing the diluted products may separate over a period of time, they do remain permanently redispersible in contrast to the available band-ply lubricants not made in accordance with this process.

RECORDING SUBSTRATES AND PLASTICS

Dialkanolamine Derivatives and Halogenated Hydrocarbon Carrier

R.P. Pardee; U.S. Patent 4,071,460; January 31, 1978; assigned to Ball Brothers Research Corporation has found that certain dialkanolamines are extremely effective lubricants upon sundry substrates but especially upon those substrates having dynamic presentations thereon, such as phonograph records, discs and recording tapes.

The composition of the process which imparts to a given substrate a low coefficient of friction consists essentially of a solution of a dialkanolamine in a carrier or solvent. The dialkanolamine has the formula: $(HOR_1)_2N—R_2$, wherein R_1 is an alkylene having 2 to 3 carbon atoms, i.e., ethylene, isopropylene and propylene and R_2 is a hydrocarbon radical having 4 to 20 carbon atoms. The hydrocarbon radical, R_2, may be an alkyl, aryl, or alkaryl, that is, an alkyl-substituted aryl radical, such as tolyl. Illustrative examples of such dialkanolamines are the following: N-butyl·N,N-β,β'-diethanolamine, N-octyl·N,N-γ,γ'-diisopropanolamine, N-decyl-N,N-γ,γ'-diisopropanolamine, and N-phenyl-N,N'-β,β'-diethanolamine.

Example 1: A solution was prepared consisting of 99.97 wt % of trichlorotri-fluoroethane (Freon TF), and 0.03 wt % of Anti-Stat 273C, a commercial N,N-bis(2-hydroxyethyl)alkylamine (Fine Organics, Inc.), the alkyl moiety thereof ranging from dodecyl to tetradecyl. A clear solution resulted and was sprayed onto a phonograph test record, National Association of Broadcasters (NAB) test record No. 12-5-98, the trichlorotrifluoroethane was allowed to evaporate, and the playing surface was lightly buffed to leave a thin coating of the alkylamine thereon. The record was then subjected to playing and compared with an un-treated record to determine changes in any surface noise. For this purpose, the signal from the stylus, tracking at one gram load in the record grooves, was fed to a Tektronix 5100 Series Storage Oscilloscope for display. During the first number of playings, the coated record showed significantly less surface noise than did an uncoated record; and progressively throughout some 120 playings, the level of background or surface noise of the coated record ultimately reached the noise level that the uncoated record showed on its first playing.

Example 2: Accelerated phonograph record wear tests were conducted on NAB test records which had been treated with the compositions of the process. The results achieved from treated records were compared with the results of the wear test conducted on a cleaned test record which had not been treated in accordance with the process. The test utilized a standard type automatic record turntable rotating at 33$\frac{1}{3}$ rpm with the stylus on the tone arm adjusted to 9.5 g load on the record surface. This high stylus load was used in order to accelerate the wear process and thereby provide better discrimination among record treatments. Various compositions of the process were applied to the record surfaces in accord-ance with the procedure described in Example 1. Test results are presented in the tabulation below:

Phonograph Test Record Identification	Number of Times Played	Composition of Record Treating Solution, Trichlorotrifluoroethane +	Appearance After Test	Rating*
A	129	0.03 wt % Anti-Stat 273C	Clean surface, no visible weak particles	0
B	108	0.018 wt % Anti-Stat 273C	Practically clean, 1 or 2 visible wear particles	1
C	130	0.03 wt % Anti-Stat 273E	Practically clean, a few wear particles	2
D	128	None (control)**	Surface evenly flecked with visible but tiny white wear debris	75

*0 = clean, 100 = heavily covered with wear debris.
**Record surface was wiped with velvet swatch saturated with the trichloro-trifluoroethane only.

The above examples show that compositions of the process effectively prevent phonograph record groove wear and preserve original recorded fidelity; and the compositions markedly reduce background or surface noise when applied to a record and retard the rate of noise increase with many playings.

Tetrafluoroethylene Telomeric Compositions

R.P. Pardee; U.S. Patent 4,096,079; June 20, 1978; assigned to Ball Brothers Research Corporation describes a process for improving lubricity and wear resistance of a given substrate by applying a composition consisting essentially of low-molecular weight tetrafluoroethylene telomers, an antistatic agent, preferably in the form of a tertiary amine, and a volatile organic solvent, and removing the volatile solvent to produce a thin, dry coating upon the substrate. The compositions have been found to be most effective as preservatives for coating gramophone or phonograph records which provide marked reduction of record groove wear while substantially minimizing noise and harmonic distortion.

Example 1: A lubricating composition was prepared by adding 594 g trichlorotrifluoroethane (Freon TF) to 6 g Vydax AR, a 20 wt % dispersion of tetrafluoroethylene telomers in 80 wt % trichlorotrifluoroethane so as to provide about a 0.2 wt % concentration of the telomers in the final concentration. Generally, the average particle size of the telomers in the dispersion is about 5 microns. The resulting mixture was thoroughly agitated and allowed to settle for about 168 hours. The upper clear solution was then removed from the relatively fluffy, white sediment to yield about 570 g of clear solution. About 0.53 ml of Anti-Stat 273C, a commercial N,N-bis(2-hydroxyethyl)alkylamine was added for each liter of clear solution to provide a concentration of approximately 0.03 wt %.

The alkyl moiety of the alkylamine ranges from about dodecyl to tetradecyl. This resulting clear composition was applied to a National Association of Broadcasters (NAB) No. 12-5-98 phonograph test record by gently spraying over the surface thereof whereby evaporation removes the solvent at room temperature and results in an almost instantaneous deposition upon the record surface of a practically invisible coating thereon. The thus-treated record was thereafter carefully buffed by rubbing in the direction of the grooves employing a velvet buffing pad so as to provide a bright, lustrous finish thereon.

In order to determine both reduction of distortion and background noise of treated as compared with untreated records, tests were conducted and were determined by using a Hewlett-Packard Wave Analyzer (Model No. 3590A with a 3594A attachment). The studies which were made over the frequency range of about 500 Hz to about 13 kHz employed a "window" bandwidth of about 100 Hz. The sweep rate was set at 100 Hz, per second, the meter damping being set at a "medium" setting. The maximum input voltage was set at 0.1, reference adjustment to relative and the meter to linear dB, with a frequency range of 62 kHz. After 50 days, noise in the 3 kHz band had increased by an average of 1.4 dB compared with an average increase of 3.7 dB for untreated record subjected to the same test, and the second harmonic had changed by an average of 2 dB compared with an average change of 8 dB for an untreated record. Thus, the composition of the process when applied to the record surface demonstrates that there is a significantly lower increase in the background noise and a significantly smaller change in harmonics as compared with an untreated record surface when both records are subjected to the same number of playings.

Example 2: A lubricating composition according to the process was prepared by adding 598.8 g of trichlorotrifluoroethane solvent (Freon TF) to 1.2 g of

MP-51, a tetrafluoroethylene telomer, so as to provide a 0.2 wt % concentration of telomers in the final composition. The resulting mixture was carefully stirred for about 30 minutes and only a few insoluble particles remained which were then removed by filtering through filter paper No. 3. To the resulting clear solution there was added Anti-Stat 273C, described in Example 1 in an amount equivalent to 0.53 ml per liter of solution, so as to provide a concentration of about 0.03 wt % in the final composition. This composition was sprayed upon a phonograph record surface and carefully buffed after evaporation of solvent.

The phonograph record test equipment described in Example 1 was again employed. The mean noise level on the treated record was measured at 59.3 dB below reference level compared with 60 dB below reference for an untreated record; and the second harmonic of the 3 kHz fundamental test frequency for the treated record was measured at a mean peak height of 29 dB below reference level compared with 30.3 dB below reference for the untreated record. Thus, it was demonstrated that there is no significant increase in background noise or change in harmonics due to the application of the lubricating composition of the process to a new record.

Methyl Alkyl Siloxanes for Video Disc Lubricant

N.V. Desai and R.J. Himics; U.S. Patent 4,159,276; June 26, 1979; assigned to RCA Corporation describe a method for purifying methyl alkyl siloxane lubricants. U.S. Patent 3,833,408 describes the application of methyl alkyl siloxane compositions of the formula

$$(CH_3)_3SiO \underset{\underset{R}{|}}{\overset{\overset{CH_3}{|}}{-(SiO)_x-}} Si(CH_3)_3$$

wherein R is an alkyl group of 4 to 20 carbon atoms and x is an integer, as lubricants for conductive video discs comprising a molded plastic disc having audio and video information in the form of geometric variations in a spiral groove in the surface of the disc. These discs are coated first with a conductive material which acts as a first electrode, then with a dielectric layer and a final layer of lubricant. A metal tipped stylus acts as a second electrode of a capacitor and the information signals are monitored by the stylus which notes changes in capacitance between the stylus and the disc surface as the information, in the form of depressions, passes beneath the stylus.

Further developments in this system have produced a video disc which is made of a conductive plastic material, e.g., a PVC copolymer resin containing sufficient amounts of conductive carbon particles so that the disc can provide capacitance readout, while the plastic resin surrounds the carbon particles providing a dielectric surface layer on the conductive particles. This development has eliminated the need for separate coatings of metal and dielectric on the plastic disc.

Video discs are also being developed which do not require a conductive surface or a grooved surface, the stylus being maintained in synchronization with the information pattern by means of electrical signals rather than the groove walls.

These changes in the materials used for the video disc have somewhat changed the requirements for the lubricant and in certain respects the commercially available methyl alkyl siloxane lubricant (General Electric Company, SF-1147), wherein R in the above formula is a decyl radical and x is about 2 to 7, is now unsatisfactory.

This material contains an antioxidant compound which is added as

A portion of this antioxidant reacts with residual hydride groups in the methyl alkyl siloxane lubricant, to form chemically bound compounds of the type

The presence of the unbound, free antioxidant compound can cause changes in the video disc surface with time, contributes to inhomogeneities in the disc surface from one lot to another, particularly in the wetting and lubrication dynamics during playback of the disc, and thus its presence is undesirable. However, in removing the unbound antioxidant, the basic structure of the methyl alkyl siloxane lubricant should remain the same, particularly with respect to molecular weight and molecular weight distribution of the methyl alkyl siloxane molecules. Further, a suitable method for removing antioxidant must not further contaminate the siloxane lubricant with difficult-to-remove solvents or other material. This method purifies the methyl alkyl siloxane lubricant, removes chemically unbound antioxidant but without adversely affecting the lubricant or the chemically bound antioxidant so as to improve the performance of this class of lubricants for the video disc application.

The process comprises a multistep extraction of the methyl alkyl siloxane lubricant with approximately equal volumes of acetone. The siloxane and acetone mixture is stirred together for some period until the antioxidant is dissolved in the acetone. The stirring is controlled so that only one phase is apparent. The mixture is then allowed to stand until two phases are visible, about 4 to 20 hours, and the siloxane lubricant and acetone phases are separated. The extraction is repeated in like manner until all of the free antioxidant has been removed from the oil layer.

Generally, two extractions are sufficient to remove all the unbound antioxidant completely. A single extraction will remove the bulk of the unbound antioxidant, but minor amounts are still present even when a large excess of acetone is used.

Acetone is the preferred solvent because it is a very effective solvent for the antioxidant, inexpensive, and is readily separated from the siloxane lubricant. After the extraction step, the acetone and siloxane will form distinct layers. If the ambient temperature is low, the lubricant layer can appear hazy due to the presence of small amounts of acetone, but this can be improved by increasing the temperature of the mixture to normal room temperatures.

In order to remove residual acetone from the lubricant layer, the mixture is heated at about 80°C under vacuum. The bulk of the acetone can be removed at 80°C using a rotary evaporator with a water aspirator, for example, and evacuation can be continued at about 80°C under 10 mm Hg pressure when all the acetone will be removed.

The above procedure produces a methyl alkyl siloxane having the same molecular weight and molecular weight distribution as the feedstock and which contains chemically bound antioxidant but is free of chemically unbound antioxidant. In addition, the process purifies the methyl alkyl siloxane, particularly removing catalyst residues. The losses are low and about 99.5% of the starting material is recovered.

The chemically unbound antioxidant-free methyl alkyl siloxane lubricant can be applied to a video disc by evaporation, spin coating or spray coating from solution. For example, a layer about 200 to 300 A thick of the lubricant can be applied by spray coating from a 1% solution of the siloxane in heptane or isopropanol at a loading of from about 0.2 to 2.0 wt % of the solution of the methyl alkyl siloxane.

Soaps and Esters from Alpha-Olefin Acids for PVC Lubricant

According to a process described by *H.C. Foulks, Jr.; U.S. Patent 4,107,115; August 15, 1978; assigned to Emery Industries, Inc.* soaps and ester-soaps of high molecular weight branched- and straight-chain aliphatic monocarboxylic acids which are obtained from α-olefins containing 22 or more carbon atoms are useful lubricants for structural resins. High molecular weight acids useful for the preparation of the soaps and ester-soaps are obtained by the free radical addition of a short chain monocarboxylic acid to C_{22+} olefin or by ozonization of the C_{22+} olefin. The products of this process provide excellent internal-external lubrication for PVC homopolymers and copolymers.

The following examples illustrate the process and all parts and percentages are given on a weight basis unless otherwise indicated.

Example 1: To obtain mixed acid products useful in the preparation of the soaps and ester-soaps of this process equal parts of C_{30+} α-olefin (Gulf C_{30+} olefin fraction, MP 160° to 167°F, containing 78 wt % C_{30} and higher olefins) and pelargonic acid were fed into the top section of a countercurrent absorber while a stream of oxygen and carbon dioxide containing approximately 1.5 to 2% ozone was fed into the bottom section. The rates of flow of the O_3/O_2 gas stream and the olefin feed were adjusted so that the C_{30+} α-olefin absorbed as much ozone as possible in passing through the absorber and so that all but trace amounts of ozone were removed from the oxygen. The temperature in the absorber was maintained in the range 65° to 85°C. The effluent gases were scrubbed with water to remove organic vapors and particulate matter and then passed through a catalytic furnace where organic matter was oxidized to carbon dioxide and water. The gas was then dried and recycled.

The ozonide was removed from the bottom of the absorber and passed into a decomposition vessel containing a heel of pelargonic acid, 0.25% sodium hydroxide based on weight of ozonide and previously decomposed ozonide to serve as a diluent. The decomposition vessel was maintained at a temperature of 95°C while adding oxygen containing 1% ozone and the ozonide added over a 2-hour period. When the addition was complete the decomposition was continued for 2 additional hours before transferring to an oxidation reactor. The oxidation was carried out in the presence of manganese acetate tetrahydrate (0.1% based on the C_{30+} olefin) in an oxygen atmosphere. The time for oxidation was 4 hours.

The mixed oxidation product was then stirred with 0.5% phosphoric acid (75%) for 15 minutes and an activated bleaching clay (Filtrol Grade No. 1) added with additional stirring. The mass was filtered to remove the manganese salts of phosphoric acid and the filter aid and then stripped of pelargonic acid under reduced pressure using a Vigreaux column. The stripping was conducted at 230°C and during the final stages the pressure was reduced to 0.5 torr. A portion of the mixed acid product, crystallized from glacial acetic acid, was analyzed by gas-liquid chromatography of the methyl esters employing a modification of ASTM Test Method D 1983-64T. A Hewlett Packard Model 7550 chromatograph equipped with a 6' x ⅛" stainless steel mesh Diatoport S was used. The instrument was programmed for an 8°C per minute temperature rise over the range 75° to 333°C with a helium flow of 15 ml per minute and 50 psig. The mixed acid product (equivalent weight 586; 7 to 8 Gardner color) had the following compositional analysis:

Acid	Weight Percent
C_{9-21}	10.27
C_{22}	3.85
C_{23}	5.14
C_{24}	3.26
C_{25}	6.83
C_{26}	3.08
C_{27}	11.57
C_{28}	2.83
C_{29}	12.54
C_{30}	1.72
C_{31}	10.53
C_{32}	1.29
C_{33}	8.13
C_{34}	0.89
C_{35}	6.00
C_{36+}	11.95
Total	99.88

Example 2: A predominantly α-methyl branched high molecular weight mono-carboxylic acid was prepared by charging a glass reactor with 200 g of an α-olefin mixture containing greater than 85 wt % C_{22-88} olefins (Gulf C_{22+} α-olefin fraction, MP 127°F), 326 g propionic acid and 8 g di-tert-butyl peroxide. The system was flushed with nitrogen and a slight nitrogen flow maintained while the reaction mixture was heated at reflux for about 4 hours. At the completion of the reaction unreacted propionic acid was removed under vacuum at 200°C. 225 g of the C_{25+} α-methyl monocarboxylic acid having an acid value of 48 was recovered.

Example 3: A reactor was charged with a mixture of 300 g of the C_{22+} olefin of Example 2 and 200 g of pelargonic acid (Emfac 1202 pelargonic acid). A stream of oxygen containing 3% ozone was continuously bubbled in below the surface of the liquid at a rate of 24 scfh at 4 psig so that approximately 35 g ozone was being charged per hour. The temperature of the absorber was main-tained above the titering point of the reaction mixture with vigorous agitation to insure intimate contact with the ozone and the progress of the reaction fol-lowed by analyzing the off-gases. Ozonolysis was terminated when ozone ab-sorption dropped below 15%. The ozonides were oxidatively cleaved by the dropwise addition of the ozonide mixture into a vessel containing 100 g pelar-gonic acid and 0.75 g sodium hydroxide over a period of about 90 minutes.

The reaction mixture was vigorously agitated and maintained at about 95°C while bubbling in a stream of oxygen containing 1% ozone at a rate of 2.4 scfh. When the addition was complete, stirring was continued for an additional 90 minutes while bubbling in the O_3/O_2 mixture. The ozone generator was then turned off. Manganese acetate tetrahydrate (1.5 g) was added and the temperature of the reaction mixture raised to 120°C while bubbling in pure oxygen with stirring. After 3½ hours the oxidation reaction was complete and the mixed oxidation product was stripped of pelargonic acid by heating to 245°C while pulling a vacuum of 25 torrs on the system. The mixed acid product contained approxi-mately 80 wt % C_{21+} monocarboxylic acids.

Example 4: The metallic soap of mixed C_{29+} monocarboxylic acids obtained by the ozonization of an α-olefin mixture containing greater than 75 wt % olefins having 30 or more carbon atoms in accordance with the procedure of Example 1 was prepared by double decomposition. The sodium salt of the acid was first prepared by adding 0.1 equivalent of the mixed acids (recrystallized from 5:1 methanol) to an aqueous solution containing 0.1 equivalent sodium hydroxide and maintained at 85°C. The reaction mixture was stirred at 90°C for 30 min-utes and 0.1 equivalent calcium chloride dissolved in 1,000 ml water added with agitation. The calcium soap immediately precipitated from solution and was re-covered by filtration. After thoroughly washing with water to remove the so-dium chloride the soap was dried at 65°C. The resulting soap of the mixed high molecular weight acids contained 3.4 wt % calcium, had a negligible acid value and melted at 134° to 143°C.

Example 5: Employing a procedure similar to that described in Example 4 the calcium soap of a high molecular weight α-methyl branched acid obtained by the addition of propionic acid to a C_{30+} olefin mixture was prepared. 0.5 equiva-lent of the high molecular weight α-methyl branched acid was first converted to the sodium salt by neutralization with 0.5 equivalent sodium hydroxide. The sodium salt was then converted to the insoluble calcium soap by the addition of an aqueous solution containing 0.5 equivalent calcium chloride. The precipitated calcium soap was washed until there was less than 0.1% sodium chloride in the filtrate and dried at 65°C. The soap contained about 2 wt % calcium and melted between 122° and 128°C.

Example 6: To demonstrate the ability of the products of Examples 4 and 5 to function as lubricants for PVC the calcium soaps were incorporated in the following standard pipe formulation: PVC resin (Geon 101-EP), 100 parts; tin mercaptide stabilizer, 2 parts; titanium dioxide, 3 parts; acrylic processing aid, 4 parts; and lubricant soaps, 0.5 part.

The ingredients were blended in a Henschel high speed mixer and the resin evaluated in a Brabender plasticorder, a convenient laboratory evaluation tool which measures the flow properties of the resin against time. Evaluation conditions were as follows: resin charge, 55 g; No. 6 roller head; temperature, 195°C; and rotor speed, 60 rpm. Test results obtained are set forth below and compared with an unlubricated control resin. Times are given in minutes.

Lubricant Soap	T_s*	Torque m-g	T_p**	Torque m-g
4	18.5	550	19.75	3100
5	31	600	34.5	3250
None	1.25	650	2.75	3800

*Time to start of fusion.
**Time to fusion peak.

It is evident from the above data that the soaps of this process are effective lubricants for PVC and extend the fusion time of PVC resins.

Examples 7 through 9: A series of ester-soaps having varying calcium contents were prepared employing the high molecular weight α-methyl monocarboxylic acid obtained by the free radical addition of propionic acid to a C_{30+} olefin mixture. The ester-soaps were prepared by simultaneously reacting the monocarboxylic acid, tripentaerythritol and calcium hydroxide at 220° to 230°C in the presence of 0.03 wt % dibutyltin oxide catalyst while removing the water of reaction. Reactant charges (in equivalents) and properties of the resulting ester-soaps were as follows:

	7	8	9
Reactants			
α-Methyl monocarboxylic acid	1	1	1
Tripentaerythritol	0.75	0.5	0.25
Calcium hydroxide	0.25	0.5	0.75
Properties			
Wt % calcium	0.53	1.07	1.46
Acid value	20.8	18.5	16.6
Hardness	117	132	85

Percent calcium was determined by ashing and atomic absorption (Perkin Elmer Model 303) and hardness measured in accordance with ASTM D 1321-61T.

The ester-soap products were compounded with a typical PVC resin formulation as follows: PVC resin (Diamond Shamrock PVC-40, inherent viscosity 0.83),

100 parts; acrylic processing aid, 4 parts; tin mercaptide stabilizer, 2 parts; and epoxidized soy, 1 part.

The resin formulations were then evaluated in the Brabender machine (56 g sample; 160°C; No. 6 rotor head at 60 rpm). All of the ester-soaps proved to be effective lubricants for the PVC resin and extended the fusion time beyond that obtained with an unlubricated control resin and an identically formulated resin lubricated with 0.5 phr of a commercially available wax product which contains about 2 wt % Ca and is derived from montan wax and 1,3-butylene glycol. For example, the resin containing ester-soap 8 had not started to fuse in 20 minutes time whereas the resin containing an identical amount of the commercial product started to fuse (T_s) in 8 minutes at 825 m-g torque and had the fusion peak (T_p) at 10'30" at a torque of 3,300 m-g.

Lubricant for Thermoplastic Materials

A process described by *J. Boussely, M.-M. Chandavoine, M. Chignac, C. Grain and C. Pigerol; U.S. Patent 4,098,706; July 4, 1978; assigned to Sapchim-Fournier-Cimag, France* relates to lubricating systems for thermoplastic materials, which comprise at least two lubricating agents selected from 1,2-propane diol dibehenate, thiodiglycol dibehenate, diethyleneglycol dibehenate, triethyleneglycol dibehenate, trimethylolpropane tribehenate, dipentaerythrite pentabehenate, 1,4-butanediol dibehenate, glyceryl tribehenate, pentaerythrite tetrabehenate and one metallic salt of behenic acid.

The esters of the process, whether used alone or in association, have been found to be superior to the previously known lubricating agents with respect to one or more of the characteristics generally attributed to such agents. For instance, the compounds of the process increase the quality of the synthetic material, by improving either the internal lubricating effect or the external lubricating effect, or even both effects.

Furthermore, they have a low degree of toxicity and, when incorporated in a resin which is to be used for manufacturing containers, they have high resistance to extraction from the resin into the contents of the container. These findings are important because they are closely related to the problem of providing containers for food and drink and the possible pollution of the latter by the lubricated polymer from which the container is made.

Finally, the compounds are not volatile and are stable under the influence of heat and air, even at high temperatures, which is very important with regard to the problems of manipulation and pollution in the workshops where the various operations are carried out.

Magnetic Field Reactor for Preparation of Lubricant Dispersions

V.V. Kafarov, A.V. Kuramzhin, D.M. Bodrov, E.I. Obelchenko, A.A. Chernyavsky, M.S. Yaschinskaya, D.D. Logvinenko, O.P. Shelyakov, K.L. Tsantker and L.B. Gladilina; U.S. Patent 4,149,981; April 17, 1979 describe a method of producing plastic and liquid lubricants, which will accelerate the process of producing these materials.

According to the process, the dispersion is accomplished in a vortical bed of ferromagnetic particles formed under the action of a rotating magnetic field. Subsequent steps of melting the dispersion, moisture removal, cooling, deaeration and homogenization can also be included.

The proposed method of producing plastic and liquid lubricants makes it possible to reduce by several thousands of times, as compared to the known batch processes, and by several dozens of times, as compared to the known continuous processes, the duration of chemical reactions between the components of lubricants and duration of dispersing these components. The method ensures continuous run of technological processes with high productivity, decreases by 10 to 20% specific consumption of expensive components and by 2 to 3 times consumption of energy and lowers operating temperatures and pressures.

The proposed method is accomplished in a reactor representing a length of a nonmagnetic pipeline around which a system of windings is arranged creating a rotating magnetic field. Nonequiaxial ferromagnetic particles are placed in the reactor; the particles under the effect of the rotating magnetic field are set in a compound motion: each particle travels in the direction of field rotation and simultaneously rotates precessionally about its smallest axis at a speed of 10,000 rpm. Ferromagnetic particles, operating as elementary mechanical stirrers, create a vortical layer filling the whole operating volume of the reactor, and, at the same time, emit acoustic and ultrasonic oscillations of a wide frequency spectrum.

In addition, under the action of an alternating magnetic field, ferromagnetic particles emit magnetostrictive oscillations. Eddy currents, arising in the particles as in electric conductors, give rise to rapidly alternating magnetic and electric fields. Due to the combined action of all the above-cited factors, an intensive stirring and dispersion of the components takes place in the working zone of the reactor, at the same time the components are fed into the reactor continuously and at a given ratio. The duration of treating the components in the reactor, even when the saponification reaction (for soap lubricants) takes place, does not exceed several seconds at temperatures no more than 70° to 90°C under atmospheric pressure.

The product obtained in the reactor is discharged continuously and delivered to the subsequent stages of treatment, if necessary. Thus, for example, when preparing sodium or lithium soap lubricants, the dispersion obtained is heated at a temperature of about 160° to 250°C for producing a melt of thickeners in oils with a regular structure; then water is evaporated, deaeration performed, and the product is cooled at a rate ensuring prescribed crystalline structure of the lubricant. After that the lubricants are subjected to homogenation to improve their rheological properties. The ferromagnetic particles are retained by the magnetic field in the working zone of the reactor and do not contaminate the product.

The best effect can be obtained when using nonequiaxial ferromagnetic particles with the ratio between their large and small size within the range from 6 to 20. Various ferromagnetic metals and alloys both magnetically soft and hard, such as carbonaceous steel, nickel, cobalt-nickel alloys, and the like, can be used for preparing particles.

Example 1: To obtain plastic lithium soap lubricant, nonequiaxial ferromagnetic particles, from magnetically soft carbonaceous steel, with a ratio between large and small size equal to 9 to 11 and with a surface covered by a polyethylene layer are placed into a 0.5-liter reactor fitted with an electric inductor having an active power of 1.7 kW and supplied from three-phase alternating-current mains at 380/220 V and 50 Hz. The inductor is switched on and a magnetic field is established inside the reactor, the magnetic field rotating at a speed of 3,000 rpm. The components are fed into the reactor with the help of a metering device at 76°C, the flow rates of the components being as follows:

Technical-grade stearin	44.8 kg/hr
10% aqueous solution of lithium hydroxide	36.2 kg/hr
Mineral oil with a viscosity of 7 cs at +50°C	392.0 kg/hr
5% solution of diphenylamine in the same mineral oil	27.0 kg/hr

The soap-oil dispersion formed at a flow rate of 500 kg/hr, from the reactor with the rotating magnetic field, is delivered with the help of a metering pump into a thermal unit. The soap-oil dispersion is melted in the thermal unit at +220°C under 15 kg/cm². The product leaving the thermal unit goes to an evaporator where a rarefraction of 150 to 220 mm Hg is maintained. Due to a sharp pressure drop, the moisture from the product is completely removed. The product temperature falls down to +150°C. After the evaporator, the product with the help of a second measuring device is fed into a scraper cooler where it is cooled down to +40°C, and then the product passes through a filter and a slotted homogenizer where it is treated under 100 to 120 kgf/cm². After that the product is discharged. The lubricants obtained (465 kg/hr) have the following characteristics:

Ultimate strength at +50°C	4 gf/cm²
Viscosity at -50°C and deformation rate 10 s⁻¹	6,450 poises
Free alkali content as calculated for NaOH	0.08%
Drop point	178°C
Oxidability (in mg of KOH)	0.13
Colloidal stability	24.2%
Evaporativity	18.8%
Mechanical impurities	None
Water content	None
Corrosive action on copper plates at +100°C for 3 hours	None

Example 2: To obtain plastic lithium soap lubricant, similar to that described in Example 1, a 2-liter reactor fitted with an inductor having an active power of 7.5 kW and the same type of ferromagnetic particles are used. The flow rates of the components are as follows:

Technical-grade stearin	190 kg/hr
10% aqueous solution of lithium hydroxide	153 kg/hr
Mineral oil with a viscosity 7 cs at 50°C	1,660 kg/hr
5% solution of diphenylamine in the same mineral oil	120 kg/hr

The soap-oil dispersion (2,123 kg/hr) formed in the reactor is treated by following the procedure described in Example 1; lubricant oil is obtained (1,980 kg/hr) with characteristics close to those given in Example 1.

SPECIALTY LUBRICANTS AND FLUIDS

Railway Lubricating Oil

A process described by *J.L. Thompson, S.S. Deluga, J.W. Harnach and E. Shamah; U.S. Patent 4,131,551; December 26, 1978; assigned to Standard Oil Company* relates to lubricating oils of high dispersancy-detergency and high alkalinity reserve for use as crankcase lubricant in marine and heavy duty diesel, such as railway diesel engines.

The compositions are improved lubricant compositions comprising (A) a lubricant mineral base oil, (B) a Mannich condensation reaction product comprising the reaction product of an alkylphenol, a polyamine and formaldehyde, (C) an alkaline earth metal salt of a Mannich condensation reaction product comprising the reaction product of an alkylphenol, formaldehyde and a polyamine, (D) an alkylbenzene sulfonate of an alkaline earth metal of low total base number, (E) an alkaline earth metal salt of bisalkylphenol sulfide, (F) a chlorinated paraffin, and (G) a small amount of a polydimethylsiloxane.

Example 1: Preparation of (B) Mannich Condensation Product — A stirred reactor is charged with 0.4 mol of nonylphenol over a period of 7 hours. About 1 mol of boron trifluoride, BF_3, is blown into the phenol while maintaining the temperature below 175°F. The resultant BF_3 complex has a boron content of about 1%. 100 g of the BF_3-nonylphenol complex is added to 1,100 g of polybutene having an average molecular weight of about 900, diluted with solvent-extracted 5W oil, with stirring, at about 100° to 125°F. After stirring the reaction mass at 100° to 125°F for about one hour, the reaction mass is neutralized. The reaction mass is then heated to about 500°F while excess volatiles are stripped therefrom with inert gas.

The polybutyl phenol amine condensation reaction product is prepared by charging the stirred reactor with 1,200 g of the polybutyl phenol produced in the preceding step, together with 775 g of tetraethylenepentamine and the temperature is adjusted to 80°F or less. Then there is added 710 g formaldehyde. After the formaldehyde addition, the reaction mixture is rapidly heated to about 320°F, while blowing with an inert gas to remove volatiles. The stripped reaction mixture is then filtered and the filtrate is diluted with solvent and ready for use.

Example 2: Preparation of (C) Calcium Salt of Mannich Reaction Product — 8.0 mols of nonylphenol in a diluent oil were added to a flask under a nitrogen blanket. 4.0 mols of ethylenediamine were added at a rate to keep the flask below 300°F. The mixture was heated to 300°F for one hour. The mixture was cooled. Antifoam agents and diluent oil was added. 3.0 mols of calcium hydroxide in 400 ml of diluent oil were added to the mixture. The reaction mixture was heated to 190°F for one hour, then heated to 300°F to remove water which was blown with nitrogen. The mixture was cooled and filtered to a clear product.

Example 3: Preparation of Sulfonate (D) — 1.070 g of benzene is charged into a reaction vessel and heated with steam. 17.4 g of aluminum chloride is slowly added to the benzene and the mixture is stirred until a complex agent reaction mixture is completed, approximately one-half hour. Into this mixture is mixed 870 g of a polypropylene which has a molecular weight of from 400 to 600.

The polypropylene is added at a rate so that the addition is complete in about 20 minutes. At the end of the addition, the reaction is continued for another 20 minutes. At the end of this time, the mixture is heated to approximately 250°F and is blown with nitrogen or steam to remove benzene, unreacted polymer, and light alkylates. The heavy alkylate is recovered. Approximately 720 g of polymer alkylate is produced. The sulfonation of the alkylate is done by mixing in a jacketed vessel the alkylate and approximately an equal amount of 22% oleum over a time period of about 1.5 hours. During this mixing step, the temperature of the mixture is not allowed to exceed 95°F.

Upon completion of the mixing, the mixture is allowed to react for approximately one hour at a temperature not greater than 130°F. At the end of this time, the mixture is diluted with 250 g of water to form a concentration of sulfuric acid in the aqueous layer of less than 85%. The mixture is allowed to settle and separate into a lower sulfuric acid layer and an upper sulfonic acid product. The separation is substantially complete in approximately 20 minutes.

To prepare the calcium overbased sulfonate, 1.38 mols sulfonic acid, 300 ml xylene, 929 mols calcium oxide and 24.7 mols methanol are placed in a reaction vessel. Into this mixture, carbon dioxide and ammonia are bubbled at 80°F. The carbonation is continued for approximately one hour. At the end of this time, the temperature of the reaction vessel is increased to 250°F and the reaction mixture is blown with an inert gas to remove the xylene, the methanol and unreacted carbon dioxide and ammonia. The mixture is filtered and the overbased calcium sulfonate is recovered. Overbasing technology is well known and variations in base number are readily achieved.

Example 4: Preparation of Calcium Alkylphenol Sulfide (E) — To a 5-liter flask fitted with a stirrer and Dean Stark trap was added the following: 157 g (0.71 mol) nonylphenol; 784 g (2.99 mols) dodecylphenol; 886 g (239 g/mol phenols) SX-5W oil; 184 g (2.49 mols) calcium hydroxide; and 129 g (4.03 mols) sulfur. The mass was heated to 360°F and held there for two hours. Volatiles are removed by heating to 460°F while blowing with a small stream of nitrogen. Filter-aid was added and the product was isolated by filtration. The base number was 123.

Improved Test Lubricant Formulations

	Test Oil, % by wt			
	1	2	3	4
Mannich additive (B)	1.87	3.0	2.7	3.5
Ca Mannich salt (C)	1.1	1.6	1.6	1.6
Calcium sulfonate (D)	5.0	2.0	2.0	2.0
Bisphenol sulfide (E)	2.7	4.2	4.1	2.0
Chlorinated paraffin (F)	0.13	0.2	0.13	0.20
Siloxane polymer (G)	5 ppm			
Base oil (A)	balance			

120 Hour Caterpillar 1-G Test Results

	Test Oils			
	1	2	3	4
Top groove fill, %	37	6	46	19
CRC demerits				
Weighted carbon	91	18	72	89
Weighted lacquer	46	90	109	135
Weighted total	137	108	151	224

EMD 2-567 Engine Tests

Silver Corrosion Demerits	. . . . Test Oils.			
	1	2	3	4
Right bearing No. 1	28	22	29	42
Left bearing No. 2	46	28	34	46
Average	37	25	32	44

Alkalinity Retention

Tests	ASTM Test	Alkalinity . After Test. .		Increase in Alkalinity Retention, %
		1	2	
Cat. 1-G	D-2896	4.8	8.3	61
	D-664	2.3	3.7	73
EMD 2-567	D-664	1.1	2.1	40

Example Inspection of Test Oil 2

Gravity	24.1°F
Viscosity	
SUS at 100°F	998
SUS at 210°F	79
Index	72
Flash point	470°F
Pour point	+5°F
TBN, D-664	8.4
TAN, D-664	0.8
TNB, D-2896	10.0
Chlorine	850 ppm
Calcium	0.29 wt %
Sulfur	0.32 wt %
Nitrogen	0.07 wt %
Sulfated ash	0.92 wt %
Zinc	5 ppm
Boron	5 ppm
Magnesium	10 ppm
Potassium	1 ppm
Silver disc four ball test	
at 350°F (scar diameter)	1.52 mm
at 500°F (scar diameter)	1.64 mm

Test oil 2 shows, in these data tables, a better performance. In each test the performance of test oil 2 is significantly better than the comparison oils. The composition of test oil 2 has been successfully tested in diesel engines operating in commercial service, and in admixture with other railway oils.

The Caterpillar 1-G and EMD 2-567 are well known tests in which diesel engines are run under various operating conditions of temperature, speed and load to test lubricant performance. The Caterpillar 1-G and EMD 2-567 are the standard tests for the examination of railway diesel lubricants. ASTM D-2896 and D-664 are potentiometric titration methods for determination of basicity in petroleum oils. D-2896 uses perchloric acid as a titrant and D-664 uses hydrochloric acid or potassium hydroxide. Results are expressed as total base number.

Bearing Lubricant

R. Baur; U.S. Patent 4,151,102; April 24, 1979; assigned to Swiss Aluminum Ltd., Switzerland describes a synthetic bearing lubricant which prevents wear even under very high loads and is such that when it contaminates the rolling lubricant the rolled product is not stained. The 100 pbw of the lubricant comprises: 91.5-95.3 pbw of polyisobutene; 3.7-7.5 pbw of at least one ester of a C_{2-5} alcohol with an α-hydroxy-carboxylic acid; and 0.6-1.2 pbw of at least one unsaturated C_{13-19} carboxylic acid. The manufacture of the lubricant takes place simply by mixing the components in accordance with the required composition. The mixing operation can be made easier by warming the viscous components.

Extensive plant trials with various lubricating systems have shown that the lubricant of the process can be used to full advantage in all known lubricating systems (e.g., closed circuit systems and various open circuit systems with ball, roller and cone bearings). It has been found particularly advantageous to employ the lubricant in oil mist and oil droplet/compressed air lubricating systems. It was found that the friction which is found to occur with ring bearings on starting up cold, can be avoided to a large extent with the lubricant according to the process.

Trials have shown that the tendency of the rolls to stick can be reduced without impairing the advantages accrued from the process, if the polyisobutene is completely or partly replaced by polymethacrylate dissolved in mineral oil, or by a mixture of polymethacrylate and kerosene dissolved in mineral oil. In the latter case, the ratio of polymethacrylate solution:kerosene should be approximately 2:1 to 1:2, preferably 1.2:1 to 1:1.2. The kind of mineral oils containing polymethacrylate are commercially available products (e.g., Viscoplex SV 36, Rohm GmbH), and are in general used as lubricant additives to lower the stock point and to raise the viscosity index.

The polybutenes used to form the hydrodynamic lubricant film are likewise commercially available products (e.g., Indopol L 10 and Indopol H 100, Amoco Chemicals). The viscosity of the lubricant can be altered over a relatively large range simply by mixing in polybutenes of various chain lengths. For example, by altering the ratio of mixing of the two abovementioned Indopols, at 60°C a viscosity range of 10 to 5,000 cs can be obtained.

Temporary Rust-Proofing Lubricant for Steel Plates

A process described by *T. Sakurai, S. Shimada and Y. Kamimura; U.S. Patent 4,113,635; September 12, 1978; assigned to Nippon Steel Corporation* is directed to rust-proof lubricant compositions comprising partial esters of pentaerythritol with fatty acid, as the chief constituents, rust-preventatives, lubrication-improving agents, surface active agents, etc., having melting points of 30° to 60°C, which can be applied on steel plates without the necessity of any solvent or heat-drying, and which may be adapted for the continuous coating of steel plates.

The reason why the partial ester of pentaerythritol of a fatty acid is selected is based mainly on the following facts. The partial ester of pentaerythritol of a fatty acid has hydroxyl groups together with ester groups, and their configuration is of the tetrapod type, so that the adsorption strength to the metal surface is strong and the sliding lubrication is remarkably excellent as compared with gly-ceride, for example, palm kernel oil, and commercial metalworking oils. More-over most of the partial esters of pentaerythritol of a fatty acid have melting points of less than 60°C, and their viscosity in melt is lower, being different from those of the polymers. With these properties, they can be easily applied on steel plates, and can be used for the continuous coating operation of steel plates by the usual coating method, such as roll coating, or spray coating.

Example: The coating compositions of this process listed in Table 1 are inde-pendently liquidized by heating to 70°C, and applied continuously on coils of degreased coil rolled steel plates (0.8 mm thick) each in an amount of 1 to 2 g/m² by spray coating. Immediately on applying the spray coating, the thickness of the coating is made uniform by use of hot rolls maintained at a temperature of 75° to 80°C, and cold air is sprayed without delay into the steel plates which are then taken up into a coil.

The thus-coated plates were subjected to a pressing formability test, and the re-sults are shown in Table 1. As controls, samples for the test are prepared by applying a commercial metalworking oil and two lubricants of the solid film type, one of which requires the use of a solvent, and the other of which is com-posed of a soap. Results are shown in Table 2.

Table 1: Performance of Coating Lubricant Compositions of This Process

No.	Fatty Acid or Its Source	Monoester	Diester (%)	Triester
A	Stearic acid	10	86	4
B	Stearic acid	10	86	4
C	Fish oil fatty acid	31	44	25
D	Fish oil fatty acid	31	44	25
E	Coconut oil fatty acid	20	10	70
F	Coconut oil fatty acid	20	10	70
G	Hydrogenated beef tallow fatty acid	5	80	15
H	Hydrogenated beef tallow fatty acid	5	80	15

No.	Pentaerythritol Partial Ester	Dicyclohexylamine Rust Preventive (%)	Other Additives (%)	
A	95	5	None	—
B	85	5	Polyoxyethylene alkyl ether phosphate	10
C	85	5	Mineral oil	10
D	85	5	Tricresyl phosphate	10
E	95	5	None	—
F	85	5	Trichloroethyl phosphate	10
G	95	5	None	—
H	85	5	Polyoxyethylene alkyl ether phosphate	10

No.	Formability (mm)	Rust Prevention*	Degreaseability**
A	163	oo	oo
B	Draw fit	oo	oo
C	162	oo	oo
D	Draw fit	oo	oo
E	160	oo	oo
F	167	oo	oo
G	168	oo	oo
H	Draw fit	oo	oo

*,**See footnotes after Table 2.

Table 2: Performance of Coating Composition on the Market

Composition	Film Type	Formability	Rust Prevention*	Degreaseability**
Metal working oil***	Liquid	105	y	x
Solvent-type	Solid	148	x	x
Fatty acid soap-type	Solid	73	o	oo

*A weathering test was carried out for 15 days in an atmosphere of 80% humidity at 35°C. oo = less than 0.1% rust; o = 0.1 to 5% rust; y = 6 to 30% rust; and x = more than 31% rust.

**Samples prepared by applying the lubricants independently were tested after they had been allowed to stand for 1 month in an atmosphere of 80% humidity at 40°C.

***Commercial drawing lubricant.

A cleaner solution containing 2% of a degreasing agent, Fine Cleaner No. 353, is sprayed onto the samples at the pressure of 1 kg/cm^2 for 2 minutes.

The degreaseability is estimated by measuring percentage of a water-wettable area after the washing. oo is 100%; o is 81 to 99%; and x is below 80%. The pressing test for determining the formability was carried out using steel disks (blanks), 480 mm diameter, cut out from their respective lubricant-coated steel plates.

The formability is evaluated by the height at which the blank steel is broken by pressing with a spherical heat punch having a diameter of 200 mm, using a 150-ton press tester (blank holder pressure 10 ton).

Centrifugal Compressor Refrigeration Lubricant Based on Castor Oil

A process described by *G.C. Gainer and R.M. Luck; U.S. Patent 4,159,255; June 26, 1979; assigned to Westinghouse Electric Corp.* comprises a halocarbon refrigeration system, using improved lubricants. A lubricant composition is provided which is a mixture of high viscosity and low viscosity fluids, which in combination have a low affinity for R-12, dichlorodifluoromethane, and yet provide excellent chemical and thermal stability in a refrigeration environment. This lubricating composition blend minimizes parasitic losses during refrigeration running, by minimizing the amount of R-12 dissolved in the lubricating blend, and therefore lost to the chilling function of the system.

The lubricating composition comprises a mixture of 100 parts of chemically and thermally stable castor oil, and a low viscosity blending fluid additive, having a viscosity of up to 335 SUS at 100°F, which is soluble in castor oil, and chemically and thermally stable in the presence of halocarbon refrigerants. The blending fluid is selected from pentaerythritol esters of saturated fatty acids, dipentaerythritol esters of saturated fatty acids, alkylated diphenyl esters, neopentyl esters, and their mixtures.

Castor oil has a very high viscosity, approximately 1,555 SUS at 100°F, making it completely unsuitable as a compressor lubricant. Castor oil, however, is relatively inexpensive and has extremely low affinity for R-12. The low viscosity blending fluids described above, when added to castor oil, can reduce the mixture viscosity to about 600 SUS at 100°F, which is suitable for use in centrifugal compressors. While the low viscosity fluids are themselves relatively soluble in

R-12, the mixture of them with castor oil exhibits a very low affinity for R-12, good lubricating qualities, low wear rates, and chemical and thermal stability in the presence of R-12. In addition, and very importantly, the additives are inexpensive and commercially available.

High Density Metal-Containing Lubricants for Oil Well Drilling

A process described by *G.L. Hurst; U.S. Patent 4,076,637; February 28, 1978; assigned to Tyler Corporation* relates to metal-containing greases of a high density which are useful in the down-hole drilling operations of the oil industry, for example.

According to the process, a solid metal is heated until it reaches a liquid state and is then thoroughly admixed into a carrier fluid which is thermally stable at the temperature of the molten metal. It was discovered that the metal disperses into fine globules which are held in suspension without any substantial tendency to coalesce back into larger particles. Furthermore, upon sufficient cooling, the globules harden and remain dispersed in the carrier fluid. The finely divided metal particles are homogeneously dispersed throughout the carrier fluid in the form of spherules of micron and submicron size. When lubricants are used as carrier fluids, the presence of the spherical metal particles causes an increase in the density of the lubricant without any appreciable loss of the lubricating qualities of the lubricant itself.

The discrete spherical metallic particles suspended in the carrier fluid are separated from the carrier fluid by common washing, dissolving or evaporation techniques resulting in a metal powder consisting of the abovedescribed metal spherules.

Example 1: 9 parts of Wood's metal (50% bismuth, 25% lead, 12.5% tin, and 12.5% cadmium by weight) and 1 part of a silicone-based grease sold as Dow Corning No. 111 were deposited in a suitable container. The container was positioned on a hot plate and heated until the contents reached 80° to 100°C. The Wood's metal, which has a melting point of 70°C, became molten. The metal and grease were then thoroughly admixed using a blender type mechanical mixer. The liquid metal vanished into and was seemingly enveloped by the silicone grease. The resulting suspension contained discrete spherules of Wood's metal and did not display any tendency to precipitate. The average size of the discrete metal spherules was less than about 2 μ. The metal-containing grease was then cooled and the resulting density was determined to be about 5 g/ml.

Example 2: 9 parts of Wood's metal and 1 part of a silicone-based grease sold as Dow Corning No. 41 were placed in a suitable container. The container was heated until the contents reached approximately 80° to 100°C at which point the Wood's metal became molten. Upon thoroughly admixing the contents, the surprising phenomena of Example 1 occurred, that is, the liquid metal appeared to disappear into the grease carrier fluid. The resulting average particle size of the metal globules was less than 1 micron. Upon cooling, the resulting metal-containing grease displayed a density of approximately 5 g/ml.

Example 3: The procedure of Example 1 was repeated substituting a hydrocarbon-based grease sold as Micro Lube (Micro Lube Inc.) for silicone grease. The petroleum-based grease completely enveloped the liquid metal and the suspension appeared to be substantially stable. This experiment demonstrated that the un-expected suspension of a molten metal when admixed with a lubricant carrier fluid does not depend upon synthetic lubricant bases, but occurs readily when petroleum based carrier fluids are employed. The resulting metal-containing grease had a density of approximately 5 g/ml.

Nuclear Service Lubricant

A process described by *D.N. Palmer and J.E. Davison; U.S. Patent 4,128,486; December 5, 1978; assigned to Combustion Engineering, Inc.* relates to anti-seize/antigall lubricants for nuclear service at elevated temperatures, in particular to the material applied at metal to metal interfaces to prevent seizure or fusion of one surface of the metal to the other.

The process involves the use of ^{11}BN, as an antiseizing compound to be applied to threaded metal parts prior to their joining. This compound is particularly ef-fective for elevated temperature nuclear reactor applications. For such usage, the ^{11}BN is, of course, formulated in a carrier composition so that the solid lub-ricant can be thinly but evenly placed between the threadedly engaged members. In some case, the carrier composition includes constituents that function as a secondary lubricant or as a lubricant assisting material.

An antiseize/antigallant type lubricant must possess a continuous network of lub-ricant molecules or molecule agglomerates over the substrate surface in order to function in a satisfactory manner in gallant applications. The continuous network or film structure must be capable of withstanding very high point stress pres-sures and relatively high compressive forces without film rupture, and must also yield or slide to impart lubricity. Film adhesion, network coherency, and film strength in unbindered lubricants are controlled by many variables including dis-persion homogeneity and surface wetting. Thus, the proper addition of film-form-ing aids (surfactants/dispersants) and matrix extenders is most desirable in the development of a useful antiseize/antigallant lubricant.

Thus, extenders, supplementary lubricants, etc., suitable for practice of this proc-ess, should exhibit a high level of stability in the range of 500° to 700°F. In par-ticular, the materials should exhibit a vapor pressure below 10^1 atm at 650°F, they should be free of nuclear poisons, and they should decompose into ma-terials that will not create severe corrosion problems for their environment. In total, only a few classes of nonvolatile materials, all polymers, have been found that are suitable for nonvolatile carrier substances. They are ^{11}B carborane poly-siloxanes, polyphenyl ethers, polyphenyl siloxanes/polyalkylaryl-siloxanes, and dimethylpolyalkalene ether copolymers of methyl phenyl poly-siloxanes.

In the instance of each of the above polymers, the polymer should be a nonreac-tive fully polymerized substance.

A multiplicity of formulations are contemplated within the preferred ranges for ^{11}BN/^{11}B carborane polysiloxane formulations depending upon the use of the lubricant and the lubricant form desired.

The following are examples of typical formulations which have the following characteristics; Nuclear Grade 400° to 650°F; 1,000 to 5,000 R/hr; 90% gamma,

10% fast neutron, 2 to 3 MeV energy spectrum; 1 to 3 years service; ^{11}BN and/or polysiloxanes (phenyl- or methylphenylsiloxanes) and/or ^{11}B carborane methylphenyl polysiloxanes and polyphenyl ethers.

Stable Oil-to-Grease Emulsion	Function	Weight Percent
1a—multiple vehicle system		
^{11}BN (nonturbostratic hexagonal)	solid lubricant	9.5–23.5
Phenylpolysiloxane A*	vehicle extender	22.5–33.0
Phenylpolysiloxane B**	vehicle extender	22.5–33.0
^{11}B-carborane methyl polysiloxane***	thickener/lubricant	45.0–10.0
GE SF-1066†	wetting agent/dispersant aid	0.5–0.5
1b—multiple vehicle system		
^{11}BN (nonturbostratic hexagonal)	solid lubricant	9.5–23.5
Polyphenyl ether††	vehicle extender	20.0–30.0
Phenylpolysiloxane B**	vehicle extender	25.0–36.0
^{11}B-carborane methyl polysiloxane***	thickener/lubricant	45.0–10.0
GE SF-1066†	wetting agent/dispersant aid	0.5–0.5
2a—simple vehicle system		
^{11}BN (nonturbostratic hexagonal)	solid lubricant	11.5–25.5
Polyphenyl ether††	vehicle extender	43.0–64.0
^{11}B-carborane methyl polysiloxane***	thickener/lubricant	45.0–10.0
GE SF-1066†	wetting agent/dispersant aid	0.5–0.5
2b—simple vehicle system		
^{11}BN (nonturbostratic hexagonal)	solid lubricant	12.5–26.5
Polysiloxane A* or B**	vehicle extender	42.0–63.0
^{11}B-carborane methyl polysiloxane***	thickener/lubricant	45.0–10.0
GE SF-1066†	wetting agent/dispersant aid	0.5–0.5

*Similar to 550 silicone oil (Dow Corning).
**Similar to 710 silicone oil (Dow Corning).
***Similar to Analabs (Olin) Dexsil 300 carborane polysiloxane, ^{11}B content, 100%, ^{10}B content less than 10 ppm, or Uscarsic (Union Carbide).
†Dimethyl polyalkylene ether copolymer of a methylphenyl polysiloxane (General Electric).
††6 ring.

Penetrating Oil Composition

P.J. Gibbons; U.S. Patent 4,113,633; September 12, 1978 describes a penetrating oil which comprises the following components:

	Percent
Mono- and dimethylbenzenes	12–35
Lower alkanols (C_{1-4})	12–30
Antioxidants and rust-reducing agents	1–4
Detergents	0.2–1.0
V.I. improvers	0–4
Paraffin and naphthalene bright-stock base qs to 100%	

The mono- and dimethylbenzenes are chosen from toluene and the xylenes. Any of the xylenes or mixtures thereof may be used. While either toluene or xylene may be used as the methylbenzene component, a mixture of the two is preferred with a toluene/xylene ratio of about 2.5 to 1.

The alkanol component may be any lower alcohol such as methanol, ethanol or a propanol. For tax and economy reasons a blend of methanol and isopropanol is preferred in a ratio of 1 to 5 respectively. Such a ratio also blends well with the methylbenzenes.

The bright oil base is a blend of natural paraffin oils, pretreated for high-pressure, low-viscosity lubricant use, such as available commercially for transmission oils.

The antioxidants and reducing agents are those commonly used in automobile lubricants with tert-di-butyl-cresol and similar sterically hindered aromatics preferred as the prime antioxidants and augmented with zinc dialkylthiophosphates as a combined derusting agent and antioxidant. This antioxidant mixture may also be augmented with alkyl succinic acid. Up to about 3.5 to 4% of combined antioxidants and rust looseners are sufficient but as little as 0.5% will suffice and about 2% is preferred. The zinc alkylthiophosphate and hindered aromatics should be included for their combined loosening power.

Examples of compositions below are in parts by weight.

	. . . . Examples.				
	1	2	3	4	5
Toluene	12.5	16.6	21.0	25.0	16.6
Xylene	4.0	5.5	6.5	8.0	5.5
Isopropanol	10.0	13.5	17.0	20.0	13.5
Methanol	2.1	2.8	3.5	4.2	2.8
Benzyl acetate (odorant)	1.2	1.6	2.0	2.4	1.6
Tert-dibutyl-p-cresol	0.8	0.7	0.6	0.4	0.7
Alkyl succinic acid	0.4	0.3	0.25	0.2	0.3
Zinc dialkyl thiophosphate	1.0	0.9	0.7	0.65	0.9
Polymethacrylate	1.9	1.6	1.4	1.1	–
Methyl silicone polymer	1.0	0.9	0.25	0.2	--
Lube oil bright stock to 100%					

The formulation of Example 2 has been found to provide the best combination of properties for use in automotive shops and for householder service. For plumbing, where greater penetrating activity is desired the formulation of Examples 3 and 4 is preferred. Example 5 is adequate for service where the parts will have to be cleaned and refitted after separation.

Gas Bearing Gyroscopes

B.H. Baxter; U.S. Patent 4,076,635; February 28, 1978; assigned to British Aircraft Corporation Limited, England describes a lubricant for use in gas bearing gyroscopes. In a gas bearing gyroscope the gyroscope rotor is supported by means of a gas bearing. These bearings are usually of the type wherein the rotation of the various bearing components generates the required gas pressure to maintain a working clearance between the bearing surfaces. At low rotational speeds, therefore, the gas pressure is insufficient to maintain the working clearance and rubbing contact occurs between the bearing surfaces.

According to the process, the lubricant includes a colloidal form of a metal oxide hydrate, the metal being chosen from those elements in the first transition period of the Periodic Table, and a fatty acid having a chain length in the region of 16 to 24 inclusive carbon molecules, the fatty acid being combined with surface regions of the colloidal particles of the metal oxide hydrate.

Preferably the lubricant is prepared by forming a suspension, comprising a quantity of a colloidal form of the metal oxide hydrate and a suspending medium, adding to the suspension the fatty acid in a proportion of not more than 20% by weight of the mass of suspended colloid, dispersing the fatty acid with respect to the colloidal particles in suspension such that the fatty acid combines with a surface region of the colloidal particles to form a lubricant layer on the particles, and subsequently removing the suspending medium.

The preferred method of preparing the lubricant is as follows: The iron(III) oxide hydrate is precipitated from an aqueous solution of ferric ions, by the addition of a suitable alkali precipitant, in this example ferric chloride, while the pH value of the solution is maintained within the range 6.0 to 8.0.

The precipitated solid product is then washed in water until free from chloride ions, or other anions.

The precipitant is then transferred to a nonaqueous medium by washing with a suitable organic solvent such as acetone until free of water followed by further washing with a volatile hydrocarbon, such as heptane, until free of acetone.

The iron(III) oxide hydrate is dispersed in the heptane at this stage and can be stored in this form as a suspension if necessary.

The metal oxide hydrate content of the suspension is subsequently determined by analysis so that the required quantity of fatty acid, up to 20% by weight as previously mentioned, can be added to the suspension. The resulting mixture is dispersed to a stable colloidal suspension by use of high frequency vibrations.

Finally the stability of the final product, i.e., the dispersed lubricant, is established by analysis of the infrared spectrum of the dispersion.

The lubricant is applied to the surface to be lubricated in the heptane dispersed form or it may be further diluted by a halogenated solvent such as chloroform. In either case, the solvents are allowed to evaporate depositing a lubricant residue on the surfaces.

Lubricant for Processing of Molten Glass

A process described by *E.S. Nachtman and R.G. Hitchcock; U.S. Patent 4,119,547; October 10, 1978; assigned to Tower Oil & Technology Co.* relates to high temperature lubricants and more particularly to a lubricant composition of the type in which water is present as a major component.

In general, such high temperature lubricants have been provided in the form of an aqueous emulsion in which the water represented the continuous phase and a lubricant the discontinuous phase. Difficulties have been experienced with the use of lubricant compositions of the type described in that an excessive amount of smoke and mist is generated when used in such high temperature manufacturing processes and the protection between the work piece and tool is insufficient, especially in glass working operations.

This process deviates from conventional practice in the production of a high temperature lubricating composition and makes available a number of unique operating characteristics, such as reduced smoke or mist generation during use and less fire hazard, while providing the desired barrier film between the work and tool.

It has been discovered that the characteristics of the type described can be achieved by reversal of the status of the components wherein the lubricating component is embodied as a liquid and/or solid in the continuous phase, while the aqueous phase is embodied as the dispersed phase. Thus, the hot surfaces to be cooled and lubricated are first engaged by the continuous phase containing the lubricating component or components to wet out the surfaces before an aqueous hydrophilic and lubricant repellent phase can form on the hot surfaces of the tool or work. This is especially effective in the processing of molten glass which is a highly hydrophilic material but it is equally effective as a high temperature lubricant in the processing of metal and other materials at high temperature.

The desired reversal of phases is achieved by the use of an oil-soluble surfactant or combination of surfactants where final HLB number is low, to stabilize the composition. Representative of such oil-soluble surfactants are lipophilic tertiary amides in which the lypophilic group is a long chain alkyl such as a C_{8-18} alkyl, as represented by octyl to octadecyl, and the like, a polycyclic group such as cholesterol, aryl alkyl and derivatives thereof in which the alkyl is from 1 to 10 carbon atoms and the aryl is phenyl, naphthyl, and the like. Surfactants having a final HLB number below 7 and preferably below 5 are suitable for use. A lipophilic tertiary amide of the type described is Lubrizol 5162 (Lubrizol Corporation).

The continuous lubricant phase may consist, for example, of a low molecular weight polymer formed of such polymers as isobutene, high molecular weight alcohols or polyols which are immiscible in water, as well as petroleum based oils of varying viscosity, such as naphthenic based oils. The continuous lubricant phase can be formulated to contain a solid which provides high temperature lubrication under the conditions described without raising hazards or the generation of smoke or mist. Representative of such solid lubricants are graphite, molybdenum disulfide, mica, boron nitride and potassium iodide.

Example 1: Glass Mold Lubricating Composition –

    (a)   17.5% by weight isobutene polymer, having an average molecular weight of 340 (Poly Vis OSH),

    (b)   20.0% by weight isobutene polymer, having an average molecular weight of 1,350 (Lubrizol 3156),

    (c)   10.0% by weight naphthenic base oil (100 SUS at 100°F),

    (d)   7.5% by weight graphite (1 to 5 microns, ashless),

    (e)   40.0% by weight water, and

    (f)   5.0% by weight alkanol surfactant (Lubrizol 5162).

Example 2: Glass Ring Lubricating Composition –

    (a)   47.5% by weight isobutene polymer, having an average molecular weight of 340

    (b)   5.0% by weight alkanol amide surfactant (Lubrizol 5162),

    (c)   7.5% by weight graphite, and

    (d)   40.0% by weight water.

In Examples 1 and 2, the solid lubricant and surfactants are first added to the carrier along with any other lubricant component. Then the water is added with mixing. The water or aqueous phase is thus dispersed as the discontinuous phase in a continuous phase of the lubricant component.

Automatic Transmission Fluid

E.F. Outten, J. Ryer and J.E. Williams; U.S. Patent 4,116,877; September 26, 1978; assigned to Exxon Research & Engineering Co. have found that mineral oil containing at least a seal swelling amount of the combination of an organophosphite having the general formula $(RO)_3P$ where R is a hydrocarbyl group containing from 8 to 24 carbons, preferably tris(nonylphenyl)phosphite and a phenol of the general formula

where R is a hydrocarbyl group of from 8 to 24 carbons, preferably nonylphenol; the organophosphite having a weight ratio to the phenol from 4:1 to 1:4, preferably 3:1 to 1:3 optimally about 3:2, surprisingly enhances the elastomer compatibility of the oil, as well as providing the essential components from which a hydraulic fluid of enhanced elastomer compatibility can be formulated.

Examples 1 through 9: Elastomer compatibility of a nitrile seal with the combination of the process is shown by the results of a test wherein various combinations are admixed into a power steering fluid and the elastomer seal immersed in the test fluid for 70 hours at 149°C. The results are shown in Table 1.

Table 1

Ex. No.	Total Weight Percent of Combination	Tris(Nonylphenyl) Phosphite	Nonylphenol	Tensile Strength (psi)
1	0.0	0.0	0.0	527
2	0.5	0.5	0.0	518
3	0.5	0.0	0.5	425
4	0.5	0.2	0.3	1,156
5	0.5	0.3	0.2	1,208
6	1.0	1.0	0.0	668
7	1.0	0.0	1.0	1,069
8	1.0	0.7	0.3	1,746
9	1.0	0.3	0.7	1,865

The test fluids were prepared by mixing the elastomer compatibility agent to be tested in a power steering fluid (PSPF) blend of Solvent Neutral 150 mineral oil and a naphthenic solvent oil which contained about 0.7 wt % of a commercial ashless dispersant/antioxidant, 1.35 wt % of a commercial multifunctionalized V.I. improver and 0.1 wt % of a commercial friction modifier. Tensile strength was measured according to ASTM D1414-72 after fluid aging for 70 hours at 149°C according to ASTM D471-72.

The marked improvements in elastomer compatibility provided according to this process are clearly shown by comparing Examples 1 through 3 with Examples 4 and 5 and Examples 1, 6 and 7 with Examples 8 and 9.

The PSPF blend as detailed above was mixed with tris(nonylphenyl)phosphite and a commercially available hindered phenol to evaluate their elastomeric compatibility properties. The results with a nitrile rubber showed a blend of phosphite and phenol such as is taught in the British Patent 1,282,652 to be inferior in elastomer compatibility to the process combination, as seen in Table 2 wherein Elongation to Break measurements were carried out according to ASTM D1414-72 after fluid aging as above per ASTM D471-72.

Table 2

Concentration of Phosphite-Phenol .... Agents in PSPF (vol %)*		Tensile	Elongation
Tris(nonylphenyl) Phosphite	Hindered Phenol	Strength (psi)	to Break (%)
0.5	0.5**	420	27
0.5	0.5***	480	23

*PSPF as described above.
**Ethyl Anti-oxidant 702 (Ethyl Corp.), an alkylated tert-butyl methylene bridged bisphenol.
***Ethyl Anti-oxidant 728 (Ethyl Corp.), an impure alkylated tert-butyl methylene bridged bisphenol.

Boron Dispersant for Automatic Transmission Fluid

E.J. Friihauf and D.L. Murfin; U.S. Patent 4,080,303; March 21, 1978; assigned to The Lubrizol Corporation describe lubricant compositions comprising (1) a viscosity improving agent, (2) a boron-containing dispersant, and (3) an ester of an aromatic carboxylic acid having 6 to 10 carbon atoms in the aromatic nucleus and a total of 6 to 40 aliphatic and alicyclic carbon atoms as well as additive concentrates for making such lubricant compositions.

Preferably, (1) is a carboxy-containing interpolymer in which some carboxy radicals are esterified and the remaining carboxy radicals are neutralized with an amino compound, and (2) is a borated, acylated polyamino compound, having an acyl group containing at least 50 carbon atoms. These lubricant compositions are useful as automatic transmission fluids.

Examples 1 through 9: To make the following lubricant compositions, a master blend is prepared containing 98 parts by weight of a conventional mineral oil automatic transmission fluid base stock, 0.76 part by weight of a borated succinic acid-polyamide dispersant prepared in the manner described in U.S. Patent 3,087,936 and 1.34 parts by weight of an esterified/amine-treated styrene/maleic anhydride copolymer prepared in the manner described in U.S. Patent 3,702,300. To this blend is added 2 parts by weight of the aromatic diesters listed in the table below.

Example	Aromatic Ester
1	Di(allyl)phthalate
2	Di(n-butyl)phthalate
3	Di(capryl)phthalate
4	Di(n-hexyl)terephthalate
5	Di(isononyl)isophthalate

(continued)

Example	Aromatic Ester
6	Di(undecyl)phthalate
7	Di(octadecyl)-1,8-naphthalene dicarboxylic acid
8	Di(ethylhexyl)phthalate
9	Isooctyl 4-tetrapropylbenzoate

Each of the preparations made in Examples 1 through 9 exhibit seal swell properties appropriate for an automatic transmission fluid in laboratory tests.

HYDROTREATING AND PROCESSING OF SPECIALTY OILS

Upgrading Lubricating Oil Stock

R.F. Bridger, C.A. Audeh and E.-A.I. Heiba; U.S. Patent 4,090,953; May 23, 1978; assigned to Mobil Oil Corporation describe an improved process and means for forming lubricating oils which are highly resistant to deterioration, e.g., oxidation and sludge formation, upon exposure to a highly oxidative environment.

The process comprises contacting a lubricating oil stock, such as, for example, from a Midcontinental U.S. crude or an Arabian Light crude in a flow reactor or under conditions comparable to those existing in a flow reactor with elemental sulfur in amounts of from 0.025 to 0.2% by weight of the oil stock in the presence of a catalyst material selected from the group consisting of alumina, silica, an aluminosilicate, a metal of Groups II-A, II-B, VI-B or VIII of the Periodic Table of Elements, an oxide of a metal of Groups II-A, II-B, VI-B, or VIII, a sulfide of a metal of Groups II-A, II-B, VI-B or VIII, clay, silica combined with an oxide of a metal of Groups II-A, III-A, IV-B or V-B and combinations thereof.

The elemental sulfur may be provided for the treatment, if desired, by a sulfur precursor, such as, for example, H_2S, an organosulfur compound, i.e., added or naturally occurring, or combinations thereof. The naturally occurring organosulfur compound may be utilized if present in the lubricating oil stock in a quantity providing greater than about 0.125 wt % sulfur. When such an organosulfur compound is the source of elemental sulfur herein, it can serve for generation of sulfur in situ. The catalyst materials for use in this process serve to assist the extrusion of naturally occurring sulfur from the lubricating oil stock and, if sufficient organosulfur compounds are present therein, dehydrogenation is enhanced.

The basic test procedure employed in evaluation of product yield from the process is described by Dornte in *Industrial and Engineering Chemistry*, 28, 26-30, (1936), modified as below indicated.

It is interesting to note that performance of a lubricating oil in the below described test method is indicative of that oil's performance in the field. The test is conducted in an air circulation apparatus of the type described by Dornte. A tube containing 30 g of lubricating oil (with or without additive) is placed in a heater thermostatted at 162°C. Air is circulated through the oil sample at a rate of 5 l/hr. Metal surfaces are provided for oil contact to act as oxidation accelerators. The metal surfaces employed include: iron wire, analytical grade, Washburn and Moen No. 15 gage, wound into a coil approximately ⅝ inch o.d. and

$2\frac{5}{8}$ inches long to give a surface area of approximately 15.3 in^2; copper wire, electrolytic, B and S gage No. 18, 6.2 inches long; and a lead (tin-free) square 0.25 x 0.25 inch cut from $\frac{1}{16}$ inch thick sheet.

Another test method used herein is the standard Rotary Bomb Oxidation Test (RBOT) designated ASTM D2272. Each sample tested in the RBOT was blended with a standard commercial additive package prior to testing.

The lubricating oil stock (150 SUS Arabian Light) used in the following examples was conventionally refined by distillation, followed by furfural extraction and methyl ethyl ketone dewaxing. It has the following physical properties and furfural extraction conditions.

Furfural dosage, vol %	180
Tower temperature, °F, top	185
Tower temperature, °F, bottom	140
Gravity, °API	30.9
Pour point, °F	0
Flash point, °F	410
Sulfur, wt %	0.63
Nitrogen, wt %	0.0029
Aniline point, °F	210
Viscosity, SUS at 100°F	152
Viscosity index	103
ASTM color	$1\frac{1}{2}$

Example 1: A 50-g quantity of the above oil stock, without treatment in accordance with the process, was subjected to the RBOT test and a 30-g quantity of the untreated oil was subjected to the abovedescribed Dornte test. The results of the tests are recorded in the table below for comparison purposes with tests conducted on the same oil stock treated by the process (Examples 2 through 5).

Example 2: A 450-g quantity of the above oil stock was mixed with 0.2 wt % of elemental sulfur (based on weight of oil) and allowed to flow through a reactor containing 10 g of $\frac{1}{16}$-inch extrudate zeolite X (i.e., NaX). The reaction temperature was maintained at 175°C, the reaction pressure was maintained at 25 psig and the LHSV was 1 hr^{-1}. The oil was then cooled to room temperature and residual corrosive sulfur was removed by stirring the oil with finely divided sodium hydroxide (50 g) for 16 hours. After removal of the sodium hydroxide by filtration, the oil was tested in the RBOT and Dornte tests, and shown to have improved oxidation properties, as demonstrated in the table below. The flow reactor used was a 15 ml downflow reactor.

Example 3: A 450-g quantity of the above oil stock was mixed with 0.1 wt % of elemental sulfur (based on weight of oil) and allowed to flow through the reactor and under the reaction conditions of Example 2. The oil was then washed with sodium hydroxide, filtered and tested as in Example 2. The results showing improved oxidation properties are recorded in the table below.

Example 4: A 450-g quantity of the above oil stock was mixed with 0.05 wt % of elemental sulfur (based on weight of oil) and allowed to flow through the reactor and under the reaction conditions of Example 2. The oil was then washed with sodium hydroxide, filtered and tested as in Example 2. The results show-ing improved oxidation properties are recorded in the table below.

Example 5: A 450-g quantity of the above oil stock was mixed with 0.025 wt % of elemental sulfur (based on weight of oil) and allowed to flow through the reactor and under the reaction conditions of Example 2. The oil was then washed with sodium hydroxide, filtered and tested as in Example 2. The results show-ing improved oxidation properties are recorded in the table below.

Example 6: In order to demonstrate the beneficial effect of the flow process over that of the art whereby the sulfur contacting is conducted in a stirred batch reactor (i.e., as in U.S. Patent 3,904,511), a 450-g quantity of the above oil stock was thermostatted under nitrogen at 175°C in the presence of 10 g of the zeolite catalyst used in the above Examples 2 through 5. Elemental sulfur (2.25 g) was added and the oil was stirred at 175°C for 1 hour while hydrogen sulfide was evolved. The oil was cooled to room temperature and residual corrosive sulfur was removed by stirring the oil with finely divided sodium hydroxide as in Ex-amples 2 through 5. The product was filtered and tested as in Examples 2 through 5 and the results of the tests are recorded in the table below.

Example 7: Further, in demonstration of the beneficial effect of the flow proc-ess over that of the art whereby the sulfur contacting is conducted in a stirred batch reactor (i.e., as in U.S. Patent 3,904,511), a 450 g quantity of the above oil stock is thermostatted under nitrogen at 175°C in the presence of 10 g of the zeolite catalyst used in the above Examples 2 through 5. Elemental sulfur (0.45 g) is added and the oil is stirred at 175°C for one hour while hydrogen sulfide is evolved. The oil is cooled to room temperature and residual corrosive sulfur is removed by stirring the oil with finely divided sodium hydroxide as in Examples 2 through 5. The product is filtered and tested as in Examples 2 through 5 and the results of the tests are recorded in the table below.

Ex. No.	RBOT (min)	Dornte (hr)*
1	268	43.3
2	360 ⎫	86.3 ⎫
2**	–	90.9
3	395 ⎬ avg 394	59.5 ⎬ avg 75.8
4	400	79.0
5	420 ⎭	63.4 ⎭
6	343 ⎫	65.2 ⎫
6**	345 ⎬ avg 329	77.3 ⎬ avg 62.5
7	300 ⎭	45.1 ⎭

*Time required for absorption of 1 mol O$_2$/kg oil at 162°C with Fe, Cu and Pb present.
**Repeated.

Simultaneous Dewaxing and Hydrogenation of Feedstock

C. Olavesen, B.M. Sankey and J.B. Gilbert; U.S. Patent 4,124,650; November 7, 1978; assigned to Exxon Research & Engineering Co. describe an improved proc-ess for the production of low pour point, high V.I., stable synthetic oils from feedstocks derived by thermally and noncatalytically polymerizing a mixture of linear C$_{6-20}$ carbon atom mono-α-olefins. The improvement comprises simultan-eously saturating residual olefinic bonds and removing at least a portion of the wax from the feedstock by contacting the feedstock and hydrogen with a hy-drogen form mordenite catalyst thereby producing a stable, high V.I., synthetic oil of low pour point. The catalytic metal component of the catalyst is selected from Groups VI or VIII metals, particularly Pt or Pd.

In the example below, a raw, untreated thermal, noncatalytically polymerized synthetic oil (TPO) feedstock is contacted with a noble metal on a hydrogen form mordenite catalyst, in the presence of hydrogen and the results are compared with conventionally hydrotreating the same TPO feedstock as well as treating a paraffinic distillate derived from a natural crude oil with a noble metal on a hydrogen form mordenite catalyst.

A commercially available hydrogen form mordenite, Zeolon 900-H (Norton Chemical Company), was treated with ammonium nitrate solution and calcined to produce a hydrogen mordenite containing less than 0.1 wt % residual sodium. Platinum in an amount of 0.5 wt % of the total catalyst (dry basis) was added to the mordenite by aqueous impregnation with a solution of chloroplatinic acid followed by washing and calcination in air at 950°F. The platinum-containing catalyst was then reduced by treating with hydrogen at 700°F and 1,350 psig hydrogen pressure for 4 hours prior to use. The catalyst was used to treat various petroleum derived, sulfur-containing feedstocks for more than 1,000 hr prior to the experiment with the TPO.

The feedstock used in this experiment was a raw or untreated TPO boiling in the range of from 700° to 1025°F which was produced from a C_{10-14} mixture of linear, mono-α-olefins derived from steam cracking a paraffin wax. This feed was passed over the platinum-containing mordenite catalyst in the presence of hydrogen.

Another sample of the same TPO feed was conventionally hydrotreated over a nickel molybdate on alumina catalyst and a light, conventional lube oil distillate was catalytically dewaxed over a noble metal on hydrogen form mordenite catalyst. The data for the TPO feed and products are shown in Table 1, while Table 2 contains the data for the light lube distillate.

The data in Table 1 clearly illustrate the simultaneous saturation in olefin content and reduction in pour point (wax content) with little loss in V.I. when the TPO was treated in accordance with this process. It also illustrates the increase in wax content which occurs when the TPO is conventionally hydrogenated. Comparing the data in Table 2 with that in Table 1 illustrates the unexpected selectivity (yield-pour point relationship) of the hydrogen form mordenite with the TPO and the dramatic loss in V.I. with the crude oil distillate.

Table 1: Conventional Hydrogenation of TPO Compared with Simultaneous Hydrogenation/Catalytic Dewaxing of TPO

	TPO Feed	Conventional Hydrogenation*	Simultaneous Hydrogenation/ Catalytic Dewaxing**
Yield on feed, LV %	100	100	60
Viscosity, cs at 100°F	19.8	20.8	29.4
Viscosity, cs at 210°F	4.3	4.5	5.4
Viscosity index	139	143	131
Density, kg/dm³ at 15°C	0.842	0.838	0.848
Pour point, °F (ASTM)	+45	+70	-35
Cloud point, °F (ASTM)	+60	+88	-20
Aromatics, wt % (silica gel)	9.5	5.2	4.8
Bromine number	16.4	0.1	0.4

*Catalyst—Cyanamid Aero HDS-9A (nickel molybdate on alumina). Reactor conditions— 800 psig H_2, 2,000 scf H_2/bbl, 1 vol/hr/vol, 550°F.

**Reactor conditions—1,350 psig H_2, 5,000 scf H_2/bbl, 575°F, 0.17 vol/hr/vol, product stripped at 175°F, 10 mm Hg. Catalyst is 0.5 wt % Pt on hydrogen-mordenite.

Table 2: Catalytic Dewaxing of Petroleum Derived Feedstock

| | . SAE 5 Grade Raffinate. . | |
	Feed*	Product
Yield on feed, LV %	100	48
Viscosity, cs at 100°F	17.5	41.8
Viscosity, cs at 210°F	3.7	5.3
Viscosity index	104	39
Density, kg/dm³ at 15°C	0.864	0.901
Pour point, °F (ASTM)	79	−25
Cloud point, °F (ASTM)	−	−
Aromatics, wt % (silica gel)	−	−
Bromine number	−	−

*Derived from Leduc (Western Canadian) crude, boiling range 610° to
850°F at atmospheric pressure.

The following were the conditions used: reactor conditions—1,350 psig H_2, 0.5 v/h/v, 5,000 scf H_2/bbl, 550°F; product stripped at 300°F and 10 mm Hg. Catalyst—hydrogen mordenite; Na content reduced to less than 0.1 wt % by ammonium nitrate treating and calcination, impregnated with palladium chloride solution to give 0.25 wt % palladium on catalyst, calcined in air at 1000°F for 1 hour and reduced at 650°F and 1,350 psig H_2 for 4 hours; catalyst pretreated for 150 hours with sulfur containing distillate.

Hydrogenation of Low Molecular Weight Polyisoprene

S. Yasui and H. Sato; U.S. Patent 4,122,023; October 24, 1978; assigned to Sumitomo Chemical Company Limited, Japan have found that the hydrogenation of certain low molecular weight polyisoprenes affords synthetic saturated oils, of which fractional distillation products have a wide variety of flow characteristics suitable for various uses including lubricating oils and cosmetics and some of them are quite similar to squalene in physical properties.

According to the process, synthetic saturated oils are produced by hydrogenation of low molecular weight polyisoprene having the 1,4 structure of at least 70% in the main chains and a number average molecular weight of about 290 to 3,000.

The low molecular weight polyisoprene suitable as the starting material may be produced by conventional procedures. For instance, such polyisoprene is obtainable by polymerization of isoprene in the presence of an α-olefin using a catalyst composition comprising an organometallic compound and a nickel compound with or without an electron donor as described in Japanese Patent 115,189/1974. The molecular weight of the polymer to be produced can be readily regulated by controlling the amounts of the α-olefin, the organometallic compound, the nickel compound and the electron donor.

The low molecular weight polyisoprene thus produced may be separated as liquid polymer from the reaction mixture by a conventional separation procedure. For instance, the catalyst for polymerization is deactivated by treatment with methanol, ethanol, propanol, n-amyl alcohol, water or the like and then eliminated by washing with an aqueous solution of acid (e.g., hydrochloric acid, sulfuric acid, nitric acid, formic acid, acetic acid, oxalic acid). The resultant mixture is neutralized with an aqueous alkaline solution, washed with water and then concentrated under reduced pressure for removal of the solvent, whereby the liquid polymer is obtained.

Hydrogenation of the liquid polymer thus obtained may be carried out by treatment with hydrogen in the presence of a hydrogenation catalyst, usually at a temperature of about 50° to 350°C for about 1 to 100 hours under a hydrogen pressure of about 5 to 300 kg/cm². The treatment may be carried out in the presence or absence of an inert solvent such as alcohols (e.g., methanol, ethanol), ketones (e.g., acetone, methyl ethyl ketone), aliphatic hydrocarbons (e.g., heptane, hexane, pentane, cyclohexane) or their mixtures. As the hydrogenation catalyst, there may be used any conventional one such as nickel (e.g., Raney nickel, nickel on diatomaceous earth, Urushibara nickel), palladium and platinum. After completion of the hydrogenation, the catalyst and the solvent are removed from the reaction mixture by usual methods, and the distillation of the reaction mixture under reduced pressure affords the hydrogenated product of the liquid polymer.

The thus obtained hydrogenated liquid polymer, i.e., the synthetic saturated oil of the process, has a broad molecular weight distribution, comprises polymers ranging from low molecular weight ones to high molecular weight ones and shows generally the following physical properties:

Appearance: colorless, transparent, odorless
Boiling point at 760 mm Hg: $\geqslant 250°C$
Specific gravity at 20°C: 0.80 to 0.92
Refractive index, $n_D{}^{20}$: 1.44 to 1.50
Inherent viscosity at 25°C: 5 to 10^5 cp

The main components in such a hydrogenated liquid polymer are hydrogenated polyisoprenes substantially representable by the formula:

$$
\begin{array}{cc}
CH_3 & CH_3 \\
| & | \\
R_1-CH_2CHCH_2CH_2\!+\!CH_2CHCH_2CH_2\!\!\xrightarrow{}{}_{\overline{n}} R_2
\end{array}
$$

wherein R_1 is hydrogen or alkyl having 1 to 8 carbon atoms, R_2 is hydrogen, ethyl or isopropyl and R_1 and R_2 are not hydrogens at the same time and n is an integer of 3 to 40, preferably 4 to 20. When, for instance, the hydrogenated liquid polymer is produced by hydrogenation of the liquid polymer according to the method described in Japanese Patent 115,189/1974, its major components are the ones represented by the above formula wherein R_1 is hydrogen and R_2 is ethyl or isopropyl. Further, the hydrogenated liquid polymer produced by hydrogenation of a liquid polymer obtained by polymerization of isoprene in the presence of lithium or C_{1-8} alkyl lithium may comprise as its major components the ones represented by the formula, wherein R_1 is hydrogen or C_{1-8} alkyl and R_2 is hydrogen.

Furthermore, the hydrogenated liquid polymer obtained by this process contains as the major components the ones represented by the formula wherein R_1 is hydrogen and R_2 is isopropyl and, when subjected to rectification and gel permeation chromatography, affords the substances shown in the table on the following page.

n	Molecular Weight*	Viscosity at 25°C (cp)	BP at 0.15 Torr (°C)	Specific Gravity (d^{20})	Refractive Index (n_D^{20})
3	316	8.0	143	0.8051	1.4490
4	380	15.6	172	0.8092	1.4529
5	450	28	195	0.8125	1.4560
6	525	45	213	0.8160	1.4585
7–9	670	105	230–270	0.8208	1.4630
10–12	840	250	280–320	0.8263	1.4683

*Determined by the vapor pressure osmometry method.

The hydrogenated liquid polymer may be separated by a conventional procedure such as fractional distillation into the initial fraction (30°C ≤ BP/1 mm Hg ≤ 150°C) having a low viscosity, the middle fraction (150°C < BP/ 1 mm Hg ≤ 450°C) having a medium viscosity and the residual matter (450°C <BP/1 mm Hg) having a high viscosity. The fraction having a molecular weight smaller than about 290 irritates the skin slightly and is unsuitable for cosmetic use and has a low flash point unsuitable for lubricating oil. Therefore, the fraction having a molecular weight smaller than 290 is not included in the process.

These fractions are applied to various uses such as machine oils for precision machines (e.g., watches, measuring instruments, telephones), engine oils for automobiles and lubricating oils for jet planes and propeller planes depending on their viscosities and flash points. In these uses, they may be used alone or in combination with conventional additives such as viscosity index improvers, flow point depressants, anticorrosive agents and carbonization inhibitors.

As known, cosmetics are generally prepared by admixing together oil-soluble materials such as vegetable oils (e.g., beeswax, vegetable wax, cetyl alcohol, stearic acid, lanolin, castor oil, olive oil), mineral oils (e.g., paraffin, liquid paraffin, vaseline, ceresine) and animals oils (e.g., squalene), water-soluble materials such as ethanol, glycerol, propylene glycol, polyethylene glycol, methylcellulose, hydroxyethylcellulose, polyvinyl alcohol, polyvinyl pyrrolidone, tragacanth gum and acacia gum, surfactants, coloring materials such as inorganic pigments (e.g., zinc stearate, ultramarine, titanium oxide, talc, kaolin), organic dyes and natural coloring matters, antioxidants, perfumes and water.

The hydrogenated liquid polymer obtained by this process may be used as oil-soluble materials in the cosmetics in the form of milky lotions, creams, stick pomades and the like. Since they are already hydrogenated, no deterioration in quality will be caused in those cosmetics.

Hydrogenation of Olefin Polymers

G.F.S. Debande, R.N.M. Cahen and J.F.J. Grootjans; U.S. Patent 4,101,599; July 18, 1978; assigned to Labofina SA, Belgium describe a process for producing high quality white oils, especially white oils exhibiting a relatively low iodine value.

The process for hydrogenating a liquid polymer derived from olefin units containing 4 carbon atoms comprises the step of hydrogenating the polymer at a hydrogen pressure of between about 20 and 120 kg/cm², and at a temperature of between about 130° and 250°C in the presence of a catalyst comprising palladium on an alumina support exhibiting a total pore volume of at least about 0.25 ml/g, wherein 35 to 50% of the pore volume are provided by small pores the sizes of which are distributed around a mean size of less than 300±50A and 25 to 35% of the pore volume are provided by large pores the sizes of which are distributed around a mean size of above 300±50A, with the mean size of the large pores being at least 1.5 times the mean size of the small pores.

The liquid hourly space velocity may be between about 0.25 and 4 hr⁻¹. A suitable hydrogen:polymer ratio is between about 250 and 6,000 Nl/l. Preferably, the reaction is effected under reaction conditions which are sufficient to yield a hydrogenated product, the iodine value of which is below 0.26.

The obtained hydrogenated products have been successfully tested using the BP acid test, DAB VII test and FDA tests, to determine their usefulness as white oils and medicinal oils.

Two-Step Alkylation Process

C.D. Kennedy and G.E. Nicks; U.S. Patent 4,148,834; April 10, 1979; assigned to Continental Oil Company describe an improved process for preparing a synthetic hydrocarbon lubricant composition, consisting essentially of di-long-chain alkyl-substituted aromatic hydrocarbons. The improved process is directed to a two-step alkylation process, using linear monoolefins as the alkylating agent, and in which the first alkylation is conducted using HF as the catalyst, wherein the improvement comprises using aluminum chloride or aluminum bromide as the catalyst for the second alkylation step. Use of aluminum chloride or aluminum bromide, instead of HF, in the second alkylation step results in a product having better low temperature properties.

Example 1: (A) HF Alkylation — Benzene (8,000 g) was added to a polyethylene reactor equipped with a stirrer, dropping funnel, and a vent. The benzene was cooled to 5°C in an ice bath, and 4,000 g of HF were condensed into the reactor. The admixture was stirred and held at 5°C while 3,000 g of 1-dodecene were added over a period of about one hour.

After the olefin addition was complete, the ice bath was removed and the HF allowed to evaporate overnight. When the reaction system was free of HF, the crude alkylate was washed with a 5% aqueous NaOH solution at about 90°C, allowed to settle and the excess caustic solution removed. The crude alkylate was then distilled to remove benzene.

(B) AlCl₃ Alkylation — Six mols of the monoalkylbenzene of step (A) were added to a creased flask equipped with stirrer, thermometer, condenser, and dropping funnel. To the flask was added a trace of HCl as a promoter and the AlCl₃ (5% by weight based on the olefin used). One mol of 1-dodecane was then added in about one hour while the temperature was held at 50°C. After the olefin addition was completed, the admixture was stirred for an additional 30 minutes, allowed to settle, and the sludge drawn off. The crude alkylate was washed with an aqueous 5% NaOH solution and then subjected to a distillation to remove unreacted monoalkylbenzene. The bottoms fraction was the desired product. The properties of the product are shown in Table 1.

Example 2: This example is comparative and shows the results obtained using HF as the catalyst in both the first and second alkylation steps. The first alkylation step was the same as in Example 1. The second alkylation step was similar to the first alkylation step except that the monoalkylate of the first step was used in place of benzene. The amounts were equivalent on a molar basis. After washing, the crude alkylate was subjected to distillation to recover the desired product as the bottom fraction. The properties of the product are shown in Table 1.

Table 1

| | | Pour Point ($°$F) | V.I. | . . . . . .Viscosities, cs. . . . . . | | |
Product	meta/para			$-40°$F	$100°$F	$210°$F
Ex. 1	22	-75	108	4,612	21.53	4.20
Ex. 2	0.39	-75	88	15,898	31.70	5.01

From the V.I. and $-40°$F viscosities, it is readily apparent that use of $AlCl_3$ in the second step gives a product having superior properties.

Example 3: This example illustrates the process and shows the results obtained using the process of Example 1 with a different feedstock in the second alkylation step. The feedstock was a mixed C_{10-14} alkylbenzene which was prepared by alkylating benzene with a C_{10-14} α-olefin mixture using HF as the catalyst. The monoalkylbenzene had the following homolog distribution: 0.2 wt % (C_{10} alkyl)benzene, 1.0 wt % (C_{11} alkyl)benzene, 25.9 wt % (C_{12} alkyl)benzene, 32.4 wt % (C_{13} alkyl)benzene, and 40.5 wt % (C_{14} alkyl)benzene. The product contained 3.0 wt % alkyl-substituted tetrahydronaphthalenes in conjunction with the mono-alkylbenzenes.

The procedure used was the same as that in (B) of Example 1, except the mol ratios of 4:1 and 2:1, of monoalkylate to benzene, were used. Run A used a ratio of 4:1 while Run B used a ratio of 2:1. The trialkyl-substituted tetrahydronaphthalene content and the physical properties of the bottom product are shown in Table 2.

Table 2

| | | Pour Point ($°$F) | V.I. | . . . . . . Viscosities (cs) . . . . . . | | |
Run	Percent TTHN*			$-40°$F	$100°$F	$210°$F
A	5.0	-70	119	10,788	34.65	5.78
B	<5.0	-65	127	10,366	33.88	5.84

*Trialkyl-substituted tetrahydronaphthalenes.

Thermal Inhibition of α-Oligomer Oils

T.A. Schenach; U.S. Patent 4,085,056; April 18, 1978; assigned to Bray Oil Co. Inc. has found that if an α-olefin oligomer oil is treated with an alkyllithium compound or base of comparable strength and the product is subsequently treated with alkyl iodide, a composition is formed with viscosity properties similar to those of the starting oligomer oil, but with dramatically improved thermal stability. This composition may be used as is in the formulation of lubricant products.

Alternately, it may be added to an untreated α-olefin oligomer oil, in which case it functions as an inhibitor to reduce the thermal decomposition of the oligomer oil.

Thermal stability tests were carried out in a 500 ml round-bottom flask, fitted with a heating mantle, a nitrogen inlet tube, and a reflux condenser. 50 g of the oil to be tested were charged to the flask and heated to 680°F (360°C) under a slow bleed of nitrogen sufficient to exclude air without removing volatile cracking products. The degree of decomposition was evaluated by the decrease in viscosity of the oil after one hour at 680°F. In some cases, a trap was inserted between the flask and the condenser in order to collect low boiling decomposition products. These were recombined with the oil remaining in the flask after the heating period. This procedure was found to be more severe, the viscosity losses being considerably greater than if the trap was not employed.

Example: A hydrogenated decene oligomer oil having a bromine number of 0.2 and a kinematic viscosity of 19.56 cs at 100°F (37.8°C) and containing approximately 75% hydrogenated decene trimer and 25% hydrogenated decene tetramer was heated for one hour under nitrogen at 680°F. At the end of the test, the viscosity had dropped to 13.56 cs at 100°F (a 30.67% loss).

225 g of the above oligomer oil were charged to a three-neck round-bottom flask equipped with a stirrer and a nitrogen atmosphere, and 80 ml of a 2.29 M solution of n-butyl lithium in n-hexane slowly added. The mixture was stirred at ambient temperature (80°F) for ten minutes and then slowly warmed to 200°F at which point the mixture was slightly hazy and most of the n-hexane had distilled off. It was allowed to cool back to room temperature, and 33 g of methyl iodide were carefully added. A thick white precipitate formed and heat was evolved. The mixture was stirred for 15 minutes, and then 100 ml of water cautiously added to hydrolyze residual alkyllithium compounds. After thorough mixing, the batch was allowed to stand for separation of the water layer. It was then water-washed, dried with $MgSO_4$, and stripped to 280°F under nitrogen. After clay treatment, the product was a pale yellow oil with a 100°F viscosity of 19.24 cs, which showed on the thermal stability test a viscosity of 14.76 cs, a 23.29% loss.

Processing of Sulfonic Acids

A process described by *D.S. Bosniack and P.F. Korbach; U.S. Patent 4,110,196; August 29, 1978; assigned to Exxon Research & Engineering Co.* relates to the processing of sulfonic acids and is particularly concerned with converting such acids into a hydrocarbon oil of superior oxidation stability.

Thus, it has been found that sulfonic acids can be converted into a hydrocarbon oil of superior oxidation stability by transforming the sulfonic acids into compounds selected from the group consisting of ammonium sulfonates, substituted ammonium sulfonates, and sulfonic acid esters; and hydrotreating the resultant sulfonic acid derivatives in the presence of a hydrotreating catalyst. The hydrotreating process frees the parent hydrocarbons that originally constituted the organic portion of the sulfonic acids by cleaving the carbon-sulfur bond of the sulfonic acid derivatives. These parent hydrocarbons are recovered from the reaction products of the hydrotreating step and comprise a hydrocarbon oil product that possesses an oxidation stability substantially greater than other hydrocarbon oils of comparable composition and viscosity and therefore will normally exhibit a longer service life than such oils when used as or as a component of motor oils, transformer oils, cutting oils, quench oils, process oils and the like.

The sulfonic acids are transformed into ammonium sulfonates or substituted ammonium sulfonates by reacting the acids with ammonia or a substituted ammonia compound such as ammonium hydroxide, methylamine, ethylamine or the like. The acids are transformed into sulfonic acid esters by reacting them with a compound, such as thionyl chloride, that will transform the acids into their corresponding hydrocarbon sulfonyl chlorides. The sulfonyl chlorides are then reacted with an alcohol to form the desired sulfonic acid esters. Alternatively, the sulfonic acid esters may be prepared by heating the sulfonic acids with a trialkyl phosphate. Before the sulfonic acid derivatives are subjected to hydrotreating, it may be desirable to decrease their viscosity by mixing them with an inert diluent oil.

The process provides an economical and ecologically acceptable procedure for converting sulfonic acids into a useful and desirable hydrocarbon product that has properties superior to oils of similar composition and viscosity prepared in conventional ways.

In the process depicted in Figure 8.1, a feed stream containing sulfonic acids that are to be converted into a hydrocarbon oil of superior oxidation stability is introduced through the line **10** into the contacting zone in a reactor or similar contacting vessel **11**. The feed stream will normally contain a mixture of differing types of sulfonic acids recovered as by-products in the acid treating of various petroleum fractions produced during the refining of crude oil. The feed, however, may be a pure sulfonic acid and need not necessarily be derived from the refining process. The feed may be composed entirely of sulfonic acids or may contain the acids in combination with other organic or inorganic compounds. Preferably, the feed stream will be a mixture of oil-soluble, aromatic sulfonic acids produced during the manufacture of medicinal or technical grade white oils.

Figure 8.1: Conversion of Sulfonic Acids into Hydrocarbon Oils

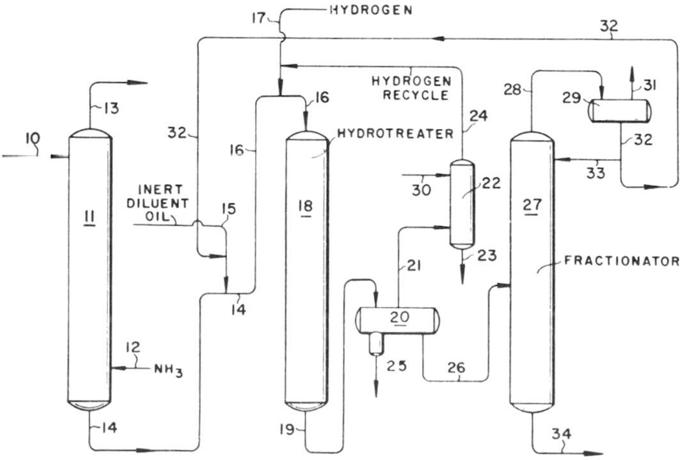

Source: U.S. Patent 4,110,196

The feed stream is introduced into the top of contacting vessel **11** and moves downwardly through the contacting zone where it comes in contact with an up-flowing stream of gaseous ammonia introduced into the bottom of the vessel through line **12**. If desired, the contacting zone may be provided with spray nozzles, perforated plates, bubble cap plates, packing or other means for promoting intimate contact between the gas and liquid.

As the sulfonic acids in the feed stream pass downwardly through the contacting zone, they are neutralized by the rising ammonia and transformed into ammonium sulfonates. The amount of ammonia injected into vessel **11** will normally be slightly in excess of that needed to neutralize all of the sulfonic acids. Any excess ammonia is removed overhead from the vessel through line **13** and may be recovered for reuse or transferred to downstream units for further processing. A bottoms stream containing the ammonium sulfonates produced in the contacting zone is removed from vessel **11** through line **14**.

The purpose of the abovedescribed neutralization step is to transform the sulfonic acids into compounds that are susceptible to hydrotreating. Studies indicate that sulfonic acids themselves cannot be effectively hydrotreated because they tend to weaken the structural support of the hydrotreating catalyst, which in turn results in the production of fines that contaminate the hydrotreated product. This observed catalyst deterioration makes any proposed commercial process that includes the direct hydrotreating of sulfonic acids unfeasible. Studies further indicate that sodium sulfonates and similar metallic salts of sulfonic acids cannot be effectively hydrotreated. These metallic sulfonates appear stable in the presence of hydrogen, conventional hydrotreating catalysts and standard hydrotreating conditions and therefore will not yield the desired hydrocarbon oil product.

It has been found that ammonium sulfonates, substituted ammonium sulfonates such as methyl ammonium sulfonate, methyl ethyl ammonium sulfonate, dimethyl ammonium sulfonate and the like, and sulfonic acid esters will undergo hydrotreating. Apparently, the hydrotreating process cleaves the carbon-sulfur bond of these sulfonates and esters and thereby results in the production of the parent hydrocarbons that originally constituted the organic portion of the sulfonic acids. It has been found that these parent hydrocarbons comprise a hydrocarbon oil that exhibits a higher stability toward oxidation than can otherwise be obtained from other hydrocarbon oils of similar composition and viscosity.

In the process depicted in Figure 8.1, the sulfonic acids in the feed to vessel **11** are transformed into ammonium sulfonates by reacting them with gaseous ammonia. It will be understood that the sulfonic acids may be reacted with any nitrogen-containing compound or mixture of such compounds that will transform the acids into ammonium sulfonates or substituted ammonium sulfonates. Such a nitrogen-containing compound will have the generalized formula:

$$R_1 - \underset{\underset{R_3}{|}}{\overset{\overset{R_2}{|}}{N}} : [HOH]_x$$

where x is 0 or 1 and R_1, R_2 and R_3 may be the same or different and may be, for example, (1) hydrogen atoms; (2) straight or branched chain aliphatic groups or substituted aliphatic groups having from 1 to 16 carbon atoms such as methyl,

ethyl, 1-propyl, 2-propyl, butyl, hexyl, heptyl, benzyl, octyl, decyl, and 1,1-di-methyldodecyl radicals; (3) cycloaliphatic groups having from 3 to 10 carbon atoms such as cyclopropyl, cyclopentyl, cyclohexyl, methylcyclohexyl and per-hydronaphthyl radicals; and (4) aryl groups having from 6 to 12 carbon atoms, such as phenyl tolyl, biphenyl, and butylbenzene radicals.

Specific examples of nitrogen compounds or mixtures of such compounds that may be used are ammonia and substituted ammonia compounds such as am-monium hydroxide, methylamine, dimethylamine, trimethylamine, ethylamine, butylamine, hexylamine, cyclohexylamine, benzylamine, octylamine, decylamine, N-butyl-N-phenylamine, a mixed branched chain isomeric, 1,1-dimethyl C_{12-14} primary aliphatic amine composition (Primene 81-4) and the like. The ammonia and substituted ammonia compounds will add to the hydrogen atom of the sul-fonic acids to form the corresponding ammonium sulfonates or substituted am-monium sulfonates, which are subsequently subjected to hydrotreating.

Before the neutralization step effluent containing the ammonium sulfonates or substituted ammonium sulfonates is subjected to hydrotreating, it may be ad-vantageous to mix the effluent with an inert diluent oil. One example of a situa-tion in which a diluent oil should normally be used is when the effluent is so viscous that it cannot be easily pumped and may cause plugging of the catalyst bed in the hydrotreater. The inert diluent oil will normally be saturated hydro-carbon or a mixture of saturated hydrocarbons such as pentane, hexane, heptane, octane, isooctane or higher molecular weight hydrocarbons. An aromatic com-pound or mixture of aromatic compounds may, however, be used as the diluent oil if it is inert to the sulfonates and the hydrotreating process.

If a diluent oil is used, a sufficient amount is normally mixed with the sulfonates-containing stream to produce a solution that is composed of from 20 to 80 vol % diluent oil. In some cases it may be advantageous to add the diluent oil before the stream containing the sulfonic acids is neutralized, especially in situations where the feed is viscous and difficult to pump.

Referring again to the process depicted in Figure 8.1, the bottoms stream from vessel **11**, which contains ammonium sulfonates, is passed through line **14** into line **16** where it is mixed with hydrogen gas injected into line **16** via line **17**. If desired, the bottoms may first be mixed with an inert diluent oil injected through line **15** into line **14**. The mixture of diluent oil, bottoms and hydrogen gas is then passed into hydrotreater **18**. The hydrotreater contains a fixed bed of stan-dard hydrotreating catalyst in the form of extrudate or pills. The catalyst will normally be composed of nickel-cobalt-molybdenum, nickel-molybdenum, co-balt-molybdenum or similar catalytically active compounds supported on inert silica-alumina.

The mixture of diluent oil, sulfonates-containing bottoms and hydrogen is passed downward through the catalyst bed under standard hydrotreating conditions, conditions that will be familiar to those skilled in the art. In the presence of the hydrotreating catalyst, the hydrogen reacts with the ammonium sulfonates to cleave their carbon-sulfur bonds and produce sulfonic acid parent hydrocarbons, ammonia, hydrogen sulfide and water.

The hydrotreater effluent containing, among other substances, sulfonic acid par-ent hydrocarbons, diluent oil, ammonia, hydrogen sulfide, unreacted hydrogen and water is withdrawn from the hydrotreater through line **19**, cooled and passed to separator or similar vessel **20** where the gases in the hydrotreater effluent are allowed to separate from the liquids and are withdrawn through line **21** and passed to scrubber or similar device **22**. Here unreacted hydrogen is separated from the other gases, primarily ammonia and hydrogen sulfide, by passing the mixture of gases upward through a downflowing scrubbing liquid injected into the top of the scrubber via line **30**.

The scrubbing liquid absorbs the ammonia and hydrogen sulfide and is withdrawn from the bottom of the scrubber through line **23** and passed to downstream units for regeneration or further processing. Any solvent, including water, that will selectively absorb hydrogen sulfide and ammonia in the presence of hydrogen may be used as the scrubbing fluid. The hydrogen is withdrawn from the scrubber through line **24** and recycled to the hydrotreater via lines **17** and **16**.

The water in separator **20** will normally have a greater specific gravity than the oil phase and is removed via line **25** and passed to downstream units for further processing. The oil phase, which consists of, among other substances, the sulfonic acid parent hydrocarbons formed during hydrotreating and diluent oil, is withdrawn from the separator through line **26** and fed to fractionator **27** where the parent hydrocarbons are separated from the diluent oil and other components. The diluent oil will normally have a boiling point lower than the initial boiling point of the parent hydrocarbons and therefore will normally be removed from the top of the fractionator through line **28** along with gases and other lower boiling constituents. The fractionator overhead is cooled and passed to distillate drum **29** where the gases are taken off overhead through line **31** and passed to downstream units for further processing.

The liquid, which will normally be pure diluent oil, is withdrawn from distillate drum **29** through line **32**. A portion of this liquid may be returned as reflux to the upper portion of the fractionator through line **33**. The remaining liquid may be recovered or recycled to the process via line **15**.

The parent hydrocarbons are removed in liquid form from the bottom of fractionator **27** via line **34** and recovered as the hydrocarbon oil product. Studies indicate that this hydrocarbon oil product possesses an oxidation stability substantially greater than other hydrocarbon oils of comparable composition and viscosity, which are prepared in other ways. This superior oxidation stability makes the hydrocarbon oil product an excellent base stock for blending with other oils to form hydrocarbon mixtures that normally exhibit longer than normal service lives when used as transformer oils, motor oils, heat transfer oils, cutting oils, quench oils, stain resistant rubber extender oils, cable oils, plasticizers and the like.

SYNTHETIC ESTER COMPOSITIONS

Mixtures of Complex and Monomeric Esters

K. Schmitt, J. Disteldorf and W. Flakus; U.S. Patent 4,155,861; May 22, 1979; assigned to Studiengesellschaft AG, Germany describe a formulation in which a quantity of a complex ester based on dicarboxylic acid, preferably branched, and hexanediol or trimethyl hexanediol, is added to a monomeric ester of a branched dicarboxylic acid.

Tables 1 through 4 list the characteristics of lubricants according to the process in relation to their percentage contents of complex ester. For example, by the addition of 1 to 10% of a specified complex ester to the monomeric ester oil listed, lubricants are obtained which can be used to particular advantage for the lubrication of transmissions, and, in addition, for the preparation of wide-range motor oils of SAE classes 5W/20, 5W/30 or 10W/40, which are thus usable also as driving fluid for high-vacuum pumps, and as industrial oils, and finally they can be used for the ATF field. Higher percentages of complex esters result in lubricants meeting requirements for extreme pressure, gear service and for hydraulic processes.

In the individual columns of Table 1, the applications of the individual mixtures are stated. The advantage that can be achieved by the process consists in the fact that merely by the addition of different quantities of a single complex ester to one and the same monomeric ester oil, high-performance lubricants are obtained for practically all important applications.

Although trimethyladipic acid octyl decyl ester (a diester) is given in all the tables as the monomeric ester, the effect described is nevertheless also obtained when a monomeric ester is used which is based on branched glutaric acid or branched succinic acid, as for example, monomethyl glutaric acid, dimethyl glutaric acid, monomethyl succinic acid, dimethyl succinic acid, monomethyl malonic acid, dimethyl malonic acid, etc.

The complex esters to be used are prepared in the following manner. The listed quantities of a dicarboxylic acid ester, a diol, and 0.05 to 0.1% of the total quantity of tetraalkyl titanate (generally tetraisopropyl titanate) are condensed at temperatures of 150° to 250°C under nitrogen shielding, and with the yielding of a quantity of monoalcohol equivalent to the amount of diol used. The removal of the last volatile components is performed in vacuo. The complex esters mentioned in Tables 1 through 4 possess the following characteristics.

Complex Ester 1: This ester was prepared by the reaction of 1.02 mols trimethyladipic acid dimethyl ester and 1.0 mol hexanediol-1,6, according to the general instructions. The characteristics of the complex ester are: pour point °C, +6; flash point °C, 304; and MW, 3,300.

Complex Ester 2: This ester was prepared by the reaction of 1.5 mols of trimethyladipic acid dimethyl ester and 1.0 mol of hexanediol-1,6, according to the general instructions. The characteristics of the complex ester are: viscosity at 100°F in cs, 396; viscosity at 210°F in cs, 35.75; pour point °C, –10; flash point °C, 285; and MW, 1,030.

Complex Ester 3: This ester was prepared by the reaction of 1.02 mols of trimethyladipic acid dimethyl ester and 1.0 mol of trimethylhexanediol-1,6, according to the general instructions. The characteristics of the complex ester are: viscosity at 210°F in cs, 735; pour point °C, +7; flash point °C, 316; and MW 2,815.

Complex Ester 4: This ester was prepared by the reaction of 1.5 mols of trimethyladipic acid dimethyl ester and 1.0 mol of trimethylhexanediol-1,6, according to the general instructions. The characteristics of the complex ester are: viscosity at 100°F in cs, 341.5; viscosity at 210°F in cs, 137.1; pour point °C, 0; flash point °C, 305; and MW, 1,640.

Table 1: Properties of Ester Oil Formulations According to Their Percentage
Content of Complex Esters

Complex ester 1, %	0	4	7	27
Trimethyladipic acid octyl decyl ester, %	100	96	93	73
Viscosity, cs				
at 100°F	12.74	18.18	22.73	126.4
at 210°F	3.25	4.43	5.30	21.53
Viscosity index	141	175	169	143
Pour point, °C	-73	-65	-59	-35
Flash point, °C	224	226	230	243
Noach value, %	14.7	13.5	11 4	7.8
Applications*	A	B	C	D

*Lubricant A is for transmissions, refrigeration machines, internal combustion
engines; B is the same as A and also as driving fluid for vacuum pumps; C is
the same as A and also for gears; and D is extreme-pressure gear lubricant,
and as a hydraulic fluid.

Table 2

Complex ester 2, %	22.5	81.5
Trimethyladipic acid octyl decyl ester, %	77.5	18.5
Viscosity, cs		
at 100°F	25.09	189.9
at 210°F	5.35	21.38
Viscosity index	155	125
Pour point, °C	-53	-17
Flash point, °C	233	248

Note: Applications are the same as in Table 1, but with modified specifications.

Table 3

Complex ester 3, %	4	7	27
Trimethyladipic acid octyl decyl ester, %	96	93	73
Viscosity, cs			
at 100°F	17.6	21.1	83.3
at 210°F	4.23	4.98	14.25
Viscosity index	168	171	144
Pour point, °C	-69	-66	-49
Flash point, °C	230	233	239

Note: Applications are the same as in Table 1, but with modified specifications.

Table 4

Complex ester 4, %	4	7	27
Trimethyladipic acid octyl decyl ester, %	96	93	73
Viscosity, cs			
at 100°F	15.30	17.58	41.8
at 210°F	3.81	4.37	7.91
Viscosity index	163	175	147
Pour point, °C	-71	-65	-56
Flash point, °C	230	235	243

Note: Applications are the same as in Table 1, but with modified specifications.

Trimethylolpropane-Dimerized Fatty Acids

K. Koch and W. Breitzke; U.S. Patent 4,113,642; September 12, 1978; assigned to Henkel Kommanditgesellschaft auf Aktien, Germany describe the development of a high viscosity neutral complex polyester lubricant produced by esterifying a mixture of (a) a branched alkanepolyol having 2 to 4 primary hydroxyls and 4 to 10 carbon atoms, (b) polyacids selected from the group consisting of dimeric fatty acids and trimeric fatty acids produced by the dimerization of unsaturated fatty acids having 16 to 18 carbon atoms, and (c) alkanoic acids having 6 to 16 carbon atoms, in such proportions that the acid number of the polyester is 0.3 or below, the hydroxyl number of the polyester is 0.5 or below, and the proportion of the hydroxyl groups being esterified by the alkanoic acid is from 50 to 90%.

Owing to their excellent properties, namely, their high viscosity, their low pour point and their favorable viscosity-temperature behavior, the neutral complex esters in accordance with the process are eminently suitable for use as lubricants, particularly for transmission fluid and lubricants for two-stroke piston engines.

The complex esters may constitute the sole oil base in the finished lubricant, or they may be mixed as a mixture component with other products which are already known for this purpose. When used as mixing components in lubricant and transmission oils, any optional quantity ratios may be mixed which are determined exclusively by the required properties such as viscosity-pour point, and viscosity-temperature behavior. However, the content of complex esters will not usually be below 10%, and preferably not below 30% by weight in the finished product.

Examples 1 through 7: Production of the Neutral Complex Esters – 268 g (2 mols) of trimethylolpropane, 565 g ($\sim$1 mol) of dimeric fatty acid (a mixture of $\sim$95% by weight of dimerized fatty acids, $\sim$4% by weight of trimerized fatty acids, and $\sim$1% by weight of nonpolymerized unsaturated fatty acids, the starting unsaturated fatty acid being a mixture of olefinically unsaturated fatty acids having 16 to 18 carbon atoms), and 632 g ($\sim$4 mols) of C_{6-12} saturated fatty acids (a mixture of $\sim$5% by weight of C_6 fatty acids, $\sim$45% by weight of C_8 fatty acids, $\sim$45% by weight of C_{10} fatty acids, and $\sim$5% by weight of C_{12} fatty acids as obtained from the distillation of coconut fatty acid), were heated to 200°C in an autoclave provided with a water separator under passage of nitrogen, the nitrogen acting as a carrier gas to flush out the water of reaction.

A mixture of powdered tin 1.5 g and p-toluene sulfonic acid 1.5 g, was used as an esterification catalyst. Toward the end of the reaction, further esterification was carried out at the same temperature, but under reduced pressure ($\sim$70 torrs). After cooling to 120°C, 1.5 g ($\sim$1% by weight) of activated bleaching clay were added and the mixture was again heated to 200°C and the surplus monocarboxylic acid was distilled off in vacuo.

The acid number of the product of esterification (the proportion esterified with monocarboxylic acid was 67%) was 0.28. The product had a viscosity of 628 cs at 37.8°C (100°F), and 57 cs at 99°C (210°F). The viscosity index was 164 and the pour point was –38°C. The complex esters given in Table 1 were produced in conformity with the above method.

Table 1: Neutral Complex Esters Produced

Ex. No.	Polyol* (1 mol)	Dimer (mol)	Monocarboxylic Acid (mol)	Proportion Esterified Therewith (%)	Viscosity (cs) 37.8°C	Viscosity (cs) 99°C	V.I.	Pour Point (0°C)	Acid No.
2	T	1** (0.35)	C_{6-12}FS*** (2.3)	77	213	25	156	-50	0.2
3	T	1 (0.25)	C_{6-12}FS (2.5)	83	113	16	152	-59	0.2
4	T	2† (0.3)	C_{6-12}FS (2.4)	80	147	18	146	-41	0.24
5	T	2 (0.5)	Isopalmitic (2.0)	67	890	71	160	-28	0.28
6	N	1 (0.5)	C_{6-12}FS (1.0)	50	613	54	157	-22	0.25
7	P	1 (0.5)	C_{6-12}FS (3.0)	75	896	77	170	-20	0.27

*T is trimethylolpropane; N is neopentyl glycol; and P is pentaerythritol.

**Dimeric fatty acid mixture with proportions given in Example 1.

***Fatty acid mixture from the prerun of the coconut fatty acid distillation having proportions as given in Example 1.

†Dimeric fatty acid mixture with ~75% by wt dimerized fatty acids, ~22% by wt trimerized fatty acids, and ~3% by wt of nonpolymerized unsaturated fatty acids, likewise of olefinically-unsaturated fatty acids having 16 to 18 carbon atoms.

Example 8: Use — Aging tests at high temperatures were carried out with a lubricating oil manufactured on the basis of the complex esters in accordance with the process, and with a commercially available lubricating oil. In addition, the compatibility of those oils with various seal materials also was tested. A commercially available single grade oil of the class SAE 80 and a multigrade oil, in accordance with the process of the specification SAE 80 W-90 were used in the tests. The composition of the oil in accordance with the process was as follows: 95.5% by weight of the product of Example 4 (complex ester of 1 mol of trimethylolpropane, 0.3 mol dimeric fatty acid and 2.4 mols of C_{6-12} prerun fatty acid), 6.5% by weight of a commercially available transmission oil additive (Anglamol 99, Lubrizol).

The characteristic data of the oil in accordance with the process given in Table 2 show that the additive used did not contain any agent for lowering the pour point and any V.I. improver.

Table 2: Characteristic Data of the Lubricating Oils Tested

Characteristic Value	Process	Commercially Available
Kinematic viscosity, cs		
at 37.8°C	147	115.4
at 98.9°C	18	11.5
Dynamic viscosity, cp		
at -26.1°C	25,000	solid
Viscosity index	146	94
Pour point, °C	-41	-19
Acid number	0.4	2.8

The lubricating oils to be tested were heated to 160°C and 200°C, respectively, for 8 hours in a glass flask, a quantity of air being conducted through the flask during this period of time at a rate of 10 l/hr. The changes in the viscosities and acid numbers were determined from the samples aged at 200°C.

The viscosity change at 99°C (210°F) for the process oil was +33.2% and +52.7% for the commercially available oil. The increase in acid number for the process oil was 1.3% and 3.4% for the commercially available oil.

The lubricant in accordance with the process exhibited a substantially smaller degree of aging than the commercially available product.

Trimethylolpropane-Isopalmitic Acid-Alkanoic Acid

K. Koch and H. Kroke; U.S. Patent 4,144,183; March 13, 1979; assigned to Henkel Kommanditgesellschaft auf Aktien, Germany describe the development of a branched-chain aliphatic ester oil consisting essentially of a full ester of a branched-chain aliphatic polyol having from 2 to 6 primary hydroxyl groups selected from the group consisting of alkanepolyols having from 3 to 6 carbon atoms with a mixture of (A) α-branched-chain alkanoic acids having the formula

$$\begin{array}{c} R_1 \\ {\diagdown} \\ {}CH{-}COOH \\ {\diagup} \\ R_2 \end{array}$$

where R_1 and R_2 are alkyl having from 1 to 19 carbon atoms and the total number of carbon atoms in the acid is from 14 to 22, selected from the group consisting of (1) acids derived from the oxidation of α-branched alcohols formed from normal alcohols by the Guerbet synthesis and (2) an acid of the formula

$$\begin{array}{c} CH_3(CH_2)_8{-}CH{-}(CH_2)_8CH_3 \\ | \\ COOH \end{array}$$

and (B) straight-chain alkanoic acids having from 8 to 10 carbon atoms, in such a ratio wherein at least one of the primary hydroxyl groups is on average at least 80% esterified by the α-branched-chain alkanoic acids and at least one of the primary hydroxyl groups is on average at least 40% esterified by the straight-chain alkanoic acids. The ester oils are outstandingly suitable both alone, and in admixture with other products already known for this purpose for use as lubricants and as hydraulic fluid, on account of their extremely favorable properties with regard to viscosity, behavior in the cold and thermostability. When used as a mixture component in lubricants and hydraulic fluids, any desired mixing proportions can be selected, which are determined exclusively by the values required with respect to working behavior, pour point and viscosity-temperature behavior.

Example: The full esters utilized for testing for behavior to cold and viscosity-temperature behavior were prepared from the polyols and the mixture of branched-chain carboxylic acids and straight-chain carboxylic acids as given below by the method outlined above of heating an excess of about 1.05 mols of the acid mixture for each mol equivalent of hydroxyl groups in the polyol in the presence of a p-toluene-sulfonic acid to a temperature of about 125°C while removing the water produced by the reaction. The esters were recovered by washing the reaction mixture with methanol. In the table, A is neopentylglycol, B is trimethylolpropane, C is pentaerythritol, D is isopalmitic acid, obtained by oxidation of the 2-hexyl-decanol formed from n-octanol by oxidation in the Guerbet synthesis, and E is a mixture of straight-chained alkanoic acids having 8 to 10 carbon atoms. The values obtained during the tests are given in the following table.

Full Ester	Pour Point (°C)	Viscosity (cs) 100°F	210°F	V.I.
A + 1D and 1E	<−60	16.6	3.73	124
B + 0.8D and 2.2E	<−60	31.37	5.85	143

(continued)

Full Ester	Pour Point (°C)	. . Viscosity (cs) . .		V.I.
		100°F	210°F	
B + 1D and 2E	-62	34.0	6.1	139
B + 1.36D and 1.64E	<-60	36.6	6.32	135
B + 1.46D and 1.54E	<-60	40.1	6.66	131
B + 1.78D and 1.22E	-56	42.8	6.97	133
C + 3.58D and 0.42E	-45	69.1	10.0	139

From the above table, the extremely favorable properties for technical use of the ester oils according to the process with reference to behavior to cold and of viscosity-temperature behavior can be clearly noted. In spite of their relatively high viscosities and their favorable viscosity-temperature behavior (viscosity Index), the products have an extremely low pour point of well below -30°C.

COMPANY INDEX

The company names listed below are given exactly as they appear in the patents, despite name changes, mergers and acquisitions which have, at times, resulted in the revision of a company name.

Asahi Denka Kogyo KK - 212
Atlantic Richfield Co. - 91, 93
Ball Brothers Research Corp. - 354, 356
Ball Corp. - 311
BASF AG - 110
BASF Wyandotte Corp. - 279, 285, 337
Borg-Warner Corp. - 297
Bray Oil Co., Inc. - 388
British Aircraft Corp. - 375
British Petroleum Co. Ltd. - 316
Carpenter Technology Corp. - 307
Chevron Research Co. - 64, 66, 72, 75, 77, 78, 218, 233, 236, 241, 243, 262, 319, 320, 321
Ciba-Geigy AG - 167, 206
Combustion Engineering, Inc. - 373
Continental Oil Co. - 237, 252, 387
Deutsche Texaco AG - 291, 328
Diamond Shamrock Corp. - 293
Dixie Yarns, Inc. - 346
Dow Chemical Co. - 34, 97, 105, 166
Dow Corning Corp. - 352
Edwin Cooper and Company Limited - 22, 38, 57, 208, 246
Edwin Cooper, Inc. - 177
Elco Corp. - 211
Emery Industries, Inc. - 278, 288, 290, 359
Entreprise de Recherches et d'Activities Petrolieres (E.R.A.P.) - 135

Ethyl Corp. - 22, 36
Exxon Research and Engineering Co. - 16, 18, 25, 26, 38, 78, 108, 112, 119, 122, 124, 128, 138, 139, 144, 153, 157, 160, 170, 185, 194, 205, 227, 249, 253, 264, 272, 302, 322, 378, 382, 389
FMC Corp. - 269
GAF Corp. - 274
General Electric Co. - 221, 349
B.F. Goodrich Co. - 350
Gulf Research & Development Co. - 113, 164, 326
Henkel Kommanditgesellschaft auf Aktien - 396, 398
Institut Francais du Petrole - 31, 142
Kao Soap Co., Ltd. - 343
Labofina SA - 70, 330, 386
Liquichimica Robassomero SpA - 6, 86
Lubrizol Corp. - 30, 42, 45, 46, 47, 56, 77, 87, 175, 229, 244, 298, 299, 379
Maruzen Oil Co., Ltd. - 84
Merck & Co., Inc. - 282
Milliken Research Corp. - 3, 347
Mobil Oil Corp. - 7, 9, 11, 88, 90, 171, 188, 189, 190, 192, 196, 200, 217, 231, 232, 236, 266, 271, 294, 295, 317, 323, 380
National Research Laboratories - 282
Nippon Oil Co., Ltd. - 254, 331
Orobis Limited - 100
Orogil - 24
Pennwalt Corp. - 200

Phillips Petroleum Co. - 111, 140, 215
PPG Industries, Inc. - 301
RCA Corp. - 357
Rhone Poulenc Industries - 142
Rocol Limited - 303
Rohm and Haas Co. - 48, 148, 163
Rohm GmbH - 152
Sapchim Fournier-Cimag - 363
Shell Internationale Research
  Maatschap IIJ BV - 326
Shell Oil Co. - 110, 150, 207, 284,
  312, 314, 324
SKF Industries, Inc. - 308
Société Nationale Elf Aquitaine - 134
Standard Oil Co. - 366
Standard Oil Co. (Indiana) - 40, 53,
  63, 71, 101, 181, 183, 197, 261

Studiengesellschaft AG - 393
Sumitomo Chemical Co., Ltd. - 384
Swiss Aluminum Ltd. - 369
S.A. Texaco Belgium NV - 119, 256
Texaco, Inc. - 33, 60, 85, 93, 115, 118,
  131, 202, 232, 259, 265
Toa Nenryo Kogyo KK - 25
Tower Oil & Technology Co. - 376
Tyler Corp. - 372
Union Carbide Corp. - 276
Uniroyal, Inc. - 238
U.S. Secretary of Agriculture - 220, 266
U.S. Secretary of the Air Force - 268,
  334
Wacker-Chemie GmbH - 343
Westinghouse Electric Corp. - 371
Wyman-Gordon Co. - 292

INVENTOR INDEX

Abdul-Malek, A. - 63
Adams, J.H. - 319, 320
Andress, H.J., Jr. - 200, 295
Aoki, K. - 212
Askew, J.D., Jr. - 232, 246
Audeh, C.A. - 317, 380
Bagrov, G.N. - 310
Bailey, W.W. - 326
Bakker, N. - 66
Baldwin, B.A. - 215
Baur, R. - 369
Baxter, B.H. - 375
Bekasova, N.I. - 305
Bell, E.W. - 220
Berlin, A.M. - 305
Birke, A.H. - 284
Bodrov, D.M. - 363
Bollinger, J.M. - 48
Boslett, J.A. - 211
Bosniack, D.S. - 224, 389
Boussely, J. - 363
Braid, M. - 171, 231, 236, 266, 271
Brannen, C.G. - 63
Breitzke, W. - 396
Bridger, R.F. - 190, 217, 232, 380
Broeckx, W.P. - 119
Brois, S.J. - 18, 25, 26, 38, 144
Bronstert, K. - 110
Brown, E.D., Jr. - 222
Brownawell, D.W. - 16
Bryant, C.P. - 46
Burnop, V.C.E. - 78
Cahen, R.N.M. - 386

Carswell, R. - 166
Caruso, G.P. - 324
Caspari, G. - 181, 261
Chafetz, H. - 33
Chandavoine, M.-M - 363
Chapelet, G. - 134, 135
Charlston, B.S. - 303
Chauvel, B. - 142
Chernyavsky, A.A. - 363
Chesluk, R.P. - 232
Chibnik, S. - 9
Chignac, M. - 363
Chou, K.J. - 202
Christian, J.B. - 334
Cier, R.J. - 217
Claiborne, J.L. - 346
Clarke, G.A. - 322
Clason, D.L. - 42
Cohen, C. - 31
Cohen, J.M. - 42
Colclough, T. - 205, 249
Crawford, J. - 178
Crawford, W.C. - 93
Crocker, R.E. - 321
Crook, M.F. - 57
Cusano, C.M. - 60, 115, 131
Dahlberg, J.R. - 301
Davis, B.T. - 22, 38, 57
Davis, K.E. - 30, 56
Davis, R.H. - 294
Davison, J.E. - 373
Dawans, F. - 142
Debande, G.F.S. - 386

De Clippeleir, G. - 70
DeJovine, J.M. - 90, 93
Deluga, S.S. - 366
Desai, N.V. - 357
DeVries, D.L. - 90, 93
de Vries, L. - 72, 75, 77
Disteldorf, J. - 393
Donahue, E.T. - 266
Dounchis, H. - 269
Dreher, J.L. - 321
Dulog, L.G. - 119
Durand, J.-P. - 142
Duteurtre, P. - 24
Eckert, R.J.A. - 110
Edmisten, W.C. - 63
Elliot, R.L. - 108, 122, 144,
    157, 160
Elliott, J.S. - 22, 38
Engel, L.J. - 124, 138, 139
Fenton, W.N. - 352
Flakus, W. - 393
Forsberg, J.W. - 77
Fossati, F. - 6, 86
Foulks, H.C., Jr. - 359
Frangatos, G. - 188, 189
Friihauf, E.J. - 30, 379
Fusco, A.M. - 347
Gaetani, F.J. - 60
Gainer, G.C. - 371
Galluccio, R.A. - 148
Gardiner, B., Jr. - 122
Gardiner, J.B. - 112, 124, 128,
    138, 139, 144, 157, 160
Gateau, P. - 142
Gibbons, P.J. - 374
Gilbert, J.B. - 382
Gladilina, L.B. - 363
Goldfain, V.N. - 310
Goletz, E., Jr. - 16
Gordeeva, G.N. - 310
Gorman, H.R. - 301
Grain, C. - 363
Grasshoff, H.D. - 291, 328
Green, M.J. - 303
Gribkova, P.N. - 305
Gribova, I.A. - 305
Grier, N. - 282
Grootjans, J.F.J. - 386
Gutierrez, A. - 25, 26
Hamblin, P.C. - 167
Hanauer, R.H. - 163
Handa, T. - 212
Harnach, J.W. - 366

Harris, H.A. - 312
Harting, G.L. - 322
Haugen, H. - 259
Hayashida, S. - 84
Heiba, E.-A.I. - 380
Heilman, W.J. - 113
Heilweil, I.J. - 11
Hellmuth, W.W. - 202
Henderson, C.C. - 232
Hermans, J.C. - 256
Hill, M.W. - 112
Himics, R.J. - 357
Hirota, T. - 343
Hitchcock, R.G. - 376
Hoke, D.I. - 47
Hori, T. - 84
Hotten, B.W. - 64, 218
Hovemann, F. - 110
Howlett, R.M. - 38
Huber, P. - 343
Hurst, G.L. - 372
Jacobson, N. - 119
Jahnke, R.W. - 298, 299
Jain, S.C. - 292
Jaruzelski, J.J. - 253
Jayne, G.J.J. - 208, 246
Jones, L. - 34
Jones, R.E. - 115
Jost, H. - 152
Kafarov, V.V. - 363
Kamimura, Y. - 369
Karll, R.E. - 53
Karpen, W.L. - 307
Katabe, K. - 343
Katoh, H. - 212
Kazintsev, T.I. - 310
Keil, J.W. - 352
Kelyman, J.S. - 34
Kennedy, C.D. - 387
Kenney, H.E. - 266
King, J.P. - 200
Kinoshita, H. - 331
Kiovsky, T.E. - 150
Kisselow, A.W. - 316
Kistler, J.-P. - 272
Kluger, E.W. - 3
Knoche, H. - 135
Koch, K. - 396, 398
Komarova, L.G. - 305
Korbach, P.F. - 389
Korshak, V.V. - 305
Krasnov, A.P. - 305
Krenowicz, R.A. - 252

Kresge, E.N. - 119
Kress, P.J. - 311
Kroke, H. - 398
Krongauz, E.S. - 305
Kuntz, I. - 128
Kuramzhin, A.V. - 363
Ladenberger, V. - 110
Lampelzammer, H. - 343
Landis, P.S. - 173, 232, 236
Lang, A. - 134
Larkin, J.M. - 33
Lee, D.A. - 211
Lee, G.A. - 311
Lee, R.J. - 53
Lenack, A.L.P. - 185
Leonardi, S.J. - 271
Liston, T.V. - 241, 243
Login, R.B. - 337
Logvinenko, D.D. - 363
Longo, J.M. - 302
Lonstrup, T.F. - 16
Loveless, F.C. - 238
Lowe, W. - 233, 236, 241, 243, 262
Luck, R.M. - 371
Lynch, T.J. - 113
Machleder, W.H. - 48
MacKenzie, K.A. - 326
Mailey, E.A. - 200
Makino, N. - 331
Malec, R.E. - 36
Mamatsashvili, G.V. - 305
Marcantonio, A.F. - 310, 311
Marie, G. - 134, 135
Marin, P.-D. - 272
Martin, W.H. - 276
Mauer, G.L. - 282
Maxwell, J.F. - 279, 285
McIntosh, G.A. - 293
McIntyre, M.H. - 7
Messerly, J.W. - 350
Metro, S.J. - 227
Michaelis, K.P. - 206
Miley, J.W. - 3
Miller, E.L. - 252
Miller, H.N. - 18, 153
Mitacek, B. - 111
Miyagawa, T. - 254
Moore, R.J. - 110
Morris, C.A. - 292
Murfin, D.L. - 379
Murphy, W.R. - 323
Nachtman, E.S. - 376
Nassry, A. - 279, 285

Nehmsmann, L.J., III - 274
Newkirk, D.D. - 337
Nicks, G.E. - 387
Nishihara, A. - 212
Nnadi, J.C. - 7, 11, 88, 90
Norman, S. - 22
Nudenberg, W. - 238
Numair, S.J. - 16
Obelchenko, E.I. - 363
O'Brien, J.P. - 208
Okamoto, N. - 25
Okorodudu, A.O.M. - 173, 192
Olavesen, C. - 382
Oravitz, J.L., Jr. - 301
Outten, E.F. - 378
Palmer, D.N. - 373
Papay, A.G. - 177
Pappas, J.J. - 119
Pardee, R.P. - 354, 356
Pavlova, S.-S.A. - 305
Peditto, A. - 6, 86
Pellegrini, J.P., Jr. - 164
Petrenko, A.V. - 310
Petrillo, V. - 6, 86
Pigerol, C. - 363
Pilz, H. - 152
Pindar, J.F. - 42
Pirson, E. - 343
Popoff, I.C. - 200
Powers, W.J., III - 85
Reick, F.G. - 222
Reinhard, R.R. - 265
Rhodes, R.B. - 110
Richardson, A. - 100
Ripple, D.E. - 45, 175
Rogan, J.B. - 101
Rohde, R. - 215
Rowe, C.N. - 323
Rumierz, J.R. - 308
Ryer, J. - 18, 112, 378
Sabol, A.R. - 71
Sakurai, T. - 212, 369
Sankey, B.M. - 382
Sato, H. - 384
Scattergood, R. - 264
Schenach, T.A. - 388
Schick, J.W. - 294
Schmidt, A. - 167
Schmitt, K. - 393
Schmitt, K.D. - 190, 196
Schoedel, U. - 152
Seki, H. - 254
Sekiya, M. - 331

Shamah, E. - 366
Shaub, H. - 170, 194, 227
Shelyakov, O.P. - 363
Shimada, S. - 369
Shipman, J.J. - 350
Shringarpurey, S.K. - 282
Sillion, B. - 31
Simak, P. - 110
Smith, D.M. - 244
Smith, H.A. - 105
Smith, R. - 293
Smith, W.L. - 34
Snyder, C.E., Jr. - 268
Soucy, R.J. - 223
Soula, G. - 24
Sowerby, R.L. - 229
Stambaugh, R.L. - 148
Staples, T.L. - 97
Steckel, T.F. - 87
Steger, J.J. - 302
Stuebe, C.W. - 46
Sturwold, R.J. - 278, 288, 290
Su, T.K. - 3
Sugiura, K. - 254
Swakon, E.A. - 197
Sweeney, W.M. - 118
Swinney, B. - 249, 264
Tamborski, C. - 268, 334
Taylor, B.W. - 326
Telegin, V.D. - 310
Thayer, H.I. - 164
Thir, B. - 337
Thompson, J.L. - 366
Tomoda, Y. - 212
Traver, F.J. - 349
Trepka, W.J. - 140
Tsantker, K.L. - 363
Vanderlinden, A. - 70

Van Doorne, G.C. - 330
van Hesden, J.W. - 297
VanVleck, G.T. - 349
Vasiliev, J.N. - 310
Verdicchio, R.J. - 274
Vinogradov, A.V. - 305
Vinogradova, O.V. - 305
Vlasova, I.V. - 305
Vorobiev, V.D. - 305
Waddey, W.E. - 194
Wainwright, P. - 303
Wakim, J.M. - 207
Waldbillig, J.O. - 131
Warne, T.M. - 183
Weetman, D.G. - 259
Wenzel, F. - 152
West, C.T. - 40
White, S.A. - 101
Willette, G.L. - 163
Williams, D.A. - 166
Williams, J.E. - 378
Williams, M.A. - 278
Williams, R.P. - 215
Wimmer, F. - 343
Winans, E.D. - 25
Wirth, H.O. - 206
Witzel, B.E. - 282
Wood, D.L. - 110
Woods, D.R. - 208
Woods, W.W. - 237
Wulfers, T.F. - 314
Wulk, P.K. - 316
Yaffe, R. - 265
Yamamoto, R.I. - 60
Yaschinskaya, M.S. - 363
Yasui, S. - 384
Yoto, M. - 212
Zielinski, J. - 18

U.S. PATENT NUMBER INDEX

4,066,560 - 349	4,080,307 - 167	4,094,801 - 77
4,067,817 - 288	4,081,385 - 60	4,094,802 - 24
4,068,056 - 138	4,081,386 - 229	4,096,077 - 197
4,068,057 - 139	4,081,387 - 175	4,096,078 - 265
4,068,058 - 139	4,081,388 - 24	4,096,079 - 356
4,069,162 - 112	4,081,389 - 264	4,096,898 - 350
4,069,163 - 266	4,081,390 - 100	4,097,386 - 236
4,070,295 - 153	4,081,445 - 256	4,097,388 - 268
4,070,370 - 22	4,082,680 - 111	4,098,705 - 212
4,071,459 - 31	4,083,791 - 38	4,098,706 - 363
4,071,460 - 354	4,083,792 - 88	4,098,707 - 188
4,071,548 - 25	4,084,737 - 301	4,098,708 - 46
4,072,618 - 295	4,085,053 - 181	4,098,709 - 163
4,072,619 - 166	4,085,055 - 142	4,098,710 - 122
4,073,737 - 108	4,085,056 - 388	4,100,080 - 320
4,073,738 - 110	4,086,170 - 70	4,100,081 - 321
4,075,112 - 330	4,086,171 - 110	4,100,082 - 42
4,075,113 - 330	4,086,172 - 236	4,100,083 - 323
4,075,393 - 288	4,086,173 - 7	4,100,084 - 85
4,076,634 - 305	4,088,585 - 307	4,100,085 - 86
4,076,635 - 375	4,088,587 - 262	4,100,086 - 33
4,076,636 - 90	4,089,790 - 320	4,100,187 - 38
4,076,637 - 372	4,089,791 - 259	4,101,427 - 170
4,076,639 - 217	4,089,792 - 236	4,101,428 - 178
4,076,672 - 343	4,089,794 - 157	4,101,429 - 284
4,077,892 - 189	4,089,854 - 312	4,101,430 - 271
4,077,893 - 150	4,090,953 - 380	4,101,432 - 192
4,077,992 - 347	4,090,971 - 47	4,101,599 - 386
4,079,012 - 224	4,092,254 - 190	4,102,798 - 18
4,079,013 - 317	4,092,255 - 135	4,104,177 - 324
4,080,302 - 294	4,093,578 - 310	4,104,178 - 292
4,080,303 - 379	4,094,799 - 93	4,104,179 - 205
4,080,304 - 105	4,094,800 - 183	4,104,180 - 78

4,104,181 - 173
4,104,182 - 202
4,105,571 - 194
4,107,054 - 237
4,107,058 - 322
4,107,059 - 200
4,107,061 - 278
4,107,115 - 359
4,108,783 - 45
4,108,784 - 46
4,108,785 - 288
4,110,196 - 389
4,110,227 - 337
4,110,233 - 326
4,110,234 - 238
4,113,633 - 374
4,113,635 - 369
4,113,636 - 124
4,113,637 - 31
4,113,639 - 16
4,113,640 - 314
4,113,642 - 396
4,113,725 - 11
4,115,282 - 328
4,115,284 - 331
4,115,285 - 297
4,115,286 - 215
4,115,287 - 249
4,115,288 - 196
4,116,872 - 298
4,116,873 - 72
4,116,874 - 254
4,116,875 - 11
4,116,876 - 25
4,116,877 - 378
4,116,917 - 110
4,118,331 - 299
4,119,547 - 376
4,119,551 - 265
4,119,552 - 57
4,120,803 - 36
4,120,804 - 34
4,120,887 - 25
4,122,021 - 238
4,122,022 - 326
4,122,023 - 384
4,123,369 - 252
4,123,371 - 84
4,123,372 - 232
4,123,373 - 26
4,124,513 - 265
4,124,514 - 265

4,124,650 - 382
4,125,470 - 352
4,125,472 - 171
4,125,479 - 232
4,127,138 - 118
4,127,491 - 222
4,127,492 - 6
4,127,493 - 22
4,128,486 - 373
4,128,488 - 87
4,129,508 - 30
4,129,509 - 282
4,129,510 - 244
4,129,512 - 316
4,130,492 - 302
4,130,494 - 227
4,130,496 - 190
4,131,551 - 366
4,131,552 - 310
4,131,553 - 40
4,131,554 - 87
4,132,656 - 90
4,132,657 - 274
4,132,659 - 218
4,132,660 - 334
4,132,661 - 131
4,132,662 - 290
4,132,663 - 113
4,134,844 - 93
4,134,845 - 207
4,134,846 - 48
4,135,034 - 346
4,136,040 - 93
4,136,041 - 185
4,136,042 - 164
4,136,043 - 30
4,136,044 - 266
4,136,047 - 101
4,136,048 - 97
4,137,183 - 261
4,137,184 - 66
4,137,185 - 144
4,137,186 - 71
4,138,346 - 279
4,138,347 - 93
4,138,348 - 291
4,138,349 - 221
4,138,370 - 160
4,139,417 - 134
4,139,480 - 128
4,140,642 - 272
4,140,643 - 30

4,140,834 - 311
4,141,844 - 265
4,141,845 - 265
4,141,846 - 246
4,142,865 - 119
4,142,979 - 256
4,142,980 - 53
4,144,178 - 343
4,144,180 - 200
4,144,181 - 157
4,144,183 - 398
4,144,247 - 246
4,145,298 - 140
4,146,487 - 308
4,146,488 - 276
4,146,490 - 269
4,146,492 - 115
4,147,640 - 208
4,147,641 - 48
4,147,666 - 206
4,148,737 - 241
4,148,738 - 243
4,148,739 - 243
4,148,834 - 387
4,148,970 - 293
4,149,980 - 63
4,149,981 - 363
4,149,982 - 211
4,149,983 - 282
4,149,984 - 152
4,151,099 - 285
4,151,102 - 369
4,152,278 - 220
4,153,562 - 253
4,153,563 - 231
4,153,564 - 9
4,153,565 - 236
4,153,566 - 18
4,153,567 - 3
4,155,858 - 319
4,155,860 - 223
4,155,861 - 393
4,156,061 - 119
4,157,971 - 265
4,157,972 - 64
4,158,633 - 177
4,159,252 - 303
4,159,255 - 371
4,159,276 - 357
4,159,898 - 31
4,159,956 - 77
4,159,957 - 77

4,159,958 - 75   4,160,739 - 148   4,161,452 - 148
4,159,959 - 77   4,161,451 - 233   4,161,475 - 56
4,159,960 - 189

NOTICE

Nothing contained in this Review shall be construed to constitute a permission or recommendation to practice any invention covered by any patent without a license from the patent owners. Further, neither the author nor the publisher assumes any liability with respect to the use of, or for damages resulting from the use of, any information, apparatus, method or process described in this Review.

FUEL ADDITIVES FOR INTERNAL COMBUSTION ENGINES 1978
Recent Developments

by Maurice William Ranney

Chemical Technology Review No. 112
Energy Technology Review No. 30

The search for improved fuel additives has taken on new and significant dimensions as the world begins to react to the far-reaching consequences of its limited petroleum resources. With the attendant ever-increasing costs for merchandise transport and personal travel to and from work, truly new and innovative ideas and processes are required.

While this is but one approach toward stretching the available energy supplies, successful research for more effective additives, which really increase mileage of the fuels in question, will provide many new business opportunities in the next few years. Fuel additives manufacture and sales can be expected to show well above normal growth in the next decade.

The magnitude of the problem is compounded by concern for environmental conditions and the quality of life which continues to be the focal point for consumer activity and governmental legislation around the world.

These changes in urgency, concern and direction are readily evident in the recent U.S. patent literature. This book, presenting over 200 processes and hundreds of formulated fuel compositions brings much of the recent proprietary efforts together, thus providing a comprehensive picture of technological development in this field.

A partial and condensed table of contents follows. Numbers in ( ) indicate the number of processes per topic.

1. **DETERGENTS AND ANTI-ICING ADDITIVES (79)**
Multipurpose Carburetor Detergents
Maleic Polyamines
Polyamine Mannich Bases
Succinamic Acids
Fatty Acids +
  Alkanolamine Sulfonates
Ester-Type Dispersants
Borates & Phosphonates
Multipurpose Additives
Nitroketonized Amides
Substituted Morpholines
Telomeric HC + P Compounds
Intake Valve
  Cleanliness Agents
Phenylthydrazines
Hydrocarbylamines
Sodium Carbonate
Mineral Oil Fractions

2. **FLOW IMPROVERS AND POUR POINT DEPRESSANTS (22)**
Ethylene-Based Polymers
Hydrogenated Polybutadiene
Acrylate Polymers +
  4-Vinylpyridine
Succinamides
Fatty Amines
Wool Fat

3. **OXIDATION AND CORROSION INHIBITORS (25)**
Alkylphenols & Imidazolines
Guanidine Carbonate
Malonic Esters
Alkanolamides +
  Dicarboxylates
Olefin-Thionophosphine
  Reaction Products
Hindered Phenols and
  Amino Compounds
Hexahydropyrimidines
Dimer Acid Mixtures

4. **OCTANE IMPROVERS AND ANTIKNOCKS (16)**
Methylmethoxypropane
Manganese, Citrate &
  Alkyl Tin Compounds
Stabilized Me Carbonyls
Catalyst Plugging Preventers
Benzotrifluoride

5. **COMBUSTION AIDS & FUEL COMPOSITIONS (22)**
Combustion Efficiency Aids
Di-**tert.**-Butyl Peroxide
Encapsulated Hydrogen
Controlled Fuel Wettability
  Agents
Emulsified Fuels
Surfactant Blends
Isobutane Oxidation
Methanol Treatment

6. **ANTISTATS, BIOCIDES DYES & OTHERS (32)**
Polysulfone Copolymers
Metal Salicylates
Organoboron Compounds
Disazo Dyestuffs
Demisting Agents
  for Jet Fuels
Dispersants for Diesel Fuels
Carbon Inhibitors
Dehazing Composition

ISBN 0-8155-0709-7

300 pages

CORROSION INHIBITORS 1979
RECENT DEVELOPMENTS

by J.S. Robinson

Chemical Technology Review No. 132

There are both multipurpose and highly specialized corrosion inhibitors described in this new work. Corrosion is an ever-present problem in industry, but it is clear from the nearly 300 processes and techniques detailed here that research is making substantial strides in supplying products to overcome it. The book comprises chemicals, such as sequestering agents and oxygen scavengers; and physical barriers, such as coatings.

The author has arranged the processes according to their most significant end use, as shown below. The first chapter, for instance, covers substances used to overcome the problems encountered in water engineering. It encompasses both antiscalants and corrosion inhibitors, the latter often being required during scale removal operations.

It should be emphasized that some inhibitors serve in several capacities—the benzotriazole derivative found under lubricants may well be ideal for circulating water systems.

Chapter headings and **examples of some** important subtitles follow below. The number of processes per topic is shown in parentheses following chapter headings.

1. CIRCULATING WATER SYSTEMS (69)
Cooling Water
  Phosphonomethylamino Carboxylates
  Phosphorus-Free Inhibitors
  Inhibitor of Low Pollution & Toxicity
  Inhibitor in Seawater Coolant
Maleic-Furan Copolymers for Boilers
Hydrazine Compound as Oxygen Scavenger
Scale Removal and Acid Cleaning
Inhibitor for Hydrofluoric Acid
Heat Exchanger Agents
Municipal Water Supplies
  Pyrophosphate-Zinc Combination
  Coated Chlorine Evaporator Chamber

2. OIL WELL AND REFINERY USE (39)
Alkynoxymethylamines in Drilling Fluids
Amine Bisulfites in Water Flooding
Alkylpyridines in Water Flooding
Pipelines and Tanks
  Macrocyclic Tetramine Films
Acid-Gas Treatment
  Cu-S-Monoethanolamine Formulation
Antifoulant and Anticorrosive Processing
Quaternary Ammonium Demulsifiers

3. CONSTRUCTION MATERIALS (34)
Corrosion-Resistant Concrete and Gypsum

Rust Treatment—Primers, Pigments, Resins
  Silicone-Acrylic-Polyurethane Coatings
  Topcoated Phosphated Bolts, Nuts, Washers
  Carboxymethylated Derusting Agents
Circuit Breaker Phosphate-Chromate Coat
Multilayered Wax Coating for Marine Use
Inhibiting Shellfish and Algae Adhesion

4. FUELS AND LUBRICANTS (47)
C_{21}-Dicarboxylic Acid Motor Fuel Formula
Aminoalkylpropanediols in Motor Fuels
Halogen Treatment of Motor Fuels
Hydraulic Fluids
Benzotriazole Metal-Working Fluid
Residual Fuel—Mg-Si-Mn Combination
Tetrahydrobenzimidazole Lubricant
P_2S_5 Adducts as Lubricating Oils
Silicone-Perfluorocarbon Polymer Grease

5. INORGANIC TREATMENT OF METAL (50)
Phosphatizing with Alkylolamine Additive
Quaternary Amino Polymer Pickling Bath
Halogenated Alkynoxymethylamine Pickle
Polymeric Electrodeposition
Zn-Li Silicate-Latex Coating
Zn-Al Hot Dip Coating
Metal-Urea Chromating Composition
Hydrophobic Silicon Oxide Layers
Carbide-Reinforced Superalloys
Vacuum-Tight Metal-to-Ceramic Seals

6. ORGANIC TREATMENT OF METAL (29)
Tannin-Phosphate-Ti-Fluoride Formula
Alkanolamine Mist or Spray Inhibitor
Dicyclohexylammonium Pelargonate
Ascorbic Acid Compositions
Corrosion-Inhibiting Rubber
Radiation-Polymerized Coatings
1,2-Fused-1,3-Dinitrogen Heterocyclics
Sulfur Dioxide-Schiff Base Adducts

7. ADDITIONAL APPLICATIONS (29)
Detecting and Evaluating Corrosion
Alkyltin Tarnish Protectives
Inert Gas for Storage Protection
3-Component Synergistic Hydrocarbons
Reaction Vessels & Process Equipment—
  Chromium Dioxide Synthesis
  Acrylonitrile Plants
  Nuclear Reactors
Coal- or Gas-Carrying Pipelines
Noncorrosive Solid Detergent
Inhibition of Toothpaste Tube Swelling
Polyphosphonic Acid Sequestering Agent

ISBN 0-8155-0757-7

306 pages

ANTIOXIDANTS 1979
RECENT DEVELOPMENTS

by William Ranney

Chemical Technology Review No. 127

Antioxidants prolong the useful life of countless products including plastics, elastomers, resins, lubricants and foodstuffs.

Once oxidation of an organic substrate begins, a chain reaction of autocatalytic oxidation ensues, ultimately causing gross changes in properties. An antioxidant will inhibit the initial oxidation or suppress the subsequent chain reaction. When two antioxidants with diverse actions are combined, the overall synergistic effect can be more than merely additive.

Over the years the antioxidant field has progressed from simple phenols to those of much more complex and sophisticated structures having a high specificity.

The 256 processes in *Antioxidants* detail manufacturing technology and compositions applicable to many fields. The partial table of contents below gives chapter headings and **examples of some** subtitles. The number of processes per topic is in parentheses.

1. **POLYOLEFIN RESINS (90)**
   Oligomeric Bisphenol Derivatives
   4-Hydroxydiphenyl Sulfoxide
   Piperidine Compounds
   4,4'-Bipiperidylidene Compounds
   Multisubstituted 4-Piperidinol Compounds
   Halogenated Benzotriazoles
   Diacyl Dihydrazides
   Polyhydrazides
   Pyrrolidone Dicarboxylic Acids & Esters
   Hydroxyaryl Hydantoins
   Mercaptoquinazolone
   Phosphorus-containing Compositions
   Alkoxylated Phosphoronitrilic
     Compounds
   Esters of Phosphinodithioic Acids
   Ni Organophosphate and Benzophenone
   Zn and Ni Compounds Combined
   Metal Salt of Piperidine Carboxylic Acid
   Aerosol for Surface Protection
   Glass-Fiber-Reinforced Polypropylene
   Flame-Retarded Expandable Polystyrene

2. **POLYVINYL CHLORIDE RESINS (19)**
   Alkyltin Mercaptocarboxylic Acid Ester
   Organic S and Hydrocarbyltin Compounds
   Organotin Halides
   Mg Diketone Complex and Mercaptide
   Phosphorylated Phenol/Phenol Ester
   Carboxylic Acid Amide Stabilizer

3. **POLYESTER, POLYCARBONATE,
   POLYURETHANE PLASTICS (30)**
   Benzotriazole-Benzoxazole
     Bichromophore
   Aryl Esters of Heterocyclic Acids
   Carbodiimides
   Cyclic Diphosphites and Silanes
   Substituted Acrylonitrile Compounds
   Esterified Hindered Phenols
   Guanidine Compounds
   Polyether Polyols
   Polyaryl Ether Ketones
   Cationic Dyeable Nylon
   Unsaturated Cycloacetal Resin

4. **ELASTOMERS (45)**
   Enamines and Other Antiozonants
   Phenolic Antioxidants
   Tricyclopentadiene-Phenolic Products
   Thioalkylidene-Bisphenols
   Bis(4-Anilinophenoxy) Esters
   UV-Curable High-Vinyl Polybutadienes
   Copolymerizable Antioxidants
   Dialkyl Vinylphenols
   Nonmigratory Sulfonyl Azides
   4,4-Bis(N-Pyrrolidinyl)Diphenylmethane
   Metal Carboxylates for Halobutyl Rubber
   Chelating Agent as Metal Deactivator
   Alkylphenyl Propylene Glycol Phosphite

5. **PETROLEUM PRODUCTS AND
   SYNTHETIC LUBRICANTS (30)**
   Mineral Oil Lubricants
   Naphthylamines and S Compounds
   S-containing Ni Complexes
   1-Butoxy-1-(1-Naphthoxy)Ethane
   Treatment of Asphaltite Resin
   Ammonia-Containing Gasoline
     Compositions
   Polyester Lubricants

6. **FOOD PRODUCTS (35)**
   Nonabsorbable Antioxidants
   Polymeric N-Substituted Maleimides
   General Antioxidants
   3,4-Dihydroxyphenylalanine Derivatives
   Dialkyl Pentaerythritol Diphosphite
   Purification of Rosemary and Sage
   Chroman Compounds and Ascorbic Acid
   Sorbic Acid from Vinyl Butyrolactone
   Antioxidant Dispensing System

7. **OTHER ANTIOXIDANTS (7)**
   Herbicidal Thiolcarbamate Sulfoxide
   Preservatives for Dental Products

ISBN 0-8155-0747-X

372 pages

BOILER FUEL ADDITIVES FOR POLLUTION REDUCTION AND ENERGY SAVING 1978

Edited by R.C. Eliot

Energy Technology Review No. 33
Pollution Technology Review No. 53
Chemical Technology Review No. 120

Problems of incomplete combustion, soot deposition, and sludge formation are common to a greater or lesser degree with all hydrocarbon fuel oils, especially those sold for use in heavy duty installations, e.g., furnaces and boilers in power plants, large office buildings, hospitals, etc. The problem of particulate emissions is particularly acute, as these carbonaceous contaminants are currently a violation of local anti-pollution laws in many communities.

This book reflects the findings of investigations conducted for the U.S. Environmental Protection Agency and the Federal Energy Administration with the purpose of determining the effectiveness of chemical additives not only in reducing smoke and particulate emissions, but also as a means of increasing overall boiler efficiency, and thus conserving fuel. A survey of the recent patent literature is included.

Various kinds of compounds have shown merit as combustion improvers, including surfactants, organometallics, and low-molecular-weight polymers. Alkaline-earth and transition metal additives in concentrations of 20 ppm to 50 ppm of metal in the fuel oil were effective in reducing particulate emissions. For added versatility the organo-metallic compound is usually blended with dispersants and other ingredients to make a multipurpose product. Judicious application of these "combustion catalysts" should enable the user to cut fuel costs, while lowering air pollution at the same time. A partial table of contents follows here:

1. **MECHANISMS OF CHEMICAL CONTROL IN HYDROCARBON COMBUSTION**
   Pollutant Formation
   Residual Oils vs. Distillate Oils
   Products of Incomplete Combustion
   Emissions from Fuel Impurities
   Theoretical Basis for
     Controlling Combustion-Generated
     Emissions with Chemical Additives
   Increasing the Active Oxidant
     Concentration
   Inhibiting Formation of
     Unreactive Combustible Materials
   Promotion of Atom Recombination
   Retardation of O-Atom Production

Promotion of Pyrolysis Reactions
EPA Investigation of Residual Oil Firing

2. **EMISSION-REDUCING ADDITIVES**
   Cobalt Naphthenate
   Iron Naphthenate
   Manganese Naphthenate
   Barium Naphthenate
   Iron + Barium Naphthenates
   Calcium Naphthenate
   Methyl Cyclopentadienyl
     Manganese Tricarbonyl
   Iron Cyclopentadienes
   Barium Sulfonate
   Barium Ethyl Hexoate
   Iron Carbonyl
   Toluene
   Diglyme
   Ethyl Alcohol
   Hexyl Alcohol
   Naphthenic Acid
   Ethylhexoic Acid
   Proprietary Additives
   MC-7 Soluble (3.6% Mn, 3% Ba)
   Rolfite 404 (2% Mn)
   Improsoot
   Watcon 130
   Others

3. **FUEL-HANDLING ADDITIVES**
   Stabilizers
   Dispersants
   Flow Improvers
   Pour Point Depressants
   Corrosion Inhibitors
   Antistatics
   Metal Deactivators

4. **VARIOUS COMBUSTION AIDS**
   Calcium-Based Montmorillonite
   Ferrous Picrate + Nitroaliphatics
   Smoke Suppressants
   Organo-Metallic Phosphates
   Metal Carboxylates
   Oil-Soluble Ammonium Salts
   Preventing Sublimation of Arsenic

5. **POST-FLAME TREATMENTS**
   Powder-Type Additives
     for Flue Gas Injection
   Cold-End Treatment
   Reduction of Slag Build-Up
   Activated Manganese Compound

ISBN 0-8155-0729-1

230 pages

REPROCESSING AND DISPOSAL
OF WASTE PETROLEUM OILS 1979

by L.Y. Hess

Pollution Technology Review No. 64
Chemical Technology Review No. 140

Recovery and judicious disposal of waste oil have become financially rewarding practices by reason of resource conservation and environmental protection.

Dirty and contaminated lubricating oil still has high energy values. It can be re-refined or used as a feedstock for making other petroleum products. Industrially, it can be reclaimed to nearly original quality by simple equipment.

Sufficiently worthwhile disposal practices include road oiling, combining with fuel oil, use as auxiliary fuel in municipal incinerators, and landspreading with decomposition by suitable microorganisms.

This book is based on the latest government research reports and U.S. patents which contain practical technological process information. A condensed table of contents follows here.

1. SOURCES AND CHARACTERISTICS
Generation and Disposal Figures
Automotive Service Centers
Commercial Truck Fleets
Railroad Service Centers
Aviation Service Centers

2. RECOVERY PROCESSES
Acid/Clay Process
Extraction Processes
Distillation

3. PROPRIETARY PROCESSES (118)
Treatment with Chemicals
Solvent Extraction
Adsorption
Distillation & Hydrogenation
Filtration
Other Separation Techniques
Application of Electric Fields

4. RESEARCH PROJECTS
Innovative Techniques
Ion Exchange Percolation
Chelation
Comprehensive Waste Oil Facility

5. DISPOSAL PRACTICES
Incineration and Use as Fuel
Environmental Aspects

6. DISPOSAL OF RECOVERY RESIDUES
Waste Products Generated
Acid Sludge
Caustic Sludge
Spent Clay
Distillation Bottoms
Characterization of Wastewaters
Marine Waste Oil Processing Facility
The Re-Refining Facility
Environmental Assessment

7. DISPOSAL AND RECYCLING OF AIR FORCE WASTE OILS
Generation of Waste POLs
Kelly AFB Survey
Andrews AFB Survey
Available Disposal/Recycle Techniques

8. COMBUSTION STUDIES OF WASTE POLs
Air Force Experimental Studies
Combustion Test Equipment
Energy Recovery
Waste Oil Burn-Off in Coast Guard
    Power Plants
Waste Lube Oil Added to Diesel Fuel

9. MUNICIPAL INCINERATOR FUEL
Physical and Combustion Properties of
    Waste Oil
Heating Requirements
Phenomena Occurring in a Refuse Bed
Transferring Heat Flux
Mixing Waste Oil Directly into Refuse
Utilizing Auxiliary Burners
Impact on Air Quality

10. DISPOSAL BY LANDSPREADING
Land Disposal Practices
Microbiological Studies
Shell Project Description and Objectives
Oil Decomposition Rate and
    Effect on Fertilizer
Metals Contents of Soils
Microbial Action
Rainfall Runoff
Union Carbide Study
Combined Oil/Nitrate Application
Summary

ISBN 0-8155-0775-5

322 pages

ETHYL ALCOHOL PRODUCTION AND USE AS A MOTOR FUEL 1979

Edited by J.K. Paul

Energy Technology Review No. 51
Chemical Technology Review No. 144

This book is a companion volume to our *Methanol Technology and Application in Motor Fuels*. Ethanol is available from so-called "renewable raw materials," i.e., agricultural crops such as corn, grain, potatoes, sugar beets and sugar cane, or other biomass and suitable garbage.

In this book an attempt has been made to present an economic assessment of possible modes of preparation of ethanol from various forms of biomass, natural resources and their waste materials or by-products. A chapter on current technology is also included. The present and potential availability of biomass from sugar crops, grains and grasses and silvicultural forms is considered.

Current crop production, proposed crops grown specifically for energy production and crop wastes and residues are discussed. Finally, to determine the actual practicality of fueling motor vehicles with ethanol, either 100% or in blends, several sets of engine test data are reviewed. The results seem favorable, 10-20% ethanol blends performing very similarly to straight gasoline with slight gains in octane rating and mileage.

1. NEAR-TERM POTENTIALS
Alcohol Fuel in Automobiles
Costs of Biomass-Based Fuels
Resources Required
Government Role

2. ECONOMIC ASSESSMENT
Mitre Study
Comparison of Process Economics
Raphael Katzen Study
California Energy Commission Study

3. ETHANOL FROM MUNICIPAL WASTES
Enzymatic Hydrolysis of Cellulosic Wastes
Plant Conversion Requirements

4. AVAILABILITY OF SUGAR CROPS
Sugarcane
Sweet Sorghum
Sugar Beets
General Route from Sugar Crop to Fuel

5. AVAILABILITY OF GRAINS AND GRASSES
Criteria for Crop Selection
Grain and Grass Crops

Plant Species Relative to Production for Energy
Grasses and Grains Presently Grown in the U.S.
Documented Plant Candidates for Expanded Utilization

6. AVAILABILITY OF WOOD BIOMASS
Silvicultural Energy Farms
Utilization of Forest Residues
Mill Residues

7. ETHANOL PRODUCTION TECHNOLOGY
Fermentation
Distillation
Cellulose Conversion
Dartmouth College and Bureau of Solid Waste Management
U.S. Army—Natick Laboratories
University of California at Berkeley (Wilke)
University of Pennsylvania (Humphrey) and General Electric Company
Purdue (Tsao)
Gulf Oil Chemicals Company
Various Methods

8. NEBRASKA GASOHOL PROGRAM
Background
Formation of Agricultural Products Industrial Utilization Committee
Two Million Mile Road Test
Consumer Acceptance
Food and Fuel for the Future
Feasibility Studies
Formation of National Gasohol Commission

9. ETHANOL USE IN BRAZIL
Alcohols as Fuels in Brazil
Prospects of Alcohols as Fuels
Otto Engines with Ethanol Blends
Otto Engines with Straight Ethanol
Energetics and Economics of Cassava and Sugarcane Fuel Ethanol

10. MOTOR VEHICLE OPERATION WITH ETHANOL AND ETHANOL BLENDS
Changes in Fuel/Air Ratio
Emission and Fuel Economy
Performance Test Data
Dual Fuel Diesel Application
EPA Gasohol Test Program
BERC Testing Program
Brazilian Engine Calibration Test Program

ISBN 0-8155-0780-1

350 pages

METHANOL TECHNOLOGY & APPLICATION IN MOTOR FUELS 1978

Edited by J. K. Paul

Chemical Technology Review No. 114
Energy Technology Review No. 31

This book contains detailed descriptive information relating to methanol production technology from unusual sources, the utilization of methanol as an automotive fuel, and the conversion of methanol into gasoline. The first chapter is an overview and serves as an introduction to the subject.

The next three chapter are feasibility studies and discuss the production of methanol from coal, solid waste, and natural gas. This book does not discuss in any detail the production of methanol from liquid hydrocarbons, since one of the reasons for the future utilization of methanol in automobiles is to lessen the reliance upon petroleum hydrocarbons. Regarding natural gas, methanol can be produced from natural gas at a distant location, and shipped as a liquid at potentially less cost and less danger than the shipping of liquefied natural gas (LNG). Regarding coal, the cost of methanol appears to be comparable on an energy-equivalent basis to the production of synthetic gasoline and substitute natural gas from coal.

The next two chapters relate to the use of methanol/gasoline blends in automobiles and the use of 100% methanol as a vehicle fuel. Small amounts of methanol can be added to gasoline for use in current engines, however, the use of 100% methanol as a vehicle fuel does require engine changes, including modifications to the carburetor, and the necessity for a greater heat supply to the intake manifold.

The last two chapters relate to the production of gasoline from methanol, and these two chapters are based primarily on the efforts being conducted by the Mobil Corporation in this direction.

This book is based on federally funded studies, U.S. patents and other sources. It does convey the impression of considerable progress in methanol technology.

1. OVERVIEW
Methanol Production
Extraction
Fermentation
Syntheses
From Synthesis Gas
Oxidation of Hydrocarbons
By Irradiation of CO_x
Organic Feedstocks
Inorganic Origins
Raw Materials
 Nonrenewable Sources
 Renewable Sources
Methanol as a Fuel
Combustion Emissions
Suitable Engines

Electrochemical
 Oxidation of Methanol
Human Toxicity:
 Visual Impairment &
 Blindness from Methanol

2. METHANOL FROM COAL
Coal Gasification and
 Methanol Synthesis
DuPont Feasibility Study
Sasol Type Process Study
Badger Conceptual Design

3. METHANOL FROM SOLID WASTE
Conversion of Municipal
 Solid Waste
Pyrolysis Procedures
Suitability Specifications
 for Municipal Wastes
Conversion of Wood Wastes

4. ALASKAN METHANOL
Synthesis & Conversion Plants
Pipeline (Alyeska)
Pumping Power Requirements
Other Transportation
Other Transportation
Methanol Crude Separations
 at Los Angeles
Cost of Supply Gas
Slug Flow Interface
Overall Cost Estimates

5. METHANOL/GASOLINE BLENDS
5%, 10%, 15%, 20% Methanol
Missouri U. Study
Simulation Procedures
Engine Adjustments
Bartlesville Energy Research
 Center Study
Vehicle Optimizations Needed
Performance Mapping
U. of California Study
Engine Configurations
 and Operations
Optimum Amounts in Blends

6. 100% METHANOL MOTOR FUEL
Engine Modifications
Fuel-Air Injection Systems
Cold Start Difficulties
Thermochemical Engines

7. CHEMICAL CONVERSION: METHANOL TO GASOLINE
Mobil Oil Corp. Process
Fixed Bed Pilot Plant
Vehicle Studies with Synthetic
 Gasoline from CH_3OH

8. METHANOL TO GASOLINE
Proprietary Processes
Mobil Oil Corp. Patents
Ethyl Corp. Patents
Others

ISBN 0-8155-0719-4

470 pages

ENHANCED OIL RECOVERY

Secondary and Tertiary Methods 1978

Edited by M. M. Schumacher

Chemical Technology Review No. 103
Energy Technology Review No. 22

This is a book about greater oil recovery by succeeding methods; the possible techniques and their present or future capabilities for extracting more petroleum from oil fields after primary production.

These methods are also applicable to fields which were abandoned in previous decades when there was a plentiful supply of energy. Sometimes less than one-third of the crude oil in place was recovered, when only the original reservoir pressure was allowed to drive the oil into the producing well.

The concept of enhanced oil recovery applies to a whole collection of methods which are becoming increasingly profitable as energy becomes more expensive and the price of oil goes up. Each method appears to have its own claimed unique capability to extract the most oil from a particular reservoir.

After a decline in pressure has caused the oil recovery to become uneconomic, oil production can be increased by immiscible gas injection and waterflooding. These traditional "secondary" methods are now being supplemented by "tertiary" recovery methods including miscible fluid displacement, microemulsion (micellar) flooding, cyclic steam injection, controlled *in situ* combustion (fireflood) and other related techniques which are either thermal or miscible methods, thereby reducing the surface tension between oil and driving fluid.

Most of the data in this book are based on federally funded studies—a partial and condensed list of contents follows here:

1. DEMAND & SUPPLY
Primary Production
Enhanced Oil Recovery (EOR)
Promise & Potential of EOR
GURC Forecast
Oil Prices vs. EOR Development

2. SECONDARY METHODS
Formation Energy
Pressure Maintenance
Waterflooding
Treatment of Injection Waters
Immiscible Gas Injection

3. TERTIARY RECOVERY—
MISCIBLE AGENTS
HC-Miscible Flooding
CO_2-Miscible Flooding
Polymer-Augmented Waterflooding
Micellar Flooding

Performance & Problems
Costs vs. Performance

4. TERTIARY RECOVERY—
THERMAL TECHNIQUES
Cyclic Steam Injection
Limitations & Economics
Steam Drive
in situ Combustion
Necessity for Materials Research
Future Projects

5. VARIOUS METHODS
Oil Recovery by Nuclear
Explosions
Nuclear Stimulation
Seismic & Pollutant Effects
Underground Reservoir Construction

6. SELECTION OF
APPROPRIATE TECHNIQUES
Screening Methods
Potentials & Profits

7. MATERIALS NEEDED
Chemicals Used
Water Supplies
Energy Requirements
Research Needs

8. ENVIRONMENTAL ASPECTS
Chemical Hazards
Effects on Groundwater
Emission of
Steam & Gases

9. STATE OF THE ART
Field Tests Made
Field Test Activity in Texas
Project Developments
Classification by
Location & Methods

10. EOR PROJECTS FOR
RECOVERY OF VISCOUS OILS
The "Heavy Oil" Problem
Solvent Stimulation
in California
Solvents & Explosives
to Recover Heavy Oil
in Kansas
Thermal Recovery of
Viscous Oils

11. THE FUTURE OF EOR
Constraints to Development
Necessity for Federal
Participation
Industry's Choices

ISBN 0-8155-0692-9

207 pages